Lecture Notes in Networks and Systems

Volume 298

The series "Lecture Notes in Networks and Systems" publishes the latest developments in Networks and Systems—quickly, informally and with high quality. Original research reported in proceedings and post-proceedings represents the core of LNNS.

Volumes published in LNNS embrace all aspects and subfields of, as well as new challenges in, Networks and Systems.

The series contains proceedings and edited volumes in systems and networks, spanning the areas of Cyber-Physical Systems, Autonomous Systems, Sensor Networks, Control Systems, Energy Systems, Automotive Systems, Biological Systems, Vehicular Networking and Connected Vehicles, Aerospace Systems, Automation, Manufacturing, Smart Grids, Nonlinear Systems, Power Systems, Robotics, Social Systems, Economic Systems and other. Of particular value to both the contributors and the readership are the short publication timeframe and the world-wide distribution and exposure which enable both a wide and rapid dissemination of research output.

The series covers the theory, applications, and perspectives on the state of the art and future developments relevant to systems and networks, decision making, control, complex processes and related areas, as embedded in the fields of interdisciplinary and applied sciences, engineering, computer science, physics, economics, social, and life sciences, as well as the paradigms and methodologies behind them.

Indexed by SCOPUS, INSPEC, WTI Frankfurt eG, zbMATH, SCImago.

All books published in the series are submitted for consideration in Web of Science.

More information about this series at http://www.springer.com/series/15179

Michael E. Auer · Kalyan Ram Bhimavaram ·
Xiao-Guang Yue
Editors

Online Engineering and Society 4.0

Proceedings of the 18th International Conference on Remote Engineering and Virtual Instrumentation

Springer

REV International Conference on Remote Engineering and Virtual Instrumentation

Editors
Michael E. Auer
Carinthia University of Applied Sciences
Villah, Austria

Xiao-Guang Yue
Wuhan University of Technology
Wuhan, China

Kalyan Ram Bhimavaram
Electrono Solutions Pvt. Ltd.
Bangalore, Karnataka, India

ISSN 2367-3370 ISSN 2367-3389 (electronic)
Lecture Notes in Networks and Systems
ISBN 978-3-030-82528-7 ISBN 978-3-030-82529-4 (eBook)
https://doi.org/10.1007/978-3-030-82529-4

This Springer imprint is published by the registered company Springer Nature Switzerland AG
The registered company address is: Gewerbestrasse 11, 6330 Cham, Switzerland

Always Sebastian Schreiter did an excellent job to edit this book.
REV2022 will be held at British University of Egypt in Cairo.

Michael E. Auer
REV General Chair

Kalyan Ram Bhimavaram
IAOE President, REV2021 Co-chair

Xiao-Guang Yue
IETI Hong Kong, REV2021 Co-chair

Preface

The REV Conference is the annual conference of the International Association of Online Engineering (IAOE) and the Global Online Laboratory Consortium (GOLC).

REV2021 on "Online Engineering and Society 4.0" was the 18th in a series of annual events concerning the area of remote engineering and virtual instrumentation.

In a globally connected world, the interest in online collaboration, teleworking, remote services and other digital working environments is rapidly increasing. In response to that, the general objective of this conference is to contribute and discuss fundamentals, applications and experiences in the field of online and remote engineering, virtual instrumentation and other related new technologies like cross-reality, data science and big data, Internet of Things and Industrial Internet of Things, Industry 4.0, cloud technologies, cyber security, and M2M and smart objects. Another objective of the conference is to discuss guidelines and new concepts for engineering education in higher and vocational education institutions, including emerging technologies in learning, MOOCs and MOOLs, and open resources.

REV2021 has been organized in cooperation with the International Engineering and Technology Institute (IETI), Hong Kong, and the Technical University of Tallinn, Estonia, as an online event from February 24 to 26, 2021.

REV2021 offered again an exciting technical program as well as networking opportunities. Outstanding scientists and industry leaders accepted the invitation for keynote speeches:

- **Jeff Kodosky**

  Co-founder of NI and The Father of LabVIEW

- **Javier García-Zubía and Gustavo R. Alves**

  University of Deusto, Spain, and ISEP Porto, Portugal

- **Raphael Rohde**

  Manager Technology Development, PHOENIX CONTACT GmbH & Co. KG
- **Krishna Vedula**

  Executive Director, Indo Universal Collaboration for Engineering Education (IUCEE)
- **Gabriel Xiao-Guang Yue**

  Rector Advisor of European University Cyprus and College Advisor of RICE, RMUTR, Thailand
- **Mark Easley**

  University Marketing Manager and MCU Applications Specialist at Texas Instruments

It was in 2004 when we started this conference series in Villach, Austria, together with some visionary colleagues and friends from around the world. When we started our REV endeavor, the Internet was just 10 years old! Since then, the situation regarding online engineering and virtual instrumentation has radically changed. Both are today typical working areas of most of the engineers and are intrinsically tied to

- Internet of Things
- Industry 4.0 and Industrial Internet of Things
- Cloud Technologies
- Data Science
- Cross and Mixed Reality
- Remote Working Environments
- Online & Biomedical Engineering

to name only a few.

With our conference in 2004, we already tried to focus on the upcoming use of the Internet for engineering tasks and the opportunities as well as challenges around it. And as we can see today, it was the right decision.

The REV 2021 Conference takes up the following topics in its variety and discusses the state-of-the-art and future trends under the global theme "Cross Reality and Data Science in Engineering":

- Applications and Experiences
- Artificial Intelligence
- Augmented and Mixed Reality
- Big Data
- Biomedical Engineering
- Cyber Physical System
- Cyber Security

- Collaborative Work in Virtual Environments
- Data Science
- Evaluation of Online Labs
- Human Machine Interaction and Usability
- Internet of Things and Industrial IoT
- Industry 4.0
- M2M Concepts
- Networking, Edge and Cloud Technology
- Online Engineering
- Process Visualization and Virtual Instrumentation
- Remote Control and Measurements
- Remote and Crowd Sensing
- Smart Objects
- Smart World (City, Buildings, Home, etc.)
- Society 4.0 (Impact of COVID on Engineering Education)
- Standards and Standardization Proposals
- Teleservice and Telediagnosis
- Telerobotics and Telepresence
- Teleworking Environments
- Virtual and Remote Laboratories

Special thematic tracks have been held on the following topics:

- Non-traditional Labs and Lab Networks in the IoT and Industry 4.0 Domain
- IoT—Smart Materials and Systems
- Biomedical Engineering and Computing

Submission types have been accepted:

- Full Paper, Short Paper
- Work in Progress, Poster
- Special Sessions
- Workshops, Tutorials

All contributions were subject to a double-blind review. The review process was extremely competitive. We had to review near to 200 submissions. A team of over 90 program committee members and reviewers did this terrific job. Our special thanks go to all of them.

Due to the time and conference schedule restrictions, we could finally accept only the best 65 submissions for presentation or demonstration. The conference had over 140 participants from 29 countries from all continents.

We thank our Estonian colleagues from the Technical University of Tallinn and Videal Productions OÜ for conducting the online conference.

Committees

General Chair

Michael E. Auer	Founding President and CEO of the IAOE, CTI Frankfurt/Main New York, Vienna, Bangalore

REV2021 Co-chairs

Kalyan Ram Bhimavaram	IAOE President and Electrono Solutions Pvt Ltd, India
Xiao-Guang Yue	International Engineering & Technology Institute, Hong Kong

Program Co-chairs

Dominik May	University of Georgia, USA, and IAOE Vice President
Doru Ursutiu	Transylvania University of Brasov and IAOE Past President, Romania
Rita Y. M. Li	Hong Kong Shue Yan University, Hong Kong

International Advisory Board

Abul Azad	President, Global Online Laboratory Consortium, USA
Denis Gillet	EPFL, Lausanne, Switzerland
Bert Hesselink	Stanford University, USA
Zorica Nedic	University of South Australia
Teresa Restivo	University of Porto, Portugal
Hamadou Saliah-Hassane	Université TÉLUQ, Montreal, Canada
Cornel Samoila	University of Brasov, Romania

Franz Schauer	Tomas Bata University, Czech Republic
Tarek Sobh	University of Bridgeport, USA
Claudius Terkowsky	TU Dortmund University, Germany
Vasant Honavar	Penn State University, USA
Valerie Varney	RWTH Aachen, Germany
Krishna Vedula	IUCEE, India

Technical Program Co-chairs

Abul K. M. Azad	Northern Illinois University, DeKalb, IL, and GOLC President, USA
Sebastian Schreiter	IAOE, France

IEEE Liaison

Manuel Castro	UNED, Madrid, Spain

Workshop and Tutorial Chairs

Andreas Pester	The British University in Egypt, Cairo
Valerie Varney	RWTH Aachen, Germany

Special Session Chairs

Alexander Kist	University of Southern Queensland, Australia
Teresa Restivo	University of Porto, Portugal

Publication Chair and Web Master

Sebastian Schreiter	IAOE, France

International Program Committee

Akram Abu-Aisheh	Hartford University, USA
Yacob Astatke	Morgan State University, USA
Gustavo Alves	Polytechnic of Porto, Portugal, and IAOE Vice President
Nael Bakarad	Grand Valley State University, USA
David Boehringer	University of Stuttgart, Germany
Michael Callaghan	University of Ulster, Northern Ireland
Manuel Castro	UNED, Madrid, Spain
Torsten Fransson	KTH, Stockholm, Sweden
Javier Garcia-Zubia	University of Deusto, Spain
Denis Gillet	EPFL, Lausanne, Switzerland

Contents

Virtual and Remote Laboratories

Work-in Progress: Development of an E-learning FPGA Platform Following the IEEE SA Std. 1876 - 2019 Standard for Networked Smart Learning Objects for Online Laboratories

Marcelo Daniel Berejuck[1(✉)], Jhennifer Cristine Matias[1], Emilio Werner[1], and Hamadou Saliah-Hassane[2]

[1] Federal University of Santa Catarina, Florianópolis, Brazil
marcelo.berejuck@ufsc.br
[2] Université TÉLUQ, Montreal, Canada

Abstract. Low budget is a common problem faced by South American universities and many other universities and vocational education institutions in developing countries in Africa. The learning of Hardware Description Languages, like VHDL or Verilog, is compromised due to the lack of FPGA (FPGA is an acronym of Field Programmable Gate Array) development boards for undergraduate students from those institutions. It is an issue that can be addressed with hardware remote Labs: a solution that has been adopted by institutions offering undergraduate courses via distance education. The hardware remote lab development for this purpose requires planning of the technology that will be adopted, to prevent problems for the whole system, such as secure access, system retrieving and recovering, or data storing. Our proposal is the development of an e-learning FPGA platform that will be guided by good practices complying with the IEEE SA Std.1876 - 2019. We noticed that so far, there are only a few remote laboratories dedicated to the learning of FPGA. No one of them complies with the recent IEEE SA Std. 1876 - 2019 for Networked Smart Learning Objects for Online Laboratories. At the end of the project, we are expecting the delivery of a fully functional FPGA platform for e-learning.

Keywords: Online laboratory · Remote laboratory · FPGA · IEEE SA Std. 1876 - 2019

1 Introduction

Remote laboratories, or remote labs, are defined as automatized hardware equipment that works under the control of specialized software to allow access to realize didactic experiments through the Internet [1]. In other words, a remote Lab is a real laboratory which is controlled and monitored remotely [5]. They can

M. E. Auer et al. (eds.), *Online Engineering and Society 4.0*, Lecture Notes in Networks and Systems 298, https://doi.org/10.1007/978-3-030-82529-4_1

be categorized as interactive and non-interactive labs. Interactive labs allow the users to change parameters or configuration of the hardware experiment; on the other hand, the non-interactive execute an experiment with fixed settings and configurations most of the time in what is called in batch mode.

Examples of interactive remote labs. to name only a few are: NetLab at the University of South Australia (Australia) [7], eLab3D at the *Politécnica de Madrid* (Spain) [6], LaboREM at the University of Pau and Adour Province (France) [8], RELLE at Federal University of Santa Catarina (Brazil) [2], and VISIR at the *Blekinge Tekniska Högskola* (Sweden) [9].

Remote labs are usually available through a Remote Lab Management System (RLMS), which are software platforms capable of supporting the hardware experiment as well as technical support documents. An essential feature of an RLMS is the front end that will be available to the users, because it must be simple and intuitive to capture the users' attention. VISIR (Virtual Instrument Systems in Reality) provides the most popular front end in the e-learning communities. It is an Open Lab Platform, which is an architecture for opening hands-on instructional laboratories for remote access 24/7 with preserved context [5]. It was initially conceived to deal with basic electrical circuits and analog electronic interfaces. Developed by the Blekinge Tekniska Högskola (BTH) in Sweden together with National Instruments in USA and Axiom EduTech in Sweden, they targeted to disseminate an online laboratory concept created at BTH using open source technologies in collaboration with other universities and organizations [9]. Nowadays, several remote Labs have adopted VISIR, like PILAR project.

PILAR project [5] aims to interconnect VISIR "nodes" into a common platform in order to make use of the individual services provided by each node and create federated services through an international consortium. The nodes at the PILAR consortium are from Blekinge Tekniska Högskola from Sweden, *Universidad Nacional de Educación a Distancia* and *Universidad Católica de Deusto* from Spain, *Instituto Politécnico do Porto* from Portugal and *Fachhochschule Kärnten* from Austria.

PILAR project is an example of how important it is to follow standards and good practices in the development of a system. The interconnection of those nodes implies the need for some standards to let more efficient and secure will be the connection of the whole system. IEEE addressed this issue and established a working group to organize standards focus on Online laboratories. As a result, in 2019 IEEE Standards Association approved the IEEE SA Std. 1876 - 2019 [10]. This standard cover issues related to "methods for data storing, system retrieving, and accessing online laboratories as smart and interactive learning objects". "The standard also defines methods for integrating online laboratories as smart learning objects in learning environments and learning object repositories".

This paper introduces a research proposal for the development of an e-learning FPGA platform that will be guided by good practices complying with the IEEE SA Std. 1876 - 2019 and will be stored as a validated use case. The FPGA boards are usually expensive, which can be an issue for several universities and educational institutions. In this context, we understand that the present research proposal will fulfill this gap, offering access to FPGA boards through the remote lab concept, and according to IEEE SA Std. 1876 - 2019. Next sections will give an insight into the components, from hardware and software, as well as the resources and efforts that must be employed on this development.

2 An Overview of IEEE SA 1876-2019 Standard

The IEEE SA Std. 1876-2019 for networked smart learning objects for online laboratories was sponsored by the IEEE Education Society. This standard establishes the relationship between all the components (hardware, software, and learning environments) in order to ease the design and implementation of pedagogically driven online laboratory activities. This standard is designed to ease the design, the implementation, and the usage of pedagogically oriented online laboratories as smart learning objects and their integration in learning environments and learning object repositories [10].

Figure 1 illustrates the implementation layers of online Labs as smart learning objects that was defined in the standard. The first layer standardizes an online Lab as a Service (LaaS), which can be personalized at the second layer. The second layer describes an online Lab as a Learning Object (LO), which can be integrated in various learning environments, including Massive Open Online Courses (MOOC), Learning Management Systems (LMS), learning resource repositories, and mobile applications [10].

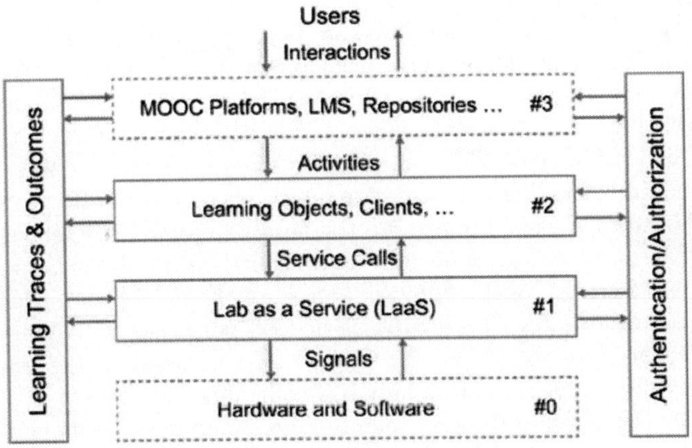

Fig. 1 Conceptual layers and normative information [10]

Next section, we will introduce the hardware and software components that must be developed to build the proposed FPGA platform. All components will we described and related to the conceptual layers depicted in Fig. 1.

3 Hardware and Software Components

Figure 2 illustrates a complete e-learning scenario, from the end-user up to the remote hardware. The end-user will require FPGA experiments from the e-learning platform. For this reason, we understand that the user interface must be an application running on a personal computer (PC), in which the user should be developing his bitstream with FPGA manufacturer tools that will be loaded by the hardware platform. The user application is related to the level #3.

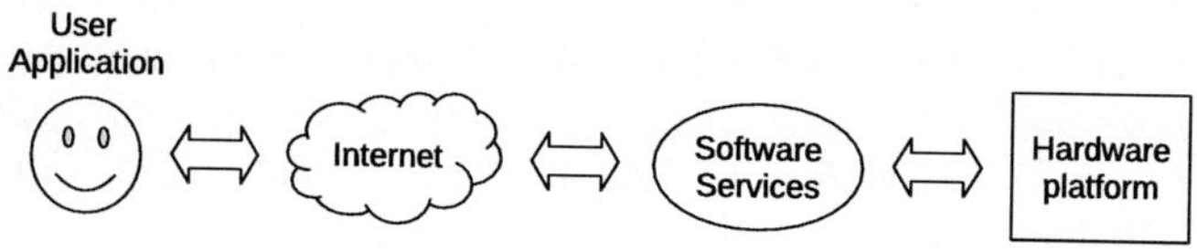

Fig. 2 Overview about the hardware and software to be developed

The software services depicted in Fig. 2 refer to the development of Learning Objects and services for LaaS (Lab as a Service) and the components generated will be related to the #2 and #1 levels. The hardware developed for the platform, which includes standards FPGA boards, will be related to the level #0. All of these components will be described in the following subsections.

3.1 Layer #0—Hardware Platform

Figure 3 illustrates the minimum hardware needed to build the remote Lab proposed in this document. The FPGA development boards are usually made with some hardware resources. The board shown in Fig. 3 has m LEDs, n Seven Segment Displays, and k generic sensors. We acquired an FPGA board from Digilent®, called "Nexys A7-50T" for our prototype. Is is also possible to build an extension board, with different hardware interfaces, and connect it to the FPGA board (see bottom of Fig. 3) to expand the number of hardware interfaces. However, this extension board can be done later after the first platform prototype.

The bitstream sent by the user will configure the FPGA through the JTAG cable. The adoption of a JTAG cable is needed when the FPGA development board does not have the JTAG interface inside it. Most of the development boards available in the market already have this interface included.

There are two ways to the user monitor the experiment: visualizing the real-time images captured by a video camera, or receiving information by RS232-USB cable; both ways will be managed by Application Programming Interfaces (APIs) in the PC.

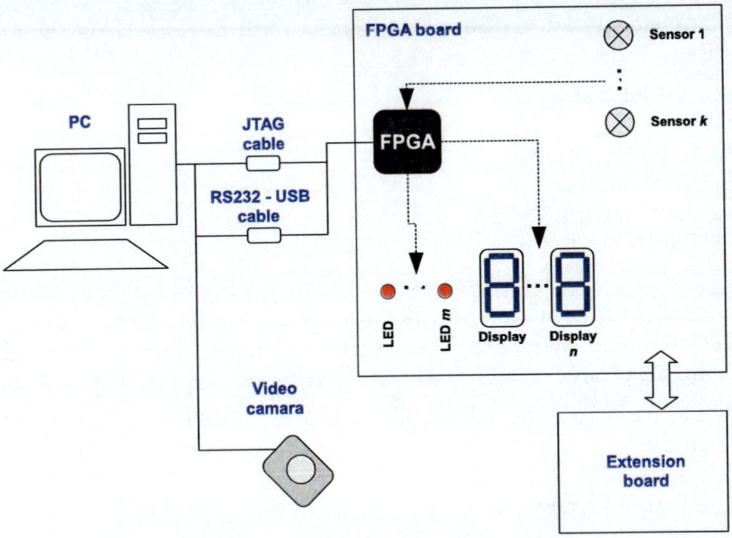

Fig. 3 Minimum hardware required for the remote Lab

3.2 Layer #1—Laboratory as a Service (LaaS)

The set of Application Programming Interfaces that will control the access to the hardware platform is an implementation of the layer #1 of IEEE SA Std. 1876-2019. A LaaS consists of a set of well-defined services that enable interoperability with external applications and internal functionalities. This point must be carefully discussed later, because components developed by the user, such as VHDL components, can also be considered services with its internal functionalities. So far, we will take into account the hardware available in the development board that will be used in the project to define internal functionalities. Other services will implement the communication with the user and the authentication/authorization. Figure 4 shows four services that are required by IEEE SA 1876-2019, and three services classified as optional.

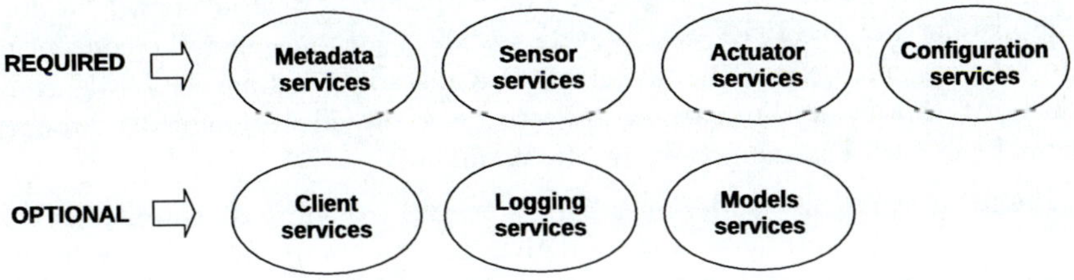

Fig. 4 Required and optional services for the LaaS

The implementation of the require services will be done using the following methods:

- getSensorMetadata();
- getSensorData();
- getActuatorMetadata();
- setActuatorData();
- getConfiguration(); and
- setConfiguration().

The metadata methods return a description of the online lab, its mechanisms and services. The information provided by this service shall be sufficient to define a user interface and the related client-server communication. Sensors and actuators have "setX" and "getX" methods to write and read data, respectively.

As optional, the following methods can be developed:

- getClients();
- getExperiment();
- getLoggingInfo(); and
- getModels().

Client service is a list of client applications providing a user interface to the online lab may be provided by the online lab itself. Logging service provides a method to retrieve logged user actions. This service can be implemented as Logging and Alarms functionality.

Models service relates various information about the connected equipment can be sent to the client application. For example, a 3D (or 2D) graphical model describing the connected equipment could be defined in the form of a Virtual Reality Markup Language (VRML) file. Similarly, a mathematical model describing the connected equipment can be sent in the form of dynamical equations, that could be used to simulate the equipment at the client side. Both these models could be read or modified through the means of virtual sensors/actuators.

3.3 Layer #2—Learning Objects and Clients

The proposed remote lab must be pedagogically sound. Within the context of this standard, the remote lab should fit in learning frameworks as a Learning Object (LO). To reach this target, the development of this layer will be done using as reference the IEEE Std 1484.12.1-2002, which specifies the conceptual data schema that defines the structure of a metadata instance for a learning object. At this layer the learning objects must provide pedagogical support. Some of the most common resources are as follows:

- The lab assignment: guides and digital formats to report the lab results;
- The course content: class lessons, digital documents, or multimedia components;
- The external Content: external multimedia components used as support for the students; and
- The assessment: tests with questions about the lab activities.

A complete pedagogical document must be prepared to address the most relevant topics related to hardware synthesis language.

4 Final Remarks and Expected Results

This research proposal embodies different skills to acquire and some tools to be developed. The challenges to overcome include hardware, firmware, and software issues at different abstraction layers, and we understood that there is a lot of research and development work to carry on. Our research teams from Laboratory at Distance (L@d) at TÉLUQ University in Montreal, Canada and from the Remote Experimentation Laboratory (RExLab) of Federal University of Santa Catarina (UFSC), Brazil are working together on this research proposal. The teams carefully analyzed the good practices available on the IEEESA Std.1876 - 2019 and we believe that it is possible to design and deploy a fully functional prototype complying with most of the mandatory requirements of the mentioned standard as a use case. We are optimistic because we understand that we will not need to expend too much time discussing implementations approach, once the IEEE Standard teamwork already did it and presented it to the scientific community as the IEEE SA 1876 -2019 standard.

References

1. Orduña P et al (2015) An extensible architecture for the integration of remote and virtual laboratories in public learning tools. IEEE Rev. Iberoamericana de Tecnologias del Aprendizaje 10(4):223–233
2. Simão JP, de Lima JP, Saliah-Hassane H. da Silva JB, Alves JB (2020) Laboratórios online móveis em um ambiente de experimentação colaborativo. In: Proceedings of XIV Brazilian symposium of collaborative systems, pp 11–20. https://sol.sbc.org.br/index.php/sbsc/article/view/9945
3. García-Zubía J (2015) Archimedes remote lab for secondary schools. In: 3rd experiment international conference (exp.at 2015), pp 60–64
4. Harward VJ (2008) The iLab shared architecture: a web services infrastructure to build communities of internet accessible laboratories. Proc IEEE 96(6):931–950
5. Garcia-Loro F, Cristobal ES, Castro M (2019) PILAR: a federation of VISIR systems for analog electronics. In: 5th experiment international conference (exp.at 2019), pp 260–261
6. López S, Carpeño A, Arriaga J (2015) Remote laboratory eLab3D: a complementary resource in engineering education. IEEE Rev. Iberoamericana de Tecnologias del Aprendizaje 10(3):160–167
7. Stefanovic M, Tadic D, Nestic S, Djordjevic A (2015) An assessment of distance learning laboratory objectives for control engineering education. Comput Appl Eng Educ 23(2):191–202
8. Luthon F, Larroque B (2015) LaboREM–a remote laboratory for game-like training in electronics. IEEE Trans. Learn. Technol. 8(3):311–321
9. Gustavsson I, Zackrisson J, Håkansson L, Claesson I, Lagö T (2020) The VISIR project: an open source software initiative for distributed online laboratories, m. 01, vol 2
10. IEEE Standard for Networked Smart Learning Objects for Online Laboratories (2019). https://standards.ieee.org/standard/1876-2019.html

On the Development of a Unified Online Laboratory Framework

Johannes Nau(✉), Johannes Richter, Detlef Streitferdt, Karsten Henke, Robert Niklas Bock, and Andreas Mitschele-Thiel

Technische Universität Ilmenau, Ilmenau, Germany
{johannes.nau,johannes.richter,detlef.streitferdt,karsten.henke, robert-niklas.bock,mitsch}@tu-ilmenau.de

Abstract. This work focuses on the requirements analysis of a modular framework to simplify developing and integrating new or existing remote experiments and laboratories. Besides this technical view, this paper also gives an organizational view on developing such a framework and managing the corresponding modules, which can also be developed by a third party. On a technical side, we provide the requirements to ultimately define the interface between different modules, enabling easy integration on different abstraction levels.

The work's basis is the review of individual remote laboratories and existing systems of the past two decades and the author's collective experience. In the spirit of uniting the remote laboratory community, we will follow the IEEE 1876 standard wherever possible and extend it to make our vision of an easy to use, integrable, and extendable framework possible.

Keywords: Virtual and remote laboratories · Standardization · Requirements engineering · Collaboration

1 Introduction

Remote laboratories allow access to various real experiments. When the laboratory simulates the experiment, it is called a virtual laboratory. Both types are a subgroup of online laboratories. For this paper's scope, we use the terminology of online laboratories as experiments that resemble real experiments. This terminology excludes learning objects that for example represent mathematical relationships or provide access to software. We also like to focus on online laboratories deployed for teaching purposes. The idea of such laboratories for "hands-on" training of students has been around for about 20 years.

Many of these laboratories are developed from the ground up. Some solutions, known as Remote Laboratory Management Systems (RMLS), have been developed to simplify the development of laboratories. However, these predefined solutions often restrict the type of experiment that can be managed or require the developer to handle most of the experiment, including communication themselves, only providing high-level functions like user management.

M. E. Auer et al. (eds.), *Online Engineering and Society 4.0*, Lecture Notes in Networks and Systems 298, https://doi.org/10.1007/978-3-030-82529-4_2

Currently, there is some effort in the community to standardize the external interface of online laboratories. We believe that this is a way to reduce the effort to develop, maintain, and extend an online laboratory. Our research aims to provide a set of loosely coupled components with a well-defined interface to allow better reuse of components in an online laboratory.

2 Online Laboratories and Management Systems

To develop a unified approach for a remote laboratory framework, we first analyzed existing remote laboratories, RLMS, architecture proposals, and standards.

We have analyzed 300 papers selected from over 2000 papers identified by searching the *IEEE Xplore* digital library. The first selection was made manually based on the paper's title, in case it was obvious the paper would not fit because the title does not suggest the description of an online laboratory or RLMS. At the second level, the selection was based on the abstract of the paper, respectively.

We will only reference the most important and exemplary papers in this section because of the tremendous number of papers.

2.1 Online Laboratories

During the literature review, we have developed a taxonomy to classify remote laboratories given in Fig. 1.

The first distinction of an online laboratory can be made based on the execution type of experiments. This classification is well known and presented in

Classification based on			Reference
Type of Execution	Virtual		[3]
	Hybrid		[4]
	Remote	Realtime	[5]
		Batch	[14]
Type of Laboratory	Mechanical		[9]
	Chemical		[11]
	Electronic		[40]
	Embedded Systems		[5]
	Robotics		[41]
	Manufacturing		[13]
	Radio Communication		[12]
Autonomy of Laboratory	Self-contained		[7]
	Remote coupled		[5]
Level of Integration	Execution only		[14]
	Integrated Development and Analysis Environment		[5]
User Interface	Remote Desktop		[40]
	Website		[14]
	Proprietary		[16]

Fig. 1 Taxonomy of online laboratories with exemplary references on the right

various papers (e.g. [1,2]). A virtual online laboratory is one where the whole experiment is just a simulation presented to the user. In a remote online laboratory, the user has control over a physical setup. The latter one is further divided into real-time and batched laboratories. A real-time laboratory allows continuous control over the experiment, while a batched laboratory queues parameterized experiments for execution on the remote hardware. A mix of a remote and a virtual laboratory is called a hybrid lab.

An example of a virtual laboratory is described in [3] simulating industrial robots in a web environment. Examples of batched laboratories are multiple instances of *VISIR* [4]. A typical hybrid laboratory is *GOLDi-Labs* [5], where either (or both) the controller or the physical hardware model can be simulated or operated in reality.

The type of laboratory is another classification criterion. Most of the laboratories we analyzed allow either the execution of control algorithms on various embedded systems (e.g. a coupled tank system [6]) or make measurements on either a digital or analog electronic hardware [4,7]. Some laboratories offer physical experiments, mostly resembling mechanical experiments like various motions on a pneumatic rail [8] or a material characterization experiment [9]. There are also various experiments in the photovoltaic domain [10]. Some more specialized laboratories offer chemical experiments [11], experiments on radio communications [12], or include manufacturing methods in a remote laboratory [13].

A third classification decides whether the online laboratory is merely a remote interface to an experiment that could theoretically be operated locally, or the online laboratory is part of the experiment and necessary to operate the experiment. We call the last mode "remote coupled laboratory". An excellent example of such a remote coupled laboratory is *GOLDi-Labs* [5].

Another differentiation can be made between simple experimental laboratories (e.g. [14]), and online laboratories with integrated development tools. These tools help to either plan, develop, analyze the experiment (e.g. [5]), or to collaborate with students and teachers (e.g. [7]). Specifically mentioning *RALFie* [15], it follows a different approach, where the experiment is not hosted in a centralized form dictating the final setup. The user creates the setup in collaboration with other users. Then, experiments will run on the self-designed hardware solutions.

The provided user interface is the last differentiation of this paper's analysis. Some solutions are based on remote desktop solutions, while the vast majority uses a web interface. Nevertheless, we also found some solutions where the student needs a proprietary client software. This solution is especially appealing if the remote laboratory has some advanced control or feedback mechanism. An example of such an advanced control mechanism is [16], where the authors provide haptic feedback to a beam bending experiment.

2.2 Management Systems and Remote Laboratory Networks

Library of Labs (LiLa) [17] is a now-discontinued project hosting various engineering and natural sciences experiments. The project was a collaboration between eight universities and three companies from around Europe. In November 2011, the project listed 41 remote experiments and 218 virtual laboratories [18]. One of the central features of the project is its scheduling system to book experiments.

SAHARA [19] is a software system to manage and run virtual experiments developed by the University of Technology Sydney (UTS), Australia. This software was developed and has been used by the *Labshare consortium* [20] to share laboratory setups between Australian educational facilities. The system has three components, the web interface, the scheduling subsystem, and the rig client. The web interface controls the system and gives access to remote experiments. The rig client interfaces with the actual hardware setup and provides a software abstraction of the experiment. The scheduling subsystem controls the concurrent booking of individual rigs.

WebLab-Deusto [21] is an open-source RMLS developed by the University of Deusto, Spain. Its key features are user access control, scheduling access to experiments, and sharing experiments between different institutions. The supported experiments can range from interfaced real-world setups to virtual laboratories used for educational purposes. As a commercial spin-off, *LabsLand* [22] sells consulting services and experiment setups based on *WebLab-Deusto*.

GoLab is an initiative of the University of Twente, Netherlands. It originated from the identically named *GoLab-Project* as an ecosystem for innovative learning technologies [23]. It hosts the platforms *Golabz* and *Graasp*, with *Golabz* being used for sharing and support and *Graasp* being the authoring and learning platform of the project. In October 2020, the project lists 53 remote laboratories and 562 virtual laboratories. The *Next-Lab* project extended the scope to include tools for inquiry-based science education [24].

REMLABNET [25] proposed an elaborate setup of services and software systems to host remote laboratories and integrate them into classroom activities. Besides features like experiment scheduling, sharing, and user access control, the proposed system aims for collaborative use cases. The proposed classroom includes interactions between the users, like voice chat and passing control over the experiment to individuals. The project had two successive publications describing the further development and integration with *Go-Lab* [23].

Lab2Go [26] proposes a web-based index of existing remote and virtual laboratories. The paper suggests an ontology outlining the setups and their preferred ways of accessing them to collect and browse information about different approaches towards remote laboratories. With the use of an open-source ontology Wiki, the project aims to host a comprehensive list of remote and virtual laboratories.

The *iLab Shared Architecture* [27] proposes a unifying architecture and middleware to build and host remote laboratories. Key features are single sign-on systems with appropriate user and rights management to control access to laboratories. The project aims for an architecture suitable for a wide range of fields and audiences.

VISIR [4] is a widely adapted remote laboratory for wiring and measuring electrical circuits. The designed circuit is then checked for validity and modeled on actual hardware, where the desired measurements are taken and reported back to the user. The authors highlight the superiority of their approach over purely simulated approaches, as the user experiences noisy real-world measurements instead of calculated results without any deviation.

DigiLab4U [1] aims to incorporate new technologies from domains like the Internet of Things (IoT) and Industry 4.0 into their remote experiments. This approach focuses on interconnected machines and sets them apart from typical, directly controlled setups.

Gateway4Labs [28] is an architecture and open source project for remote laboratories federation. The project aims towards a middleware solution to communicate between remote labs and content or learning management systems (LMS). Having an abstract layer between the components minimizes the development effort needed for developing specific plugins allowing for exchange between the laboratories and the LMS.

2.3 Standards and Conventions Regarding Remote Laboratories

There have been various approaches to standardize interfaces and architectures for remote laboratories. In [29], the authors proposed a new "smart device" paradigm, which envisions a remote laboratory as an intelligent device. This device handles all necessary work and offers services to a thin client like a browser.

The *Sharable Courseware Object Reference Model* (SCORM) [30] is a widely accepted set of definitions, enabling the exchange of online learning modules. These modules include remote laboratories as well as other e-learning approaches. Through its widespread support among learning management systems, users can host courses and laboratories and track their users' progress on different platforms.

The *Experience-API* [31], xAPI for short, is a successor of SCORM. As SCORM was tailored for desktop applications and single point of data recording, the need arose to track the users' learning progress across multiple nodes and collect information from more than one course. The goal of the xAPI is to record and transmit user interactions and study progress across a wide set of learning and training systems in order to provide all users with a holistic learning and teaching experience.

Learning Tools Interoperability by IMS Global Learning Consortium (IMS-LTI) is a set of definitions, standards, and references, which define how learning tools can be integrated into learning management systems [32]. Systems are divided into platforms and tools. Platforms like learning management systems use tools to supplement their functionality. Tools are external applications or services providing problem-oriented functionalities. IMS-LTI builds on OAuth [33] for authentication and uses the JavaScript Object Notation for information interchange. As a complete ecosystem of services, IMS-LTI implements student assignments, recording and grading student performance, linking of knowledge bases, as well as general user and role management.

Most recently, the IEEE Std 1876-2019 [34] has been published. The standard is viewing online laboratories as layer structure consisting of the hardware, Laboratory as a Service (LaaS), learning objects, and platform layers. It defines the LaaS layer, required methods, and provides an application profile for the IEEE Std 1484.12.1-2002 [35] called Online Laboratories Learning Object Metadata (OL LOM). The support of Learning Analytics and activity tracking is given through the use of xAPI. IEEE Std 1484.12.1-2002 defines the data model describing learning objects. The XML Binding for this model is defined in IEEE Std 1484.12.3-2020 [36].

3 Requirements for a Unified Remote Laboratory Framework

The last section showed that we find various experiments offered in remote laboratories. Some of them are developed from the ground up, while others use existing software solutions. The most popular ones—in our opinion—are *VISIR* and *WebLab Deusto*. Both, however, have the shortcoming of not supporting all experiment setups that are possible.

Even the existing standard only supports remote coupled laboratories as configurable laboratories, where the experiment will be configured before execution. However, the implementation as a laboratory configuration does not allow the user to connect existing but physically separated online laboratories with each other. Such an implementation also does not allow to run mixed experiments with simulations on the client and real hardware.

One possible application where such a remote coupled experiment is of enormous value are cases where students will get only a part of the necessary hardware to operate an experiment, while costly or big mechanical models are accessible remotely. In this scenario, the hands-on experience and the remote laboratory character will blend into a more immersive experience for the student. As the network latencies will become noticeable at some point, mechanisms to cope with this latency should be part of the framework.

The framework needs to implement state of the art authentication and authorization mechanism to meet the need for security. This requirement is particularly important because the framework will deal with the users' highly personalized learning data. The online laboratory user will be safe by design, as there is no physical threat to the user. However, the equipment might require protection, especially when dealing with non-predictable latencies between sensors and actuators. For example, a limit switch should stop the driving motor independently from the actual controller output to avoid damage to the model.

Apart from this shortcoming, the implementation of the existing standards (IEEE 1876 and IEEE 1484.12) is still missing. As to our knowledge, the existing software solutions, namely *WebLab Deusto* and *VISIR*, are currently not implementing the standard. We see a significant benefit in standardized software and ready to use components and systems to streamline online laboratory development. Therefore, we analyzed the requirements necessary to build such a system of components (framework).

Potential users of the framework are teachers, engineers, and developers aiming towards the creation of an online laboratory. To visualize different types of framework users, we used the "Mort, Elvis, Einstein"-Paradigm [37]. We came up with the following stereotypical personas:

- Mort. A teacher who is not knowledgeable in computer engineering. He wants to bring an existing experiment online with the least amount of effort.
- Elvis. Someone who wants to have great control over the deployed online laboratory. He can do some programming if necessary. However, he is happy to adopt most existing solutions and only modifies parts that do not work to his expectations.
- Einstein. He is foremost a developer. He is skeptical about the solutions already provided and optimizes everything in the online lab.

The framework's functionality will be used to implement an online laboratory or even a RLMS. All experiments we reviewed in Sect. 2 should be implementable with this unified framework, which means that all hardware currently used in an online laboratory and the different deployment options must be supported. As the implementation of all combinations and possible setups seems impossible, especially since more options might be available in the future, an extensible modular approach seems more viable than a monolithic approach.

From the perspective of Mort and Elvis, the framework should not only consist of interfaces without any implementation. It should allow the implenentation of a typical online laboratory setup from the hardware up to the client. Mort wants the system to support various environments so the online laboratory will run on the hardware and software already present. The framework itself should be platform agnostic whenever possible.

Elvis and Einstein are able to add missing parts like a hardware abstraction for their custom hardware or a custom authentication schema. For them, implementing against the interface should be straight forward to minimize development time. Elvis will benefit from a fine-grained architecture because this maximizes the reusability of existing modules. For example, a coarse module implements a whole experiment, including hardware abstraction, controller programming, model protection, and communication: When changing the hardware, a developer needs to re-implement everything. However, if all those functions are modules themselves, only the hardware abstraction needs to be adapted.

As Einstein will be implementing most of the modules himself, the interfaces he implements against need to be clearly described so that the resulting system stays compatible with the rest of the framework. Einstein might handle cases the original framework design did not consider, so the interface needs to be extensible.

We would like to mention that the standards reviewed in Sect. 2 share the same spirit as our thoughts. However, we see a benefit in introducing more semi-standardized interfaces withhin our architecture. Naturally, we want to keep the interfaces already defined by the standard. Also, interfaces defined by popular RLMS like Weblab Deusto should be taken into account.

From Mort's perspective, such a framework is not ideal because selecting and integrating the right modules still needs to be done. Therefore, we plan to develop a configuration tool based on a product line perspective for online laboratories. This configuration tool allows the user to select various options for the online laboratory he wants to implement. The tool will check against different constraints to guide the user through the selection process and advises on what to select. The proposed tool functions comparable to a car configurator. Even though we are focusing on the online laboratory's technical side, the same tool may suggest learning scenarios supported by the individual online laboratory.

The restrictions for the product line will be based mainly on the supported interfaces. There may be different extensions for an interface – or even completely different interfaces, effectively limiting the modules that can be integrated with each other. Even if an API's interface seems to be identical on a functional level, the protocol of the interface with its timing and internal state transitions is different for each module and can even vary over different versions. For example, in an abstract interface for receiving and sending sensor values, the called methods may be the same. However, on a model connected with a high-speed bus, they may be called significantly more often than on a low-speed bus. Thus, one model could be used in a closed-loop control while the other could not. Formalizing such restrictions and providing a way to measure or define them easily will be part of future work.

4 High-Level Architecture

Figure 2 gives an overview of a prototypical architecture that could be implemented using the framework. The remote laboratory in the example consists of a microcontroller-board and an electromechanical hardware model. Both have an interface unit connected to them (noted as *Controller Server* and *Model Server*), which will offer the service abstraction of both. The *Experiment Client* would be used to configure and connect to an experiment. In this setup, the client would only connect to a load balancer that offers an IEEE 1876 compatible LaaS API. Also, the *Experiment Client* embeds a *Development Client*—this would be a code editor to program the *Microcontroller Board.* This client would connect to a *Development Service*, which provides the compilation environment, and eventually program the controller through the *Controller Server.*

Note that every component can be further decomposed. When inspecting the *Controller Server* and the *Model Server*, some subcomponents like the maintenance module (which will keep the software updated, implements self-testing, etc.) can be reused. However, other subcomponents like the *Hardware Abstraction* need to be implemented according to the actual hardware connected. The same decomposition can be done for the load balancer, and the *Development Service*—subcomponents (like the *Authentication*) may be reused.

Other implementation based on the framework could omit the *Load Balancer* and connect directly to the LaaS API of an experiment. The experiment itself

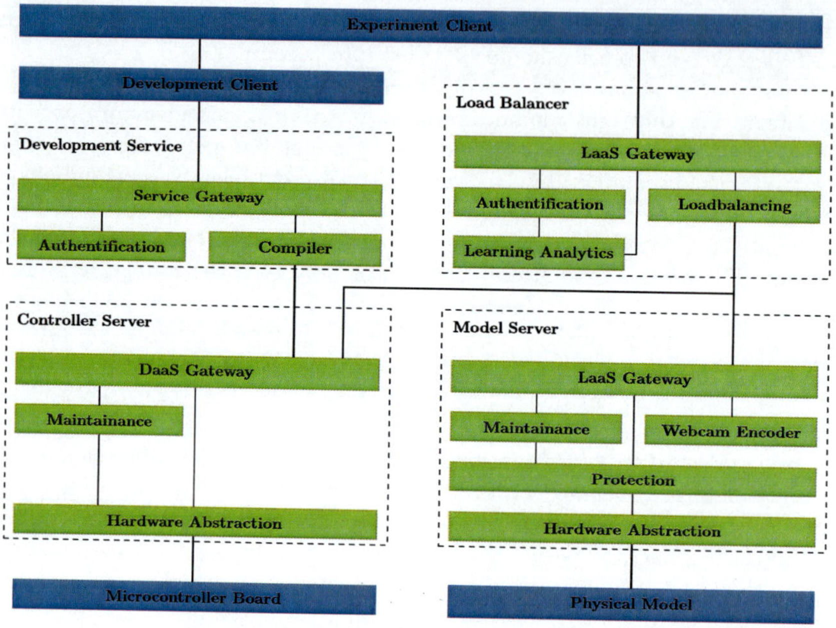

Fig. 2 Exemplary remote laboratory architecture

would then include the *Authentication* and *Learning Analytics* submodule, or use a load balancer to select the specific instances but then connect to them directly.

We also like to emphasize that some devices do not represent an experiment or laboratory by themselves. Therefore, it is not reasonable to implement a LaaS abstraction according to IEEE 1876 on them. We propose to use an abstraction we call *Device as a Service* (DaaS) where the device's capabilities will be exposed as an API. The API should be compatible with IEEE 1876 wherever possible and introduce additional features like a direct connection to other DaaS services in the same network to offer a low latency connection for sensor/actuator transmissions.

Keep in mind that this prototypical architecture lacks features like advanced user tracking, laboratory administration, and scheduling, which also need to be considered.

The first step towards a unified framework is the definition of the interfaces and protocols between services. Currently, we see the following interfaces that need to be defined without taking interfaces between subcomponents into account:

- Between different clients on the frontend
- Between client and online laboratory (IEEE 1876 is very promising)
- Between client and supporting services (like compiling, synthesizing, etc.)
- Between the framework and learning analytics (typically XAPI)

- Between the framework and Authentication provider
- Between administration panels and the framework
- The DaaS interface (might be an extended subset of IEEE 1876)
- The (physical) interface between the interface hardware and the experiment

Furthermore, each decomposition will result in additional interfaces. In the current exemplary architecture, these are mainly the interfaces between different gateways and the subcomponents. Because there are many more subcomponents than possible Gateway-Modules, it is better to reverse the definition flow. This inversion means that instead of defining modules' services, the gateway will provide a plugin type of interface for the submodules. Effectively, the Gateway defines the protocol to use rather than the other submodule.

As mentioned in the previous section, apart from the actual interface definition and implementation of the modules, a product line and configuration tool is also part of this unified framework. The architecture of the tool is independent of the actual framework. The tool is based on the description of different modules together with their interface and protocol description. For the description of interfaces, existing interface definition languages like OpenAPI [38] can be used. However, they fail to provide an adequate description of the protocol's exact timing behavior, so an extended description needs to be developed.

Based on this, we plan to develop a command line interface that can configure a desired online laboratory based on the implementers's preferences. The configuration is implemented by an executable script for each component to do the necessary tasks, like downloading or compiling the component. In essence, this will resemble a package manager like Vcpkg [39]. The same tool can be used to keep the infrastructure updated. Based on this tool, we aim to provide a web interface for the planning and configuration of an online laboratory as well.

Organizationalwise we think that a independent code versioning repository for each module is appropriate to reduce inter module dependencies. However, the product line and configuration tool should be contained in a single repository together with all scripts needed to set up a remote laboratory.

5 Summary and Future Work

Based on the literature search about individual online laboratories and RLMS, we developed and presented a plan for a unified remote laboratory framework coupled with a product line tool. The product line tool will help non-technical people to set up an online laboratory quickly. Simultaneously, the modularity will ensure high adaptability to new environments while keeping the reusability. While this tool will focus on the technical aspects of an online laboratory, it could also be augmented by didactic concepts and recommendations.

The complexity of this plan results from the need to ensure the interoperability of the different components. The first challenge of interoperability is to define clear interfaces that the components can implement. The second challenge is to keep the interfaces well-defined, including the timing behavior of the component. This challenge leads to a component-specific interface description that needs to be handled by the product line tool. Existing standards partly address the first challenge. However, additional interfaces need to be defined because of the high modularity. For this reason, both problems will be part of future work.

We are also aiming towards the design and implementation of a test environment to check for standards compliance. Even though future work's current focus is to develop the infrastructure for this framework, the development of useful components, as part of the framework, will also be included.

References

1. Pfeiffer A, Uckelmann D (2019) Open digital lab for you – laboratory-based learning scenarios in education, research and qualification. In: 5th Experiment International Conference (exp.at 2019), pp 36–41, IEEE (2019). https://doi.org/10.1109/EXPAT.2019.8876560
2. Tawfik M, Lowe D, Salzmann C et al (2015) Defining the critical factors in the architectural design of remote laboratories. IEEE Revista Iberoamericana de Tecnologias del Aprendizaje 10(4):269–279. https://doi.org/10.1109/RITA.2015.2486388
3. Dey U, Jana PK, Kumar CS (2016) Modeling and kinematic analysis of industrial robots in webgl interface. In: IEEE eighth international conference on technology for education (T4E), pp 256–257, IEEE. https://doi.org/10.1109/T4E.2016.067
4. Tawfik M, Sancristobal E, Martin S et al (2013) Virtual instrument systems in reality (visir) for remote wiring and measurement of electronic circuits on breadboard. IEEE Trans Learn Technol 6(1):60–72. https://doi.org/10.1109/TLT.2012.20
5. Henke K, Vietzke T et al (2016) Goldi - grid of online lab devices ilmenau. Int J Online Eng (iJOE) 12(04):11. https://doi.org/10.3991/ijoe.v12i04.5005
6. Bistak P (2012) Enhanced three-level hydraulic remote laboratory. In: 15th international conference on interactive collaborative learning (ICL), pp 1–4, IEEE. https://doi.org/10.1109/ICL.2012.6402210
7. Xavier MDS, Netto JFDM, Almeida TO, de Carvalho JM (2019) A proposal of a remote laboratory for pedagogical mediation in kinematics experiments. In: 2019 IEEE frontiers in education conference (FIE), pp 1–7. IEEE. https://doi.org/10.1109/FIE43999.2019.9028590
8. Andres-Gutierrez JJ, Gonzalez M, Gomez DZ (2010) Motionlab. In: IEEE EDUCON 2010 conference, pp 1499–1506, IEEE. https://doi.org/10.1109/EDUCON.2010.5492351
9. Ortelt TR, Sadiki A, Pleul C et al (2014) Development of a tele-operative testing cell as a remote lab for material characterization. In: International conference on interactive collaborative learning (ICL), pp 977–982, IEEE. https://doi.org/10.1109/ICL.2014.7017910
10. Blanchard R, Williams S, Bliss M et al (2014) Educating the world: a remote experiment in photovoltaics. In: 1st international conference on non conventional energy (ICONCE 2014), pp 28–32, IEEE. https://doi.org/10.1109/ICONCE.2014.6808676
11. Leal S, Leal JP (2013) A new chemistry e-lab experiment chemical equilibirum reaction. In: 2nd experiment@ international conference (exp.at 2013), pp 154–155, IEEE (2013). https://doi.org/10.1109/ExpAt.2013.6703050

12. Kara A, Aydin EU, Oktem R, Cagiltay N (2007) A remote laboratory for training in radio communications: Errl. In: IEEE 18th international symposium on personal, indoor and mobile radio communications, pp 1–5, IEEE. https://doi.org/10.1109/PIMRC.2007.4394833
13. Zeytunyan S, Hajian A (2014) Integrating design and manufacturing tools into ilmenau interactive hybrid online lab. In: 11th international conference on remote engineering and virtual instrumentation (REV), pp 227–230, IEEE. https://doi.org/10.1109/REV.2014.6784262
14. Winzker M, Kiessling R, Schwandt A, Paez CS, Shanab SA (2018) Teaching across the ocean with video lectures and remote-lab. In: IEEE world engineering education conference (EDUNINE), pp 1–4, IEEE. https://doi.org/10.1109/EDUNINE.2018.8451002
15. Kist AA, Maiti A, Maxwell AD (2015) Introducing ralfie — remote access laboratories for fun, innovation and education. In: 3rd experiment international conference (exp.at 2015), pp 124–125, IEEE. https://doi.org/10.1109/EXPAT.2015.7463236
16. Restivo MT, Lopes AM, Machado LDS, de Moraes RM (2011) Adding tactile information to remote & virtual labs. In: IEEE global engineering education conference (EDUCON), pp 1120–1124, IEEE. https://doi.org/10.1109/EDUCON.2011.5773287
17. RUS University of Stuttgart: Lila project. https://www.lila-project.org/content/index.html
18. The LiLa Project (2011) Lila - library of labs: dissemination of remote and virtual laboratories for natural sciences and engineering, December 2011. https://www.lila-project.org/about/presentation_neu/LiLa_Presentation.pps
19. Diponio M, Argente D. Sahara labs. https://sourceforge.net/projects/labshare-sahara/
20. The Labshare Institute: Labshare. https://web.archive.org/web/20190303121026/http://labshare.edu.au/home
21. University of Deusto: Weblab-deusto (2020). https://weblab.deusto.es/website/
22. LabsLand: Labsland: Real laboratories, on the internet (2020)
23. Golabz: Go-lab initiative — go-lab (2020). https://premium.golabz.eu/about/go-lab-initiative
24. Golabz: Next-lab — go-lab (2020). https://premium.golabz.eu/about/projects/next-lab
25. Schauer F, Krbecek M, Beno P, et al (2014) Remlabnet - open remote laboratory management system for e-experiments. In: 11th international conference on remote engineering and virtual instrumentation (REV), pp 268–273. IEEE. https://doi.org/10.1109/REV.2014.6784273
26. Zutin DG, Auer ME, Maier C, Niederstatter M (2010) Lab2go — a repository to locate educational online laboratories. In: IEEE EDUCON conference, pp 1741–1746. IEEE. https://doi.org/10.1109/EDUCON.2010.5492412
27. Harward VJ, del Alamo JA, Lerman SR et al (2008) The ilab shared architecture: a web services infrastructure to build communities of internet accessible laboratories. Proc IEEE 96(6):931–950. https://doi.org/10.1109/JPROC.2008.921607
28. Orduna P, Zutin DG, Govaerts S et al (2015) An extensible architecture for the integration of remote and virtual laboratories in public learning tools. IEEE Revista Iberoamericana de Tecnologias del Aprendizaje 10(4):223–233. https://doi.org/10.1109/RITA.2015.2486338
29. Salzmann C, Gillet D (2013) Smart device paradigm, standardization for online labs. In: 2013 IEEE global engineering education conference (EDUCON), pp 1217–1221. IEEE. https://doi.org/10.1109/EduCon.2013.6530261

30. Ball R, Burke R, Fletcher D, Hoberney A, Jesukiewicz P (2000) Sharable courseware object reference model (scorm), version 1.0. https://apps.dtic.mil/dtic/tr/fulltext/u2/a410103.pdf
31. Advanced Distributed Learning: xapi-spec. https://github.com/adlnet/xAPI-Spec
32. IMS Global Learning Consortium: Learning tools interoperability core specification 1.3 (2020). http://www.imsglobal.org/spec/lti/v1p3/
33. Oauth 2.0 — oauth. https://oauth.net/2/
34. IEEE standard for networked smart learning objects for online laboratories. IEEE Standard 1876-201910.1109/IEEESTD.2019.8723446
35. IEEE standard for learning object metadata. IEEE Standard 1484.12.1-2020. https://doi.org/10.1109/IEEESTD.2002.94128
36. IEEE standard for learning technology–extensible markup language (xml) schema definition language binding for learning object metadata. IEEE Standard 1484.12.3-2005 https://doi.org/10.1109/IEEESTD.2020.9059045
37. Atwood J (2007) Mort, elvis, einstein, and you. Coding Horror. https://blog.codinghorror.com/mort-elvis-einstein-and-you/
38. OpenAPI Initiative: Home - openapi initiative (2020). https://www.openapis.org/
39. GitHub: microsoft/vcpkg (2020). https://github.com/microsoft/vcpkg
40. Ravanasa K, Hashemian R (2014) vlab, a high speed multi-accesses parallel processing remote laboratory access for fpga design technology. In: IEEE international conference on electro/information technology, pp 377–381, IEEE (2014). https://doi.org/10.1109/EIT.2014.6871794
41. Angulo I, Garcia-Zubia J, Hernandez-Jayo U, et al (2017) Roboblock: a remote lab for robotics and visual programming. In: 2017 4th Experiment@International Conference (exp.at 2017), pp 109–110, IEEE. https://doi.org/10.1109/EXPAT.2017.7984373

GOLDi 2.0: Beyond Raw Digital Signals – Electrical Interface Emulation

Karsten Henke(✉), Johannes Nau, and Robert Niklas Bock

Ilmenau University of Technology, Ilmenau, Germany
{karsten.henke, johannes.nau, robert-niklas.bock}@tu-ilmenau.de

Abstract. The interactive hybrid remote lab *Grid of online lab devices Ilmenau* (GOLDi) has been successfully used internationally for many years as a cloud-based system for controlling electromechanical hardware models with various specification techniques. However, the complexity of digital systems is constantly increasing. To keep up with this trend, the didactic concept of remote labs is also constantly evolving. GOLDi is currently reaching its limitations on complexity and variability because all sensors and actuators are required to communicate via a few binary, not time-sensitive signals. This prevents the usage of sensors and actuators which require specific protocols, modulated or specifically timed signals. We propose a new way of transmitting complex signals including timing information by introducing interface hardware between the control unit and the physical system to emulate the output signals of one as the input signals of the other. This in turn allows for new experiment designs tailored towards more realistic setups of sensors and actuators which provide a much richer learning experience.

Keywords: Remote laboratory · E-Learning · Electrical interface emulation

1 1 Introduction

Our interactive hybrid online lab – called *Grid of Online Lab Devices Ilmenau* (GOLDi) – stands for a grid concept to realize a universal remote lab infrastructure. The remote lab is used for teaching practical lessons as well as to give hands-on experiences for the development of embedded electronics. This is not done using on-site lessons, but remotely via the Internet. This has the advantage that courses can be offered internationally worldwide and gives students from different countries, speaking different languages, the same access, and equal possibilities in the lab [1, 2].

Over the years the collection of GOLDi experiments has been constantly expanded with new electromechanical models (physical systems) as well as new control units. The sensors of all current models provide binary status information (e.g. elevator reached the upper floor), which is evaluated by the selected control unit (microcontroller, FPGA, PLD). The control unit subsequently provides binary actuator signals (e.g. movement cabin motor down on). But many of these electromechanical hardware models could be extended by additional sensors (e.g. ultrasonic module) and actuators (e.g. PWM servo) with more complex protocols (e.g. I2C, SPI, 1-wire, USART). These

M. E. Auer et al. (eds.), *Online Engineering and Society 4.0*, Lecture Notes in Networks and Systems 298, https://doi.org/10.1007/978-3-030-82529-4_3

types of sensors and actuators cannot be controlled by the current GOLDi system, because the latency within a distributed system does not allow to reproduce the time behavior of these signals and protocols.

This in turn means that the variety and complexity of new experiments are limited. We believe that more complex experiments with setups that are much closer to real-world applications provide a more holistic learning experience, especially for students who already know the fundamentals of digital circuits. Therefore, we are currently redesigning the architecture of the GOLDi infrastructure to enable these new more complex experiments in the future.

2 2 Milestones in the GOLDi Remote Lab History

With the electrical interface emulation presented in this article, we reach a new milestone in the history in the development of the GOLDi remote lab. A brief chronological overview of the milestones to date should therefore be given at this point.

2.1 First Steps – 20 Years Ago

The Integrated Communication Systems Group at the Ilmenau University of Technology has many years of experience in integrated hardware and software systems and over 20 years of experience in dealing with internet-supported teaching in the field of digital system design [3, 4]. At that time, we developed a new teaching concept, called *Living Pictures* [5] that we use in several phases of the learning process. These were highly interactive Java applets that could be used for demonstrations as well as for experimental purposes. These also served as tools in certain steps of the design process of digital systems. In the classic setup of our first remote lab, every electromechanical hardware model is connected to one specific control unit (see Fig. 1).

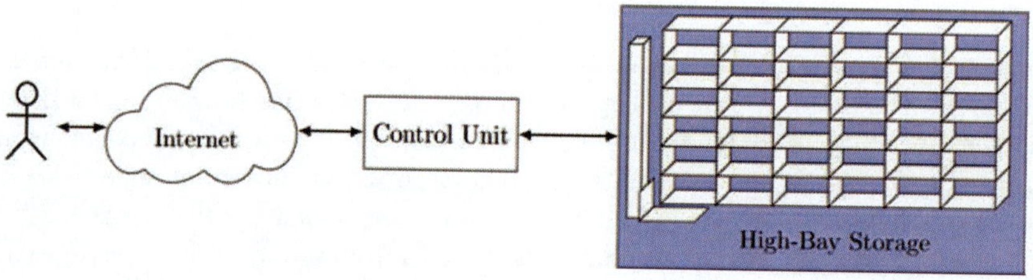

Fig. 1 Classic setup of the first remote lab

The user can control the selected control unit (e.g., a microcontroller, FPGA, or a PLC) over the internet. The sensor and actuator values and the corresponding webcam stream were transferred to the client-side running a Java-Applet.

2.2 Grid Concept of an Interactive Hybrid Online Laboratory

For a reliable, universal, and more flexible use of our remote laboratory, in 2010 we have developed the grid concept characteristic for GOLDi (see Fig. 2) [4]. Based on the flexible grid structure an experiment consists of two components: On the one hand, there are various control units (e.g. FSM, microcontroller, FPGA). On the other hand, there are electromechanical physical systems (e.g. elevator, 3-axis model, or warehouse). Each physical system can also be simulated in a virtual model. The interconnection between the control units and the selected physical hardware model during a remote lab work session (experiment) as well as the user management is done by the *Lab Server*, which also handles the webcams. These extensions offer possibilities for a flexible, scalable, and robust usage of remote labs, giving the students a high degree of freedom for doing practical work inside the lab.

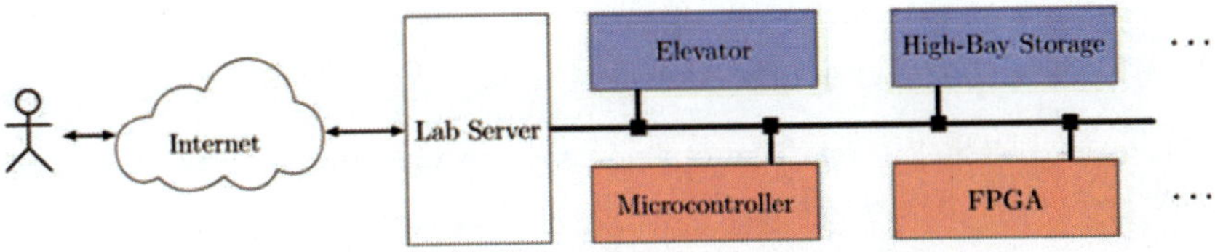

Fig. 2 Grid structure of the GOLDi remote lab infrastructure

The ongoing development of the GOLDi platform focuses on different aspects of the system, usability, reliability, and maintainability. While usability and reliability are crucial for a successful application for educational purposes, maintainability issues mainly impact future developments. As a distributed platform, GOLDi faces different challenges that are caused by the physical inaccessibility of the hardware for the core development team.

2.3 Distributed GOLDi Remote Lab Infrastructure

Our remote lab GOLDi is used for teaching practical lessons as well as to give hands-on experiences for the development of embedded electronics. This is not done using on-site lessons, but remote via the internet. This has the advantage that courses can be offered internationally worldwide and gives students from different countries, speaking different languages, the same access and equal possibilities in the lab. Since 2012, we have had the opportunity to use the remote lab infrastructure of our partners in Armenia, Georgia, and Ukraine at their home institutions within the framework of two major TEMPUS projects [4, 5].

Our first approach at the beginning of these projects was to duplicate the entire remote lab setup from the Ilmenau University of Technology (remote lab server, control units, electromechanical hardware models, remote lab infrastructure) to each partner institution. While this step allowed a simple deployment without the necessity of changes to the system, we quickly noticed that the maintenance of the whole system became a major issue. Following the idea of independent deployments at different locations, the first concept was that each partner institution should maintain its laboratory independently. The Ilmenau team would only provide support, while specially trained staff at each GOLDi location was maintaining their local deployment. This approach turned out to be very ineffective.

After installing the first GOLDi remote labs at our partner universities, we had to realize the following [6]:

- Each institution has specific modifications according to their requirements,
- Each university has different network architectures,
- Each university has different remote lab configurations,
- To have access to an experiment of the partner labs, a separate user account on the corresponding partner website is necessary.

That's why we decided to re-engineer the network of all distributed GOLDi remote labs to a GOLDi cloud system.

2.4 The GOLDi Cloud System

Available GOLDi Servers (each corresponding to a partner remote lab) are registered in the GOLDi cloud (see Fig. 3).

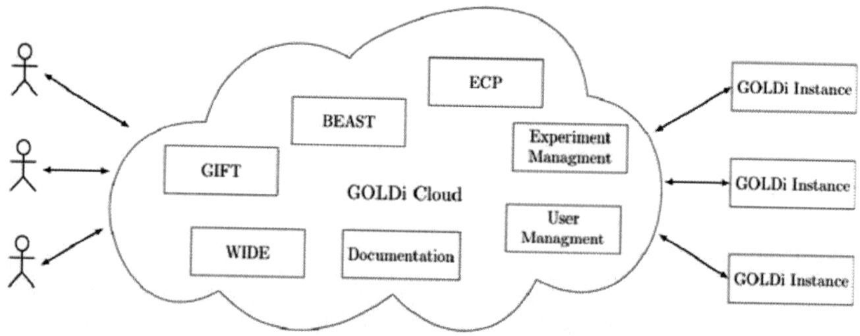

Fig. 3 GOLDi cloud architecture [8]

The new cloud structure has the following advantages [7, 8]:

- All partner universities have the same GOLDi software version,
- New functionalities are immediately available for all partners,
- Usage of all partner labs with one central goldi-labs.net user account,
- The ability to quickly change the experiment location if local hardware is unavailable or blocked by another user,
- The user can determine which lab has the lowest delay (best reaction time) regarding his current geographic position and internet connection,

GOLDi remote labs are running at ten universities in Armenia, Australia, Georgia, Germany, and Ukraine.

2.5 Revision of the Hardware Infrastructure and Maintenance Processes of the GOLDi Cloud System

Unfortunately, it turned out that there were still problems and delays when the laboratory was extended with new functionalities of the installed hardware. The firmware of the hardware components had to be updated every time a new function was added or a bug was fixed. This required each hardware component to be manually disconnected from the running lab, connected to a separate computer via USB, running the GOLDi firmware update software, and manually reconfigured at each partner institution. This procedure caused considerable difficulties for some of our partners and slowed down the overall process of GOLDi lab development. That is why we decided once again to revise the GOLDi cloud system – but this time concerning the hardware components at each distributed partner remote lab and the communication between them. From now on it was possible to update all components automatically from the distance (by the Ilmenau Team as GOLDi coordinator) without any manual intervention on site.

This revision of the GOLDi infrastructure has the advantage for our partner institutions that they can significantly reduce the maintenance effort of their labs. This is primarily due to the conversion of the existing CAN bus system for the internal communication of all hardware components within the GOLDi infrastructure to an Ethernet-based system. All this led to a reduction of the previous update process from several weeks to a few seconds [9].

With the now stable running and maintainable GOLDi instances, it was now possible to consider making the design steps more effective in a new milestone.

2.6 "Hidden" Integration of Industrial Design-Tools in the Development Process

With our remote lab infrastructure, we support a unique design process of digital control systems - independent of the selected specification technique (see Fig. 4, left side). According to the given task, this process usually consists of the conceptual formulation and the design of the control algorithm by using different non-commercial development systems to finally achieve a validated control. This can be uploaded via the *Experiment Control Panel* (ECP), running inside the student's web browser to the selected control unit in the remote lab, which will be programmed automatically in the background in the remote lab. Then the student can execute his experiment in the remote lab.

One of the problems that have emerged during the intensive usage of the GOLDi remote lab over the years is the usage of external non-commercial third-party development tools, necessary for the software- or hardware-oriented design. The installation and setup (e.g. of Atmel Studio or Intel Quartus Prime) on a private PC or laptop is quite complicated. Safe and efficient usage requires expert knowledge and the tools are not always platform-independent. Above this, these IDEs occupy several gigabytes in memory size and sometimes take hours to install. These very powerful tools, of which only a fraction of the offered functionality is needed to solve the given task, generate a large overhead only to control an electromechanical hardware model in the remote lab. Besides, high school and university students may have only little or no knowledge of microcontrollers, FPGA, compiler linker toolchain, or hardware-related languages.

That's why we developed a new tool as part of the GOLDi remote lab infrastructure – called WIDE (WEB IDE) – which is described in detail in [10]. WIDE supports all the mentioned design flows with the only requirement of a web browser. This means WIDE is running inside the ECP of the student's web browser. The WIDE backend system offers services a toolchain for each language that previously needed to be executed at the user's computer (see Fig. 4, right side). These services include mainly the compilation of source code from different languages to firmware files.

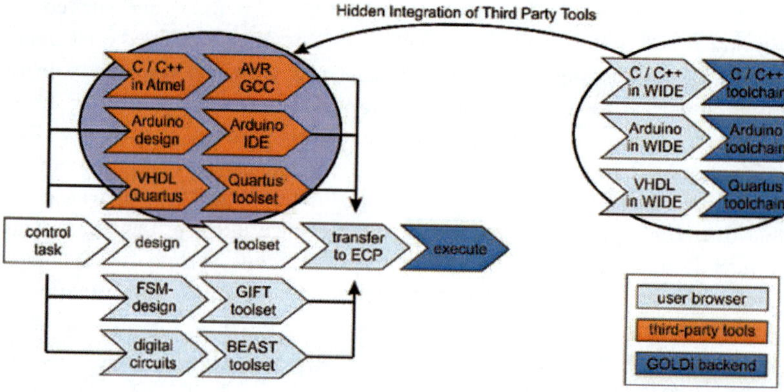

Fig. 4 Unique GOLDi design flow (left: old version; right: third party tool integration)

Due to the complete abstraction of build and synthesizing processes and the manual upload of the synthesized code to the remote lab, there are no more obstacles due to the required additional knowledge of third-party tools. This means that the focus can now be on creating functionality.

3 3 GOLDi 2.0 – the New Approach Towards Electrical Interface Emulation

However, the complexity of digital systems is constantly increasing. To keep up with this trend, the didactic concept of remote labs is also constantly evolving. GOLDi is currently reaching its limitations on complexity and variability. The sensors of all the current models provide binary status information (e.g. elevator reached the upper floor), which is evaluated by the selected control unit (microcontroller, FPGA, PLD). The control unit subsequently provides binary actuator signals (e.g. movement cabin motor down on).

Many of these electromechanical hardware models could be extended by additional sensors (e.g. ultrasonic, I2C, SPI) and actuators (e.g. PWM, I2C, SPI, USART) with more complex protocols. These types of sensors and actuators cannot be controlled by the current GOLDi system, because the latency within a distributed system does not allow to reproduce the time behavior of these protocols. This in turn means that the variety and complexity of experiments are limited.

That's why we have decided to revise and extend the architecture of GOLDi while keeping the existing functionality. The key component is an electrical interface emulation. This means that the chosen control unit (e.g. a microcontroller) can interact with our system just like it would interact with the real sensors and actuators on the remote hardware model via the corresponding protocol. In reality, however, it is directly wired to an interface unit, which emulates the real sensor/actuator. The internal state of the interface and the model is continuously synchronized.

During the development of the new architecture the following aspects are also considered and implemented:

- Compatibility with the new IEEE "Standard 1876-2019 for Networked Smart Learning Objects for Online Laboratories".
- Location-independent experiment selection, which means that it is possible to control a physical system at location B with a control unit from location A (see Fig. 5).
- Increased maintainability by reducing the number of devices that need to be maintained manually.

Assuming, a student wants to control a water level experiment by using a microcontroller to compensate the water loss caused by a consumer. He can determine the water level utilizing an ultrasonic distance sensor to finally control a pump. A common interface for such a sensor is composed of a trigger and a response signal. The microcontroller will send a pulse on the trigger line and will receive a pulse on the response line. The time delay between trigger and echo is directly proportional to the water level. This is no problem at all, as long as he does it in the hands-on lab, where the control unit is hard wired to the physical system. However, the control of such an ultrasonic sensor in the current architecture is not possible due to the latency problem which interferes with the time delay measurement. This is solved in the new architecture by emulating the sensors and actuators on the controller side. Effectively this allows us to break up the direct coupling between the microcontroller and the water level model and synchronize both parts transparently.

The new architecture will distinguish between the Control Unit (CU) and the Physical System (PS). Both the CU and the PS can be further decomposed to the CU-Entity (CU-E) and CU-Interface (CU-I) or PS-E and PS-I respectively. Here the CU-E is the actual controller and the PS-E is the actual hardware model. The CU-I and PS-I will implement the interface to our system (see Fig. 5).

In addition to sensors and actuators can also be triggered via special protocols. For example, students can control a servo on the physical system by generating the required Pulse Width Modulation (PWM) signal sequence on the microcontroller. These signals are processed by the CU-I so that the PS-I can recreate the PWM signal generated by the control unit to operate the connected servo. In this way, sensors and actuators with any digital interface (I2C, SPI, etc.) can be used in the PS-E.

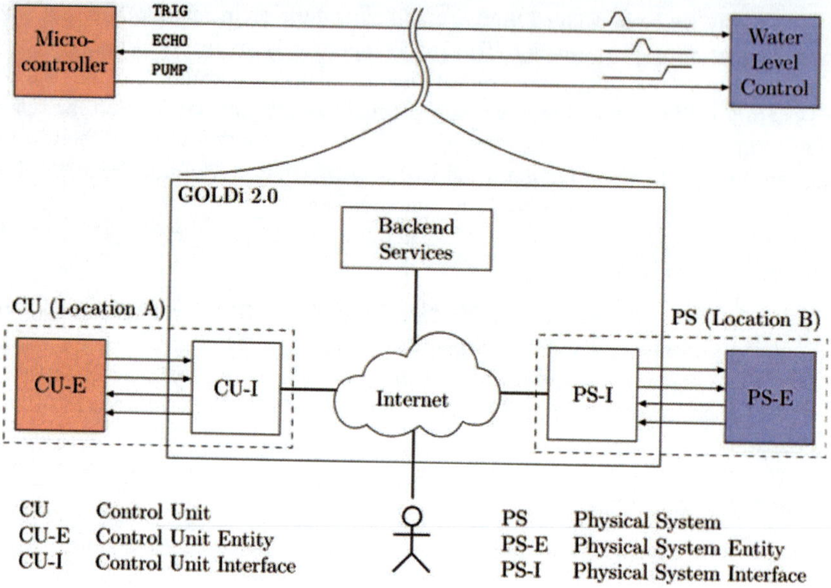

Fig. 5 Overview of the new GOLDi 2.0 architecture

4 4 Examples for New Sensor and Actuator Handling Possibilities

In the following, we will give three examples of such an Electrical Interface Emulation. For all examples, we will suppose that a microcontroller is used as the CU-E. It would also be conceivable to control a Physical System via an FPGA or a PLC.

First, we want to show how it would look like if you connect the microcontroller directly to the sensor or actuator in terms of signal behavior as seen as in the top of Fig. 5. By separating the microcontroller physically from the sensor or actuator this connection is split. The microcontroller should not notice that it is not directly connected to the sensor or actuator and that there is an internet connection between locations A and B. Consequently, the electrical interface needs to be emulated on both sides by the interface unit (see bottom of Fig. 5).

4.1 Control of an Ultrasonic Sensor

Assuming we want to continuously record the fill level in a water tank within the practical experiment *water level control*. With the direct coupling between the microcontroller and the ultrasonic sensor, the microcontroller sends a trigger pulse to the ultrasonic module [11]. This sends a delayed response signal as an answer. The duration between trigger impulse and response is a value for the filling level. Let us assume a filling level of 30 cm. The microcontroller would, for example, transform the duration to the hexadecimal value $1E_H$ and make it available for the control algorithm.

In case the microcontroller of location A should query the ultrasonic sensor at location B, the signal behavior of the interface emulation is shown in Fig. 6.

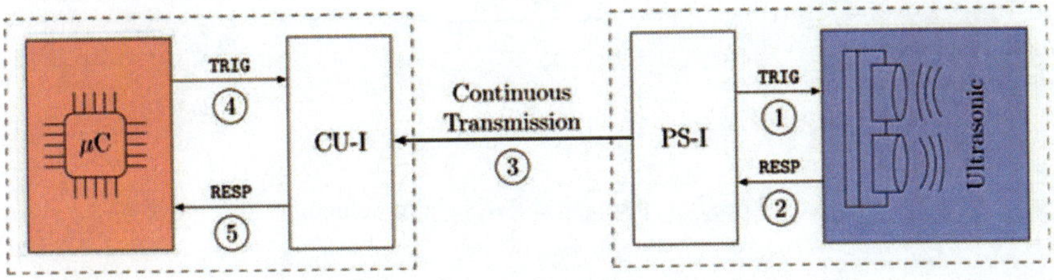

Fig. 6 Control of an ultrasonic sensor

At the hardware model *water tank* (location B) the water level is continuously measured by the PS-I controlling the ultrasonic module in a normal way. For this purpose, the PS-I sends a trigger pulse (1) and evaluates the duration until the response pulse (2) arrives. The PS-I sends the encoded time delay between these pulses continuously to the CU-I of the control unit at location A. In this case, the microcontroller is directly connected to the CU-I. To determine the water level, the microcontroller sends the trigger pulse (4) to the CU-I. The CU-I then generates a response signal (5) with the same delay that the PS-I previously transmitted. The microcontroller then calculates the value of $1E_H$, which corresponds to a filling level of 30 cm, and makes it available for the control algorithm.

In the end, it does not matter whether the PS-E and the CU-E are connected directly or through the internet using the PS-I and CU-I respectively. In any case, the microcontroller will receive the same response signal to his trigger signal. In this case, we have a continuous unidirectional data transfer within the interface emulation from the PS to the CU. The next example shows the opposite data transfer (i.e. from the CU to the PS).

4.2 Control of a PWM Servo Actuator

In this case, a servo motor should be used as an actuator to set a certain angle within a hardware model. This angle is to be controlled via PWM. The angle to which the servo arm is to be set is controlled by the width of the pulses [12]. A 50 Hz signal (20 ms period length) is common, which is between 500 µs (left position, 0°) and 2500 µs (right position, 180°) at high level and the rest of the period length at low level. Let us assume that the servo arm should be set to 90°. This means that if the microcontroller were directly connected to the servo, it would continuously generate a 1.5 ms long pulse every 20 ms.

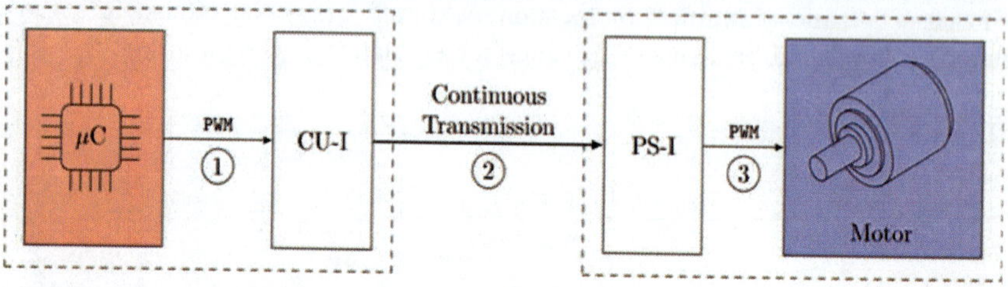

Fig. 7 Control of a PWN servo actuator

Within the new GOLDi remote lab infrastructure this would mean that the microcontroller would of course also send the 1.5 ms pulse at 20 ms intervals (1) - but this time only to the CU-I (see Fig. 7). The CU-I determines the pulse width and frequency to parametrize the PWM signal. These parameters are continuously transmitted from location A to the PS-I of location B (2). On the model side, the corresponding signal form is again generated from these parameters via the PS-I (3) and the connected servo will finally set to the desired angle position.

4.3 Control of a Real-Time Clock

The last example will demonstrate a bidirectional data exchange within the interface emulation of the new GOLDi remote lab infrastructure. A Real-Time Clock (RTC) using an SPI interface was chosen as an example [13]. An I^2C or 1-wire interface would work similarly.

If the RTC would be directly connected to the microcontroller, it would first activate the RTC via the chip select signal (CS = 0). The clock can be configured (set time, date) or read out (read time) via the clock signal (SCLK) and the data input (DIN) and output (DOUT) signals (1/3). Effectively internal registers are set or read in the device.

Exactly this signal behavior must be reproduced both on the control side via the CU-I and on the model side via the PS-I if the microcontroller is connected to a remote RTC within the remote lab infrastructure (see Fig. 8).

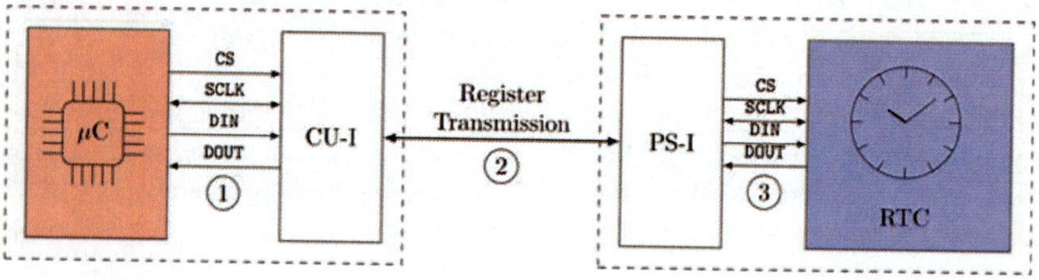

Fig. 8 Control of a real-time clock

If, for example, the clock is to be configured, the CU-I must send the corresponding values (2) to the PS-I so that the PS-I generates the corresponding signal behavior according to the SPI protocol (3). In this case, these are the RTC register contents. If the clock is to be interrogated, this must first be done by the PS-I according to the SPI protocol. A corresponding numerical value is then sent back to the CU-I (2). The microcontroller then receives the corresponding information according to the SPI protocol specification (1).

With the RTC query, the latency of the transmission comes into consideration, especially if the content is transmitted through the internet. In the simplest case, you could ignore it. However, if you want to work more precisely, you could also correct the latency dynamically by constantly determining and averaging it. Then you can proactively modify the register contents on the controller side to reflect the correct response.

5 5 Conclusion and Future Work

The new architecture and the associated electrical interface emulation will provide better coverage of the experiments that are included in the didactic concept for teaching digital systems. Therefore, we will be able to offer experiments in our remote lab, that were previously only possible in the hands-on lab. The architectural concept solves the problem of time-based protocols in high latency remote systems. However, this doesn't solve the general problem of latency in control loops, so that closed-loop control of fast systems won't be feasible in GOLDi 2.0.

With the introduction of electrical interface emulation in our remote laboratory GOLDi, we are now able to interact with sensors and actuators via signals or protocols that require a specific time behavior. This interaction is possible even if the control unit and the physical system are not located at the same place. This hidden emulation is not visible to the user. He can use the full functionality of the sensors and actuators of the physical system while being flexible in programming the control unit.

Acknowledgements. The authors would like to acknowledge the work of Pierre Helbing and Niklas Eichenberg for their contribution to the GOLDi 2.0 project.

References

1. Henke K, Vietzke T, Wuttke H-D, Ostendorff St (2016) GOLDi – grid of online lab devices ilmenau. Int J Online Eng (iJOE). 12(04):11–13. ISSN: 1861-2121
2. GOLDi labs cloud Website. https://goldi-labs.net. Accessed 5 Nov 2020
3. Sire S, Geoffroy F, Gillet D (2003) A virtual assistant for sending hints and perturbations to students based on an Electronic Laboratory Journal (eJournal). In: Proceedings of the ITHET 2003, Marrakech, Morocco, 7–9 July 2003
4. Henke K, Wuttke H-D (2003) Web-based educational tool access. In: IASTED International Conference Computers and Advanced Technology in Education – CATE 2003, Rhodes, Greece, 30 June–2 July 2003

5. Wuttke H-D, Henke K (2002) Living pictures – tool-oriented learning modules andlaboratory for teaching digital via Internet. In: Proceedings of the ICEE-2002 International Conference on Engineering Education UMIST, Manchester, Great Britain, 18–22 August 2002
6. ICo-op Tempus project Website. http://www.ICo-op.eu. Accessed 12 Nov 2020
7. DesIRE Tempus project Website. http:// http://tempus-desire.eu. Accessed 12 Nov 2020
8. Henke K, Vietzke T, Hutschenreuter R, Wuttke H-D (2016) The remote lab cloud goldi-labs.net. In: Proceedings of 13th International Conference on Remote Engineering and Virtual Instrumentation REV2016, Madrid, pp 37–42, February 2016. https://doi.org/10.1109/REV.2016.7444437
9. Fäth T, Seidel F, Henke K, Wuttke H-D, Hutschenreuter R (2018) On effective maintenance of distributed remote laboratories. In: Proceedings of the 15th International Conference on Remote Engineering and Virtual Instrumentation REV2018, Smart Industry & Smart Education, March 2018. Springer, Düsseldorf, pp. 80–89. ISBN 978-3-319-95678-7
10. Henke K, Nau J, Hutschenreuter R, Bock R, Wuttke H-D (2020) "Hidden" integration of industrial design-tools in e-learning environments. In: REV2020, Proceedings of 17th International Conference on Remote Engineering and Virtual Instrumentation, Athens, USA, pp. 44–63, February 2020
11. HC-SR04 Datasheet. https://joy-it.net/de/products/SEN-US01. Accessed 5 Nov 2020
12. How Servo Motor Works. https://howtomechatronics.com/how-it-works/how-servo-motors-work-how-to-control-servos-using-arduino/. Accessed 12 Nov 2020
13. DS3231 Datasheet. https://datasheetspdf.com/pdf/1081920/MaximIntegrated/DS3231/1. Accessed 12 Nov 2020

A Reliable Real-Time Web Interface for an Online Laboratory

Ratchatin Chancharoen(✉), Kantawatchr Chaiprabha, Keerati Tantrapirom, Pasawee Pungrasmi, Soravis Tangthavonsirikul, and Tanapoom Jitnaknan

Faculty of Engineering, Chulalongkorn University, 254 Phayathai Road, Pathumwan, Bangkok 10330, Thailand
ratchatin.c@chula.ac.th, {6130034021, 6130044321, 6130353221, 61313334521, 6131316221}@student.chula.ac.th

Abstract. The paper proposed a real-time Web interface for a physical plant that can be plugged into a commercial tele-meeting application. The Web interface is developed with responsive and interactive with a physical hardware. This results in a telepresence working environment that the students can work as a team to investigate the performance of a machine, i.e., a linear stage. The proposed system is applied to the third-year laboratory course and the feedback is positive. The working environment complies with the health measures for Covid-19 and supports a new campus life.

Keywords: Tele-laboratory · Remote laboratory · Telepresence

1 Introduction

Despite the fact that Thailand has successfully handled and controlled the Covid-19 pandemic, health measures [1] are applied around the country including an university. Remote learning strategies [2–5] are adopted in most of the universities in Bangkok, e.g., traditional face-to-face lectures are prohibited at Chulalongkorn university. Lectures are now transforming towards MOOC [6] where professors prepare their media in his/her lecture studio and their students learn the material either at online meeting room or MOOC/LMS. The purpose of this project are to develop a laboratory with a real-time Web interface and implemented it in a compulsory course in mechanical engineering curriculum in order to (1) comply with the health measures for Covid-19 and maintain the class activity during fluctuate situation that may occur (2) demonstrate an engineering solution for a contemporary issue which will inspires engineering students to continue in their career and (3) research in telepresence technology in the scope of digital technology, robotics and cyber-physical systems. This research utilizes available commercial tools, open source packages and university's infrastructure; it also develops the jigsaw that completes the system and test its performance.

The main contribution of the paper is a python-based real-time Web application that interfaces with a physical hardware. The Web application was designed with responsive and interactive features. Note that the plant is a motor control that joint-level

M. E. Auer et al. (eds.), *Online Engineering and Society 4.0*, Lecture Notes in Networks and Systems 298, https://doi.org/10.1007/978-3-030-82529-4_4

signals are communicated real-time at one millisecond updated rate. The HTML-JavaScript-SocketIO-MQTT communication link was used to give a highly responsive and interactive Web application. The application consists of (a) a client module, (b) a server module and (c) a MQTT broker. The client module is written in HTML5, JavaScript and CSS. The server module is a python program that comprises SocketIO and MQTT. The MQTT broker is also installed on the same internet server. The event-driven programming concept and small package communication are highly utilized in this approach. The activity on the Web interface triggers an event and the associated data is sent through SocketIO and MQTT to the control board; the activity on the control board, e.g. the change of the encoder's signal, will publishes the MQTT payload and the Web display will response with SocketIO engine.

The system was successfully built and used in the compulsory laboratory course. There were two groups, four students per group, during the laboratory sessions and there was only one physical plant. The proposed real-time Web interface is used to create the telepresence working environment for a laboratory.

2 The Architecture Telepresence Laboratory

The proposed laboratory can be accessed from anywhere with the internet [3, 4, 7–10] and with any browser. The students will work as a team through a tele-meeting software, ZOOM in our case, to complete their laboratory's assignment. The quality of the audio and visual (AV) are excellent with a available commercial software. The students are likely to be on campus within the university wireless network. The investigating plant is at the mechanical engineering building. The internet server, HPE SimpliVity 380 Gen10, sits within the campus network but on another building. Its CPU is Intel Xeon Silver 4210 10C/20 T 2.2 GHz with 768 GB RAM, 5 TB Storage and two NVIDIA Tesla P40 GPUs. The internet server and the university's infrastructure provide enough power to the project. MQTT is designated as a communication protocol between the server and the plant's control board, and SocketIO as a communcation between the server and the student's device. In this architecture, the server and the control board communicate with intranet IP, while the students only access the investigating plant through the internet server with public IP. This is a two-layer concept where our control board handles and controls the experimental plant; the server supervisory control the control board and interface with the students. The GUI is a Web application running on the server and there can be several GUIs running for a single plant at the same time. The set up allow us to go for a real-time Web interface controlling the physical plant which is highly responsive and interactive (Fig. 1).

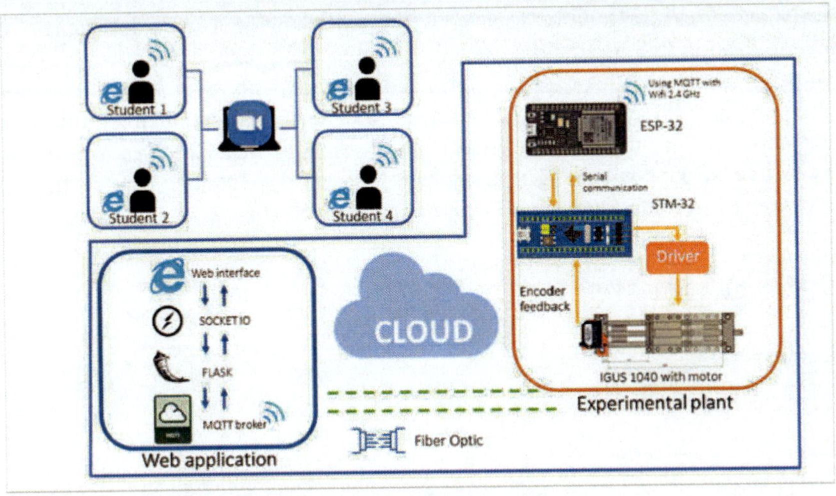

Fig. 1 Architecture of the telepresence experimentation

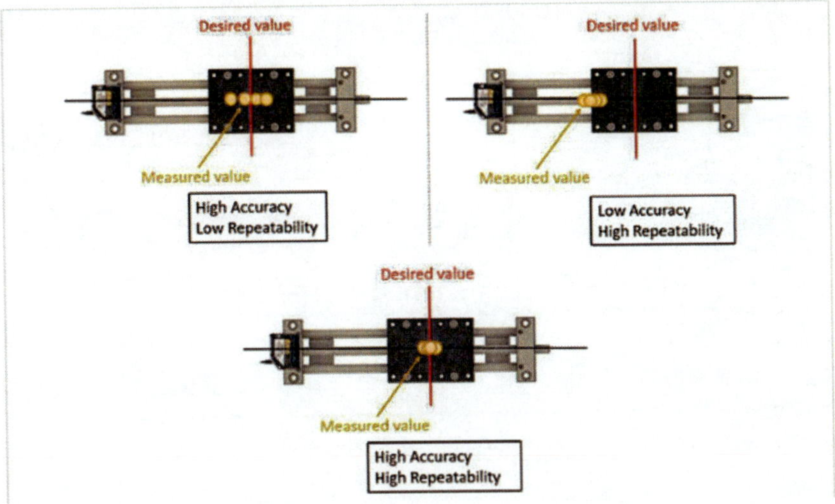

Fig. 2 Experimentation assignment

In this setup, the server is a cloud server,while the plant is a physical plant. The students can access it from anywhere within the campus wireless network.

2.1 Learning Outcome and Learning Strategy

In the experimentation, the students were assigned to work as a group to investigate the accuracy, repeatability, and resolution of an IGUS Drylin linear guide in two-week period. The expected learning outcome [3] includes (1) understanding an accuracy, repeatability, and resolution of a machine, (2) understanding a screw mechanism that converts rotation into translation, (3) understanding the proximity and range sensors, their performance and characteristics, (4) hands-on experience with an electrical motor and an encoder, (5) designing an experiment to collect, analyze, interpret a data, and draw a conclusion, (6) effective communication about their work in writing, and (7) effective working in team. The assignment was an open-end mini project where the students must develop a process to determine the result (Fig. 2).

3 A Real-Time Web Interface

3.1 Overview

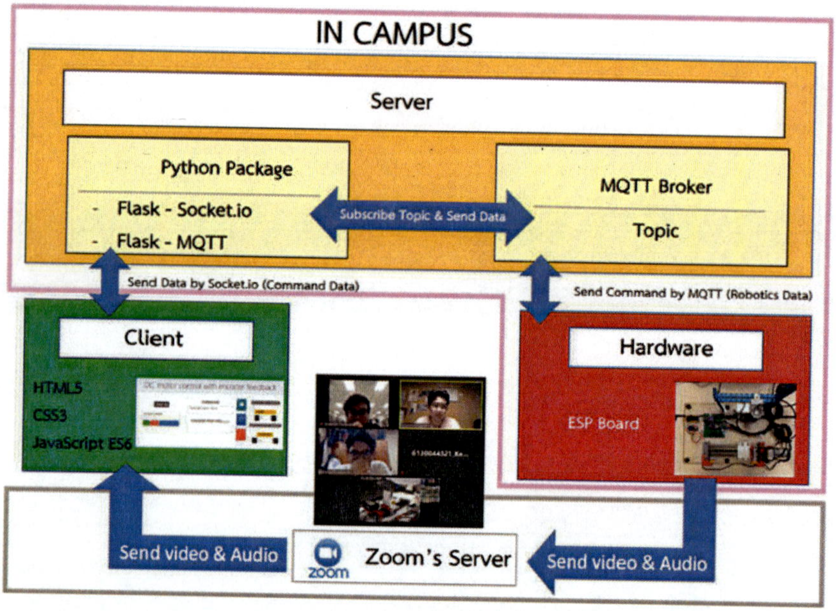

Fig. 3 Command and data flow

The data flow in the system are classified into three types, i.e., the audio and visual (AV), the robotics data, and the command data. The AV is live streaming with the available commercial software. The standard HD (720P) requires up to 1.5 Mbps bandwidth. The available commercial software has a technology to compress and effectively handle this data. The robotics data is normally a data that encodes the position or the state of a machine. The data package is small, compared to the AV, but the update rate is extremely fast. The communication latency was also a concern. The speed test within the university network is about 100 Mbps with 10 ms latency which is sufficient for this signal. But it is a shared medium thus the communication speed and latency are not guaranteed at all time. Flask, Flask-SocketIO, and Flask-MQTT are used to develop a small package communication between the server and the machine and with event-driven concept. The command data is a very small data package that requires extremely low bandwidth, but high reliability to connect Web interface (via SocketIO) and the server (via Flask-SocketIO). These signals demand a low latency and high reliability communication, not the bandwidth (Fig. 3).

A tele-meeting software can be used to handle the AV signal. In the project, we developed a framework for the communication of the robotics data and command data that can be plugged into the tele-meeting software.

3.2 Software Component

3.2.1 Flask

Flask is a Python based micro Web framework [11]. The flask core is simple and minimum but extensible. There are several extensions which selectively integrated into our project. With this characteristic, we were able to design a very small Web application to a specific purpose, e.g., highly responsive and interactive to a hardware in our case. Using the Python is also a strong point which its community is abundant, and the language is highly capable and suitable for an event-driven programming.

3.2.2 Socket.IO

SocketIO is a flask-based Python package designed for a real-time bidirectional communication which works on every platform. The socketIO is a fast communication bridge between the Python-based server and the HTML Web interface. It is a library to work with WebSocket using the event-driven concept and minimum package size.

3.2.3 MQTT

MQTT is a message transport protocol that can be used for machine to machine communication [12, 13]. It is designed to be extremely lightweight with publish/listening concept and support event-driven execution. In our project, MQTT was used to partition the communication between the Web interface and the physical hardware. MQTT comes with retained messages, last will and testament, clean session, and keep alive concepts which are really help in practical use.

3.2.4 Source Code

To design a reliable real-time Web interface for a physical plant, we need to understand computer languages, operating system, and its ecosystem. In brief, there are three languages in this project, i.e., HTML, JavaScript, and Python. The HTML (Hypertext Markup Language) defines what and how to display though the Web browser. The format is effectively defined with Cascading Style Sheets (CSS) but not mentioned in the paper. The JavaScript can be embedded inside the HTML to enable interactive features on the Web where client-side script can be implemented and most of the Web browser come with a dedicated JavaScript engine to execute it. There are React.js or Angular.js framework that can be used to develop a responsive Web application. However, we developed our real-time Web interface with Python Flask framework. The event-driven programming concept is strongly used so that a click on the Web page will generate the event which triggers everything all the way to the motor. The data package in real-time communication channel was maintained at minimum in our framework (Fig. 4).

HTML Source Code

```html
<form>
   <button class ='pos_left' >Left</button>
   <input type='number' class='position_val' >
   <button class ='pos_right' >Right</button>
</form>
```

Python Source Code

```python
@socketio.on( 'GO_LEFT')
def Go_left( 'position'):
   topic = 'plant/command'
   payload = '.....'  ## command to plant eg. Go left 1000000
   qos = 0
   mqtt.publish( topic , payload , qos )
```

Python Source Code

```python
topic = 'plant/encoder'
mqtt.subscribe( topic )

@mqtt.on_message()
def send_position_to_client(client , user_data , message)
   position = message.payload.decode()
   socketio.emit( POSITION , position )
```

Javascript Source Code

```javascript
$(document).ready(function() {
     var socket = io.connect('http://' + document.domain
                    + ':' + location.port);

   <!-- button_function-->
   var left =$('button.pos_left').on('click',function(e){
     e.preventDefault()
     let position = $('input.position_val').val()
     socket.emit('GO_LEFT', position )
   })

   <!--show_Encoder_Position-->
   socket.on('POSITION',function(position){
     var text = position;
     var $textarea = $('.Encoder_Position');
     $textarea.val(text + '\n');
   })
}
```

Fig. 4 The source codes

The HTML Source code, which is used to layout the content on the Web interface, defines a button "LEFT" with the ID 'pos_left' and a text input 'position_val'. A click on this button generates EVENT and Event handler, defined with built-in JavaScript, will be activated cosequently. The 'position_val' will be sent to the server through sockketIO with the EVENT id, "GO_LEFT".

On the server side, a Python code is also written to handle the EVENT. When the EVENT: "GO_LEFT" is triggered the server will pass through the received 'position_val' to the MQTT server which was installed on the same server. Roughly, we used three languages to passthrough the number, which entered from the Web interface, to the MQTT server. Then the physical plant, connected to the MQTT server, will response to the arriving massage. MQTT protocol is a message transport protocol with the event-driven concept and is designed to manage the messages effectively, especially in our case. When the encoder, connected to the plant control board, rotates, the board will processes the encoder's signal in real-time. The change of the encoder's position triggers EVENT then the board will passthrough the position to the MQTT server. On the server- side, Python is connected to the MQTT and passthrough the message to the client side via socketIO. The JavaScript processes this EVENT and dynamically update the text display on the Web.

3.3 Hardware

In this experimentation, the students were assigned to work as a group to investigate the (1) accuracy, (2) repeatability, and (3) resolution of a linear stage. This is only one axis of robotics system and several mechanisms were used to generate a linear motion [14]. The students were assigned to determine the overall characteristics. They had to design, experiment, plan, execute, and record their own experiment, then interpret the data, draw a conclusion, and write a technical report. The plant is designed with IGUS Drylin SLW-1040 linear guide, which is driven by D37 gear motor with encoder feedback. The linear guide is equipped with MIN and MAX industrial limit switches that will prevent the motion in the direction beyond the working space. The controller will control the motor either in open loop (Power control) or closed loop (Position controlled). Note that the investigation plant is equipped with Keyence Laser Analog Sensor, IL-065, to sense the position of the slider on the rail.

The control board are STM32 and ESP32 co-processor. Our STM32 control board is loaded with our firmware which can interface and control up to two motors with encoders and is commanded with alpha-numeric codes. The coupled ESP32 provides the internet access and control some I/Os. It is designed as a hardware interface that converts an electromechanical plant into a NC machine with internet access in a small package [11]. The hardware locates in the campus with wireless access but with security authentication that requires both an operating system and a human to pass (Fig. 5).

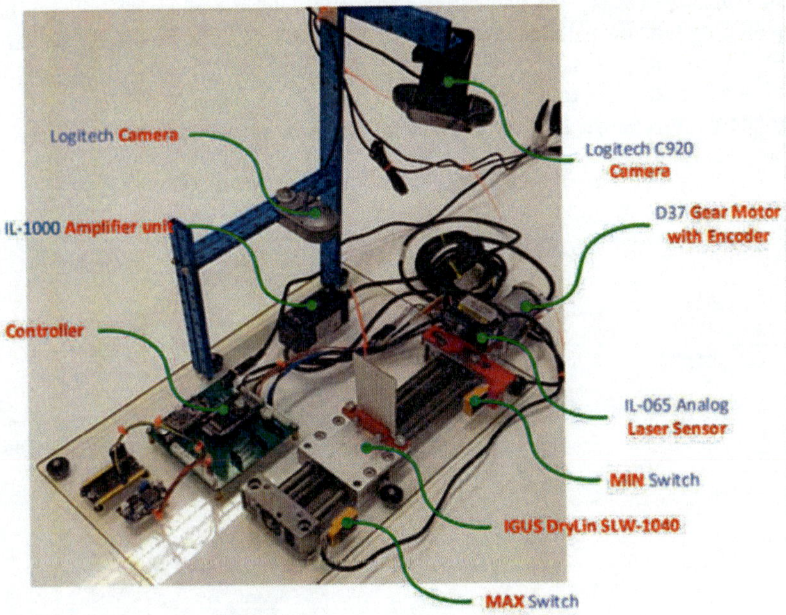

Fig. 5 The IGUS drylin linear stage

4 The Telepresence Working Environment

A typical experimental apparatus normally comes with a control panel. In the project, the control panel is designed as a Web interface which can be run on any platform. The control panel is more like a dashboard that is directly associated with a plant. The advantage of the Web-based control panel are that it is (1) customizable, (2) cost effective, (3) multiple-access, (4) remotely accessible, (5) in compliance with the health measure, and (6) its multiple information display. The software control panel outperforms the physical hardware panel, especially in the long term where a physical plant is still capable of demonstrating basic principles, but the experimentation technique is advanced, or the learning strategy is improved. Furthermore, there can be several control panels connecting to an investigation plant where each panel is designed for different laboratory's assignment. In term of cost, most students have their own tablet or laptop and university's internet access which can be used to gain the control of a plant. When the panel is shifted to the student's tablet, it is free maintenance and suits our students in digital age. The software panel also supports multiple-access where all students, not only the student in front of a physical panel, gain control. The multi information display can be implemented. Raw signal, interpreted data, information, trend can be selectively displayed. The software control panel connects to a physical plant through the digital communication channel which can be accessed from anywhere with the internet access. The feature benefit is that the Web-based control panel is demanded in the no touch society. Despite several benefits, there were concerns regarding (1) response and (2) reliability where a reliable real-time Web interface is a solution (Fig. 6).

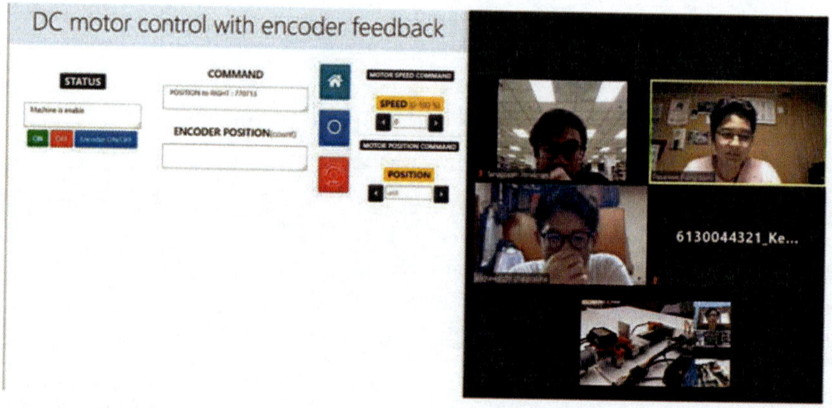

Fig. 6 The working environment

A software dashboard includes switch/button, color lamp, slider, gauge, and text which cover most of the item on a physical control board. In our project, the dashboard is designed with three columns. The first column is the general machine control including ON/OFF switch, the general text message, and the ON/OFF switch of the streaming signal, which is the encoder in our case. The second column is the display of the command is sent to the plant and the feedback signal, the encoder in this case. The plant is multiple-access where the individual student can command but all will see those live command with the current encoder position. The third column is the control panel. There are three buttons, (1) HOME, (2) ZERO, and (3) EMERGENCY. The HOME button is for commanding the stage to go to HOME position. The ZERO button is for setting the current position as the ZERO position. The subsequently position is the position relative to this position. The EMERGENCY STOP is the button for shutting off the plant in emergency case. The plant can be controlled either in VELOCITY or POSITION mode. If the number is placed in the SPEED box and the left or right is clicked, the plant will be in the velocity command and move with the power specified in the text box and in the direction as clicked. Similarly, if the number is placed in the POSITION box and the left or right is clicked, the plant is in position mode and it will JOG to the position specified in the text box relative to the current position and in the direction as clicked. The graphical user interface or GUI can be improved along way when the plant is in service as we get feedback from the user in real use.

To build the telepresence working environment, the ZOOM is used as a meeting software. We use several cameras and microphone to capture the environment in the real scene and real time, and set the screen, with the OBS, as one of the participants in the meeting software. In this way, each participant can work with his/her group, see the real scene and apparatus, command, and sense everything on the control panel. Participants were working at different places but still able to worked as a group and connected to the hardware. There was one instructor on the site to help and correct what may happen in the field.

5 The Actual Performance

The tele-operated linear stage is successfully developed and used in one of our compulsory courses in mechanical engineering curriculum, i.e., 2103360 Mechanical Engineering Experimentation and Laboratory II. The students are able to work as a team to complete their mission in the two-week period. In detail, students can actuate the motor and sense the stage position by both the encoder and the laser sensor through our Web interface from their own device while real-time monitoring the situation at the investigating plant through a video conferencing software, i.e., ZOOM. The latency is not noticed as it faster than the AV signal through the conferencing software.

In term of the learning outcomes, the expected outcomes to (1) understand accuracy, repeatability, and resolution of a machine, (2) understand a screw mechanism that converts rotation into translation, (3) understand the proximity and range sensors (6) communication are achieved either with hands-on or remote laboratory. The outcome of (4) hands-on experience with the hardware is in doubt since the students can only watch the machine from a monitor and command it through the responsive Web interface. With the proposed real-time Web interface and the proposed telepresence working environment, this outcome is achieved at some level. The outcomes of (5) designing an experiment and (7) working effectively in team are fulfilled since the students can still work as a team and work and complete the assignment in time with good quality.

The students' feedbacks were positive. (1) the assignment is challenge and attractive; (2) it is easy to understand the content; (3) the learning's content is good; including the design of experiment, the accuracy and precision, the screw mechanism, and the graphical result; (5) the investigation plant is interesting; (6) the telepresence working environment is great and the remote laboratory is inspiring; (7) the access, from anywhere with any platform, is good; (8) it is modern and technological; (9) it is easy to access and operate the plant; and (10) an instructor at the site to help when needed is good.

6 The Things that also Found Satisfied Include that

The wish list of the comments for the laboratory includes that (1) It should be more flexible to command the plant. There are many things that are not on the Web interface and thus cannot be commanded. (2) The UI/UX is confusing and should be improved. (3) The opportunity to build their own hardware. (4) The ability to control the view of the cameras remotely. (5) The exclusive access to the plant that blocks the access from anybody else so that no one can interfere during their work. (6) There should be more sensors and different type of sensors installed. (7) The assignment was unclear (8) The unit should be a single unit, millimeter. (9) There should be more than one physical plant so that two groups can work parallelly without time sharing. (10) The network connectivity was not reliable.

From the feedbacks, the telepresence working environment gets both positive and negative feedbacks [14]. The Web interface is a software, and it can be redesigned to provide a better UI/UX at low cost. A telepresence technology is still advancing and will response to their wish list related to the like real atmosphere. The data latency is already sufficient since there was no one notice about the delay. The reliability of the communication network is complained. The 5G cellular network and Wi-Fi 6 are now available, and both wireless technologies significantly will improve the reliability and latency in their network.

7 Benefits and Drawback

The highlight benefits of the proposed real-time Web interface for the laboratory are (1) more access/utilization [10]: the Web interface provides multiple-access where all the students can access, command, and observe the plant. (2) cost effective [10]: the physical control panel is replaced with the software Web interface where most students have their own tablet. The maintenance cost is reduced. (3) compatibility: the Web interface is compatible with most platform, Windows, Mac, or android. Students normally use their tablet to work and they can effectively handle data on their tablet. Digital skill is enhanced. (4) time effective: the time sharing can be applied to increase utilization. In the assignment, the students must design an experiment in investigate somethings and more than half of their time is in planning, discussion, and analysis.

(5) customization: the Web interface is software and is easier to be modified or customized. (6) comply with the health policy. Students only touch their own tablet and work in social distancing environment.

Moreover, there are opportunities to develop further for (1) another assignment but share a physical plant. (2) behavior monitoring as his/her commands are logged. Automatic assessment is possible. [15] (3) a laboratory that its objective is a remotely investigation. (4) a laboratory that investigate somethings in the micro scale. The real-time Web interface is a good choice for these events.

8 Conclusion

The experiment with a real-time Web interface is successfully developed and used in a compulsory course. It complies with a health measures: social distancing and no physical contact. The students can still work as a team to complete their mission to remotely experiment the physical plant. The drawback of the proposed technique is not noticed. The students are incubated with digital tools to carry on his/her new campus life. According to the feedback at the end, the students appreciate the technology and its value at some degrees. In term of remote laboratory, the proposed platform also opens the door to several possibilities that may disrupt the traditional laboratory as more and more benefits are discovered.

References

1. Dijkhuizen LGM, Gelderman HT, Duijst WLJM (2020) Review: the safe handling of a corpse (suspected) with COVID-19. J Forensic Leg Med 73:1–5
2. Restivo MT, Mendes J, Lopes AM, Silva CM, Chouzal F (2009) A remote laboratory in engineering measurement. IEEE Trans Ind Electron 56(12):4836–4843
3. Lowe D, Newcombe P, Stumpers B (2013) Evaluation of the use of remote laboratories for secondary school science education. Res Sci Educ 43(3):1197–1219
4. Rivera LFZ, Larrondo-Petrie MM (2016) Models of remote laboratories and collaborative roles for learning environments. In: Proceedings of 2016 13th International Conference on Remote Engineering and Virtual Instrumentation, REV 2016, pp 423–429, art no 7444517
5. Corter JE, Nickerson JV, Esche SK, Chassapis C (2004) Remote versus hands-on labs: a comparative study. In: Proceedings - Frontiers in Education Conference, FIE, 2, pp F1G-17–F1G-21
6. Guerrero M, Heaton S, Urbano D (2021) Building universities' intrapreneurial capabilities in the digital era: the role and impacts of Massive Open Online Courses (MOOCs). Technovation 99:102139
7. Neamtu DV, Fabregas E, Wyns B, De Keyser R, Dormido S, Ionescu CM (2011) A remote laboratory for mobile robot applications. IFAC (Italy) 44:7280–7285
8. Mujkanovic A, Garbi Zutin D, Schellander M, Oberlercher G, Vormaier M (2015) Impact of students' preferences on the design of online laboratories. In: IEEE Global Engineering Education Conference, EDUCON, 2015-April, pp 823–826, art no 7096067
9. Dziabenko O, García Zubía J (2015) Planning and designing remote experiment for school curriculum. In: IEEE Global Engineering Education Conference, EDUCON, 2015-April, pp 874–878, art no 7096076

10. Angulo I, Rodriguez-Gil L, Garcia-Zubia J (2018) Scaling up the lab: an adaptable and scalable architecture for embedded systems remote labs. IEEE Access 6:16887–16900
11. Aslam FA, Mohammed H, Mohd. Munir JM, Gulamgaus MA, Lokhande PS (2015) Efficient way of web development using python and flask. Int J Adv Res Comput Sci 6(2):54–57
12. Al-Fuqaha A, Guizani M, Mohammadi M, Aledhari M, Ayyash M (2015) Internet of Things: A Survey on Enabling Technologies, Protocols, and Applications. IEEE Commun Surv Tutor 17(4):2347–2376
13. Poongothai M, Muthu Subramanian P, Rajeswari A (2018) Design and implementation of IoT based smart laboratory. In: 2018 5th International Conference on Industrial Engineering and Applications, pp 169–173
14. Nam J, Do H, Kang J (2017) Investigation of friction induced vibration in lead screw system using FE model and its experimental validation. Appl Acoust 122:98–106
15. Romero S, Guenaga M, García-Zubía J, Orduña P (2015) Automatic assessment of progress using remote laboratories. Int J Online Eng 11(2):49–54

Remote Labs for Communications

Manjunath Somashekar[1(✉)], Preethi Biradar[2], Kalyan Ram Bhimavaram[2], Panchaksharayya S. Hiremath[2], and S. Arun Kumar[2]

[1] M S Ramaiah University of Applied Sciences, Bengaluru, India
[2] Electrono Solutions Pvt Ltd., Bengaluru, India

Abstract. The world is progressing at a rapid pace where everything needs to be wireless and portable. A software defined radio (SDR) is a radio communication system where the components traditionally implemented in the hardware are done by software on a computer. We use GNU Radio platform for this implementation of SDR. GNU Radio is a free and open source platform that provides various signal processing blocks in order to implement SDR's. HackRF is an open source hardware that is available for the software defined radio. In the SDR paradigm, many radio system components are implemented in software and can be reconfigured on-the-fly allowing a single piece of hardware for multi-standard, multi-band and multi-functional wireless systems. Such a paradigm change has far-reaching impact on the evolution of future wireless communications and networking systems. In this paper we discuss about various experiments that have been performed using the HackRF and GNU Radio companion where we build our SDR models.

Keywords: Software defined radio · Remote laboratories · GNU radio

1 Introduction

With the advancement in computing technologies, digital signal processing and digital communication algorithms, artificial intelligence, radio frequency (RF) hardware design, and many other elements have evolved modern communication systems into complex, intelligent, high-performance platforms that can adapt to operational environments and deliver large amounts of information in real-time, error-free. The latest step in communication systems technology is the software-defined radio, or SDR, which adopts the most recent advances in all fields to yield the ultimate transmitter and receiver. We have used GNURadio Companion, a python based visual programming interface which allows the user to use the whole tool for free and have complete control over the system that one is building. In this program we use the inexpensive HackRF while one can also use other SDR's which has to be interfaced properly. The key advantage of SDR is re-configuration, which makes it possible to completely change the radio system by means of a simple software update [1]. This feature is essential in missions where hardware replacement is not possible. SDR has various applications in even military applications, aeronautical and also civilian applications posing a wide

M. E. Auer et al. (eds.), *Online Engineering and Society 4.0*, Lecture Notes in Networks and Systems 298, https://doi.org/10.1007/978-3-030-82529-4_5

range of uses. The SDR platform provides easy access to all the layers in a wireless network system, which is not possible with traditional wireless systems.

2 Experiments

2.1 Amplitude Modulation

Amplitude Modulation is the simplest to understand and implement, in this type of modulation the transmitter uses the information signal $V_m(t)$ to vary the amplitude of carrier signal $V_c(t)$ to produce the modulated signal $V_{AM}(t)$.

$$\text{Information: } V_m(t) = V_m \sin(2\pi f_m t + \Phi);\ \text{Carrier: } V_c(t) = V_c \sin(2\pi f_c t + \Phi) \quad (1)$$

$$\text{AM: } V_{AM}(t) = \{V_C + V_m(t)\} \sin(2\pi f_c t + \Phi) \quad (2)$$

AM Signal Transmitter

1. Create a signal source block. Change the frequency to 500 Hz. This is within the audible frequency range. Change its amplitude to 0.5. In addition, change the output type from Complex to Float and hit OK. Add a Constant Source block to the signal, change the type of both blocks to Float, and set the constant for the constant source to 1.
2. Create another signal source block where the new signal source to have a frequency of 2 kHz (audible) and connect its output to an input of the modulation multiply block. This will be the carrier. And then Multiply the output of the above two signals.
3. To view the signal, first connect the output of the multiply block to a Throttle and Create a WX GUI Scope sink and WX GUI FFT sink block.
4. We can hear a buzz (Fig. 1).

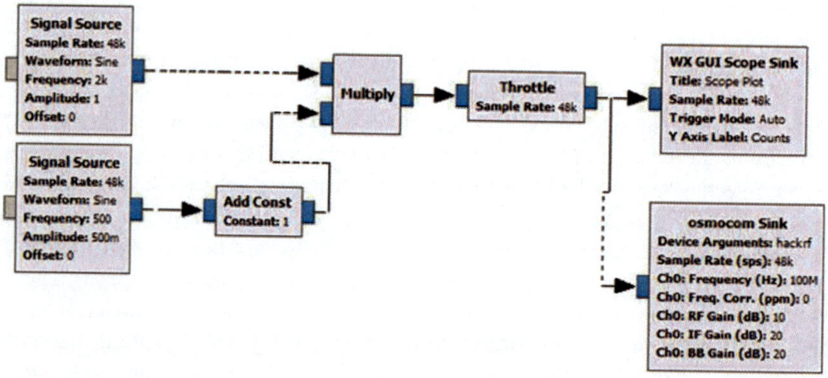

Fig. 1 GRC for AM transmitter

AM Signal Receiver. Demodulating the Modulated Signal $V_{AM}(t)$ and extracting information from it is the key function of an AM receiver. There are several methods to employ AM demodulation techniques, and in general we use the already present AM Demod Block in the GRC.

5. On a different computer, create the Audio Source Block which takes input from the microphone
6. Create a Band Pass Filter and set required sample rate. Set the low cut-off frequency to 1.2 kHz, the high cut-off frequency to 2.8 kHz, and the transition band to 100 Hz. The purpose of this filter is to eliminate other signals and noise outside the frequency range of the desired signal. Multiply the output of the above two signals.
7. Once you have verified that the modulated signal is received by the microphone/receiver, connect the output of the bandpass filter to an envelope detector, to obtain the original signal. Create a Null source block and a Max block. Change the type of the Null source and Max to Float.
8. Connect the Null source to one of the inputs of the Max block and Add a Low Pass Filter block, then set the cut-off frequency to 1 kHz and the transition width to 100 Hz.
9. To view the signal, first attach the output of the low pass filter to a. Create a WX GUI Scope sink and WX GUI FFT sink block and we can verify that it is the same as the input signal (Fig. 2).

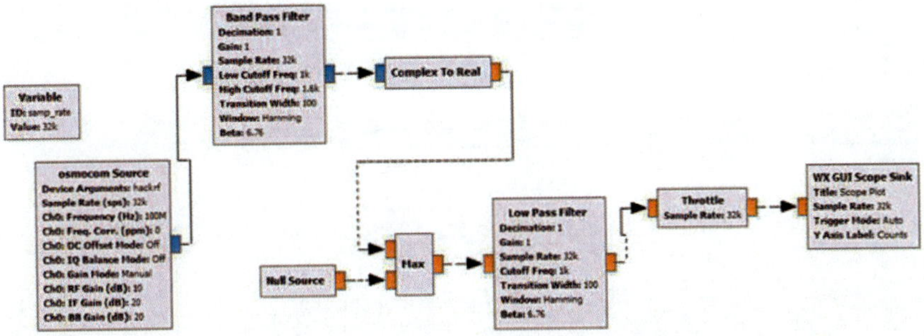

Fig. 2 GRC for AM receiver

2.2 Frequency Modulation

FM Audio Transmitter. Frequency modulation is encoding of the information in a carrier wave by varying the instantaneous frequency of the wave. In the FM radio broadcasting of an audio signal, representing the voice or music, the difference between the carrier frequency and its center frequency is proportional to that of the modulating signal. Our Goal is to transmit an audio or a wav file using Frequency Modulation.

For the implementation of the FM transmitter the modulation technique we have used is the Wide Band Frequency Modulation.

1. We have used the wav file source block which reads the raw data files in binary format from the specified .wav file. For this simulation we have used an instrumental at the bitrate of 16 kHz.
2. The signal (.wav file) is then passed through WBFM transmit block which takes a single float input stream of audio samples in the range [−1, +1] and produces a single FM modulated complex baseband output. Audio rate and quadrature rate are the sampling rate of input and output signal. Audio rate defines how many times per second input is sampled and Quadrature rate is defined as Sample rate of output stream (integer). Here the audio rate is predefined and set to 48 kHz, we set up the quadrature rates in the multiples of audio rate, 48 kHz × 4 = 192 kHz. Tau is the Preemphasis time constant (default). We set the maximum deviation at a slightly reduced rate than the default for better quality.
3. The modulated signal is then sent to the rational resampled for interpolation and decimation.
4. The resampled signal is then sent to the osmocom sink where we have to set the frequency at which the audio is transmitted. The device arguments is set to "HackRF" because we use the HackRF device (Fig. 3).

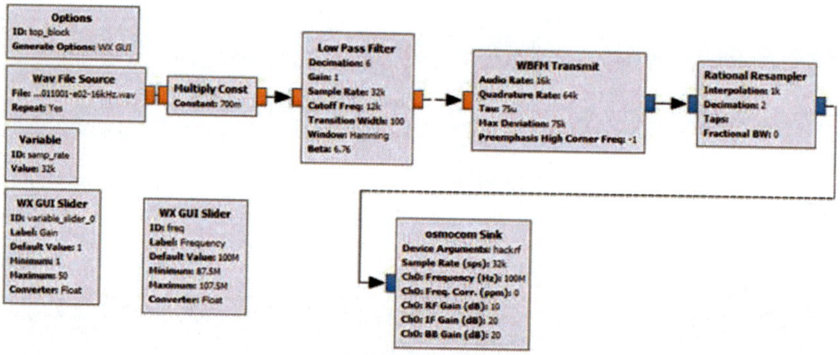

Fig. 3 GRC for FM transmitter

FM Audio Receiver. FM receiving is done by the method called as Frequency Discrimination. Our Goal is to receive the transmitted audio successfully on the HackRF device or a mobile phone.

5. osmocom source receives the transmitted modulated signal and has to be set at the same frequency rate as the transmission, the device arguments are set to "HackRF" because we use the HackRF device.
6. The received signal is given to the rational resampler where the Interpolation is 1 and decimation is 100.
7. The resampled signal is given to WBFM receive block where the quadrature rate is 48 k.
8. The demodulated signal is then given to the audio sink, where the audio is played through the computer's default speakers (Fig. 4).

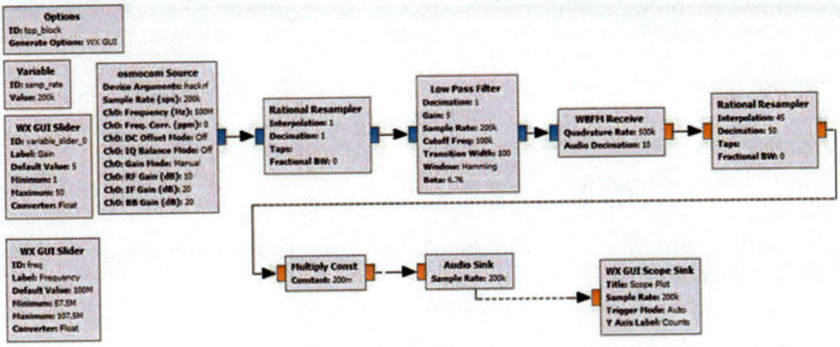

Fig. 4 GRC for FM receiver

2.3 Differential Phase Shift Keying Modulation (DPSK)

This is a type of phase modulation technique which is used to transmit data by altering the carrier wave's phase. In this, the modulated signal's phase is moved e to the element of an earlier signal. The phase of the signal tracks the low or high state of the earlier element. This kind of phase-shift keying doesn't require a synchronous carrier on the demodulator.

Image Transmitter Using DPSK

1. We use the file source block which Reads raw data values in binary format from the specified file. Here we specify the image that has to be transmitted.
2. This is then given to a packet encoder block which encodes the binary format values into data packets. In this block we set the Samples to 4 and the Bits to 1.
3. The data packets are then encoded modulated DPSK Mod block and we set the Samples the same as the packet encoder.
4. The modulated signal is then given to the osmocom sink to transmit it at a certain frequency (Fig. 5).

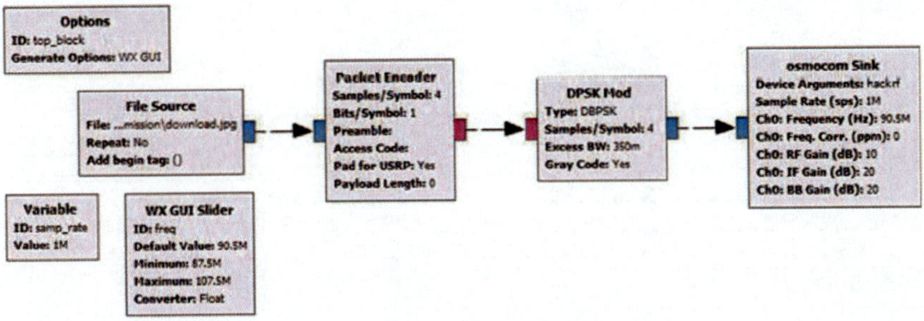

Fig. 5 GRC for DPSK image transmitter

Image Receiver Using DPSK

5. We use osmocom source to receive the transmitted signal at the same frequency.
6. The received signal is then demodulated using the DPSK Demod block, we set the Samples to 4.
7. The demodulated signal is given to a packet decoder which decodes the encoded packets.
8. This is then given to a file sink block, which stores the transmitted image over another blank image (Fig. 6).

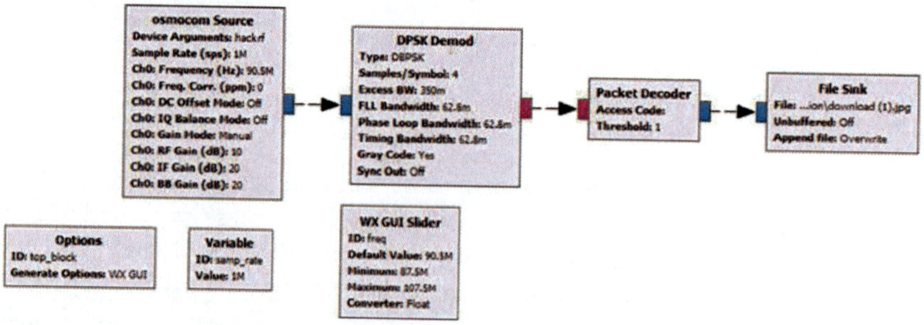

Fig. 6 GRC for DPSK image receiver

3 Results

(See Figs. 7, 8, 9 and 10).

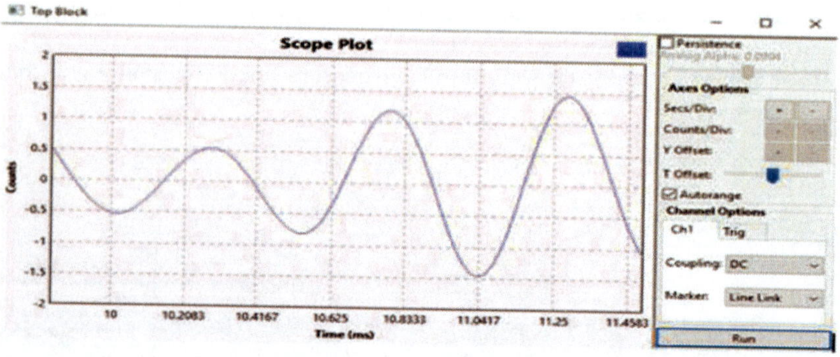

Fig. 7 Scope plot of the transmission of AM signal

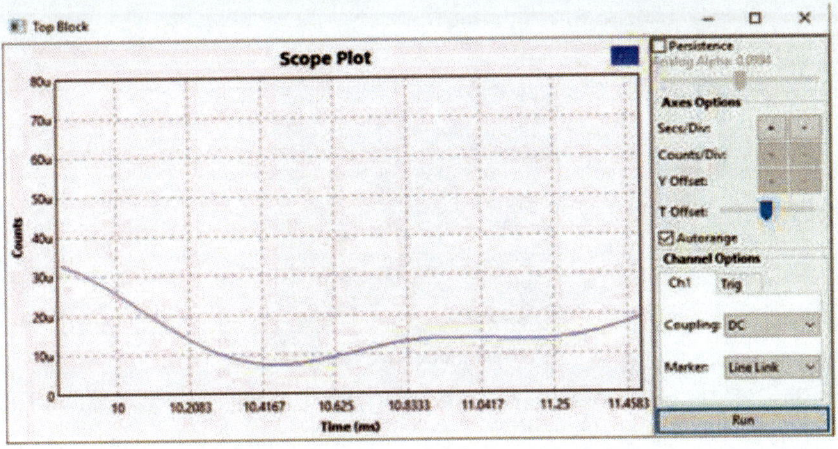

Fig. 8 Scope plot of the received AM signal

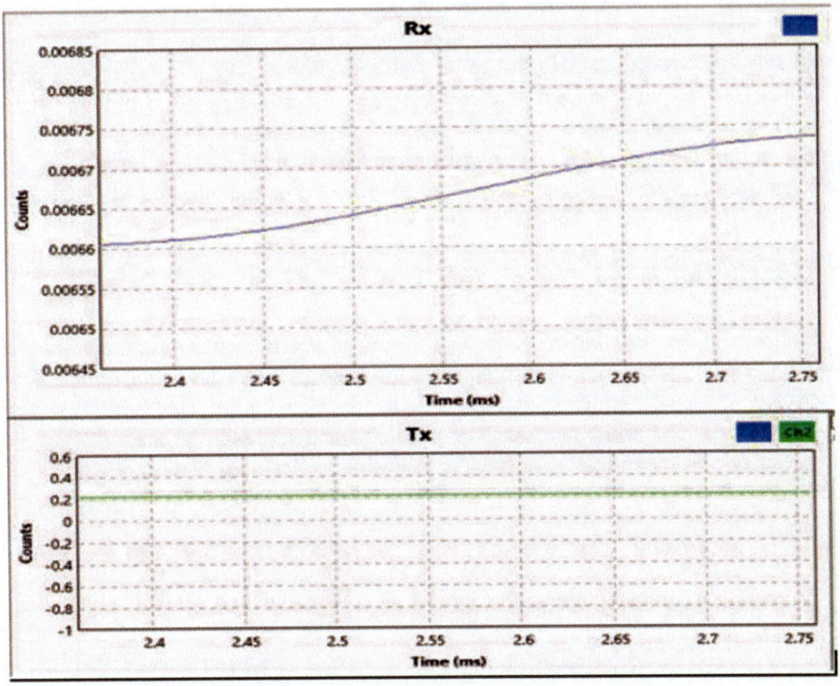

Fig. 9 Scope plot of the transmitted and received FM signal

Fig. 10 The image received after performing DPSK modulation and demodulation

4 Remote Lab Architecture

Implementation of RACL is easy and cheap. Use of Java is an element that will reduce cost and development time. Software structure of RACL consists of two main parts: Client side and Server side. Hardware of RACL has three main parts: They are Server Machine, Data acquisition card and Web cam [2]. After realizing and utilizing RACL project we depleted problems of congestion, lack of instructors and lack of space. RACL diminishes demand for buying additional Laboratory Devices by leading more effective use of available ones. RACL can be used by both students and teachers.

5 Conclusion

In this paper have seen the importance of SDR in the field of communications and looked at the inexpensive way of practicing the concepts. We have looked at different experiments and their code on the Open Sourced GNU Radio. This type of learning exposes the students to the near real-life problems and scope in the domains. Also we have seen briefly regarding the Implementation of the Remote Laboratories.

References

1. Albiol A, Corbi A, Burgos D (2017) Remote implementation of signal processing students lab. IEEE Access 5:16068–16076
2. Ugur M, Savaş K, Erdal H (2010) An Internet-based real-time remote automatic control laboratory for control education. Procedia Soc Behav Sci 2:5271–5275

Automated Testing for Sustainable Remote Laboratory System

Kyathsandra Narasimhamurthy Chandrashekar(✉), Susheen Srivatsa Chelur Nataraja, Bharath Gangarpu Chandrasekhar, Adithya Prasanna, and Suchitra Vankalkunti

Siddaganga Institute of Technology, Tumkur, India
{kcnmurthy, suchitrav}@sit.ac.in

Abstract. In this paper, performance analysis of Automated testing of different circuits in Remote laboratory are presented. Remote laboratory system for Analog Electronic circuits has more than 30 experiments. It's necessary to monitor the working condition of each experiment 24/7 for effective usage of the system. It's humanly difficult to check different types of experiments all the time. So, it's necessary to automate this testing process and inform the maintenance team for about the health condition of the circuits. The automatic testing system is aimed testing the system at regular interval and inform about the faulty circuits. Paper also presents typical analysis report of monostable multivibrator experiment considered for testing.

Keywords: Remote laboratory · Automatic testing · Monostable multivibrator

1 Introduction

Experiential learning schemes are being introduced in the Education system for better understanding of the concepts. In present situation most of the electrical engineering courses are taught by one faculty and the corresponding laboratory courses are conducted by students under the supervision of other faculty or teaching assistants. Concepts of electrical engineering can be better conveyed by conducting experiments in the classroom itself. It's also necessary that students get ample opportunity to conduct these experiments to solidify their understanding and clarifying doubts beyond class hours. Many online virtual labs portals are available, however in all these simulation-based learning, user will not be able to perform real time circuit analysis. Among many innovative ideas Remote Laboratory system is one of the effective tools. Using this any authorized user can conduct experiments available in the remote laboratory any number of times from anywhere at any time [1–7]. This will also facilitate the Professors to demonstrate the real-time implementation of the circuits in the classroom itself. In Remote laboratory system authorized user can access the hardware circuits. The circuits in the remote laboratory system need to be functional at the time of access otherwise the remote user will lose interest in the system. To have a sustainable remote laboratory system an automated testing system is essential to monitor the working condition of all the circuits.

M. E. Auer et al. (eds.), *Online Engineering and Society 4.0*, Lecture Notes in Networks and Systems 298, https://doi.org/10.1007/978-3-030-82529-4_6

The challenge in development of such monitoring system is, it should test all the circuits periodically and inform about its health status to the monitoring team. Since, mainly the hardware part has been involved it's difficult to check their correctness humanly. The paper presents an innovative "AUTOMATED-TESTING" system to monitor the working conditions of the circuits and keep updated about its health status to the maintenance team.

2 Development of Automatic Testing System

Remote laboratory system has Analog electronic circuit experiments like Rectifiers, Regulators, BJT & Opamp amplifiers, 555 Timer based multi-vibrators, Diode based wave shaping circuits, Schmitt triggers, Oscillators etc. The block diagram representation of the Remote laboratory system is shown in Fig. 1. The remote laboratory system is developed using Python, Arduino scripts and indigenously designed and developed hardware system.

The generation and measurements of waveforms for all the circuits of the Remote laboratory system is provided by Analog Discovery (AD) module a virtual instrumentation product by National Instruments. In Analog Discovery module, it's possible to generate the desired input waveform and compare the output parameters with the default values by writing the script in Analog Discovery Module.

Automated Testing is a software-oriented tool developed by interfacing with the hardware to verify the working status of all the above said circuits using appropriate test signals. As each and every circuit is unique their output waveforms are also different. It needs specific and distinct test signals to verify the testing conditions of all the circuits. The purpose of this is mainly to save the time to find the location of the fault, so that the maintenance team can fix the problem quickly.

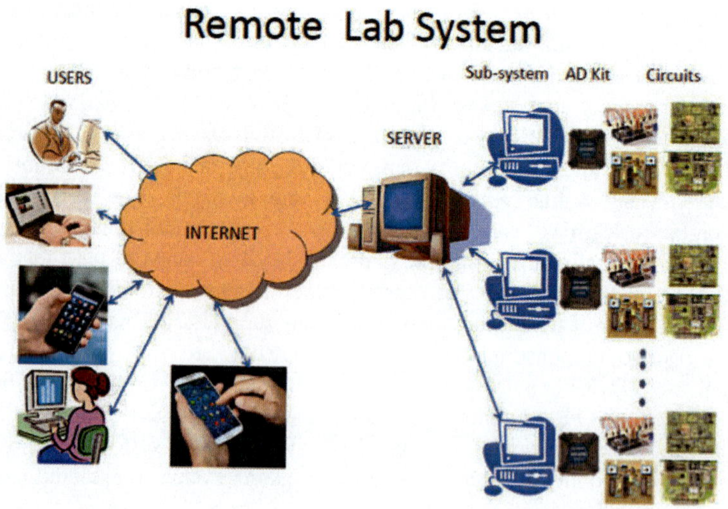

Fig. 1 Remote laboratory system

3 Behavior of the Circuit as the Testing Condition

The challenge in testing the different circuits is to how and when to decide the circuit is not functional. In this paper, its explained how the output node voltage of circuit under test is compared with a specific default value upon excited by a predefined input specifications. Both, the generation of the test patterns and the comparison of output with default values are done in AD module. For illustration purpose 555 timer based Monostable-Multivibrator experiment is chosen.

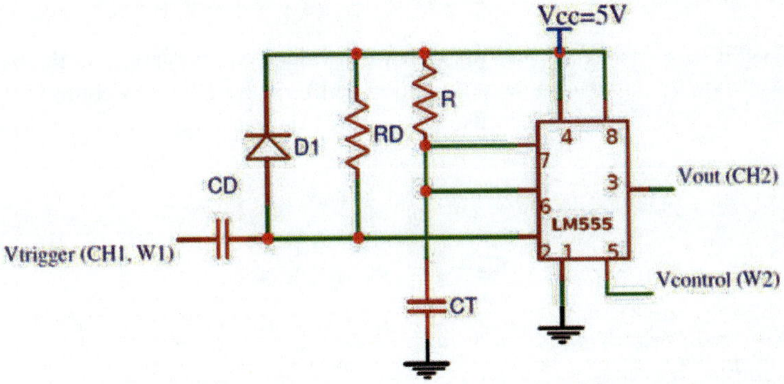

Fig. 2 Monostable Multivibrator in remote laboratory system

Circuit diagram of 555 Mono stable multi-vibrator is shown in Fig. 2. Basically in this circuit upon giving a valid trigger signal the circuit will generate a pulse of predefined width (time period). The width of the pulse is decided by the external parameters like resistor R, timing capacitor C_T and the control voltage at pin number 5 of the 555 IC. User has the access to vary the values of R and control voltage.

The time response of the circuit is shown in Fig. 3 the trigger signal shown in Fig. 3a will trigger the circuit when its amplitude is less than 1/3 Vcc (V_{TL}). When the circuit is not triggered the circuit is said to be in stable state and the timing capacitor voltage goes high (Vcc) and capacitor starts charging towards Vcc through resistor R as shown in Fig. 3 As soon as the capacitor voltage reaches 2/3 Vcc (V_{TH}) the output voltage drops to 0 V and the capacitor will also immediately discharges and settles at 0 V. The time taken be the capacitor to reach VTH from 0 V is defined as **Pulse Width** of the Mono- stable multivibrator.

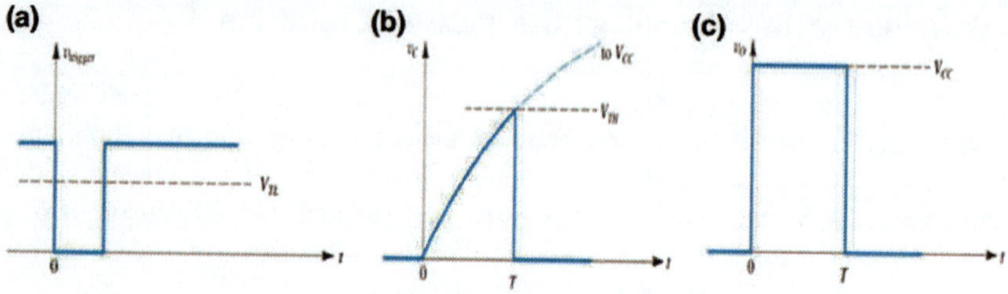

Fig. 3 (a) Vtrigger (input signal) (b) charging of the capacitor CT (c) Vout

The control voltage VTH and the resistor R values can be varied by the user. This means user has the control to vary the pulse width of the circuit. Vcontrol.

The design equations for the pulse width of the output in quasi state is given below.

Design Equations

$$V_C = V_F - (V_F - V_I)e^{-\frac{t}{RC}} \quad (1)$$

With $V_I = 0$, $V_F = V_{CC}$, at t = T, $V_C = \frac{2}{3}V_{CC}$

Equation (1) becomes,

$$\frac{2}{3}V_{CC} = V_{CC} - (V_{CC} - 0)e^{-\frac{T}{RC}}$$

Pulse width $T = \ln(3)\,RC_T \approx 1.1RC_T$

Pulse width with external voltage V_{EXT} at Vcontrol is

$$T = RC_T\, ln\left(\frac{V_{CC}}{(V_{CC} - V_{EXT})}\right) \quad (2)$$

Output will go to quasi stable state for a pre-defined period of time T whenever an input-Vtrigger goes below V_{TL} and also the output remains in the high state until the trigger is removed. Therefore, a single span of voltage is enough to make the output to go to quasi stage for a time period of T. In order to get this, the Capacitor C_D, Diode D1 and resistor R_D are being used in Fig. 2 which produces the spikes signal from the square input and Diode limits the positive signal to the amplitude of V_{Don}.

Table 1 shows the input settings and output parameters to be measured in wave generation and scope of the waveform tool to test Mono-stable multivibrator experiment.

In Remote Laboratory System the monoshot vibrator experiment consists of 3 different variations, where the Resistance R can be varied to 1 kΩ, 10 kΩ and 1 kΩ parallel with 10 kΩ. Upon Selecting the R value as 10 kΩ and C_T as 0.1 μF and considering the above input parameters, if the Mono-stable multivibrator circuit is working fine average voltage of output will be greater than 0.2 V and pulse width to be 91.6 μs.

Table 1 Details of input and scope channel settings for testing 555 timer based Mono-stable multivibrator

Experiment	Wavegen	Channel 1	Channel 2
Mono-stable Multivibrator	**Wavegen 1:** Square wave Amplitude: 3 V, Offset: 2 V Frequency: 1 kHz Duty cycle: 50% **Wavegen 2:** DC: 3.5 V (V_{EXT})	Offset: 0 V Range: 2 V/div	Offset: 0 V Range: 2 V/div
		View-Measurements-Add	
		Vertical: C1:Maximum, minimum C1: Peak to peak **Horizontal:** C1: Frequency, Positive width	**Vertical:** C2: Maximum C2: Average **Horizontal:** C2: Positive width

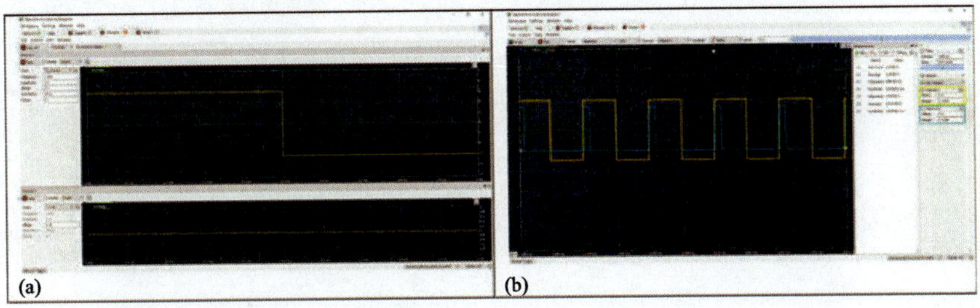

Fig. 4 (a) Trigger signal and control voltage (b) Trigger signal (yellow) and output waveform

Table 2 Parameter monitored for finding the working status of the circuit

Mono-stable multivibrator with R as 10 kΩ, C_T as 0.1 μF and V_{EXT} = 3 V			
Parameter monitored	Expected	Practical	
		Good one	Faulty one
Average voltage	>0.2 V	0.37 V	0.02 V
Pulse width	91.6 μs	90.62 μs	NA

Based on these observations made in Fig. 4 and Table 2, it's necessary to generate suitable input signal to get the output. Irrespective of the working condition of the circuits, its necessary to measure the output voltage. In the next section the process of measuring the output voltage is discussed in detail.

4 Process Flow of Automated Testing

The process flow for automatically monitoring working condition of 555-timer based Mono-stable multivibrator is shown in Fig. 5. As it is said earlier the Python code is used to provide the interface between the user and the circuit, Arduino is used to generate the desired signal pattern for switching the circuit components and routing the signals from and to the AD kit and selected circuit and conclusion will be done through the Waveform software.

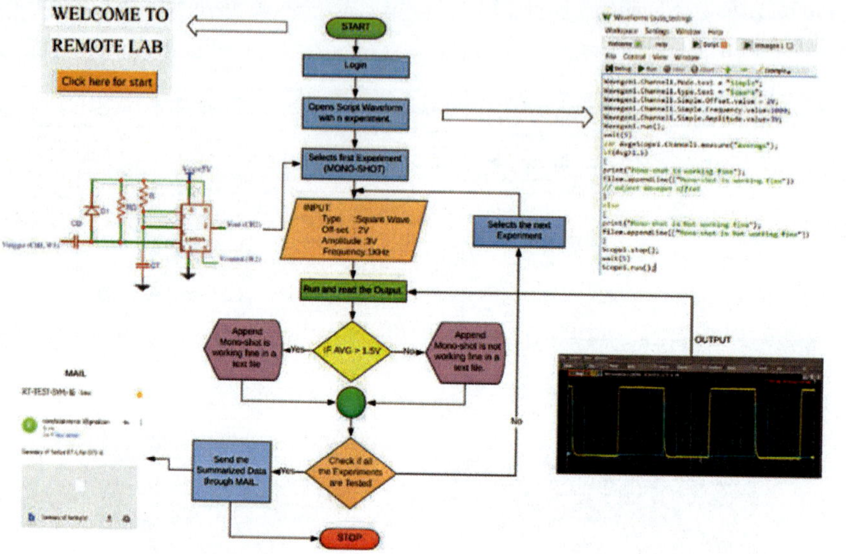

Fig. 5 Process flow of automatic testing system developed for Mono-stable multivibrator experiment

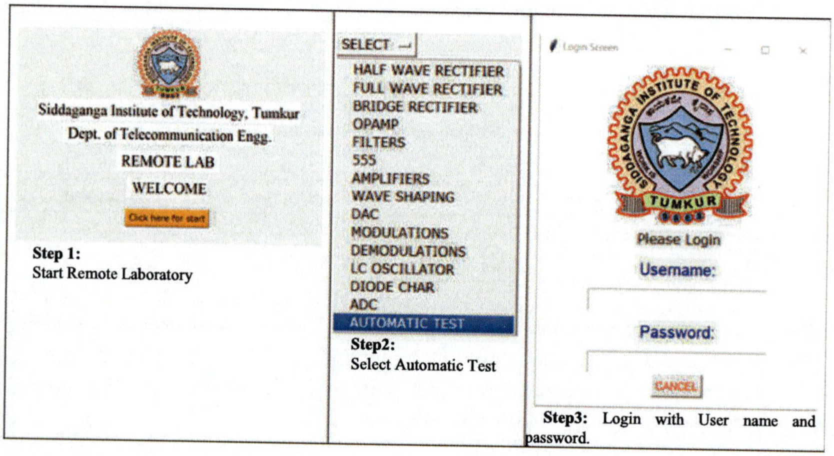

Fig. 6 Conduction steps to perform automatic testing of circuits in Remote laboratory system

The flow chart in Fig. 5 for Mono-stable multivibrator is developed keeping the trigger frequency and pulse width of the output signal. It's always challenging to write the flow chart for all experiments of the remote laboratory system. Figure 6 shows the steps to perform automatic testing. As soon as the automatic test is being selected, a new login window pops up by asking username and password. Once there is match with the username and password, the new waveform window with the script code gets opened automatically. Waveform Scripts are necessary to provide the input, read and analyze the circuit health status automatically based on the test condition given in the code.

Scripts Description

The Fig. 7 (a) shows the window of waveform script, where the initiation of the Digilent waveform input parameters, reading the output values, formulated test conditions are written in Java script to perform the test.

From the script code, objects behind the user interface can be accessed. This allows configuration using the interface and automating parts of it with script. Usually the script editor uses java-script language to create custom Math channel in Scope, custom waveform pattern in WaveGen and Meter channel function. These expect mathematical function which will be called for each sample to transform input value (data or X).

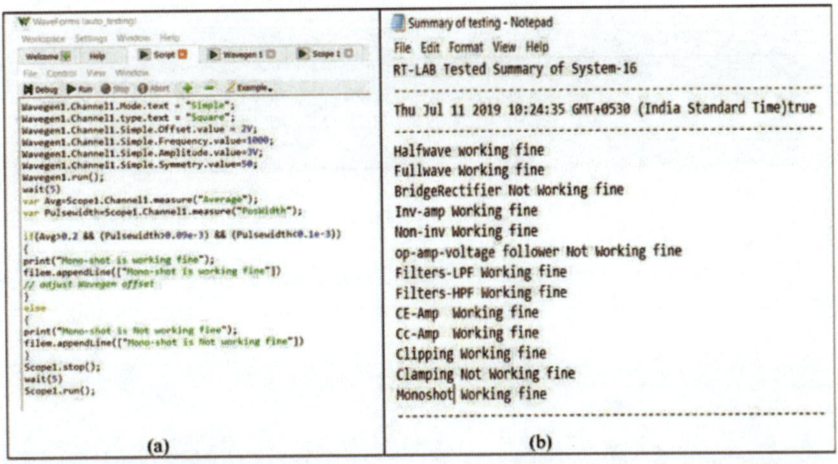

Fig. 7 (a) Script for testing the Mono stable multivibrator circuit (b) summary of the testing session

Let us have a look at the functions of each code in case of mono stable multivibrator experiment.

```
Wavegen1.Channel1.Mode.text = "Simple";
Wavegen1.Channel1.type.text = "Square";
Wavegen1.Channel1.Simple.Offset.value = 2V;
Wavegen1.Channel1.Simple.Frequency.value=1000;
Wavegen1.Channel1.Simple.Amplitude.value=3V;
Wavegen1.Channel1.Simple.Symmetry.value=50;
```

The above lines of the codes are to the set the waveforms inputs with Square wave type signal characterized by the duty cycle of 50%, amplitude of 3 V, frequency of 1 kHz with an offset of 2 V. The offset has been given since the trigger signal has to go below V_{TL} Value to make the output to go for quasi state for the predefined time.

```
Wavegen1.run();
```

Once the inputs are set, the waveform scripts start running in the background of Wavegen and the Scope upon corresponding selection of the experiment simultaneously through the python code. i.e. the python code sends the signal to the Arduino to select the corresponding experiment for testing before the waveform runs, so that the measurement of the corresponding output will be made with respect to the selected experiment only.

```
wait(5)
var Avg=Scope1.Channel1.measure("Average");
var Pulsewidth=Scope1.Channel1.measure("PosWidth");
```

wait (5) is a time lag of 5 s given for the system to settle with all the corresponding setting assigned for the interface tools.

Later we declare a two variables Avg and Pulse width to store the output Average value and the Pulse width to compare the expected value as the test condition.

```
if(Avg>0.2 && (Pulsewidth>0.09e-3) &&(Pulsewidth<0.1e-3))
{
print("Mono- stable multivibrator is working fine");
filem.appendLine(["Mono- stable multivibrator is working
fine"])
}
else
{print("Mono- stable multivibrator is Not working fine");
filem.appendLine(["Mono- stable multivibrator is Not working
fine"])}
```

The above lines of codes tell the test condition which has been implemented using if else statement in Java script. If the output average value is greater than 0.2 V and pulse width lies between the region of 0.09 to 0.1 ms for the corresponding input are considered as a good state of the circuit, else it is classified as the fault state of the circuit.

In the code if statement got satisfied, then it prints "Mono-stable multivibrator is working fine" in the console window and the filem.appendLine (["Mono-stable multivibrator is working fine"]) code will append the text "Mono-stable multivibrator is working fine" in to text file. This file is the one which has already generated with date and time at the time of the declaration of the automated test in python code which has been created though initiation of the java script in the global declaration.

Similarly, if the condition is not satisfied, then the text "Mono-stable multivibrator is Not working fine" will be appended in the text file and also will be printed on the console too. Figure 7 (b) shows the summarized data of the appended text of different experiments.

```
Scope1.stop();
wait(5);
Scope1.run();
```

Later on, the Waveforms will get stopped for 5 s to get settled for the next experiment. Where the selection of next experiment will be done by python and the Arduino and the process will go on similarly by running the scope to read the data.

Similarly, the same test process will continue until the loop is stopped for all the experiment with different test conditions and with different script codes.

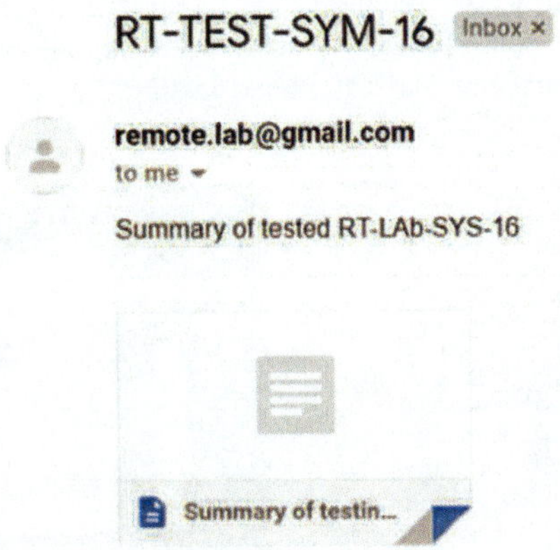

Fig. 8 Email sent by the system

Once the testing process got completing for the Experiments, the summarizing status condition will be saved in the predefined file with date and time. This summarized data will be sent automatically to the Administrator and to the maintenance team through an Email with corresponding Remote Laboratory system number. Figure 8 shows the type of mail received by the administrator. Hence, the process helps us to save the time by automating the things.

5 Advantages

Manual testing is testing of the software as well as the hardware, where tests are executed manually by the human beings. It is performed to discover bugs in hardware under development through the software interfaced in it.

In manual testing, the tester checks all the essential features of the given application or software. In this process, the software testers execute the test cases and generate the test reports without the help of any automation software testing tools which is a long-time process.

It is a classical method of all testing types and helps finding the bugs in software systems and also sometimes results poor detecting due to some factors like human error, etc...., that is why, it is generally conducted by an experienced tester to accomplish the software testing process.

In Automated Software Testing, testers write code/test scripts to automate test execution. Testers use appropriate automation tools to develop the test scripts and validate the software. The goal is to complete test execution in a less amount of time.

Automated testing entirely relies on the pre-scripted test which runs automatically to compare actual result with the expected results. This helps the tester to determine whether or not an application performs as expected.

Automated testing allows you to execute repetitive task and regression test without the intervention of manual tester. Even though all processes are performed automatically, automation requires some manual effort to create initial testing scripts. Figure 9 (a) and (b) show the manual test and automated testing.

(a) Manual Testing (b) Automated Testing

Fig. 9 (a) Manual testing (b) automated testing

As most of the part of the testing process is automated, you can have a speedy and efficient process. Also, this process can be recorded, reuse and execute the same kind of testing operations for required number of times. It can easily increase productivity because it provides fast & accurate testing result. It also supports various applications leading to check even the smallest unit.

The main intension of introducing in the **Remote laboratory system** is manual test takes lot more time to test and servicing each circuit. This **"long time lag"** service for the Remote laboratory will not give the students to know the Practical output along with the theory class on time and also due to the unseen notification by humans in manual test may miss-guide the students by getting wrong outputs in the experiment.

6 Conclusion

Automatic testing system developed to observe the working conditions of the analog circuits of Remote Laboratory system has been implemented successfully. Upon getting the report from the automatic testing system its now very easy to locate the faulty circuit and make the Remote Laboratory system available to the user with full potential. This has motivated us to develop many more experiments and add it to the existing Remote Laboratory system. This way the automated testing system helps to create a sustainable Remote laboratory and save time in testing process of all the systems.

Acknowledgements. Authors of this paper express their gratitude for all those who helped in developing the remote laboratory system. We also thank officials of Siddaganga Institute of Technology (SIT), Tumakuru, the Management, Director, CEO and Principal for their support in establishing Remote Laboratory system.

References

1. http://vlab.co.in/
2. Hesselink L et al (2003) Stanford cyberlab: Internet assisted laboratories. Int. J. Dist. Educ. Technol. 1:21–39
3. http://www.uml.edu/IT/Services/vLabs/
4. de la Torre L, Guinaldo M, Heradio R, Dormido S (2015) The ball and beam system: a case study of virtual and remote lab enhancement with moodle. IEEE Trans. Industr. Inf. 11 (4):934–945
5. de la Torre L, Sanchez JP, Dormido S (2016) What remote labs can do for you. Phys. Today 69:48–53
6. Sanchez-Herrera MRS, Mejias A, Marquez M, Andujar JM (2018) A fully integrated open solution for the remote operation of pilot plants. IEEE Trans. Industr. Inf. 15:3943–3951
7. Heradio R, de la Torre L, Dormido S (2016) Virtual and remote labs in control education: a survey. Annu. Rev. Control 42:1–10

Interactive Lab Experimentation and Simulation Tools for Remote Laboratories

Fariba Moghaddam[1(✉)], Denis Gillet[2], Aldo Vaccari[1], Christophe Salzmann[2], and Yves Piguet[2]

[1] University of Applied Sciences and Arts Western Switzerland, 1950 Sion, Switzerland
fariba.moghaddam@hevs.ch
[2] École Polytechnique Fédérale de Lausanne, 1015 Lausanne, Switzerland

Abstract. New extensions and features to the remote laboratories currently used for undergraduate courses in automatic control laboratory at the School of Engineering of University of Applied Sciences and Arts Western Switzerland are presented. This paper shows how lab experimentation and lab simulation can be linked in a complementary way. In fact, their adequate combination offers the students the opportunity to apply classroom theoretical knowledge while conducting a real experiment. Besides, the lab-based simulation tools can be used for experiment modeling and controller design. Simulation tools also facilitate analyzing and interpreting data and finally validating the results. We discuss different technical implementing trends as well as several pedagogical approaches in the way remote lab exercises are performed including real-time data acquisition, experimentation environment within user friendly interface and simulation tools.

Keywords: Remote labs · Simulation tools · Smart lab experiments · Real time · Learning analytics

1 Introduction

In science and engineering education, it is imperative that students be able to put into practice what they have learned in the classroom and to make the transition from theory to practice with confidence and effectiveness. If hands-on laboratories have long played a central role in engineering education, financial and practical constraints can seriously limit access to these activities. The importance and flexibility of distance learning, whereby teaching is undertaken remotely on digital platforms, have been largely proven, especially during the Covid-19 pandemic. Remote labs provide access to scientific experiences through a digital learning platform where students can interact with the real experiments not locally available.

There are two different ways to conduct laboratories: experimentation and simulation. They both play an important role in engineering education. Students feedbacks show that they consider remote labs as being as operative as hands-on labs. Furthermore, they seem to think that simulations are less effective than remote labs because simulations don't feel as realistic. Indeed, the realistic nature of remote labs affords

M. E. Auer et al. (eds.), *Online Engineering and Society 4.0*, Lecture Notes in Networks and Systems 298, https://doi.org/10.1007/978-3-030-82529-4_7

students the opportunity to more directly apply theories learned in the classroom to real-world, while simulations may lead students to overlook and even to miss the link between theory and application.

The goal of this work is to show how lab experimentation and lab simulation can be linked in a complementary way. In fact, their adequate combination offers the students the opportunity to apply classroom theoretical knowledge while conducting a real experiment. Besides, they can use the lab-based simulation tools for experiment modeling and controller design. Simulation tools also facilitate analyzing and interpreting data and finally validating the results.

This paper presents the remote experimentation environment together with interactive simulation tools used for automatic control laboratory sessions at the school of engineering of University of Applied Sciences and Arts Western Switzerland. We discuss different technical implementing trends as well as several pedagogical approaches in the way remote lab exercises are performed including real-time data acquisition, experimentation environment within user friendly interface and simulation tools.

2 Laboratory Experiments for Automatic Control

We suggest a classic control laboratory whose aim is to offer remote users the ability to design PID (proportional-integral-derivative) controllers and test them for example on a servomotor for speed and position control. A DC (direct-current) motor is paired with an incremental encoder to provide position and speed feedback.

The experiment remote control panel is a user-friendly interface which allows real-time data acquisition and visualization of the physical measurements such as speed, position, voltage on the remote servomotor, in open-loop and closed-loop configurations (Fig. 1). A camera offers continuous live video streaming of the experiment in addition to other remote measurements. It has been shown that the degree to which students derive realism from the remote experience is certainly a function of the user interface and visualizations used in the lab experiences.

Real-time Interaction over Internet should enable students to efficiently interact with physical distance equipment, as a tight interaction provides them with the best possible feedback to minimize drawbacks inherent to the distance, with sufficient information to reproduce the state of the distant equipment and its operational conditions.

Besides, the simulation environment is integrated to the same remote control panel application where the students interact with the remote experimentation system in real time (see Sect. 3. Interactive Tools).

The experiment remote control panel together with the simulation tools propose the following sequence of learning objectives:

1. Open-loop experimentation on the remote system to characterize its dynamic behavior, for example by performing a step response,
2. Data collection and data processing,
3. System model building through interactive curve fitting to identify the transfer function of the system,

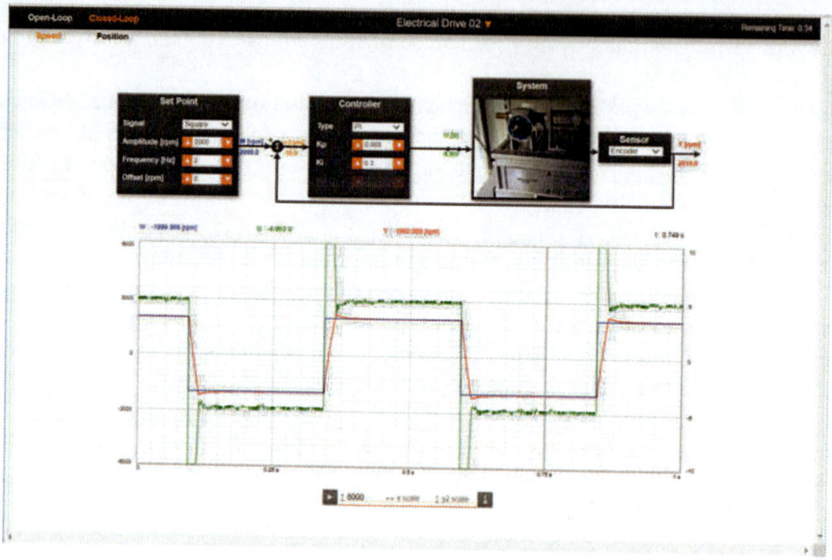

Fig. 1 Remote control panel for a servomotor speed and position control

4. Controller design based on the identified model and other given specifications,
5. Closed-loop tests to validate the calculated controller on the remote system for different set points,
6. Controller fine-tuning based on closed-loop simulation and experimentation tests,
7. Disturbance insertion on the real system and test the robustness of the controller,
8. Observation and interpretation of the results.

The proposed control panel is implemented in a MOOC platform and is actually fully dedicated to hands-on activities. From a pedagogical point of view, one deals with the extensions of e-Infrastructure for self-organized learning environments. Different learning scenarios and pedagogical contents split into theoretical, experimental and evaluation phases lead the student during the laboratory session. Based on pedagogical contents, students are able to get orientation on their own, as well as to determine and elaborate adequate learning methods for new competences and skills. The learning scenario of each session is split into short videos describing how to use the various tools (experiment description, theoretical development, control panels, …), a set of offline exercises and a set of numerical questions (quiz form) to validate user's finding and to evaluate user's understanding.

3 Interactive Tools

In addition to the control panel where students can set up the experiment conditions, supervise its proceeding and collect data, software modules provide computation and visualization support for the design of feedback controllers. Interactivity invites students to investigate different hypotheses, but without guiding them too tightly: students remain free to also try faulty designs and learn from the mismatch between their expectations and their observation of simulations and experiment results.

3.1 Purpose

The typical steps to obtain a suitable controller are as follows:

1. System modeling. Among the mathematical tools used in automatic control, transfer functions, which related the system output (measurement) to the system input (control signal) and assume linearity and time invariance, are an extremely powerful approach. They can be obtained from first principles, decomposing the system into subparts, each with their dynamic behavior described by differential or algebraic equations, with Laplace transform counterpart; or from the parameter identification of "black-box" generic models. Often both approaches are used together: a qualitative understanding of the system provides the model structure (model order, integrator or chain of integrators, delay) with unknown lumped parameters which are identified from input–output experiment data.
   Simple first- or second-order models, possibly with an integrator or a time delay, have a very simple step response students must get familiar with. In the system modeling tool, a sequence of system input and output samples acquired from the remote experiment are displayed. The student can select a range of samples where the effect of the steady-state gain and the time constant(s) are clearly visible, and then fit asymptotic lines whose position gives the model parameters. The resulting model response is displayed along with the samples to show the quality of the fit.
2. Controller design. Depending on the system, on the available measurements (system output) and actuators (system input), a controller structure must be chosen. The structure can take into account system knowledge which is not captured in the model, to enforce high or low gains in some frequency ranges and reject perturbation and measurement noise. In the case of a servomechanism, dry friction, a nonlinear perturbation which can be approximated as a constant drag when the velocity is not reversed, can be rejected with an integral effect. Once the structure is decided, controller parameters must be adjusted to obtain the desired closed-loop behavior in terms of stability, robustness and performance. To achieve a suitable compromise between contradictory goals, these criteria are evaluated simultaneously and displayed as simulations in the time domain and as frequency responses. With direct manipulation, students get a better understanding of how they are related, what can be obtained, and why they are useful.
3. Controller validation on the remote experiment. The controller which has been designed and analyzed with computed graphics can be applied to the real system. If there is a good match, students get confidence in their approach. Otherwise, the comparison triggers reflection to understand the causes of the discrepancy. Was the model suitable for its intended use? Does the real system have additional aspects which were neglected, for instance saturation of the input, or the effect of sampling or noise?

3.2 User Experience Design

The interactive tools were added to the existing user interface for our remote experiments (Fig. 2). This user interface is implemented as a web application to avoid any installation for the students. It shows the block diagram of the system, either in open loop without a controller or in closed loop with the controller. The diagram structure, the input shape and parameters, the controller structure and parameters, and the sensors to use can be selected easily with buttons, pop down menus and numeric input fields. All the settings remain visible to let the students know exactly what they are doing and to save their work with screenshots, by saving the web browser page, or printing to PDF or to a real printer. The display is especially important when two students collaborate, possibly remotely where they cannot see what their colleague does with the mouse or the keyboard, or when they request advises from an assistant. A video stream shows the experiment and provides an easy way to understand what happens. For the servomechanism, measurements above the noise level and in a suitable bandwidth are clearly visible.

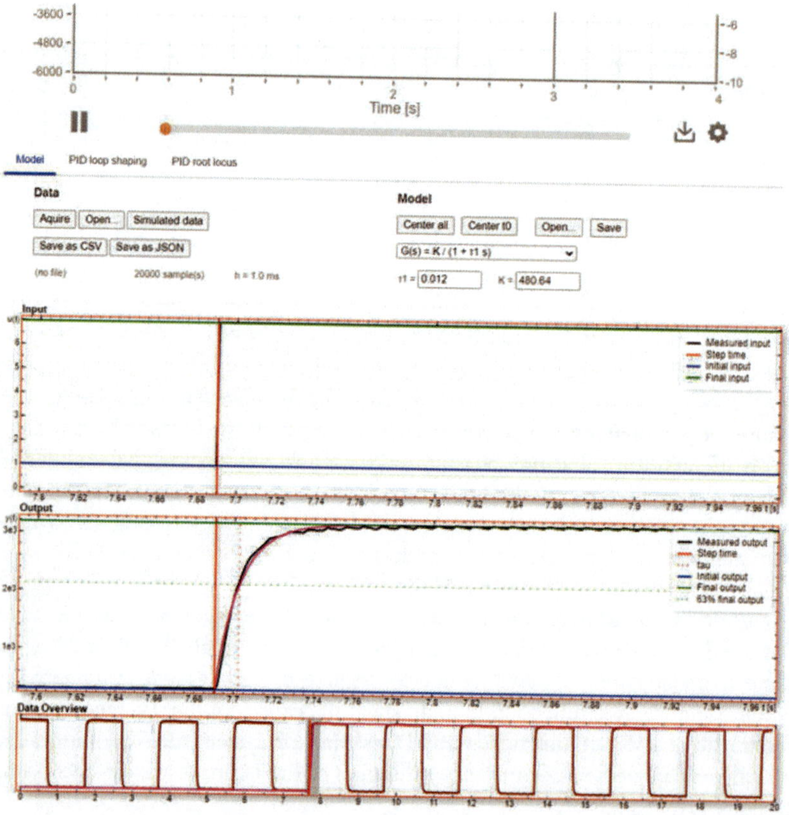

Fig. 2 Interactive tool for transfer function fit on experimental responses

Interactive tools are added below the area reserved to thc remote experiment, in different tabs. Tools correspond to different steps, as described in the previous section. Each step uses data from the previous one: acquisition sampled data from open-loop experimentation to modeling, model parameters from modeling to controller design, and controller parameters from controller design to validation on the remote experiment. Data can be saved to the local file system, which serves three purposes: students can check their content and understand clearly what happens, without hidden steps obscured by the software; they can reload data in the same tool or for the next step, maybe coming back at another time after they have thought "offline" about their work; and they can import data into other applications, such as Matlab or a word processor, for further analysis or to write a report.

The main exchange format is JSON, a text format easy to read and expressive. Figure 3 shows the model parameters corresponding to the model fit of Fig. 2. Samples acquired from the experiment can also be saved to and loaded from CSV files which are more suitable to tabular data.


```
{
    "K":480.63537749401445,
    "tau":0.012237174850223909,
    "integrator":false,
    "d":null
}
```


Fig. 3 Outcome of transfer function fit as a JSON file

3.3 Technical Aspects

The interactive tools have been developed in Sysquake, a scientific application for numerical computation and interactive visualization with a programming language mostly compatible with Matlab. Sysquake exists in two versions, native for optimal computation performance, and web-based to run locally in the browser without server support. The same interactive applications run likewise in both versions.

For the simulation tools, applications were compiled to JavaScript and integrated into the HTML page which contains the remote experiment control panel. HTML input elements, such as pop down menus and input fields, were added to permit direct parameter entry in addition to the direct manipulation of graphics, and import/export functionality was added to interface with the remote experiment on one side to fetch experimental data or set parameters, and with files on the other side to save or reload data and controller designs. Because of the time allocated to students for the remote lab experiments, this approach was preferred to a command-line interface which would be more open to different kinds of analyses and controller design methods, but would requires more learning efforts.

4 Typical Student Session

While students can freely switch between the remote experiment and the interactive tools, the expected typical path they would take could be as follows:

- Set the parameters in the remote experiment control panel to have a periodic square signal with an output significantly larger than the measurement noise, and a period significantly larger than the system time constant.
- Acquire samples in the identification tool.
- Select samples around a clean step response.
- Fit a model of adequate structure, typically a first-order model for the servomechanism controlled in speed (Fig. 2).
- Save the model to a JSON file, switch to the controller design tool, and load the model (Fig. 3).
- Design a controller with good robustness and performances. For the servomechanism controlled in speed, a PI controller (PID without derivative term) is expected to give good results. The phase margin can be observed in the Nyquist diagram, the step overshoot in the step response of the system output, and the control signal

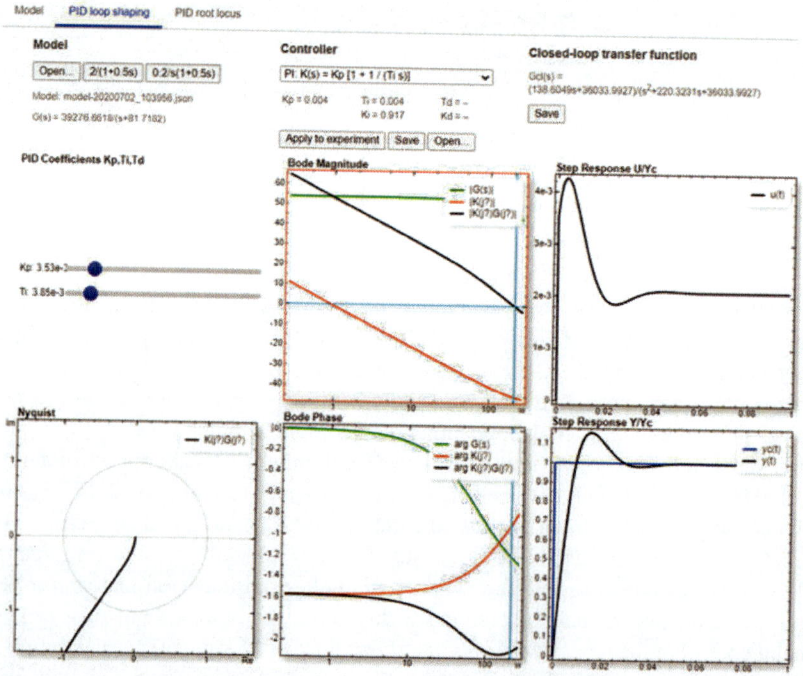

Fig. 4 Design of a PI controller with the interactive tool. The controller frequency response can be manipulated in the magnitude Bode diagram, vertically to change the PI controller gain Kp or horizontally to change the integral time Ti = Kp/Ki, or the PI parameters are changed directly with the sliders, and all the graphics are updated continuously

(system input) can be made smaller than the servomechanism input voltage admissible range (Fig. 4).

The controller parameters are applied to the remote experiment, which is switched to closed-loop mode. The setpoint signal can be changed to obtain a suitable response (Fig. 5).

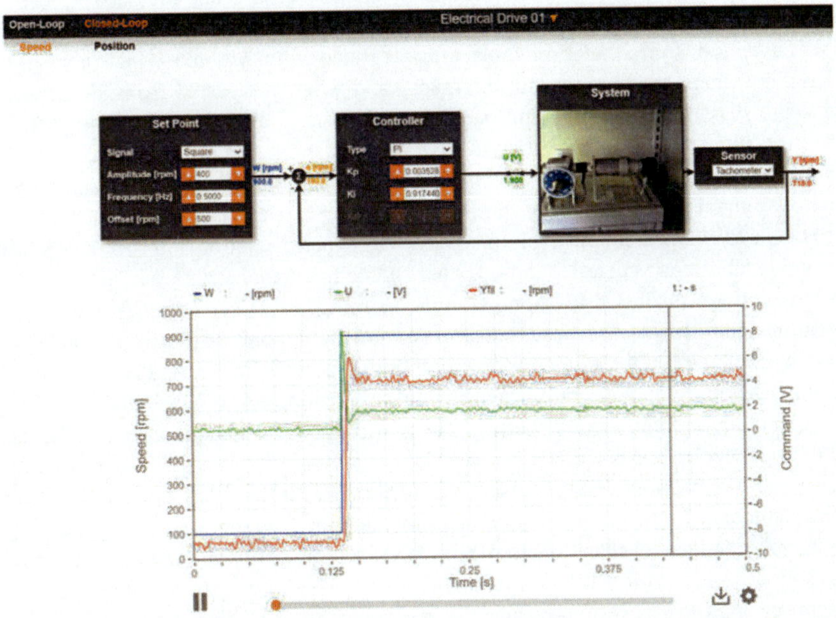

Fig. 5 Remote experimental system in closed-loop with the PI controller designed in the previous step. The Set Point and the trace scale can be adjusted to show if experience confirms the theory

5 Trends in Implementing Remote Labs

While remote experiments have been deployed for more than two decades, they are based on technologies which have evolved continuously [1]. Three aspects are worth mentioning: The seamless integration in learning platforms, which was described in previous section; the changing optimal compromise which results from processing power available on the server and the client and the communication bandwidth; and security aspects.

5.1 Integration in Learning Platforms

A good integration of remote lab experiments and simulation tools in existing learning platforms is important to give students an efficient learning experience. The software should take care of aspects which have no educative value, such as spending time learning a user interface which is used only once, or entering lengthy data at the keyboard. But the flow of information which is a key aspect of the methodology should remain clear to the students. The best approach depends not only on the topic, but also on the students and on the teaching context with respect to other courses.

At University of Applied Sciences and Arts Western Switzerland, these interactive experimentation and simulation tools are integrated with Moodle which also offers video clips where teachers explain the theory and what is expected from the students, and static PDF files. Another approach, more linear, is also possible. With MOOC platforms such as edX, the matter is divided into short sections; sessions with lab experiments the computation tools is more focused, and the intermediate results which must exchanged are managed by the platform [2].

Thanks to HTTPS support, the interactive lab experimentation and simulation tools for remote laboratories detailed in this paper can be integrated seamlessly in the Graasp platform introduced in [3] to support online learning activities. The graasp.eu digital education platform [4] has been developed with the support of the European Commission and Swiss public funds for enabling online personal, collaborative and inquiry learning. It is open to any educational institution or nonprofit organization promoting active learning and knowledge sharing. The platform has also been adapted to be used in under-connected areas [5].

Graasp has been extensively used during the COVID19 crisis not only by engineering students for blended design and computational thinking activities [6], but also by school teachers to share content with parents and children carrying out school activities at home during the confinement period. The simplicity to create activities integrating multimedia resources and the possibility to share them with students just with a secret link were the main reasons for this swift adoption.

5.2 Client–Server Architecture

For anyone observing the evolution of technologies involved in remote experiments since Internet access became more and more accessible in the nineties, it is quite obvious the environment has changed very significantly. On the client side, mostly for security reasons, Java applets, Flash Player applications, and other native browser plugins have been replaced by JavaScript web applications. The performance which can be obtained in JavaScript has increased dramatically thanks to just-in time compilers and the competition between the main browser editors. Hybrid architectures where heavy computations are performed on the server and where the client is limited to user input and result visualization, such as AJAX or Jupyter, can often be avoided. This has the benefit to reduce the cost of the server infrastructure, security and maintenance.

Communication protocols are still open and give the freedom to develop innovative solutions. Learning platforms are not standardized to a level which would permit an easy migration should it be required. This is why for this remote experimentation setup, effort has been put mostly on the remote experiment and client-side tools with data under the direct control of the client rather than based on cloud storage.

5.3 Security Aspects

The connection between the student computers (clients) and the servers connected to the lab experiments must be secured against malevolent actors. The lab experiments and server computers must be protected against unauthorized connection attempts which could steal resources, impair their usage by the students they are destined to, and possibly damage them by submitting repeated extreme commands. The student computers should be protected against malware, phishing attempts, or any content which would tarnish the lab institution. And the experimental data integrity should be guaranteed to ensure students can rely on their remote experiment outcome.

On the remote lab side, servers are connected to wired networks which are difficult to attack. The weakest side is the client computers and their Internet access: Wi-fi connections have become the norm on laptops and tablets, and the security of public wi-fi access points is often low or impossible to assess. To improve overall security, the use of TLS (SSL) encryption has become prevalent (HTTPS URLs), with more and more restrictions enforced to discourage practices known to be easily abused. Modern browsers flag pages served without encryption as unsecure, refuse unencrypted auxiliary resources accessed from encrypted pages ("mixed content"), and remember if a site has been served via HTTPS to not switch back to unencrypted HTTP in future accesses ("HTTP Strict-Transport-Security"). Cross-origin constraints restrict possible implementations further (cross-origin refers to web content in the same page obtained from different servers).

In addition to encryption, TLS guarantees the server identity: it prevents a malevolent third-party to impersonate a server, collect sensitive data sent by the client and provide them with fake replies. Identity is based on certificates with a chain of trust: browser (or the operating system) has a list of trusted parties, and thanks to public key signatures, a chain of certificates leads to the server certificate linked to its IP address. Currently, this restricts servers to fixed IP addresses.

These constraints are expected to continue tightening. For instance the DNS protocol (domain name system) used to obtain the numerical IP address from the domain name, which is still used without encryption and can be easily diverted (sometimes for legitimate reasons), is replaced by DNS-over-HTTPS in the main browsers, which can lead to other concerns. Applications such as remote lab experimentation are bound to follow closely these trends.

The interactive tools are served from a server implemented in LabVIEW to the students' browsers via HTTP 1.1 (a simple request/reply protocol where the connection is closed after one or a few requests) and WebSockets (a lasting bidirectional stream of messages starting with an HTTP handshake). Both of them are supported natively by browsers. For the reasons described above, the encrypted versions of these protocols is used, HTTPS and WSS. On WebSocket, messages are encoded in JSON, a good compromise between legibility and conciseness which is supported natively in browsers and elsewhere. Control parameters and measurement data are sent as numbers, and the video stream is sent as JPEG images in the same connection, the easiest way to ensure synchronization between video and real-time signal traces. The video quality suits well its purpose, a direct way to assess the system functioning.

Web browsers are de-facto standard environment for client applications. Client applications written in JavaScript enable students to fully interact with remote systems, through real-time video streaming, oscilloscope or other data display systems as well as control panel to specify the controller parameters. Thanks to the Smart Device (SD) API, different client applications can connect to different SD services through the API standardized in the framework of the IEEE P1876 working group [7]. As for Communication protocol, WebSocket are currently the most effective way of exchanging information between the client web application and the remote server.

6 SUDAC Experimental System

Along with the existing pool of educational remote laboratories available at University of Applied Sciences and Arts Western Switzerland and at Swiss Federal Institute of Technology, the project aims at offering a new pool of sustainable renewable energy installations operating under real weather conditions to be installed in the southern countries (Iran, Lebanon, Niger and Djibouti). The Swiss students in engineering disciplines are supposed to use the photovoltaic cells installed in the south countries to do different real measurements (angle, temperature, PV voltage output, electrical power, …) and to test control algorithms under warmer and more severe weather conditions than those typically observed in Switzerland. The proposed system (Fig. 6) offers the ability to remotely conduct experiments with real biaxial photovoltaic cells. The biaxial solar tracking system actuated by a motorized structure allows adjusting the azimuth and elevation angles to obtain the Maximum Power Point Tracking (MPPT). A camera offers continuous live video streaming of the test site in addition to other remote measurements. A solar charge controller and a battery complete the set of PV cells experimentation system.

Fig. 6 SUDAC experimental system

7 Conclusions

The results of this study point to important consensus in the affordances of both remote labs and simulations tools for science learning and critical elements of the user interface that enhance student engagement across both types of tools. An appropriate combination of lab experimentation and simulation tools offers the students the opportunity to conduct a real experiment while using the lab-based simulation tools for experiment modeling, controller designing, analyzing and interpreting data.

Acknowledgements. The work presented in this paper is carried out in the framework of the MOOLs project co-funded by the Swissuniversities Development and Cooperation network (SUDAC) program. The program seeks targeted support for collaboration between the various types of Swiss higher education institutions and their partners from the Global South in order to achieve an excellent standard of education and research on global challenges.

References

1. Saliah-Hassane H, et al (2019) Special session—online laboratories in engineering education: innovation, disruption, and future potential. In: Proceedings of the IEEE international conference on teaching, assessment, and learning for engineering (TALE), pp 1228–1232
2. Salzmann C, Piguet Y, Gillet D (2019) New tools for MOOC/MOOL to sustain continuity of experimentation in control. In: 12th IFAC symposium on advances in control education (ACE), Philadelphia, pp 254–259
3. Moghaddam F, et al (2019) Massive Open Online Labs (MOOLs): an innovative solution to achieving SDGs in the global south. In: Proceedings of 5th experiment international conference (exp. At 2019), pp 394–398
4. Rodrguez-Triana MJ, et al (November 2014) Rich open educational resources for personal and inquiry learning: agile creation, sharing and reuse in educational social media platforms. In: International conference on web and open access to learning (ICWOAL), Dubai
5. Gillet D, et al (2019) Promoting and implementing digital STEM education at secondary schools in Africa. In: Proceedings of 2019 IEEE global engineering education conference (EDUCON), pp 698–705
6. Farah JC, Moro A, Bergram K, Kumar Purohit A, Gillet D, Holzer A (September 2020) Bringing computational thinking to non-STEM undergraduates through an integrated notebook application. In: The 15th annual European conference on technology-enhanced learning (EC-TEL), Heidelberg, Germany
7. 1876-2019 - IEEE standard for networked smart learning objects for online laboratories, IEEE Education Society/Standards Committee (EDU/SC) (2019). https://standards.ieee.org/standard/1876-2019.html

Aligning Technic with Didactic – A Remote Laboratory Infrastructure for Study, Teaching and Research

Anke Pfeiffer(✉), Hadi Adineh, and Dieter Uckelmann

Hochschule für Technik Stuttgart, Schellingstraße 24, 70174 Stuttgart, Germany
anke.pfeiffer@hft-stuttgart.de

Abstract. The project Open Digital Lab for You (DigiLab4U) focuses on the creation of a net-worked laboratory-based learning environment where students, teachers and researchers can access industry related laboratory equipment focusing on the Internet of Things (IoT) and Industry 4.0 (I4.0). This paper describes the alignment of the didactical requirements with the technical design and architecture to build up a sustainable remote laboratory system for study, teaching and research. The focus is on the development of a dashboard and corresponding technical integration for remote access of an RFID measuring chamber. The remote chamber will be used in the Information Logistics study program at the Hochschule für Technik Stuttgart (HFT Stuttgart) as part of laboratory exercises and will be available to other users across universities in the future. Subsequently the initial project idea is briefly described (1), technical design according to the didactical demands are outlined (2, 3) and first evaluation results will be presented (4).

Keywords: Technical design · Remote laboratory · Engineering education

1 Introduction

Right from the start of the DigiLab4U-Project in September 2018, the focus was on the joint development of the technical infrastructure along the (media-)didactical requirements, to support learning, teaching and research processes in an online learning environment, that offers remote access to real industry-related laboratory equipment. Using the example of the development of the remote laboratory RFID measuring chamber, this paper will explain how didactic considerations can be transferred to a technical design. The measuring chamber is a real industry-related application and functions as a test environment to check the reading range or the 360° reading profile of RFID and in particular of UHF transponders (ultra-high frequency—800–1000 MHz).

In general, laboratory-based learning is an important pillar to prepare students for gaining practical knowledge and experience in the field of engineering [1]. In the project scope of DigiLab4U is currently being analysed to what extent and with which didactic conceptual design, key competences for the future working life, such as self-regulated learning or social skills in the context of lifelong learning can be addressed in lab-based learning scenarios [2, 3]. Moreover, necessary changes need to be taken into

M. E. Auer et al. (eds.), *Online Engineering and Society 4.0*, Lecture Notes in Networks and Systems 298, https://doi.org/10.1007/978-3-030-82529-4_8

account, which come with the digitization. For the engineering profession these changes are characterized in many areas by the increasing shift of value creation from the physical to the digital world, also known as IoT and I4.0 [4]. For this reason, RFID as a key technology of I4.0 should be considered in engineering education. At the same time, digitization in education and research enables new forms of cross-location networking and exchange of remote or virtual lab infrastructures. This offers universities the opportunity to reduce cost pressure and simultaneously increase the availability of lab equipment [5]. Both digital trends can offer students, teachers and researchers a broad spectrum of practical fields of action for learning, teaching and research processes [6–8]. For the organisation of those processes in the context of DigiLab4U, two questions were raised, that concern likewise the didactical and the technical perspective: Q1: What are the didactical requirements for a remote lab-based learning environment? Q2: How can these requirements be met in a technical design? Subsequently the methodical procedure, the technical design and first evaluation results will be briefly described, using the above mentioned example.

2 Didactical Requirements Analysis

For the methodical procedure a Design Based Research (DBR) approach was chosen [9]. The DBR is particularly valuable for a collaborative design development of a remote lab, where didactical requirements and technical demands should be closely interlocked. The development process is structured in the phases problem analysis/definition, design/construction and evaluation/reflection, between which interdependency exists and which makes the process iterative and the sequences variable [10].

For the *analysis phase*, an extensive requirements analysis was conducted in 2019, including nine expert interviews with professors, ten guideline-based student interviews, a literature review and a document analysis (lab scenario review, project documents). The general goal was to identify the main didactical needs and possibilities for the meaningful integration of new technologies and methods for lab-based learning in the field of I4.0 and IoT. More details can be found in Burghardt et al. [11]. Several didactical requirements were derived from the analysis, which should guide the development of the remote laboratory (see Table 1).

In the *development phase*, the challenge was to consider the didactic requirements for the technical implementation of the remote lab. This is done in close coordination and cooperation between the responsible technical and didactical researcher of the respective field. In the subsequent *evaluation phase*, the extent to which the didactic requirements could be implemented in the remote laboratory were checked (see Sect. 4).

Table 1 Six didactical requirements for the technical design of a remote lab

Requirements for the technical design	Examples of the theoretical findings of the literature analysis and document review	Examples of the expert interviews
Support theory practice transfer	• Transfer of scientific findings into practice [12] • Identifying the strengths and limitations of theoretical knowledge while operating real-world behaviours [1]	• Practical application should enable the transfer of knowledge • Access to the real lab prepares students for the professional practice and offers research opportunities
Support of self-regulated learning	• Depending on the type of learning motivation, the requirements of the current learning situation, they monitor, evaluate the progress of the learning process themselves [12–14]	• Providing a good balance between instruction and self-regulated learning • Students should be given the opportunity to learn on their own, to manage their time, to set themselves goals, methods or materials, • Can overwhelm inexperienced students
Support of learning analytics data	• Learning analytics is the measurement, collection, analysis and reporting of data about learners and their contexts and can support lab-based learning processes [8, 15] • Goal is the understanding and optimizing of learning and the environments in which it occurs [15]	• Suitable LA are rarely available, but would be desirable • LA as a learning process facilitator would be helpful to provide individual feedback • Supports the learning (progress) status review
Supports access to real industry related lab equipment	• Teaching and research requires real lab equipment [16] • access to real measurement data should be guaranteed (scenario review)	• Exchange of real lab environments offers added value for learning processes • Enables research activities
Support of the integration in a learning management system (LMS)	• The learning environment should offer equal access to learning/teaching resources and lab (project proposal)	• Integration of the remote lab into a known system simplifies access and handling • Lab-based tasks can be organised via a LMS

(*continued*)

Table 1 (*continued*)

Requirements for the technical design	Examples of the theoretical findings of the literature analysis and document review	Examples of the expert interviews
Support of cross-institutional access	• International project: international, resource-saving exchange of lab equipment [5] (project proposal)	• Exchange of real lab environments offers added value for cross-institutional student collaboration • Supporting interdisciplinary and international cooperation

3 Technical Design: Development of a Prototype for a Remote Lab

Based on the Table 1, the considerations underlying the technical design are briefly outlined in this section.

Remote Experiment User Interface: One of the main requirements was the *support of the theory-practice transfer*. For this reason, not a simulation, but a real industrial lab was made remotely accessible, to give students the opportunity to gain lab-experiences while conducting experiments in real time, collecting and analysing real measurement results as well as executing their own research projects in the field of RFID.

As second important requirement the system should support *self-regulated learning*. To gain this goal, multiple objectives were taken into account. First of all, the system should be accessible 24/7 and therefore checking the environment security and implementing robots (in some cases) is indispensable. Furthermore, the dashboard design should support the self-regulated user experience. This requirement effects mostly the user interface (UI) of the remote experiment. Figure 1 shows a UI of a sample experiment. The numbers refer to the UI elements, that were chosen for the dashboard-visualization of the remote lab and they are described next. The user is able to see the experiment information as well as the set parameter values (2) by himself or by choosing from a list. In some cases, by changing values, the user should have a chance to see the result on his screen. For example, in our case, when he changed the RFID tag in a combo list, the picture of chosen tags (3) will be shown in the screen. Based on this, the user can make his choice.

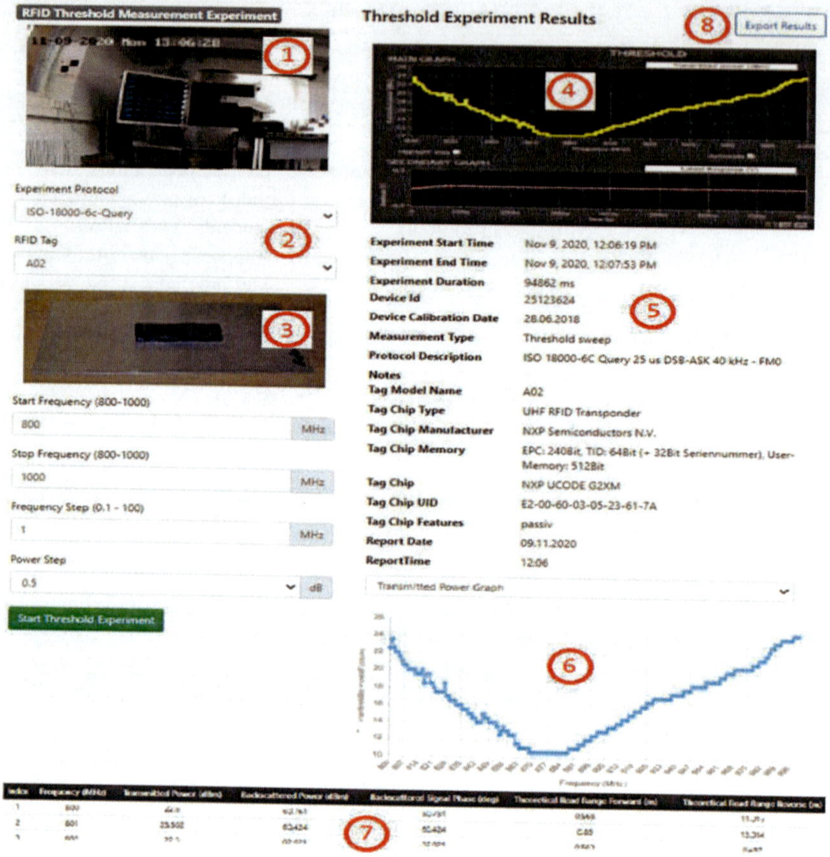

Fig. 1 HFT RFID lab, remote experiment user interface

After setting the experiment parameter, the user could start the experiment by clicking on a corresponding button. Gaining better understanding about what is actually happening inside the lab, the UI shows the user a live video (1) stream of the lab and the acting robot. This is how the user could learn more about the actual experiment materials, the instruments (e.g. RFID Measuring Chamber, Robot Arm), and the combination between them. Partial results (4) could be seen also in the UI. Once the experiment is finished, the whole results will be shown in three separated parts: experiment meta-data (5), charts (6) and a table (7). Moreover, there is an ability to download the result as an Excel file (8). This enables the user to do further analysis, archive the results and compare it with other results.

Developed Architecture and Technical Implementation: Considering the UI elements and the further mentioned requirements in Table 1, multiple services have been defined and implemented. Some of them must be shared between all laboratories of the project DigiLab4U and some others should be implemented inside each independent laboratory. Figure 2 shows the specific architecture that was developed for the remote experiment for the RFID measuring chamber.

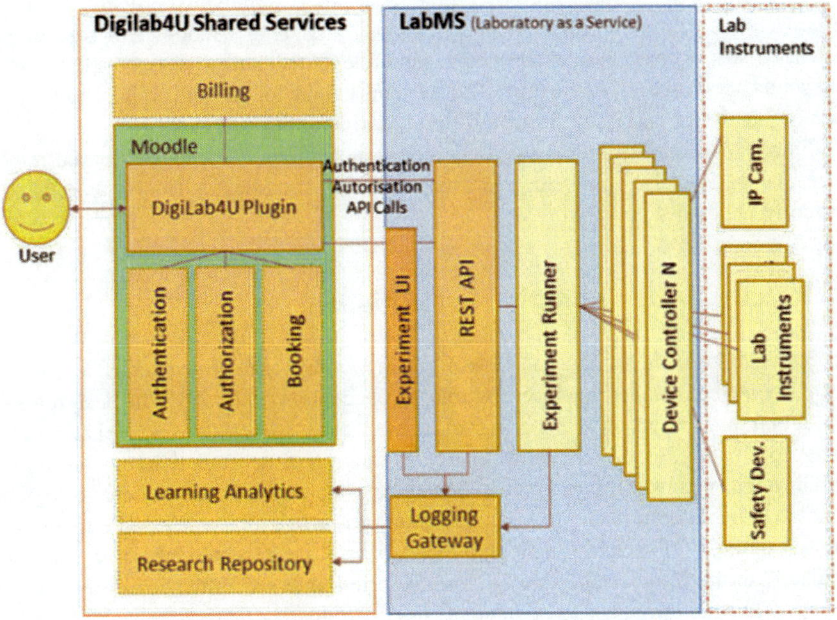

Fig. 2 Remote experiment general architecture in DigiLab4U

Learning Management System (LMS) is placed at the heart of the shared services. It enables central management, support and operations of the laboratory-based learning processes. Every user and corresponding rules should be defined in the LMS. Authentication and Authorization are also managed in this part. In addition, the booking laboratory time slots are available through the LMS as well. Moreover, cross-institutional access could be supported via federating Moodle instances from multiple institutions. In case billing and invoices are needed, the system is prepared to implement a billing-service.

Learning Analytics and Research Repository are other shared services. The first one is responsible for analysing the learning processes. Storing and manipulating research materials as well as experiment results are parts of the Research Repository responsibilities.

The Laboratory Management System (LabMS) is an application which is developed inside the DigiLab4U project and which is responsible to do experiments at laboratory side as well as managing didactic/industrial laboratory instruments and safety devices. In brief, LabMS could be seen as an application to provide the concept of "Remote Laboratories as a Service". It is aimed to simplify the communication between user and laboratory instruments. This application could be reached via HTML User Interfaces or directly via REST API. Passing through REST API, user requests could reach the

"Experiment Runner". This component is responsible for implementing experiment logic and doing it by implementing laboratory devices and sending back results to the user. It could communicate with devices, using their controller, or just by using the MQTT protocol [17]. In case of any problems, it could stop the experiment. Checking safety laboratory conditions is another responsibility of this module. In case of any security alarm, it could stop experiments, send relevant signals to devices, log the state, and send back the result to the user. These events can be logged and equally be used for learning analytics and the reflection of lab-based learning processes.

Finally, this architecture is matured enough not only to support web based remote experiments, but also to integrate Virtual Reality experiments. This functionality is currently under development.

4 Results of the Formative Evaluation

The evaluation of the remote lab took place in October and November 2020 and was not yet completed at the time of this article. In the context of the DBR the design for the evaluation was based on the method eLearning oriented walkthrough (eLOW) [18]. Center of this method builds the formative evaluation of the remote lab via an online-walkthrough, with the aim to optimize the remote lab-based learning scenario (e.g. the remote lab itself, additional learning resources and tasks). Seven selected experts from various groups of actors (students, teachers and representatives from industry) "walked" through the remote lab environment and assessed different elements of the system. The walkthrough was done via a video communication system and was guided with questions to address and connect the aspect of usability, didactical considerations and requirements with regard to industry-related needs and practices. The results of the evaluation can be summarised as follows:

Concerning the *self-regulated learning*, it was proposed from the experts, that real-time online help should be gained by implementing additional information-boxes that appear behind little question marks or that are shimmed to important words and objects. To offer the user a better orientation, they recommended to visually re-design certain UI areas: e.g. display of RFID-tag information, colour design and placement of buttons according to the current measurement phase (start, in progress, finish, new experiment), and the result representation.

Being compliant with *theory-practice transfer*, some disruptive factors would be desirable. To support this experience, some items for related tests will be added to help students to investigate on what happens if a measurement gets disrupted. *The remote-access to real lab equipment* was rated positively throughout, especially for practical applications and research reasons. Mainly the possibility to perform real time measurements independent of location and time was highlighted. In this context, it is planned to present the live operating robot in full screen in the future, to further support the real lab experience. The evaluation results for the *use of Learning Analytics* in the remote lab are unfortunately under construction and not yet available. The remote lab will be revised according to the evaluation results and will be integrated into Moodle in a next step. Next spring, it will be evaluated summatively within the scope of a case-based task in the Information Logistics study program at the HFT Stuttgart.

Acknowledgements. We thank the Federal Ministry of Education and Research of Germany (BMBF) for supporting the research project *DigiLab4U*.

References

1. Feisel LD, Rosa AJ (2005) The role of the laboratory in undergraduate engineering education. J Eng Educ 94:121–130
2. Stifterverband für die Deutsche Wissenschaft e V (2016) Hochschulbildung für die Arbeitswelt 4.0. Jahresbericht 2016. Hochschul-Bildungs-Report 2020. Edition Stifterverband, Essen
3. Gottburgsen A, Wannemacher K, Wernz J, Willige J (2019) Ingenieurausbildung für die digitale Transformation. Zukunft durch Veränderung. VDI-Studie April 2019
4. von der Heyde M, Auth G, Hartmann A, Erfurth C (2017) Hochschulentwicklung im Kontext der Digitalisierung - Bestandsaufnahme, Perspektiven, Thesen. In: Eibl M, Gaedke M (eds) Informatik 2017. Lecture Notes in Informatics (LNI), pp 1757–1772. Köllen Druck+Verlag GmbH, Bonn
5. Orduña P, Rodriguez-Gil L, Garcia-Zubia J, Angulo I, Hernandez U, Azcuenaga E (2017) Increasing the value of remote laboratory federations through an open sharing platform: LabsLand. In: Auer ME, Zutin DG (eds) Online Engineering & Internet of Things. Proceedings of the 14th, vol 22. Lecture Notes in Networks and Systems, pp 859–873. Springer
6. Parkhomenko A, Tulenkov A, Sokolyanskii A, Zalyubovskiy Y, Parkhomenko A, Stepanenko A (2019) The application of the remote lab for studying the issues of smart house systems power efficiency, safety and cybersecurity. In: Auer ME, Langmann R (eds) Smart industry & smart education, vol 47. Lecture Notes in Networks and Systems. Cham, Springer, pp 395–402
7. Guedey M, Pfeiffer A, Uckelmann D (2020) Transferring research on IoT applications for smart buildings into engineering education. In: 2020 IEEE global engineering education conference (EDUCON), Porto, Portugal, 27–30 April, 2020, pp 250–254. IEEE. https://doi.org/10.1109/EDUCON45650.2020.9125150
8. Wuttke H-D, Hamann M, Henke K (2015) Learning analytics in online remote labs. In: 3rd experiment international conference. Online experimentation: June 2nd–4th, 2015, Ponta Delgada, São Miguel Island, Azores, Portugal: exp.at 2015, 2015 3rd Experiment International Conference (exp.at 2015), Ponta Delgada, Portugal, 6 February 2015–6 April 2015, pp 255–260. IEEE, Piscataway. https://doi.org/10.1109/EXPAT.2015.7463275
9. Euler D (2014) Design-research - a paradigm under development. In: Euler D, Sloane PFE (eds) Design-based research. Zeitschrift für Berufs- und Wirtschaftspädagogik Beiheft, vol 27, pp 15–41. Franz Steiner Verlag, Stuttgart
10. McKenney S, Reeves TC (2014) Methods of evaluation and reflection in design research. In: Euler D, Sloane PFE (eds) Design-based research. Zeitschrift für Berufs- und Wirtschaftspädagogik Beiheft, vol 27, pp 141–153. Franz Steiner Verlag, Stuttgart
11. Burghardt M, Ferdinand P, Pfeiffer A, Reverberi D, Romagnoli G (2020) Integration of new technologies and alternative methods in laboratory-based scenarios. In: REV 2020 international conference on remote engineering and virtual instrumentation, pp 297–317
12. Haug A (1980) Labordidaktik in der Ingenieurausbildung. VDE Verlag GmbH, Berlin

13. Dyrna J, Riedel J, Schulze-Achatz J (2018) Wann ist Lernen mit digitalen Medien (wirklich) selbstgesteuert? Ansätze zur Ermöglichung und Förderung von Selbststeuerung in technologieunterstützten Lernprozessen. https://nbn-resolving.org/urn:nbn:de:bsz:14-qucosa2-338278. Accessed 11 Nov 2020
14. Konrad K, Traub S (2018) Selbstgesteuertes Lernen. Grundwissen und Tipps für die Praxis, 6th edn. Schneider Verlag Hohengehren GmBH, Baltmannsweiler
15. Siemens G, Long P (2011) Penetrating the fog: analytics in learning education. In: Long P (ed) Proceedings of the 1st international conference on learning analytics and knowledge, pp 31–40. ACM, New York
16. Tekkaya AE, Wilkesmann U, Terkowsky C, Pleul C, Radtke M, Maevus F (eds) (2016) Das Labor in der ingenieurwissenschaftlichen Ausbildung. Zukunftsorientierte Ansätze aus dem Projekt IngLab: acatech Studie. acatech Studie. Herbert Utz Verlag GmbH, München
17. Liu X, Zhang T, Hu N, Zhang P, Zhang Y (2020) The method of Internet of Things access and network communication based on MQTT. Comput Commun 153:169–176. https://doi.org/10.1016/j.comcom.2020.01.044
18. Jahnke I, Terkowsky C, Pleul C (2011) Wechselwirkungen hochschuldidaktischer Konzepte in fachbezogenen, Medien-integrierten Lehr-/Lernkulturen: Forschungsbasierte Gestaltung. In: Jahnke I, Wildt J (eds) Fachübergreifende und fachbezogene Hochschuldidaktik. Blickpunkt Hochschuldidaktik - Band 121, 1st edn, pp 177–192. Bertelsmann W. Verlag, s.l.

Simulation on Motion of a Trebuchet

Karthik Baburao Mulamuttal[1(✉)], Komal Kattigenahally Nagaraj[1], Gangadharan V. Kalluvalappil[2], Pruthvi Raj[2], Narasimhamurthy Kyathsandra Chandrashekar[1], and Suchitra Vankalkunti[1]

[1] Siddaganga Institute of Technology, Tumkur, India
{kcnmurthy, suchitrav}@sit.ac.in
[2] National Institute of Technology Karnataka, Suratkal, India

Abstract. In this paper, an innovative approach to analyze the performance of one of the basic mechanical system "Trebuchet" is discussed. An efficient simulation approach is developed for 'The motion of a Trebuchet' analysis. The developed simulation tool provides the access for all possible parametric variations so that the theoretical concepts are visualized easily by very simple simulation process. The process of development of the simulator is briefed in this paper. Simulations can be efficiently used to analyze the behavior of a Trebuchet system. For diverse patterns of control parameters, the system design can be tested using simulations and the most efficient system can be implemented in real time. Likewise, this paper explains in detail how the motion of a trebuchet is effectively analyzed using simulation by giving extreme conditional parameters. HTML, CSS and JavaScript tools are used to develop this simulation.

Keywords: Simulation · Trebuchet · Experiential learning

1 Introduction

Engineering requires the scientific understanding of nature and then applying it to innovate. The rapid advancements in technology have made available different tools for potential understanding and learning of engineering principles. To design a system it is necessary to have prior knowledge on the system control parameters and its behavior under different parametric conditions. Also, a user friendly and simulation tool with many parametric variations create interest in the subject matter and motivates users/students to learn the course with more confidence [1–4].

Using simulations is helpful in the following scenarios: i) when a new system is being designed and major developments in the physical layout have been demanded. ii) When a large investment is being considered for an existing developed system. iii) Simulations with animations are a very effective tool for training and educating where the animation can be the only source of understanding the complete system working. iv) Simulations have the added advantage of the ability of designing even the worst case systems, without any risk that would predict its abrupt behavior before it could be implemented in real time. v) It can be used to perform an experiment or develop designs repetitively, any number of times required. vi) There is no possibility of changes in the results at each iteration for the same values if input parameters [5–7].

M. E. Auer et al. (eds.), *Online Engineering and Society 4.0*, Lecture Notes in Networks and Systems 298, https://doi.org/10.1007/978-3-030-82529-4_9

2 Development of Simulation Tool for Trebuchet

Virtual lab system is developed to provide access to faculty and students to encourage "Experimental Learning". Experiments in Virtual lab can be accessed using laptop or smart phone and doesn't require any specific software at the user end. The prime objective is to provide remote-access to Labs in various disciplines of Science and Engineering. This Virtual Labs Would caters to Students at their UG and PG courses, to enthuse students to conduct experiments by arousing their curiosity. This would help them in learning basic and advanced concepts through experimentation, to share costly equipment and resources, which are otherwise available to limited number of users due to constraints on time and geographical distances, to provide a complete learning management system around the Virtual Labs where the students can avail the various tools for learning, including additional web-resources The advantages of using Virtual lab are user canconduct experiments at anytime from anywhere. User can conduct experiments along with peers for better learning. The outcomes of any experiment by measuring the desired parameters. Obtained analysis is used to generate report and summarize the readings with outcomes.

2.1 Trebuchet Simulation Tool Design

This section describes the development of the simulation tool for a trebuchet for virtual lab, to analyze by adding the counter mass, by varying the Length of the short arm and by changing the angle respectively. For the design of simulation in the web area we go with Hypertext Markup Language (HTML), cascading style sheets (CSS) and JavaScript (JS). HTML is the standard Markup language for documents designed to be displayed in a web browser.HTML describes the structure of a webpage semantically and originally included cues for the appearance of the document. It can be assisted by technologies such as and scripting languages such as JavaScript. CSS is a style sheet language used for describing the presentation of a document written in a markup language like HTML. CSS is a cornerstone technology of the World Wide Web, alongside HTML and JS. JavaScript enables interactive web pages and is an essential part of web applications. The simulation is developed in such a way so as to serve the purpose of actual learning as the user gives the desired inputs and the output is observed.

Trebuchet, (a form of catapult) is a siege machine that primarily throws a projectile using a swinging arm. The schematic diagram of the Trebuchet used for the development of simulator is shown in Fig. 1.

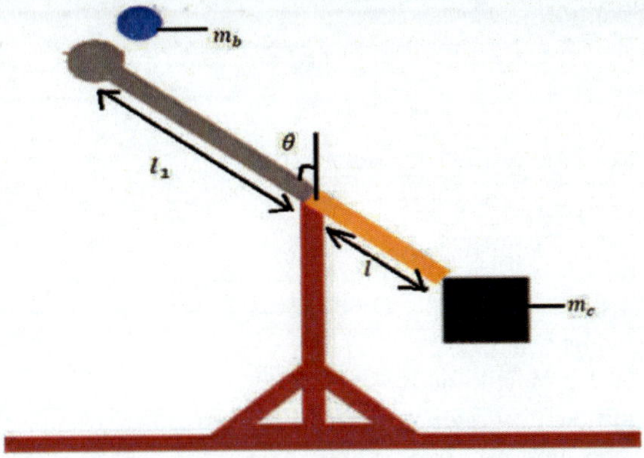

Fig. 1 Trebuchet setup for simulation

Trebuchets are large constructions made of wood and usually immobile and must be assembled on-site. The counter weight trebuchet has a long arm (load arm) that throws the projectile and a short arm (force arm) that has counter weight attached to it by hinge connection. Projectile here is a body, which is thrown near the Earth's surface and moves along a parabolic path under the action of gravity. A trebuchet is energized by lowering the long arm and raising the weighted short arm so that the projectile is thrown when the counter weight is released, on command.

The purpose of this experiment is to demonstrate the motion of a trebuchet and the resulting projectile motion, and to determine certain parameters of the projectile motion: range, height, time of flight and initial velocity.

The parameters mentioned above, vary accordingly with the counter weight, length of the short arm, and the angle of release of the projectile. During the development of the simulator for motion for Trebuchet, for few parameters certain initial conditions need to be assumed. For the selected parameters the range of values need to be provided for the users. Otherwise, a random value of the parameters will result in some output which may not be possible in real time, like entering Mass as –ve value.

The Initial Conditions and the Range of the Variables are as Follows:

1. The initial position of the arm is at 180° i.e., vertical position.
2. The length of the load arm (long arm) is fixed at 10 m.
3. The mass of the projectile is fixed at 5 kg.
4. The length of the short arm can be varied from 2.5 m to 8 m.
5. The counter weight mass can be varied from 25 to 100 kg.
6. The shot angle can be varied from 0° to 90°.

Authorized user once get the access to conduct the simulation of Trebuchet experiment can set the values of Length of the Short arm (m), Counter Mass (kg) and shot angle (degree). For the selected values of the variable parameters with assumed initial conditions the simulator gives the values of velocity, maximum height reached, range it covered and time of flight. The design equations of the Trebuchet used in the simulator to estimate all these parameters are listed below.

Mathematical Equation:

$$\text{Initial velocity } V_0 = \sqrt{\frac{2(m_c l g(1 + \cos\cos\theta) - m_b l_1 g(1 + \cos\cos\theta))}{m_b + m_c \left(\frac{l}{l_1}\right)^2}}$$

$$\text{Time of flight } T = \frac{(V_0 \sin\sin\theta) \pm \sqrt{(V_0 \sin\sin\theta)^2 + 2gh}}{g}$$

$$\text{Range} = V_0 T \cos\cos\theta$$

$$\text{Maximum height } h_{max} = \frac{(V_0 \sin\theta)^2}{2g} + h_0$$

$$h_0 = 10 + 10\cos\theta$$

m_c = Counter weight mass (kg).
m_b = Mass of the projectile body (kg).
g = Acceleration due to gravity (m/s^2).
l = Length of the short arm (m).
l_1= Length of the long arm (m).
h_0 = Height from ground level at the time of release (m).
T = Time of flight(s).
V_o = Initial velocity (m/s).
θ = Shot angle (deg).

3 Results and Discussion on Trebuchet Simulator

Analysis of Trebuchet using the Trebuchet simulator will be discussed in this section. Figure 2 shows the Trebuchet simulator's view with default length of the short arm as 2.5 m, counter weight is 25 kg and shot angle is 90°. Before pressing the start (play) button of the simulation the output parameters Velocity, height, range and time of flight are all zero.

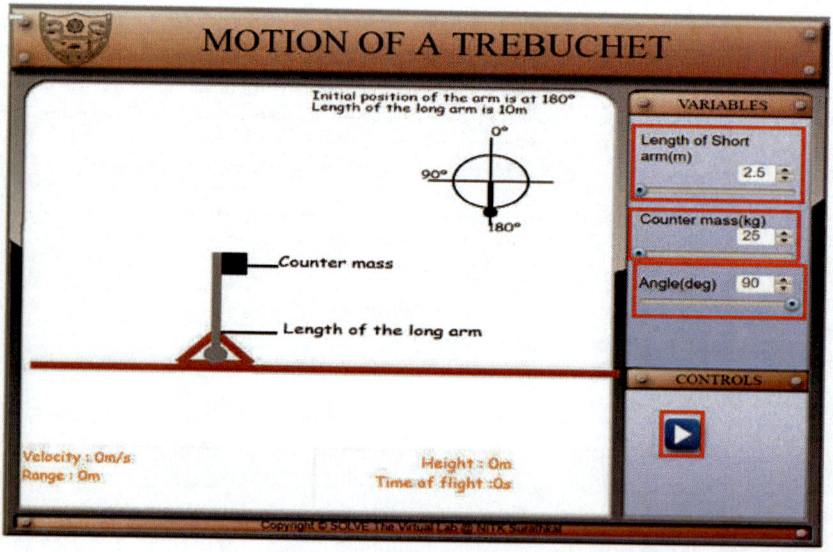

Fig. 3 Initial set-up of the Trebuchet simulator

In simulation window's zoomed view of the side view of the trebuchet is displayed. The arm length control, counter mass and release angle control sliders are given on the right side in the variables panel and controls in control panel.

The slider knobs are used to vary the variable parameters one by one to a suitable value to simulate and the play button in control panel is clicked to start the simulation. This is shown in Fig. 4.

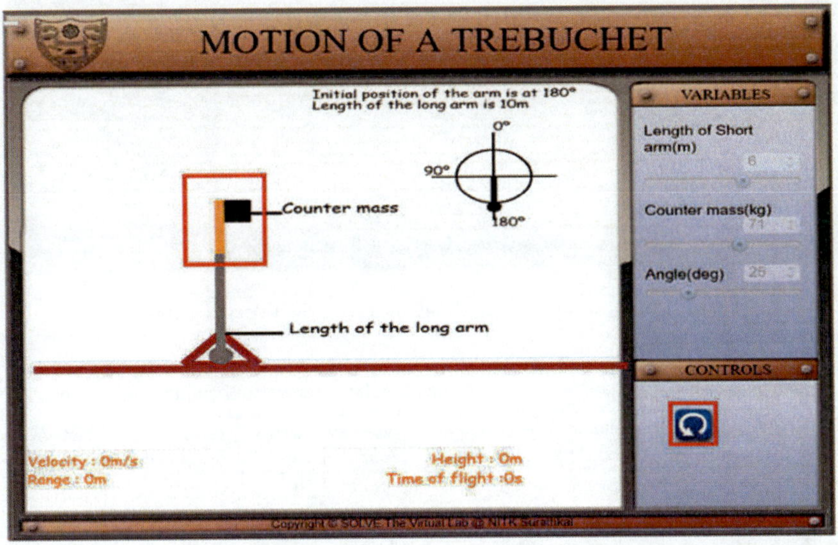

Fig. 4 Change of variable parameters

A change in the length of the Trebuchet's arm can be observed, and the animation shows the movement of the Trebuchet's arm; also a reference diagram is shown at the top to refer for the motion of the arm at the angle which is set. Then a projectile is thrown and the trajectory of the projectile is observed.

The initial velocity, height, time of flight and the range of the projectile is computed and displayed at the bottom of the simulation window as shown in the Fig. 5. It's interesting to observe that the simulator not only gives the values of the few parameters related to Trebuchet, instead it give the visualization of the entire process during the execution. This is the major difference between just using the equations to compute the parameters and making the user feel the concept involved in the topic of interest. Initially in Fig. 3 the view of the Trebuchet is different and in the end at the time of execution of the experiment it traces the entire trajectory of the mass, which makes this simulator tool unique. This type of visualization will make students curious to try for many possible combinations of variable parameters. Table 1 gives the response of the Trebuchet simulator for various input combinations. It can be observed that for first 4 readings of the Trebuchet, as the shot angle selected in 90°, the mass will fall at the base of the structure itself hence the range is 0 m.

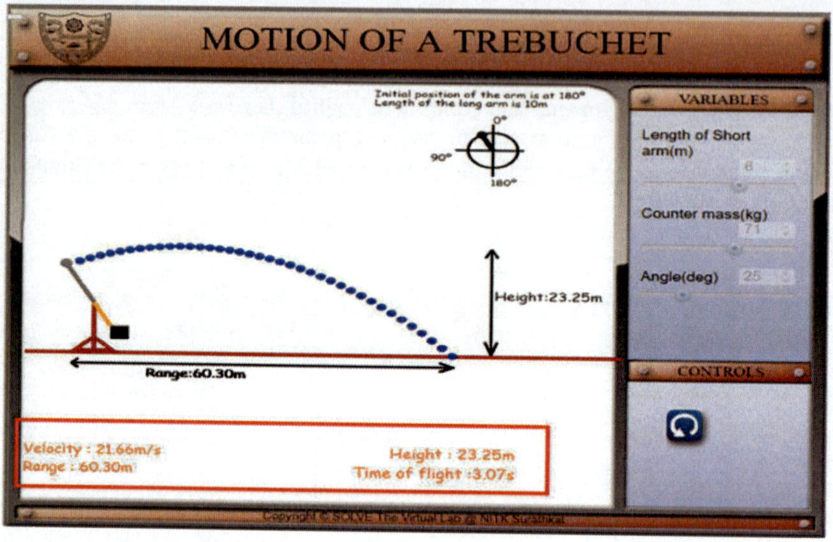

Fig. 5 Output of the Trebuchet simulator for the selected parameters

Table 1 Output parameters of Trebuchet simulator for various input combinations

Sl. No	Length of the short arm (m)	Counter mass (kg)	Angle of launch (θ)	Velocity (m/s)	Range (m)	Maximum height (m)	Time of Flight (s)
1	2.5	25	90	6.17	0	11.9	2.16
2	2.5	100	90	18.86	0	27.78	4.24
3	8	25	90	11.95	0	17.14	3.05
4	8	100	90	14.74	0	20.86	3.52
5	2.5	100	0	26.67	53.34	20	2
6	2.5	25	45	8.06	14.27	18.7	2.5
7	2.5	100	45	24.64	74.61	32.25	4.28
8	6	100	40	21.7	62.41	27.45	3.74
9	2.5	100	43	24.81	76.34	31.63	4.21
10	2.5	25	0	8.73	17.46	20	2

There are a few limitations for this experiment; in the input parameters, the counter mass weight is varied beginning from 25 kg and not from 0 kg till 100 kg. As we know the basics of a Simple balance i.e., if one end of the pan has higher weight than the other end of the pan then the pan which has higher weight comes down. According to the equation of Initial velocity which is mentioned above, if the counter mass goes below a certain value the equation takes negative values, to avoid that mass varied in the above mentioned range. It is similar in case of the length of the short arm, to avoid complex mathematical computations, the counter mass is restricted from 25 to 100 kg and length of the short arm is fixed to vary in the range of 2.5 m to 10 m.

4 Conclusion

Paper titled "Simulation on Motion of a Trebuchet" has described the interaction between the simulation of Trebuchet and the user to understand the concept of Trebuchet. User can now visualize the possible output from a Trebuchet for the given set of input parameters. This simulator will help the user to predict the input parameter settings to obtain the desired output parameters by carefully optimizing the input values over repeated simulations. Remote user can vary the length of short arm, counter mass and angle of launch and has very large combination of inputs for analysis. The analysis of velocity, range, time of flight and maximum height can be done using the simulation. The range varies from 0 to 78 m across the value of the counter mass. Analysis of these values can be helpful in the analysis of transforming energy (potential energy to kinetic energy) to designing the actual Trebuchet. There are many advantages of deploying a virtual lab in educational environment which provides a better experience for engineering students in an efficient, flexible and in a cost effective manner. User can do the analysis and learn more repeatedly conducting experiments. Students have the liberty to access the lab at anytime, from anyplace and perform at their own pace with their peer. This saves time, provides more understanding and students can spend more time with the tool to enhance their knowledge.

Acknowledgements. Authors of this paper express their gratitude for all those who helped in developing the Remote Laboratory system. We also thank officials of Siddaganga Institute of Technology (SIT), Tumakuru, the Management, Director, CEO and Principal and officials of Center for System Design, NITK, Surathkal, Mangalore, Karnataka, India for their support in making this project possible.

References

1. http://vlab.co.in/
2. Hesselink L et al (2003) Stanford cyberlab: Internet assisted laboratories Int J Dist Educ Technol 1(1):21–38
3. http://www.uml.edu/IT/Services/vLabs/
4. de la Torre L, Guinaldo M, Heradio R, Dormido S (2015) The ball and beam system: a case study of virtual and remote lab enhancement with moodle. IEEE Trans Industr Inf 11(4):934–945
5. de la Torre L, Sanchez JP, Dormido S (2016) What remote labs can do for you. Phys Today 69:48–53
6. Sanchez-Herrera MRS, Mejias, A, Marquez M, Andujar, JM (2018) A fully integrated open solution for the remote operation of pilot plants. IEEE Trans Ind Inform 1–1 (2018)
7. Heradio R, de la Torre L, Dormido S (2016) Virtual and remote labs in control education: a survey. Annu Rev Control 42:1–10

Human-Centered Design in Online Laboratories for Graduate Engineering Students

Zhongcheng Lei, Hong Zhou, Wenshan Hu(✉), and Guo-Ping Liu

Wuhan University, Wuhan 430072, China
{zhongcheng.lei,hzhouwuhee,wenshan.hu,guoping.liu}@whu.edu.cn

Abstract. To enhance user experience and provide more accessible services regarding online experimentation, a new framework of the NCSLab system which integrates various state-of-the-art technologies based on React has been designed and implemented. This paper introduces the design activity and human-centered design (HCD) principle of the new NCSLab system, and investigates the engagement of graduate students with a control background at Wuhan University in the process of design and implementation of the new version of the online system NCSLab to enhance the HCD process. The system has already been put online and applied to teaching practice at Wuhan University.

Keywords: Control engineering education · Online laboratories · Human-centered design (HCD) · Graduate course

1 Introduction

Laboratories and experimentation play a vital role in bridging theory to practice in engineering courses [1–3]. With the advancement of the Internet, Web, and communication technologies, researchers have successfully brought numerous laboratories and experimentation online to provide remote and virtual experimental services during the last two decades [4–7]. These online laboratories are complements to traditional laboratories, and more importantly, have got rid of limitations such as time reservation, fixed location, and the requirement for costly test rigs and laboratory rooms.

Online laboratories have been proved to be effective with equivalent learning outcomes [8,9], which thus have already been applied to engineering courses, for example, the PID control course [10] and the competency-based courses [11] in control engineering, and the Instrumentation and Measurements courses [12] in electrical and electronic engineering. Especially, online laboratories have proven its feasibility and effectiveness during the COVID-19 pandemic where students and equipment are physically separated due to university lockdown when other teaching methods are negatively affected and laboratory resources are not available.

M. E. Auer et al. (eds.), *Online Engineering and Society 4.0*, Lecture Notes in Networks and Systems 298, https://doi.org/10.1007/978-3-030-82529-4_10

This paper explores the engagement of students with control background in the design and implementation of the new React-based Networked Control System Laboratory (NCSLab) in the *Control Theory* graduate course. It is not always easy to involve students in the design and implementation of online laboratories, and few cases have been reported in the online laboratory related literature.

2 Evolvement of the NCSLab System

NCSLab is a web-based interactive system for online experimentation, which has been providing experimental services including remote experiments and three-dimensional (3-D) virtual experiments for over fourteen years [13–15]. With the advancement of technologies and the demand for an interactive system towards human-centered design (HCD), different frameworks and technologies have been adopted during the past decade.

With the deployment of an NGINX reverse proxy server, the coexistence of two versions of NCSLab system remains to provide experimental services for different users [16]. Currently, the two systems are identical regarding the provided experiments. Students at Wuhan University starts to use the new React-based system (https://www.powersim.whu.edu.cn/react) for experimentation, while students outside the campus may still use the old Yahoo User Interface (YUI) based system (https://www.powersim.whu.edu.cn/ncslab) as the old system has been promoted through published papers and related MOOC for years.

The two systems are summarized from different perspectives as follows.

From a technical point of view, the whole framework has been reconstructed. Table 1 listed the comparison of the used technologies regarding front-end and back-end in the YUI-based [15] and React-based system. The new system has been constructed from the scratch based on the React framework instead of the previous YUI framework which has been used for a decade in previous versions. Most previously used technologies have been replaced by open source HTML5 solutions that are compatible with modern web browsers, thus providing better user experience, for example, charting library FusionCharts which is a commercial library for the design of virtual slider and gauge has been replaced by the open source Echarts for gauge and Ant Design for the virtual slider.

From a pedagogical perspective, both of the two systems have been applied to classroom online demonstration and after-class experimentation at Wuhan University [17]. The typical courses are *Classic Control Theory* (undergraduate students in the second year), and *Control System Simulation and Computer Aided Design* (undergraduate students in the third year), *System Identification* (undergraduate students in the fourth year), and *Control Theory* (graduate students in the first year). The systems are also open for learners throughout the world on a 24/7 basis upon registration and verification through Email.

From the perspective of provided experiments, the two systems both provide remote and 3-D virtual experiments. Though the React-based system framework has been constructed from the scratch, most of the test rigs of the old system

Table 1 Comparison between YUI-based NCSLab and React-based NCSLab

Item		YUI-based system [15]	React-based system
Front-end	Framework	YUI	React
	Web technologies	FusionCharts (slider, gauge), three.js (3-D rendering), canvas.js (curve chart), WebRTC (user-side camera)	Echarts (gauge, curve chart, pie chart), Ant Design (slider, textbox), three.js (3-D rendering), WebRTC (user-side camera)
	Communication	HTTP	HTTP, WebSocket, Cross-Origin Resource Sharing
Back-end	Web server	Tomcat	Nginx
	Experiment server	Tomcat (HTTP)	Tomcat (WebSocket)
	Video server	FFmpeg, WebSocket	N/A

are migrated into the new system using React, and more and more physical and virtual test rigs are integrated into the new system.

From the viewpoint of end users, the two systems are both accessible with web browsers regardless of operating systems and web browser types. Users can design their control algorithms and customize their monitoring interface with provided widgets regardless of its implementation and without programming skills for experimentation.

3 Design Activity and Design Principle

3.1 Design Activity

Considering the design activity framework proposed in [18], the design procedure of the online system in this paper employs a *Motivation*, *Rules*, *Methods*, and *Outcomes* (MRMO) model. Figure 1 illustrates the MRMO model. *Motivation* is the specific purpose, *Rules* refers to considerations during the design process, *Methods* are actions or techniques that a designer employs to either generate or evaluate outcomes, and *Outcomes* are the results of the design activity.

The system design is carried out using the MRMO model. The motivation is to achieve a human-centered design. Rules for the proposed online system should be service-oriented, cross-platform, plug-in free and easy-access for online experimentation.

3.2 Design Principle

In this subsection, the human-centered design principle of the NCSLab system will be discussed, which has been briefly presented in [13] in the previous YUI-based system. The new framework of NCSLab system is based on React, which

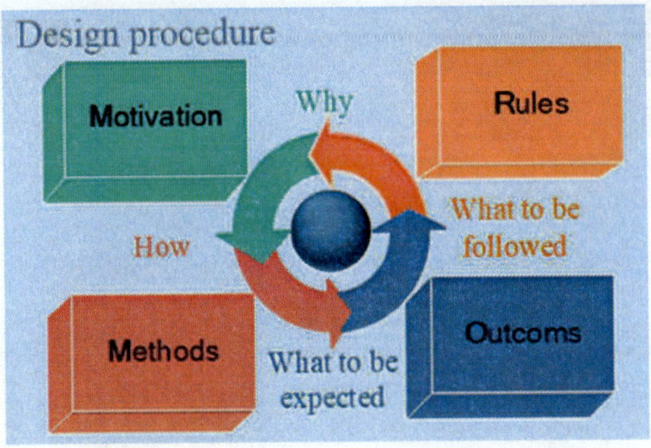

Fig. 1 MRMO model for design process

provides a modular design with components. What's more, the adopted front-end and back-end separation scheme is a perfect candidate for HCD design in which the back-end side can be ignored by students during their engagement in the design and development of the system.

HCD is vital during the design of an interactive system to make the system more usable. According to the definition of HCD by ISO standard 9241-210 [19], the HCD involves four key activities, namely, *understanding and specifying context of use; specifying the user requirements; producing design solutions; and evaluating the design*. For the design of NCSLab which is a 3-D interactive laboratory system, HCD is considered as follows.

1) *Understanding and specifying context of use*: Providing 24/7 available web-based remote and 3-D virtual experiments for users throughout the world.
2) *Specifying the user requirements*: A universal, secure, and stable online laboratory system with flexibility and interactiveness feature for all levels of students. User can not only conduct experiments, but they can also explore the system for user-defined algorithm implementation, customization of web-based interface, *etc.*
3) *Producing design solutions*: To fulfill the user requirements, a front-end and back-end separation scheme which adopts React in the front-end and Nginx in the back-end has been adopted, and various HTML5 technologies have been integrated. The engagement of students in the design process of the new system is also considered.
4) *Evaluating the design*: Apart from the design, the application of the system is crucial to verify the effectiveness of the system, during which continuous evaluation should be considered. When a new functionality has been implemented, a timely evaluation for improvement is vital. The evaluation should be conducted considering different perspectives, for example, evaluation by different users, with different courses and from different universities. Dynamic and customized questionnaires for different courses/universities are

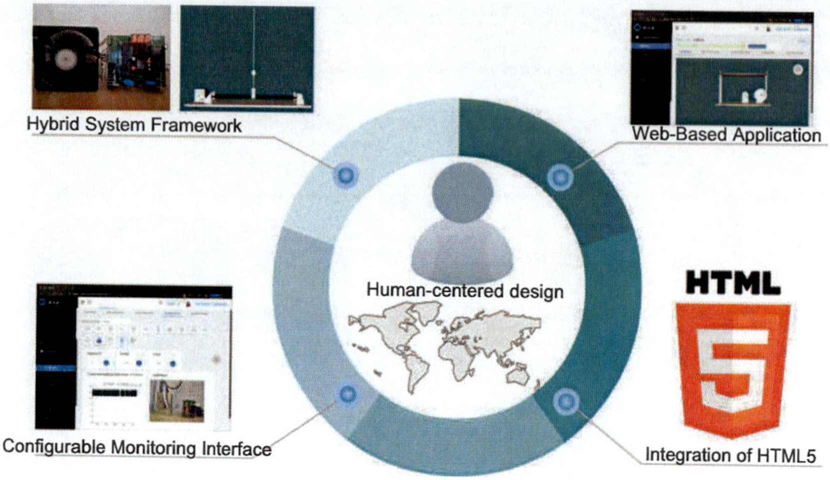

Fig. 2 Human-centered design of the proposed online system

also necessary regarding the design and implementation. Combining the "Producing design solutions" and "Evaluating the design", a "design-application-evaluation-redesign-reapplication-reevaluation" pattern has been followed to provide a better HCD system for NCSLab.

Figure 2 shows the human-centered design of the proposed online system, which is consistent with the previous versions of NCSLab [13–15]. The human-centered features of the new React-based NCSLab will be discussed in the following, in which the *Methods* and *Outcomes* of the MRMO model are included.

1) Web-Based Application: Web-based applications are more portable and less intrusive [20,21] compared with desktop applications, which gradually become flexible and powerful enabling the integration of 3-D graphics into online laboratories [6,13]. The NCSLab has been developed in a web-based interface at the very beginning. The latest web-based system can be easily accessed at https://www.powersim.whu.edu.cn/react, where users can conduct experiments at any time anywhere. The web-based system is also suitable for easy-to-follow classroom demonstrations with Internet access.

2) Integration of HTML5: From the technology point of view, 3-D technology and different 3-D rendering engines, such as Flash 3D, Unity3D, and HTML5, emerge and gradually mature. From the users' point of view, an interactive and user-friendly design would be welcomed. In this regard, the powerful Flash 3D engine, first Stage 3D and then Away 3D, was initially adopted for the development of NCSLab. However, security, crash, and frequent update warning issues brought by the Flash bothered the end users as well as the system designers, which have degraded the user experience and then restricted the popularity of the online laboratory system.

To focus on the HCD, plug-ins such as the Flash Player must be removed. Moreover, to provide 3-D virtual experiments with 3-D rendering, 3-D graphics ability should be considered when choosing proper technologies. Thus, HTML5 technologies that are plug-in free with powerful 3-D graphics ability for 3-D rendering are adopted and integrated to address above mentioned issues [15]. In the new NCSLab system, HTML5 continues to be adopted and even more HTML5 technologies such as WebSocket and Echarts are adopted.

3) Configurable Monitoring Interface: Another HCD is a configurable monitoring interface for the monitoring of online experimentation. Similar to the old YUI-based NCSLab system, the configurable feature enables the monitoring of experiments with multiple widgets provided by the system. In the new NCSLab system, new open source technologies such as Echarts and Ant Design, have been adopted. As the three.js 3-D library is powerful and open source, it remains in the new system for web-based 3-D rendering. If the user prefers to monitor a parameter in numeric, he/she can choose the crystal board, or if he/she likes to observe the trend of a parameter, the chart that depicts the real-time curve would be suitable. For the new NCSLab system, the configuration pane is based on react-grid-layout, which is flexible and widgets can be automatically organized in a proper place when dragged and placed in the pane.

4) Hybrid System Framework: Hybrid online laboratory framework allows the integration of both physical test rigs and virtual test rigs, thus, users can carry out both remote and virtual experiments within a single platform. The newly designed NCSLab system also adopts a hybrid framework where physical test rigs such as dual tank, DC motor control system and fan speed control system, and virtual test rigs such as inverted pendulum and ball and beam system are integrated for online experimentation.

4 Course Detail and HCD in the React-Based NCSLab System

Due to the limitation of the old YUI framework, and the fact that the old system is a mature setup, it is much more difficult to involve students in the design and implementation of an online experimentation system. In 2019, the new framework that adopts a front-end and back-end separation scheme and modular design based on React components has been constructed, which can provide basic functionalities.

However, the integration of more test rigs is still under construction. Considering the flexible framework and the fact that most students already have experience from their previous learning background, HCD has been applied to the *Control Theory* graduate course at Wuhan University where students are involved in the design and implementation of the new system. Moreover, it is also vital to find out how students respond to the new system after conducting online experimentation using the system, which is also a crucial means for the improvement of the system, and also a part of the HCD.

4.1 Course Detail

The *Control Theory* is a typically voluntary course for graduate students with control background at Wuhan University. Students with different learning background can choose the course when available. During the 2018–2019 Spring Semester, students from the School of Geodesy and Geomatics, School of Remote Sensing and Information Engineering, the GNSS Center, and Department of Artificial Intelligence and Automation (where the NCSLab team locates), Wuhan University had participated in the course. All of these institutes (school/research center/department) have control related disciplines such as "control theory and control engineering" and "navigation, guidance and control".

Students from the Department of Artificial Intelligence and Automation already have experience with the old NCSLab from former courses (*Classic Control Theory*, and *System Identification, etc.*) in which online demonstrations were conducted and after-class experimentation was completed for courses. Apart from that, several graduate students in the graduate course, who are also in the research team of the new NCSLab system under the supervision of the teacher who also leads the NCSLab research team, provide technical supports for the design and implementation.

In the first lecture of the *Control Theory* course, the teacher introduced the new React-based system to the students, taking the ball and beam system as an example, as shown in Fig. 3. Then the course was arranged considering HCD.

4.2 HCD in the New NCSLab System

Apart from the HCD mentioned in Sect. 3, two more different types of HCD regarding student engagement are considered during the design and application of the React-based NCSLab system.

Previously, students are passive receivers of knowledge regarding online experimentation, which is also the case with the old YUI-based NCSLab system. In the design of the new NCSLab system, students are regarded as active creators who are involved in the design of online laboratories, both during the design process and also a post-design process for evaluation which provides feedback, comments, and suggestions (questionnaires with single choice questions, fill in the blanks with predefined questions, and also open questions) regarding user experience for improvement.

Figure 4 illustrates the student actions in the design and implementation of the React-based system. Based on the YUI-based NCSLab system, students are grouped regarding different test rigs. They were supposed to design their own test rigs and integrate them into the new NCSLab system with the help of research members of the NCSLab. Though the test rigs are the same in the two versions of NCSLab, the motion control which defines the actual responses to user actions may be different as the previous YUI framework has been replaced by React. To ensure that the assigned test rig functioning well in the new system, students have to explore how to complete the definition of motion control using React.

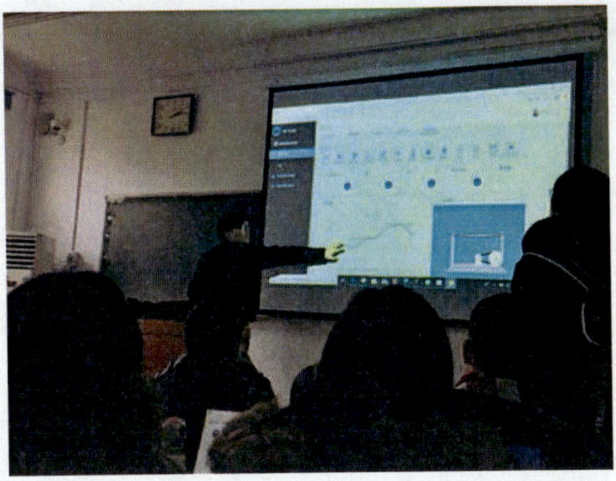

Fig. 3 Introducing React-based NCSLab to the students in the *Control Theory* graduate course in April 2019

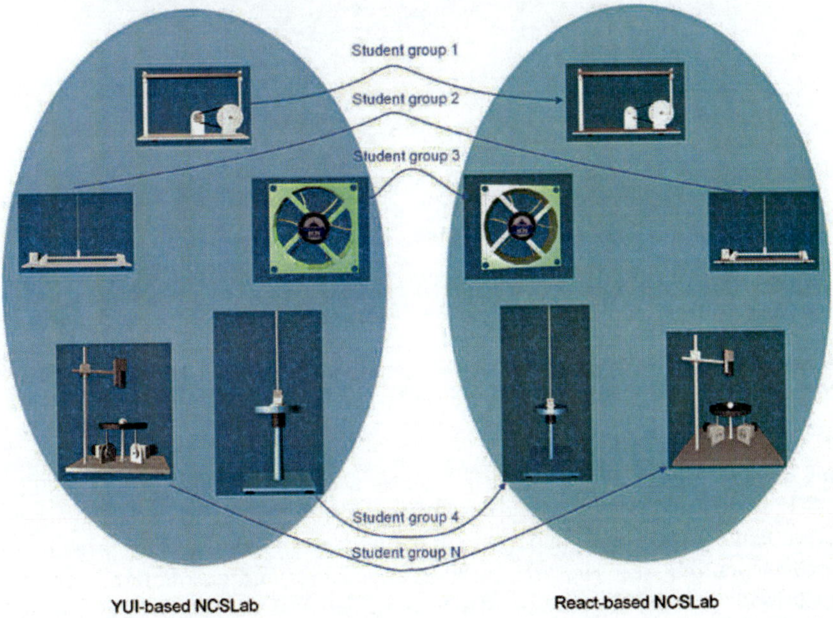

Fig. 4 Students were grouped for the integration of test rigs in the React-based NCSLab based on the YUI-based NCSLab

The upper part of Fig. 5 shows the web-based rendering definition that is supposed to be completed by students, which takes the DC motor control system as an example. Students were supposed to figure out the corresponding motion of the motor when a user changes a parameter and define the motion

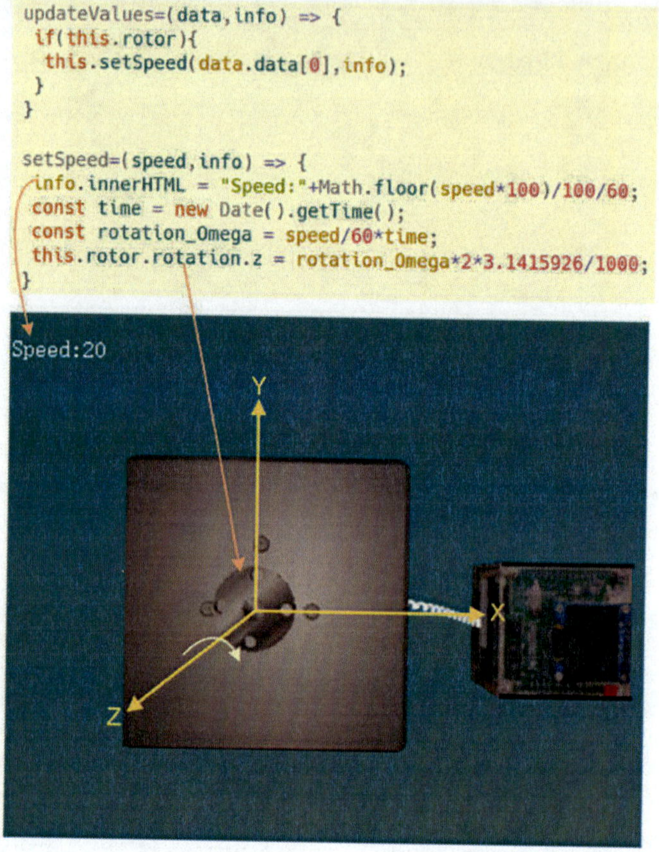

Fig. 5 Web-based rendering (motion control) definition of the DC motor control system and the corresponding implementation result

using React. The lower part of Fig. 5 illustrates the implementation result of the 3-D virtual DC motor in the web browser. Involving students in the design process could potentially help students to understand the working principle of the corresponding test rig, and their creative ideas and encountered problems may provide insight into the construction of the system, thus helping to make the system more usable.

The case in this paper is the first time to integrate the design of NCSLab system into a university course. After the completion, the new system has been put online and open for experimentation by any user. The new system has been applied to an undergraduate course *System Identification* in late 2019, in which students finished their laboratory assignments with the new React-based NCSLab system.

The limitation is that a survey regarding the course arrangement was not conducted and the feedback from students of the corresponding graduate course was not collected.

5 Conclusion and Future Work

HCD aims to achieve a more intuitive and easier to learn system for interactive use, thus the engagement of students in the process of system design and evaluation is crucial. Students can design their own test rigs through the import of the 3-D model and the definition of motion control. Through the engagement in the design of the system, students can verify their learned theoretical knowledge through practice and better understand the interconnections between the mathematical model and the 3-D model from a different perspective, and also enhance their control theory. The developed test rigs can be integrated into the new NCSLab system for online experimentation. The evaluation results and feedback from students show the new system is useful and stable, and can help students with their learning.

The COVID-19 pandemic has altered education and online learning forever. More and more universities will try to find solutions for online learning and experimentation. The new React-based NCSLab system is still under intensive construction to provide more experimental services with better user experience. In the future, more HCD will be considered and continuous evaluation will be conducted to provide continuous improvement. Future works will also consider the differences among universities to provide more tailored solutions for online experimentation.

Acknowledgments. This work was supported in part by the National Natural Science Foundation of China under Grant 62073247 and Grant 61773144.

References

1. de Moura Oliveira P, Hedengren, JD, Boaventura-Cunha J (2021) Bridging theory to practice: Feedforward and cascade control with TCLab Arduino kit. In: Gonçalves J, Braz-César M, Coelho J (eds) CONTROLO 2020. Lecture Notes in Electrical Engineering, vol 695, pp 23–32. Springer, Bragança
2. Quesada J, Calvo I, Sancho J, Sebastián R, Castro M (2020) A design-oriented engineering course involving interactions with stakeholders. IEEE Trans Educ 63(4):283–290
3. Cordeiro A, Fernão Pires V, Foito D (2018) Combining local and remote laboratories for the interactive learning of industrial automation. Comput Appl Eng Educ 26(3):675–687
4. de la Torre L, Neustock LT, Herring G, Chacon J, Garcia F, Hesselink L (2020) Automatic generation and easy deployment of digitized laboratories. IEEE Trans Ind Informat 16(12):7328–7337
5. Nafalski A, Milosz M., Considine H, Nedic Z (2020) Overseas use of the remote laboratory netlab. In: Proceedings of the 2020 IEEE Global Engineering Education Conference (EDUCON), pp 568–573

6. Galan D, Chaos D, de la Torre L, Aranda-Escolastico E, Heradio R (2019) Customized online laboratory experiments: a general tool and its application to the Furuta inverted pendulum. IEEE Control Syst Mag 39(5):75–87
7. Restivo MT, de Fátima Chouzal M, Abreu P, Zvacek S (2018) The role of an experimental laboratory in engineering education. In: International conference interactive collaborative learning, pp 644–652
8. De Jong T, Linn MC, Zacharia ZC (2013) Physical and virtual laboratories in science and engineering education. Science 340(6130):305–308
9. Potkonjak V, Gardner M, Callaghan V, Mattila P, Guetl C, Petrović VM, Jovanović K (2016) Virtual laboratories for education in science, technology, and engineering: a review. Comput Educ 95:309–327
10. Rubio E, Santana I, Rohten J, Esparza V, Martinez-Jimenez BL (2018) Remote laboratories for control education: Experience to classic PID control course. In: IEEE international conference on automation/congress of the Chilean association of automatic control (ICA-ACCA), pp 1–6. IEEE
11. Vargas H, Heradio R, Chacon J, De La Torre L, Farias G, Galan D, Dormido S (2019) Automated assessment and monitoring support for competency-based courses. IEEE Access 7:41043–41051
12. Angrisani L, Bonavolontà F, D'Arco M, Liccardo A (2020) A flexible remote laboratory with programmable device under test. Measurement 156:107584
13. Lei Z, Zhou H, Hu W, Liu GP, Deng Q, Zhou D, Liu ZW, Gao X (2021) Unified 3-D interactive human-centered system for online experimentation: current deployment and future perspectives. IEEE Trans Ind Informat 17(7):4777–4787. https://doi.org/10.1109/TII.2020.3019238
14. Lei Z, Zhou H, Hu W, Deng Q, Zhou D, Liu ZW (2018) HTML5-based 3-D online control laboratory with virtual interactive wiring practice. IEEE Trans Ind Inform 14(6):2473–2483
15. Hu W, Lei Z, Zhou H, Liu GP, Deng Q, Zhou D, Liu ZW (2017) Plug-in free web-based 3-D interactive laboratory for control engineering education. IEEE Trans Ind Electron 64(5):3808–3818
16. Lei Z, Zhou H, Ye S, Hu W, Liu GP (2020) Cost-effective server-side re-deployment for web-based online laboratories using NGINX reverse proxy. In: Proceedings of the 21st IFAC World Congress, Berlin, Germany, vol 53, no 2, pp 17204–17209
17. Lei Z, Zhou H, Hu W, Deng Q, Zhou D, Liu ZW, Gao X 3-D interactive control laboratory for classroom demonstration and online experimentation in engineering education. IEEE Trans Educ (to be published). https://doi.org/10.1109/TE.2020.3041070
18. McKenna S, Mazur D, Agutter J, Meyer M (2014) Design activity framework for visualization design. IEEE Trans Vis Comput Graph 20(12):2191–2200
19. ISO: ISO 9241-210:2010 ergonomics of human-system interaction - Part 210: Human-centred design for interactive systems (2010)
20. García-Zubia J, Orduna P, López-de Ipiña D, Alves GR (2009) Addressing software impact in the design of remote laboratories. IEEE Trans Ind Electron 56(12):4757–4767
21. Rodriguez-Gil L, García-Zubia J, Orduña P, López-de Ipiña D (2017) Towards new multiplatform hybrid online laboratory models. IEEE Trans Learn Technol 10(3):318–330

Termolabo Project: An Internet of Things Approach of Thermo-Fluids Online Laboratory

Oswaldo Andrés Vanegas Guillén[1,2](✉), José Leonardo Benavides Maldonado[3], Javier Muñoz Antón[2], Janeth Pilar Díaz Vera[1], and Carlos Hugo Lozada Cabrera[4]

[1] Universidad de Guayaquil, Avenida Delta, 090613 Guayaquil, Ecuador
{oswaldo.vanegasg, janeth.diazv}@ug.edu.ec

[2] Department of Energy, School of Industrial Engineering, Technical University of Madrid, 28006 Madrid, Spain
andres.vanegas.guillen@alumnos.upm.es, jamunoz@etsii.upm.es

[3] Universidad Nacional de Loja, 110111 Loja, Ecuador
jose.benavides@unl.edu.ec

[4] Universidad Politécnica Salesiana, 090101 Guayaquil, Ecuador
clozadacl@est.ups.edu.ec

Abstract. The present work shows the implementation of an online laboratory using an Internet of the Things (IoT) architecture for the teaching–learning of thermo-fluids systems design at the School of Industrial Engineers of the Technical University of Madrid (ETSII-UPM) and Information Technology Department at the University of Guayaquil.

The thermo-fluids laboratory is made up of an industrial water heating system made up of closed-loop piping systems, pumps, heat exchangers, actuators, sensors, and a data acquisition system based on NI-MyRio embedded system.

Project users can enter the practices oriented to the control and design of thermo-fluids through Moodle LMS. The management of the different objects that make up the laboratory is done through a Remote Learning Management System application (RLMS) developed for this purpose.

Finally, this work is validated through the implementation of an online practice of PID control.

Keywords: Control systems · Internet of Things · Online labs · Thermo-fluids systems

1 Introduction

Educational laboratories are important in the teaching–learning process as they allow students to learn from experience and put theoretical knowledge into practice [2, 3]. Due to the health emergency due to COVID-19, there are problems concerning students' mobility, preventing their use.

Currently, online laboratories are made up of remote and virtual ones, standing out three general types of a laboratory in engineering education; a saber, practical, virtual,

M. E. Auer et al. (eds.), *Online Engineering and Society 4.0*, Lecture Notes in Networks and Systems 298, https://doi.org/10.1007/978-3-030-82529-4_11

and remote; each of them with its advantages and disadvantages, as mentioned Hernández-de-Menéndez, M. et al. [5].

The present work proposes a remote online laboratory (RL) architecture integrated into a Learning Management System (LMS) through an Internet of Things (IoT) approach.

IoT technologies make it possible to improve two critical aspects of an RL: real-time communication and different devices' management. In IoT, several types of protocols are developed to be lightweight and transport messages between devices; one of them is the MQTT (Message Queuing Telemetry Transport) protocol, which is efficient in bandwidth and use of resources [1, 10].

There is a massive deployment of online laboratories [4, 9], for example, the system that proposes the control of the filling of a model of interactive tanks with the application of a classic PID control in order to maintain the level at the desired value when manipulating the closing and opening solenoid valves, this is explained in more detail in Rubio, E. et al. [8].

Another experience developed by the University of La Rioja (UNIR) in Spain applied to approximately 30,000 students geographically dispersed in different places in an educational model to carry out practices and evaluations purely online [7].

2 Termolabo Project Architecture

The architecture of the Termolabo Project is detailed below. The proposed architecture Fig. 1 has been designed to interconnect different educational laboratories through the internet.

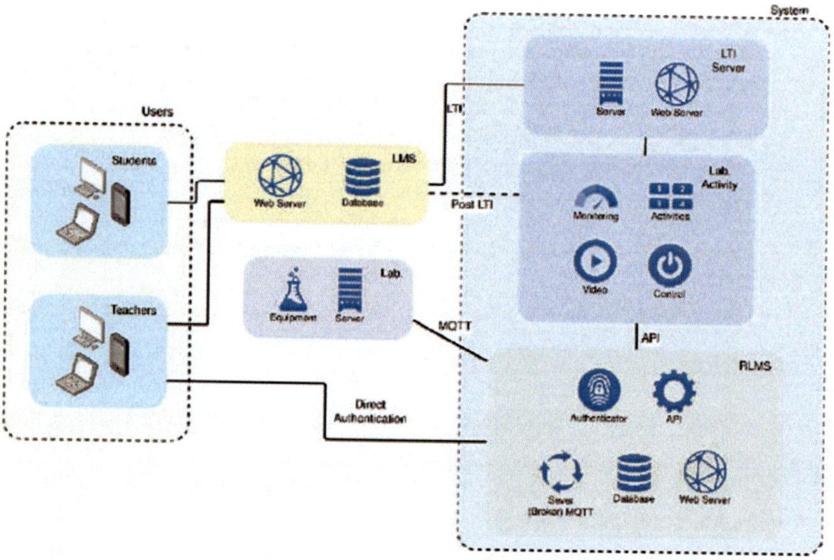

Fig. 1 Termolabo project architecture

2.1 RLMS

The Remote Laboratory Management System (RLMS) is the central core of the online laboratory system; its role is to manage the laboratory, laboratory activities, and available resources. The responsibilities of the RLMS are as follows:

1. Manage the RL.
2. Manage laboratory equipment.
3. Manage the data provided by the laboratory equipment.
4. Manage the schedules of laboratory activities.
5. Manage authorizations, authentications of the different users of the online laboratory.

The laboratory administrators and the teachers are in charge of creating the different online laboratories through the RLMS, which are virtual abstractions of their real counterparts. The resources are accessible through the MQTT network protocol, a Client/Server message transport protocol based on publications and subscriptions. This protocol's main characteristic is that it is light, efficient, and allows to minimize the users' resources and the bandwidth of the network; this protocol is widely used in Internet of Things (IoT) applications.

Through the RLMS, different laboratories can be grouped by area of knowledge, floors, buildings, and headquarters, allowing the system to be scalable. In addition to delivering the accesses and managing the topics representing the different signals that constitute the laboratory practices.

The RLMS is made up of the following parts that are explained below.

Authenticator. The authenticator is a module developed for the RLMS, which aims to identify the type of user accessing the RL resources to manage the access level. The system will act depending on the role of the user; for example, a laboratory administrator who is generally in charge of managing and maintaining the RL will have full access to all the characteristics of the resource; in the case of teachers and students, They will have access levels that will depend on the type of laboratory activity they are going to develop.

Authentication will be supported either by the information provided by the LTI (Learning Tools Interoperability) integration protocol [6], which is a communication standard between e-learning tools and systems or by an RLMS authentication service itself. In the case of LTI, it immediately provides us with the essential data that laboratory activities need for its development. It is essential to establish that each user, activity, and course will have a unique identification, which will allow the system to treat the access levels unequivocally.

API. The term "API" refers to an application programming interface. APIs allow us to communicate with other modules or sections of the system to do a job for us. In our case, we communicate with a module that will do the work of writing and reading the information we want or need from a database.

To achieve this communication at the API level, GraphQL is used, a technology developed by Facebook, which is used as a query language for the APIs, facilitating the management of data to the database. Furthermore, GraphQL has the ability to work with a subscription model, where teams can subscribe to the information they are interested in and thus be notified every time this information is updated.

Therefore, every time we need, for example, to know which equipment belongs to a specific laboratory, we will use an API developed using the GraphQL language, which will allow generating queries for the requested information.

Broker MQTT. To achieve efficient communication in terms of bandwidth and resources between the devices that are part of the laboratory, the MQTT (Message Queuing Telemetry Transport) protocol is used, which is an IoT connectivity protocol. MQTT uses a publish and subscribe model.

The MQTT Broker is a data manager who receives all the data published by the teams and decides which subscribers (students, laboratory administrators, teachers) should receive them, according to the Authenticator's rules.

Eclipse Mosquitto is used in the present work, an open-source (EPL / EDL licensed) message broker that implements the MQTT protocol versions 5.0, 3.1.1, and 3.1.

Database. In the Database all the necessary data of the users are stored, which will allow them to establish the roles of these in the virtual laboratory practices. The data stored is the detailed information of the equipment, laboratories, users, scheduling data, a grouping of laboratories, etc. The API described is used to allow the integration of the Database.

Webserver. The Web Server makes it possible for all services to be reached and for the graphical interfaces to be visible from any web browser. A web server is the one that receives the web queries that are made, typically from web browsers. Every time we visit a website or application, a web server is responsible for serving that page or web application, making a copy of it, and sending it to our browser.

Web servers can also offer web services in addition to web pages. Thus, the APIs developed will help us access our data and be reachable thanks to the implemented web server.

We use NGINX, an open-source technology, as a web server for our RLMS and labs. NGINX is written in the C programming language and is highly efficient in its service.

The occasions that the Web Server intervenes in the Termolabo Project are:

- Access from the LMS to laboratory practice.
- Direct access to the RLMS.
- Access to database queries from the API.

2.2 Virtual Laboratory Practices

As shown in Fig. 2, all activities will be accessed within a time previously chosen by the student and defined by the teacher. Access to the activity will only be guaranteed from the LMS environment; if access is attempted outside of this, access will be denied.

Upon entering the activity, teachers can create or modify possible schedules from which students can choose. Regular access to the activity is through a link placed in the respective course. This link has the same appearance as the rest of the typical LMS activities. If a student accesses practice for the first time, they will come across the scheduler, from where they will have to choose an available time, which will usually be a specific day in a specific period of time. If choosing the schedule, the space is full (places for a limit of group members), this will be indicated to the student, and they will have to select another available one. When selects an available one; you will be assigned this schedule. The student will only access the practice within the previously chosen schedule; otherwise, access will be denied.

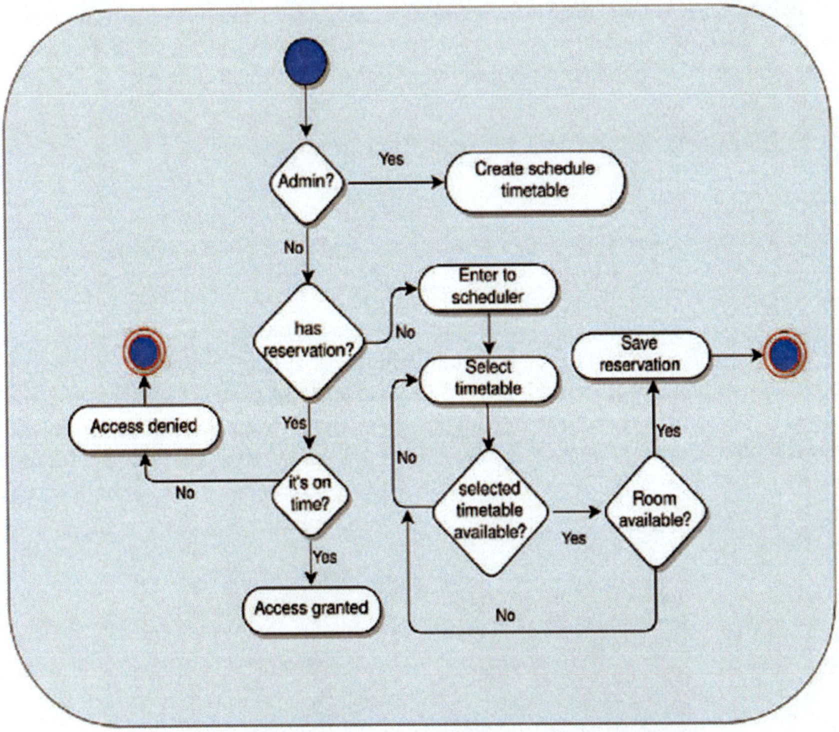

Fig. 2 Flowchart of the entry of virtual laboratory practices

2.3 Laboratories

The laboratories will house the equipment and the physical environment in which students will access remotely through the RLMS and LMS. The Laboratories have embedded data acquisition equipment and communicate using the MQTT communications protocol through their respective topics. Each integrated laboratory will have its virtual representation in the RLMS that will allow it to be managed.

2.4 LMS

LMS are web systems that allow the integration and representation of online educational content in a coordinated and structured way, allow the management of users, resources, as well as training materials and activities. Previously, online laboratories were applications with a degree of complexity, mainly due to the need to develop the management modules that LMS usually already have.

The integration of the Online Laboratory and the LMS has been achieved thanks to the specification called LTI, which allows us to integrate a learning tool (in this case, our RLMS) in an LMS learning environment. Among other things, LTI allows students and teachers to be identified in the external tool (RLMS) automatically, and that this returns the note to the LMS of the activity in question for the respective student.

The LMS used in this project is Moodle, the famous open-source platform. Moodle allows us to build a learning system, requiring few computational resources easily and with a high level of configuration through various themes and plugins.

3 Implementation of the Online Practice

Once the proposed architecture has been defined, the RL is integrated with the Termolabo Project Test Bench.

3.1 Termolabo Project Test Bench

The Termolabo Project test bench is a multicomponent system, consisting of Water in a closed-loop piping system, which has the following elements: 1 water heaters, 2 water pumps, 2 storage tanks, 4 heat exchangers of different types, 4 RTDS, 16 Solenoid valves, 2 flow meters, 1 Thermal Camera, 1 Web Camera; integrated via an embedded NI-myRio data acquisition system, which is connected to a local computer, the above can be seen in Fig. 3.

Fig. 3 Termolabo project test bench

The control of the test bench is carried out through an embedded acquisition system based on NI-MyRio and different controller boards for the management of valve opening and control of the water pumps.

3.2 Integrating the Test Bench to the RLMS

In Fig. 4, a browser screenshot of the RLMS is observed. The RLMS allows us to manage various aspects of the RL, such as access control to the RL users.

3.3 Remote Laboratory Practice

For the development of the PID control practices, a web application has been developed where students can control the plant remotely and observe its behavior in real-time through a web and thermal camera. Figure 4 shows the application developed integrated in the Moodle-LMS that was implemented for this project.

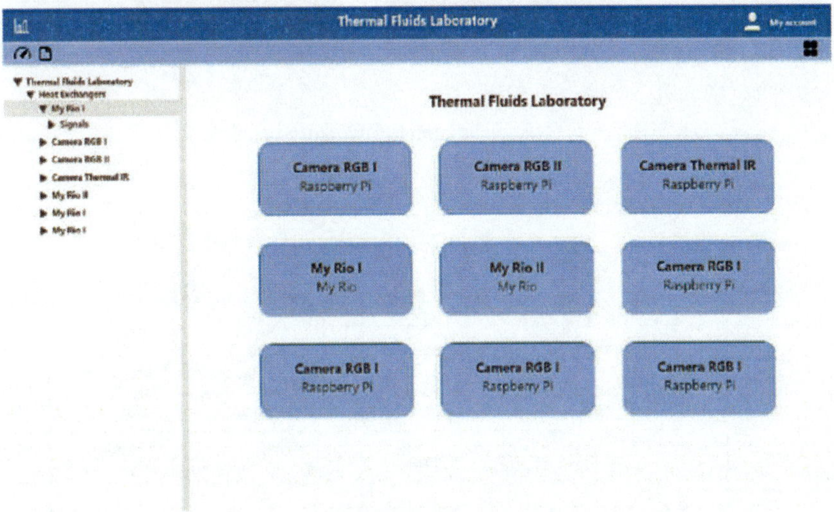

Fig. 4 Termolabo project RLMS

3.4 Implementation of the Regulation Loop

In the design and subsequent implementation of the test bench, several equipment and instruments were integrated, with the aim of carrying out the following laboratory practices: Energy Balance and Overall Efficiency, Indirect Heating/Cooling, Overall Heat Transfer Coefficient Demonstration and PID Temperature Control in the 40 L storage tank, this last practice being the one that was finally implemented in this article.

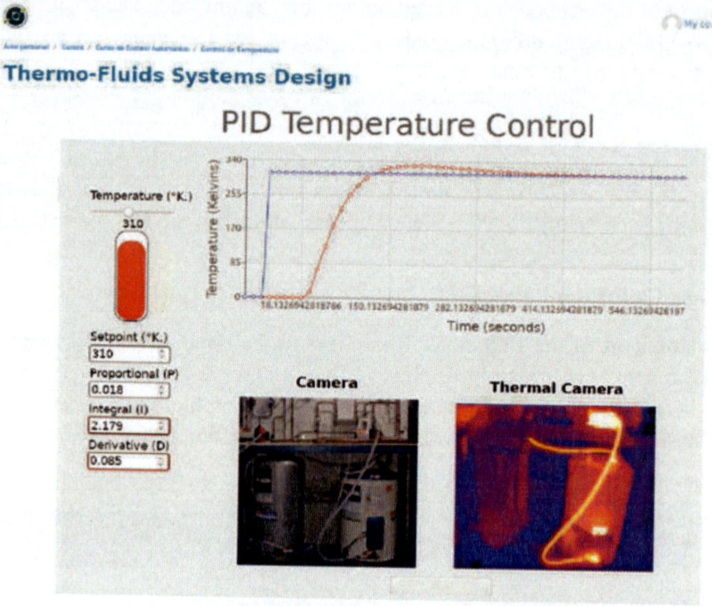

Fig. 5 Termolabo remote laboratory practice

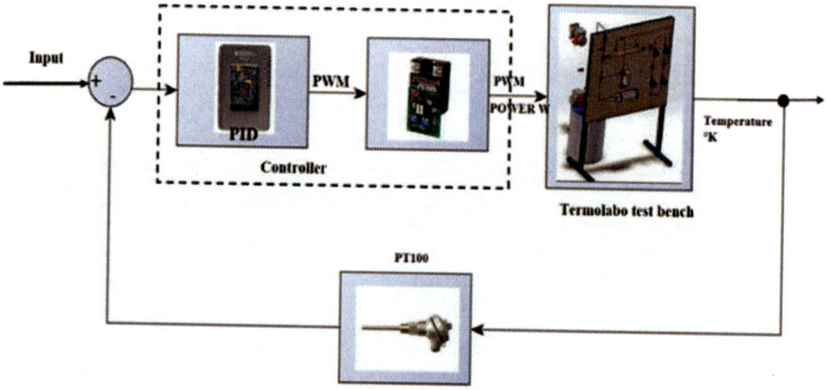

Fig. 6 Termolabo system block diagram

Furthermore, for identifying this thermal process, a control system based on a manipulated variable was proposed, in which the control of thermal resistance and a controlled output variable, which is the tank temperature, as can be seen in Fig. 5.

4 Experimental Results

The process involved is integrative; for this reason, one of the criteria for choosing the tuning method is a fast response based on Ziegler-Nichols (Z-N); and in this way, obtain the parameters of Table 1. when applying a PID controller.

Table 1 PID controller features

Parameters to control	Controller (PID)
Maximum over peak	4.22%
Stabilization time	355 seg
Rise time	68.1 seg
Error in stable state	0.00645 °K

5 Conclusions and Future Work

This work has shown a proposal for an architecture of a Remote Laboratory (RL) integrated into a Learning Management System (LMS) based on Internet of Things (IoT) technology. This approach allowed extensibility, scalability, modularity, and interoperability between heterogeneous devices that make up the RL by using the MQTT protocol.

The PID controller that was tuned with the Ziegler-Nichols method, the same one that allowed obtaining a proportional gain of 0.018, an integral of 2.17 and a derivative of 0.085. Which justifies for integrative processes and with slow dynamics, as is the case of the study plant.

Finally, it is intended in future works that this control loop, applied in this practice, is a slave loop in a cascade control that allows the manipulation of several exchangers that constitute the test bench.

References

1. Al-Fuqaha A et al (2015) Internet of Things: a survey on enabling technologies, protocols, and applications. IEEE Commun Surv Tutorials 17(4):2347–2376. https://doi.org/10.1109/COMST.2015.2444095
2. Feisel LD, Peterson GD (2002) A colloquy on learning objectives for engineering educational laboratories. Proc Am Soc Eng Educ Annu Conf Expo
3. Feisel LD, Rosa AJ (2005) The role of the laboratory in undergraduate engineering education. J Eng Educ 94(1):121–130. https://doi.org/10.1002/j.2168-9830.2005.tb00833.x
4. Heradio R et al (2016) Virtual and remote labs in education: a bibliometric analysis. Comput Educ 98:14–38. https://doi.org/10.1016/j.compedu.2016.03.010
5. Hernández-de-Menéndez M et al (2019) Virtual reality laboratories: a review of experiences. Int J Interact Des Manuf 13(3):947–966. https://doi.org/10.1007/s12008-019-00558-7

6. IMS Global Learning Consortium, I.: Learning Tools Interoperability Advantage Implementation Guide 1.3 IMS Final Release | IMS Global Learning Consortium, http://www.imsglobal.org/spec/lti/v1p3/impl/. Accessed 24 Sept 2020
7. Perales M. et al (2019) Work-in-progress: improving online higher education with virtual and remote labs. In: 2019 IEEE global engineering education conference (EDUCON), pp 1136–1139 IEEE (2019). https://doi.org/10.1109/EDUCON.2019.8725272.
8. Rubio E et al (2018) Remote laboratories for control education: experience to classic PID control course. In: 2018 IEEE international conference on automation/XXIII congress of the chilean association of automatic control (ICA-ACCA), pp 1–6 IEEE. https://doi.org/10.1109/ICA-ACCA.2018.8609754
9. Tho SW et al (2017) A systematic review of remote laboratory work in science education with the support of visualizing its structure through the histcite and citespace software. Int J Sci Math Educ 15(7):1217–1236. https://doi.org/10.1007/s10763-016-9740-z
10. Xu LD et al (2014) Internet of things in industries: a survey. IEEE Trans. Ind. Inform. 10 (4):2233–2243. https://doi.org/10.1109/TII.2014.2300753

Main Attacks and Ways to Keep Security Guidelines Updated in Remote Laboratories

Emilio Werner[1(✉)], Jhennifer Cristine Matias[1], Marcelo Daniel Berejuck[1], and Hamadou Saliah-Hassane[2]

[1] Federal University of Santa Catarina, Araranguá, Brazil
[2] Université TÉLUQ, Montreal, Canada

Abstract. Problems related to data security issues, information and access credibility are the main points encountered today for remote laboratories. Because of the use of an Internet connection, there is a need for an efficient mechanism to ensure secure and authenticated access to resources provided by the laboratories. The purpose of the research is to find solutions for the primary forms of attacks on remote labs since they are a much more efficient way in terms of accessibility and availability when compared to conventional labs. Defining which methods are adopted nowadays is the main objective that must be reached at the end of the work. Initially, a research of the current security methods of remote laboratories was carried out, which are the most frequent forms of attacks that occur today. As a result, there should be a new implementation of security-related issues. Thus, network security devices consist of one or more security functions, so that there is protection in any kind of attack. As a result, it is also expected the definition of which security methods can be adopted by remote laboratories, showing where security improvements should occur so that major attacks are avoided. Based on this context, this work intends to provide an advance in security practices to be adopted as well as ensure that the main attacks that currently occur will not affect the integrity of laboratories.

Keywords: Online laboratories · Remote laboratories · Attacks · Security

1 Introduction

Laboratories, in general, are important resources for students in all areas of knowledge. The development of research, as well as the acquisition of certain skills, can be acquired through experiments conducted in such places. However, laboratories have certain restrictions such as limitations in time and physical resources like equipment, as an increase in the number of students compared to a limited number of laboratories available in schools and universities [1].

With the constant advance of technology, there was the definition of remote laboratories, also called remote labs. These labs can be defined as equipment that is accessed through the Internet, enabling the execution of scientific experiments, making it a way of greater efficiency of access, availability and lower costs [2].

M. E. Auer et al. (eds.), *Online Engineering and Society 4.0*, Lecture Notes in Networks and Systems 298, https://doi.org/10.1007/978-3-030-82529-4_12

Despite the advantages, new problems occur with the use of the Internet for these purposes. One of the main issues encountered with the use of the Internet in remote laboratories is associated with security.

Security is related to protecting networks and computer systems from many different attacks. Among the main ones are hacking exploits, malware, data leakage, spam and Denial of Service (DoS) attacks [3].

The objective of the research is to identify a possible solution for the primary forms of attacks on remote laboratories. Ensuring the security of information and maintaining an accurate control of access to it to avoid possible frauds as well as attacks that may compromise the integrity of laboratories.

Defining which methods are currently adopted, as well as researching new ways to ensure safety, are the main objectives that must be achieved at the end of the research. In other words, in a general context, problems related to data security problems, information and access credibility are the main points found today for remote laboratories, thus doing research to address such issues to keep the security guidelines always updated.

2 Remote Laboratories

Initially, before approaching the concept of Remote Laboratories, there is a need to talk about the importance of using laboratories to improve the competencies of students in general.

According to Pereira [2] practices in laboratories are essential in teaching, as in engineering for example, since they allow students to learn effectively and develop skills necessary for their future as professionals in the area.

In the context of this, not only should theoretical learning on a subject be sufficient, but they should be allied with experiments that help to understand and enable a new way of evaluating the proposed problems as well as their possible solutions.

One of the objectives of the laboratories, in general, is to allow the training of the connectivity capacity of two different worlds: The theoretical world, seen in classrooms, and the world of practical, which happens in laboratories. That is, it should allow the development of different applications using tools to externalize this environment to the real world [4].

However, laboratories have restrictions such as time limitations on use as well as physical resources such as equipment [1]. These restrictions are aggravated by the financial resources of educational institutions.

Because of this, and especially with the significant advance of technology, there is the rise of remote laboratories. These laboratories are defined as equipment accessed remotely by users who wish to conduct scientific experiments and who do not have access to physical laboratories. Thus, by comparing both forms, remote laboratories become a much more efficient way to access and availability as well as lower costs [2].

Despite the many advantages that the use of laboratories can provide, some setbacks arise due to the use of the Internet to perform these activities. When applied in remote laboratories, these issues must be investigated and solved in order not to compromise the data used and/or generated.

One of the main problems encountered with the use of the Internet in this field is the security of both data and physical equipment. With this, there is a need for a suitable security mechanism, as seen in other areas, where secure and authenticated access to specific resources or systems are the key issues related to information security [5].

3 Main Attacks

Initially, one should understand what the main types of security attacks that occur today are. According to Pandey [5], security attacks can be divided into two major groups, passive attacks and active attacks.

3.1 Passive Attacks

They are described as attacks that have attempts to break security using observed data. The main attributes of this type of attack are as follows:

By Interception: Attacks confidentiality such as wiretaps, man-in-the-middle attacks. That is, it is considered a passive attack because the attacker does not modify the data, only uses the insecure medium to steal information [6].
By Traffic Analysis: Attacks confidentiality or anonymity. It can include tracking back on a network.

3.2 Active Attacks

For this type of attack, it requires the attacker to send data to one or both parties or block the data flow in one or both directions. The attributes are as follows:

Interruption: Availability of attacks, such as denial of service attacks.
Modification: Attacks integrity.
Therefore, network security is vital to avoid financial losses (when it comes to data security in bank transfers for example), as well as to protect users and limit damage to a company for example, as well as to ensure the security of the data transferred [3].
Manufacturing: Attacks authenticity.

4 Security of Information

Security is a highly discussed and important subject, as it is directed at protecting networks and computer systems against countless threats. Among the main ones are hacking exploits, malware, data leakage, spam and Denial of Service (DoS) attacks, as well as ensuring reliable access [3].

4.1 Safety Methods Currently Used

For a better network security, not one but several security methods should be implemented. With this, it is expected that there will be higher protection in any type of attack.

Pandey [5], simplifies network security methods in two main areas, data encryption and firewalls.

The first method, which is cryptography, is based on cyphers, in other words, mathematical functions used to encrypt and decrypt messages. The second method, called firewalls, is just groups of components that result in barriers between networks [5].

Going into detail, for Schneider, security methods are based on firewalls, Intrusion Prevention/Detection Systems (IPS/IDS), Data Leakage Prevention (DLP) and content security filtering functions (such as anti-virus). Each security device requires continuous testing to ensure that the devices are effective, accurate and productive while maintaining acceptable performance [3].

Going to another more comprehensive way, the paper entitled Internet of Things (IoT): Taxonomy of Security Attacks reports that there are several types of attacks that can lead to information compromise between IoT devices. Leading researchers to understand the risks of security breaches better, and to think about better security protections [7].

Based on this, a secure method is not the method based on offering only a security management solution, but integrated results encompassing various networks and application layer security issues [5].

4.2 New Ways to Ensure Security

One area that has been gaining emphasis is Digital Forensics (DF). Kebande's work [8], reports the concepts as well as the applications of the DF for security issues related to the Internet of things (IoT). The work in question shows the connection between these two areas, resulting in a new Digital Forensic Research Framework (DFIF-IoT). Such a generic framework for this creation is capable of supporting future IoT investigation capabilities with a degree of certainty, thus facilitating DF investigations.

Another theme that has been gaining importance in recent times is the concept and use of blockchain. Initially developed to be used with cryptocurrency, today the blockchain is also used in matters involving cloud computing and the Internet of Things (IoT). Blockchain offers a practical solution to deal with several limitations of traditional IoT applications. It can provide a secure and scalable framework for the network, so that sensitive information can be delivered without a centralized server. In addition, another advantage of Blockchain is related to reducing the operational cost of IoT [9].

5 Security Project Through the Blockchain Concept

Another way that is gaining notoriety is the use of blockchain technology as a better means of guaranteeing safe access to remote laboratories.

Blockchain can be widely applied in several services, initiated as a method to control the use and transfers of cryptocurrency, and such knowledge can be used from online payments, supply chain tracking, digital forensics and even with the sharing of health records [9].

Based on this theme, a study has already been initiated. It is currently underway to propose the use of safety techniques using blockchain mechanisms to ensure the integrity of remote labs.

The research is directly linked to the good practices available in IEEE SA Std. 1876–2019 [10], which is a standard for Networked Smart Learning Objects for Online Laboratories sponsored by IEEE Educational Society.

The mentioned standard establishes the relationship between all components (hardware, software and learning environments) to ease the design and implementation of online laboratory activities. The standard is designed to facilitate the design, implementation and use of pedagogically oriented online laboratories as smart learning objects and their integration into learning environments and learning object repositories [10].

The research teams of the Distance Laboratory (L@d) of TELUQ University in Montreal Canada and the Remote Experimentation Laboratory (RExLab) of the Federal University of Santa Catarina - UFSC in Brazil are working together on this research.

From this, the objective is to show the effectiveness of the use of blockchain techniques for a better form of data security of remote laboratories, making it an innovative fishing machine in its field. If verified, this resource can be implemented in the most diverse remote laboratories, guaranteeing another form of data protection.

6 Possible Solutions

Thus, for there to be real data security as well as the integrity of remote laboratories, not only an alternative, but a set of solutions must be developed in order to obtain the best results.

To prevent data theft, the use of a reliable Virtual Private Network (VPN) between the user's access device and the machine used in the lab makes it a way to prevent inappropriate access by users as well as ensure that data theft does not occur on the devices.

Other methods of ensuring security are investing in better ways to analyze network traffic, thus determining which data transmission or reception operations are reliable and can, therefore, be performed.

For example, the implementation of access keys in a hierarchical manner makes it possible for experiments to be supervised by a user who has a higher access key, thus being able to supervise the use of the laboratory as well as which users are performing the tests. For example, a higher access key to a teacher who accompanies students in experiments in remote laboratories dedicated to teaching some specific subject.

Another alternative capable of ensuring better security is monitoring access by users as well as making sure that only users who use a trusted connection have access to the laboratory.

The use of keys or tokens that are only enabled for a specific time can ensure that there is a greater availability of access to the laboratory.

Going in another direction, the implementation of artificial intelligence that monitors the experiments can be implemented depending on the objective of the laboratory. Thus, the AI would be able to identify possible malicious commands and end the user's session.

7 Conclusion

To finish, it is observed the great necessity to possess and to use specific technologies to obtain a better security in the remote laboratories. Since such laboratories have great importance in scientific development, especially in universities, they become better alternatives with greater accessibility and lower costs when compared to conventional laboratories.

Knowing who has access to the laboratories, as well as how these accesses were made, are the main problems to be discussed and solved. Thus, it is necessary to use new methods of protection that satisfy the security principles mentioned above.

Therefore, not only one but the set of several tools that guarantee adequate safety must be taken into consideration and worked on to make remote laboratories safe.

As well as the project using blockchain techniques to ensure a new form of security, numerous other projects can be thought of and developed. Since when talking about security, we can always implement new ways to leave a system more reliable and with greater credibility in the data presented.

No system is 100% secure when it comes to digital systems and media. However, there is the possibility of using different forms of security corroborating each other so that the main attacks are solved without compromising the laboratory.

To keep the safety guidelines up to date, initially, there should be an investment in research that brings new forms of safety, especially from other areas where they can be adapted for use in laboratories.

References

1. Salah RM, Cecil J, Atrushi D (2018) Collaborative remote laboratories for serving sciences and engineering education in iraq: rexnet project. In: ICOASE2018 - international conference on advanced science and engineering, pp 134–139
2. Pereira J (2018) Implantação e Utilização do Laboratório Remoto VISIR em Instituiçõess de Ensino Técnico, Tecnológico e Superior Revista Tecnologias na Educação 24. http://tecedu.pro.br/wp-content/uploads/2018/07/Art10-vol24-Edição-Temáticatica-VII-Junho-2018.pdf
3. Schneider D (2012) The state of network security. Netw Secur (2):14–20. https://doi.org/10.1016/S1353-4858(12)70016-8
4. Pekelman H, Gonçalves Mello A (2004) A Importância Dos Laboratórios No Ensino De Engenharia Mecânica. Congresso Brasileiro de Ensino de Engenharia, 9
5. Pandey S (2011) Modern network security: issues and challenges. Int J Eng Sci Technol 3 (5):4351–4357
6. Fernandes NC (2006) Análise de Ataques e Mecanismos de Seguran ça em Redes Ad Hoc, Ph.D. dissertation, UNIVERSIDADE FEDERAL DO RIO DE JANEIRO. http://monografias.poli.ufrj.br/monografias/monopoli10002685.pdf

7. Nawir M, Amir A, Yaakob N, Lynn OB (2017) Internet of Things (IoT): taxonomy of security attacks. In: 2016 3rd International conference on electronic design, ICED, pp 321–326
8. Kebande VR, Ray I (2016) A generic digital forensic investigation framework for Internet of Things (IoT). In: Proceedings - 2016 IEEE 4th international conference on future Internet of Things and cloud, Fi Cloud 2016, pp 356–362
9. Lao L, Li Z, Hou S, Xiao B, Guo S, Yang Y (2019) A survey of IoT applications in blockchain systems: Architecture, consensus, and traffic modeling. ACM Comput Surv 53 (1):32
10. IEEE Std 1876–2019 for Networked Smart Learning Objects for Online Laboratories 1(1). https://ieeexplore.ieee.org/document/8723446

Design and Development of an Ultra-Concurrent Laboratory for the Study of an Acid–Base Titration (ABT) at the Universidad Estatal a Distancia (UNED), Costa Rica

Carlos Arguedas-Matarrita[1(✉)], Eric Montero-Miranda[1], Laura Vargas-Badilla[1], Ronald Sánchez-Brenes[1], Eduard Rios-Badilla[1], Pablo Orduña[2], and Luis Rodríguez-Gil[2]

[1] Universidad Estatal a Distancia, San José, Costa Rica
carguedas@uned.ac.cr
[2] LabsLand, Bilbao, Spain

Abstract. The Universidad Estatal a Distancia (UNED) is the public distance-education university in Costa Rica. Since to 2019, the UNED only had available developments and laboratories focused on Physics and Engineering to improve the process of experimentation in distance education. In 2020, the UNED begins to venture into the development of laboratories in the field of chemistry with the peculiarity of working with irreversible processes through the progress of ultra-concurrent laboratories with the support of the global network of real educational laboratories available online, LabsLand. The first development in the field of ultra-concurrent laboratories between UNED and LabsLand was the Acid–Base Titration Laboratory (ABT) at UNED. The laboratory consists on the application of the analytical chemistry technique known as acid–base titration, which is performed by measuring the amount of a reagent of known concentration - called a *standard* substance – that is consumed by an unknown sample called *analyte*. Titration methods are used in many routine analyses because they are fast, accurate, convenient, and can be easily automated. Slow addition of the titrant from a burette to a known volume of the analyte solution takes place, until the reaction reaches a chemical neutralization called the “equivalence point”.

Keywords: Ultra-concurrent laboratory · Chemistry · Online education · Acid–base titration · LabsLand

1 Introduction

Remote Laboratories (RL) combine *hardware* and *software* technologies that allow students and teachers to perform experimental activities through the *internet*, just like being at the laboratory near the actual equipment. The main difference lies in the distance manipulation of automated equipment, at any time and from anywhere. These

M. E. Auer et al. (eds.), *Online Engineering and Society 4.0*, Lecture Notes in Networks and Systems 298, https://doi.org/10.1007/978-3-030-82529-4_13

devices are commonly used in physics [1] and engineering [2] education and there are few applications within the chemical science, due the typical irreversible nature of the experiments of this discipline.

An alternative to chemical distance education is the development of ultra-concurrent laboratories (ucL) which are pre-recorded real experiences configured using software tools and accessed by internet. By these means, ucLs allow students to experiment the same way as using an RL. This variation of RL is a real online interaction and can be conducted by several users simultaneously. [3]

The didactic sequences involved in RL offer the opportunity to increase student's autonomy regarding experimental work, since they are no longer limited by access and availability of equipment, henceforth being able of repeating and modifying experimental activities to own the learning process [4].

An example of an ultra-concurrent laboratory used for teaching chemistry is the one that has been developed by the University of Georgia. This laboratory seeks to expand the opportunities for self-directed learning, since the students will have unlimited remote access to the environment which would utterly foster their interest in the subject by promoting learning through guided research [5]. An ultra-concurrent laboratory is based on a set of recorded experiences carried out with real equipment generating real data and the experience can be performed by a large number of people simultaneously [4].

1.1 Remote Experimentation Laboratory at UNED

Remote experimentation at UNED is quite recent, initiating with the establishment of the Remote Experimentation Laboratory in 2018. These efforts included technical collaboration and support from RExLab – Universidade Federal de Santa Catarina to deploy their developments within UNED and furthermore the development of the RL – VISIR took place with the assistance of LabsLand [6]. The latest being a pioneer project in the region for the use of RL with educational purposes.

Due to social distancing measures taken by the COVID-19 pandemics and the lack of spaces for chemical experimentation, the UNED's group in alliance with LabsLand started the developing of new RL and ucL addressing topics and techniques such as: acid–base titration (ABT), Boyle Law, Charles Law, Gay-Lussac Law and soil exchangeable acidity. (see Fig. 1).

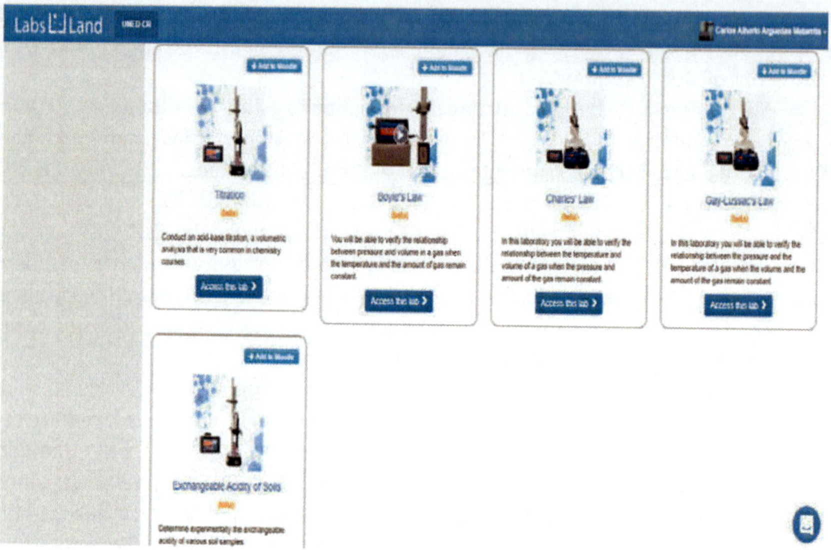

Fig. 1 New developments from the remote experimentation laboratory at UNED

This article presents the design and development process of the acid–base titration laboratory within UNED, Costa Rica. Titration is an analytical technique carried through the measurement of the amount of known concentration reagent (standard substance) that is consumed by an unknown sample (the analyte) until the reaction reaches a chemical equivalency between both species.

2 Methodology

The laboratory was elaborated jointly between UNED and Labsland. This work involved the experimental design and performance following the steps described as below:

2.1 Design of the RL

The design of the acid–base titration ucL was based on one of the most used techniques in chemical education to introduce students into the volumetric methods of analysis: the neutralization reaction. Through this analytical technique students became aware of the main concepts and components regarding experimental work in analytical chemistry, considering analyte special features within a determined matrix.

There are three basic factors to consider when working with volumetric methods: a standard solution of known concentration called the titrant, an analyte or sample and substance that allows to detect the end point of the procedure. This indicator substance can do its job by a change of color, the appearance of a precipitate or a change in the electrical potential. In acid-base titrations a standard stable solution of sodium hydroxide is commonly uscd as well as an unknown sample with acidic properties which produces hydronium ions (H_3O^+) and an indicator substance such as phenolphthalein that allows to identify the end point, where amounts of titrant and analyte are even.

In this step the equipment and instrument configuration for the appropriate development of the experiment were set up in a manner that users could corroborate the principles of an acid–base titration.

2.2 Experiment Testing

Using the instrumentation defined in the prior step, the work group conducted a series of tests in the Remote Experimentation Laboratory to adjust variables and concentrations of substances for the experiment. The equipment used is described as follows: (see Fig. 2).

2.2.1 Burette

A piece of labware used to contain and deposit the standard substance ok known concentration, in this case a sodium hydroxide solution. Unlike conventional titrations, this burette does not require high precision since an electronic drop counter is used for the estimation of the consumed volume.

2.2.2 Drop Counter

As mentioned above, this is an accessory sensor to the titration system that allows to count and register de drops of titrant added to the reaction environment. This sensor was calibrated before every experimentation.

2.2.3 pH Sensor

The pH sensor is submerged in analyte. This instrument records the variations in the reduction potential on the solution caused by the neutralization of H_3O^+ ions and hydroxide ions (OH^-).

2.2.4 Interface

Instrument that collects and transforms the registered data from the counting system and the pH sensor. This information is used to build up the tables and graphs found in downloadable files.

2.2.5 Stirring Station

This device fulfills two tasks: 1) acts as a bearing platform for all the titration system constituents and 2) it homogenizes the reaction environment contained in a beaker.

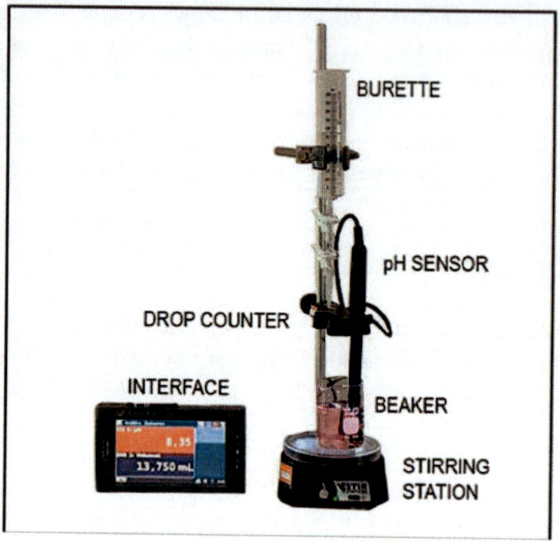

Fig. 2 Titration system used in the experiment

2.3 Experiment Recording

For the recording of the experiment, three smartphones were used over tripod holders. Each smartphone was located in an strategic place to look the procedure from three different angles: general plane (perspective), screen view of the interface and the dropping of the titrant into the beaker containing the analyte and the indicator substance.

2.4 Cloud Software Perspective

From a software perspective, this laboratory is a LabsLand ultraconcurrent laboratory [7], developed using the Open Source library weblablib [8], as well as LabsLand proprietary solutions for ultraconcurrent laboratory development. This means that LabsLand processes all the videos to suit different formats and networks and provides a web interactive experience based on web standards and without requiring any plug-in in the web browser and using exclusively secure http requests that can work through firewalls and proxies. The laboratory itself is hosted in Amazon Web Services, where a set of different technologies are used to provide highly scalable support to different loads of students to the various ultraconcurrent laboratories available in LabsLand.

3 Description of Acid–Base Titration (ABT) - RL

Users must add drops of titrant using the designated button and observe the changes in the solution, registered by the sensors and displayed in the interface, using some pre-selected parameters. The limited time of use is 60 min.

The user has the opportunity of watching the system from 2 perspectives; the first one (screen view) where they can see the changes in pH values and volume of titrant added according to the amount of drops previously programmed. As they get close to the end point of the reaction, pink starts to appear. The second angle shifts between dropping view and the perspective view, where users can get a closer look to the beaker content and the color changes.

Henceforth, as the user adds titrant to the analyte, a real-time graph is building up from the data registered by sensors, which enhances the understanding of the experience. (see Fig. 3). At the end of the experiment, users are able of downloading the raw data to perform their own graphical analysis, estimation of the equivalence point using first derivative as well as the calculation of the amount of standard substance consumed (mol) and the concentration of the analyte in the sample.

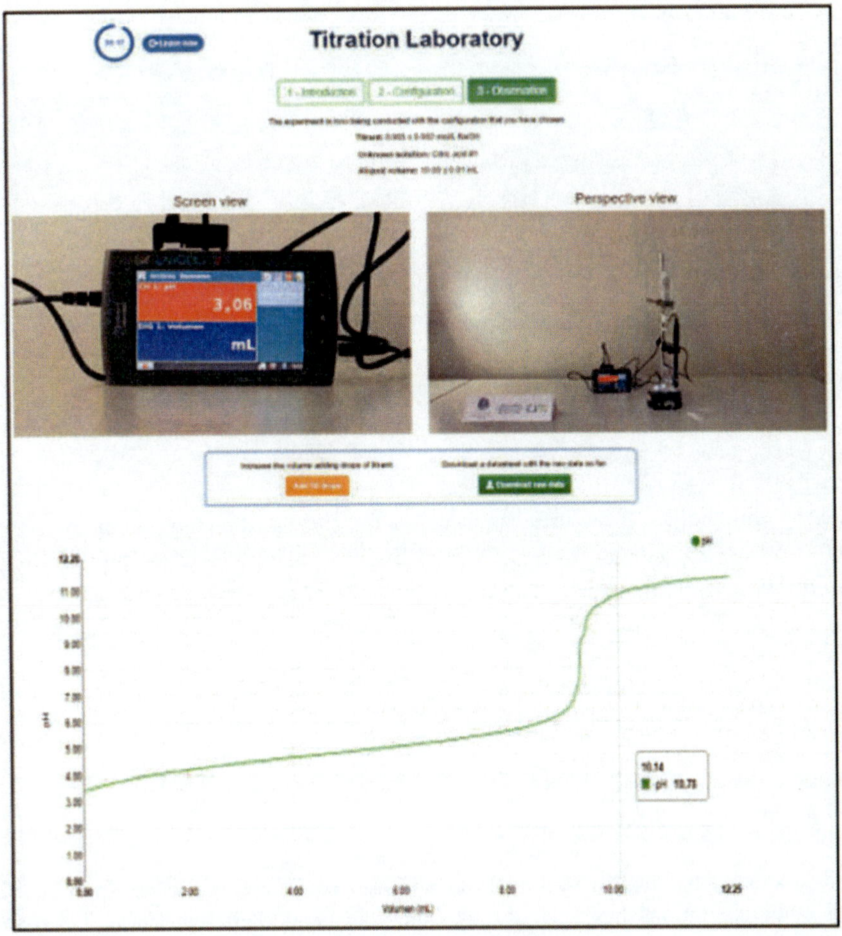

Fig. 3 User interface of the experiment

After concluding the experiment, users can repeat the experience while varying the standard solution concentration and the analyte species, choosing out of vinegar or citric acid as common analytes.

4 Example Educational Use

One of the activities that can be carried out with this laboratory is the measurement of a citric acid sample titrated with sodium hydroxide 0.065 mol/L. The user must identify the end point indicated by the color shifting and the consumed volume of titrant at that moment (see Fig. 4). Therefore, the student can replicate an experience very alike to the face-to-face laboratory work where the end point estimation is accomplished by the color change visualization, taking into account the consumed volume of titrant to the calculations.

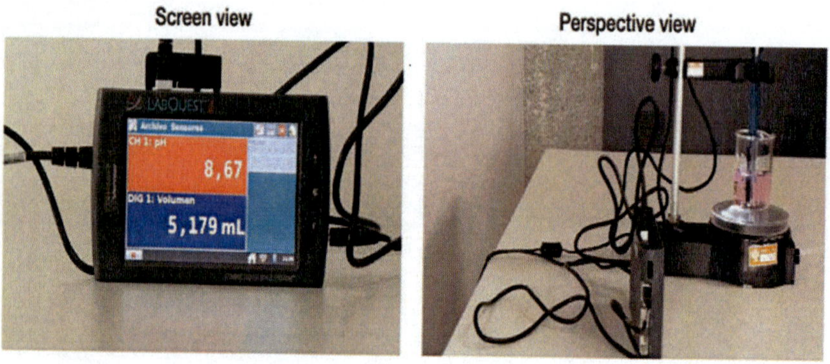

Fig. 4 View of the experiment

Applying the Eq. 1, the user can calculate the citric acid concentration (AcCitr):

$$[AcCitr] = \frac{V_{NaOH}\,[NaOH]}{V_{aliquot}} \tag{1}$$

$$[AcCitr] = \frac{5,179\ mL \cdot 0,065M}{10,00\ mL} = 0.034\ M$$

5 First Uses of RL

This laboratory has registered a total of 2363 uses so far and the access from UNED and other institutions has been through the LabsLand digital platform. LabsLand provides a truly effective handling of the RL allowing to incorporate Learning Management Systems (LMS). Moreover this company offers the ABT laboratory to third parties and the more uses UNED's laboratories have, there will be more remote laboratories that UNED and other public schools can access for free; which utterly facilitates sharing technological educational resources among institutions worldwide.

LabsLand also renders usage analytics and other tools to the developer group (UNED) and the schools and universities that have access to the equipment. In Table 1 utilization data for the ABT- ucL is shown.

Table 1 Registered uses of the laboratory in Costa Rica and Argentina

Country	Institution	Students	Uses	Total uses
Argentina	University of Buenos Aires (Basic common core)	1,043	2,081	2,154
	University of Buenos Aires (Drug quality analysis)	42	73	
Costa Rica	University of Costa Rica (UCR)	22	42	209
	UCR – West branch	12	15	
	UCR – Guanacaste	18	71	
	Colegio Científico de Alajuela (Scientific High School of Alajuela)	23	57	
	Colegio Científico de San Ramón (Scientific High School of San Ramón)	14	24	

6 Future Work and Conclusions

The acid – base titration laboratory is the first ultra-concurrent laboratory developed by UNED and LabsLand, with the educational potential of this resource, new developments focused on the teaching of experimental sciences and engineering are envisaged, specially in areas that lack of this kind of educational resources.

A very potent and versatile laboratory has been developed that gives the means to perform and acid–base titration through the internet enabling the experimental practice without being present at the laboratory. This a very significant factor in times of strong social distancing and isolation measures due to the COVID-19 pandemics.

RL are educational resources with many advantages within the distance and open education frameworks [9]. In the case of ulta-concurrent laboratories these advantages are heighten since users can work with real data that has been pre-recorded, enabling a much larger number of users. For instance, in bigger institutions with heterogenous groups of students and unrestricted access such as Universidad de Buenos Aires where the application of this instrument could be fitted based on multiple majors' profiles as those found in the first common courses.

Finally the possibilities offered by these laboratories to repeat and modify experimental practices, to carry them out at any time and place, and the fact that they are supported by different devices (cell phone, tablet, computer, among others) contributes intensively to the democratization of education [10].

References

1. Arguedas M, Concari S (2016) Remote laboratories used in physics teaching: a state of the art. In: 2016 13th international conference on remote engineering and virtual instrumentation (REV) IEEE. https://doi.org/10.1109/REV.2016.7444509
2. Samuelsen DAH, Graven OH (2016). Remote laboratories in engineering education - an overview of implementation and feasability. En LACCEI international multi-conference for engineering, education and technology. https://doi.org/10.18687/LACCEI2016.1.1.050
3. Narasimhamurthy KC, Orduna P, Rodríguez-Gil L, Susheen Srivatsa CN, Mulamuttal K. (2020) Analog electronic experiments in ultra-concurrent laboratory. In: Auer M, May D. (eds) cross reality and data science in engineering. REV 2020. advances in intelligent systems and computing, vol 1231. Springer, Cham. https://doi.org/10.1007/978-3-030-52575-0_3
4. Idoyaga I, Vargas-Badilla L, Moya N, Montero-Miranda E, Garro-Mora A (2020) El laboratorio remoto: una alternativa para extender la actividad experimental en la enseñanza de las ciencias en la universidad. Campo Universitario 1(2):4-26
5. Pokoo-Aikins GA, Hunsu N, May D (2019). Development of a remote laboratory diffusion experiment module for an enhanced laboratory experience. In: 2019 IEEE frontiers in education conference (FIE). IEEE
6. Arguedas-Matarrita C, et al (2019). Remote experimentation in the teaching of physics in costa rica: first steps. In: Proceedings of the 2019 5th experimental international conference, Madeira, Portugal
7. Orduña P et al (2018). Addressing technical and organizational pitfalls of using remote laboratories in a commercial environment. In: 2018 IEEE frontiers in education conference (FIE), pp 1–7 IEEE
8. Orduña P, Rodriguez-Gil L, Angulo I, Hernandez U, Villar A, Garcia-Zubia J (2019). WebLabLib: new approach for creating remote laboratories. In: International conference on remote engineering and virtual instrumentation, pp 477–488. Springer, Cham
9. Contreras-Mendieta JA et al (2019) Implementación de un laboratorio remoto (LR), como recurso de apoyo en un sistema de Educación a Distancia. Revista Ibérica de Sistemas e Tecnologias de Informação E 17: 923–935
10. Idoyaga I et al (2020) El Laboratorio Remoto de Validación Ácido-Base en un Curso de Química en la Universidad. In: XI Congreso Internacional sobre Aplicación de Tecnologías de la Información y Comunicaciones Avanzadas (ATICA 2020), pp 1–12

RISC-V Online Tutor

Fearghal Morgan[1(✉)], Arthur Beretta[2], Ian Gallivan[1], Joseph Clancy[1], Frédéric Rousseau[2], Roshan George[1], László Bakó[3], and Frank Callaly[1]

[1] National University of Ireland, Galway, Ireland
fearghal.morgan@nuigalway.ie
[2] Univ. Grenoble Alpes, CNRS, Grenoble INP, TIMA, Grenoble, France
[3] SAPIENTIA - Hungarian University of Transylvania, Targu-Mures, Corunca, Romania

Abstract. This paper presents the *RISC-V Online Tutor* course which provides structured, self-paced RISC-V architecture and applications training. The course browser transparently interacts with remote RISC-V hardware, implemented on an FPGA array. The course is implemented and supported by the reported vicilogic platform which provides online learning, remote FPGA prototyping and course builder. Lessons control remote hardware input signals, probe all RISC-V processor signals and overlay signal widgets on interactive course diagrams. The strategy provides a visually-rich, interactive learn-by-doing experience. The paper presents the course structure, and examples of the interactive lesson pedagogy. User experience, opinion and analytics are presented for a group of 45 users. Results indicate a high level of user satisfaction, and effective independent learning and achievement. Course availability is timely, with the growing interest in the RISC-V open-source Instruction Set Architecture, the call for training materials by RISC-V International, and the increasing demand for practical online learning systems, particularly during the Covid-19 pandemic.

Keywords: RISC-V · Online learning · Remote laboratory · FPGA · Prototyping · Assembly language · Tutor · Lesson · Course builder · Training

1 Introduction

Online training and practice initiatives such as Codecademy [1] for learning programming have been embraced by Engineering and Computer Science communities. Platforms use a range of browser technologies, emulation tools, structured learning and assessment. This paper presents the *RISC-V Online Tutor* course [2] which provides an independent-learning resource for RV32I RISC-V processor architecture and applications. Course creation and operation is supported by the reported vicilogic online learning, remote FPGA prototyping and course builder platform [3, 4]. Register to access vicilogic [5].

The vicilogic prototype builder toolsuite automatically parses and modifies a HDL hardware description language design, to connect all internal signals to a top-level register interface (which can be probed in remote FPGA hardware [6] by the course

M. E. Auer et al. (eds.), *Online Engineering and Society 4.0*, Lecture Notes in Networks and Systems 298, https://doi.org/10.1007/978-3-030-82529-4_14

browser). The user's browser interacts with the remote FPGA ARM processing system to support hardware configuration and real-time interaction with hardware. Each user course browser provides an interactive, independent training and practice environment, transparently linked to a remote RISC-V processor, automatically-allocated and configured on an FPGA [6] in the vicilogic FPGA array [3]. Lessons control remote hardware input signals, probe all RISC-V processor signals and signal widgets overlayed on interactive course diagrams (Figs. 1, 3, 6). The strategy provides a visually-rich, interactive learn-by-doing experience.

Figure 1 course browser left 'view' pane includes a subset of RISC-V components and signals, supporting the current register-register instruction lesson. The browser right pane provides guided learning course content. Signal widgets are overlayed on the view drawing to provide signal control and real-time state of all signals in the remote RISC-V hardware.

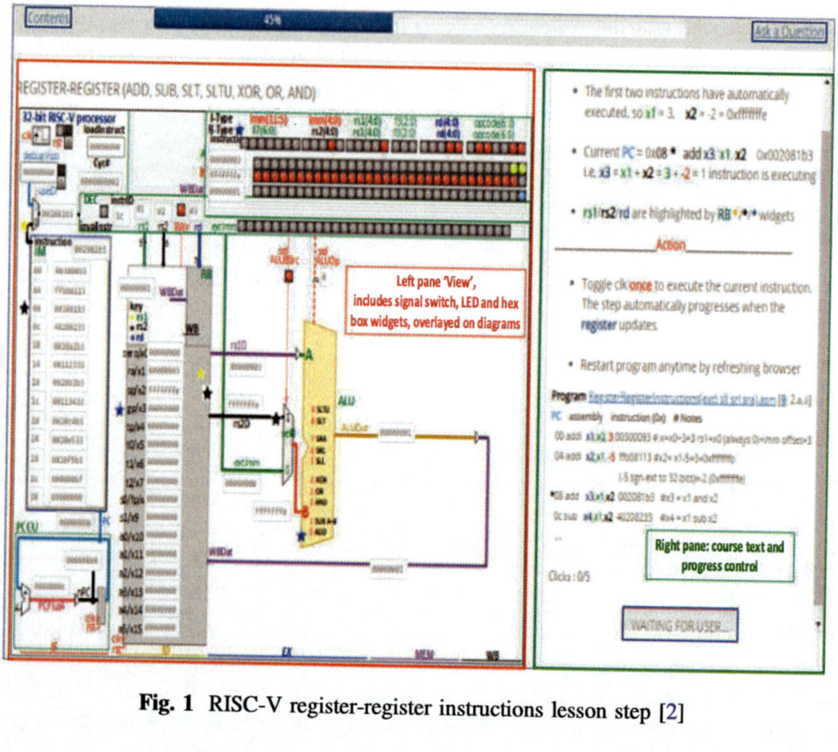

Fig. 1 RISC-V register-register instructions lesson step [2]

The paper presents the course structure, and examples of the interactive lesson pedagogy. User experience, opinion and analytics are presented for a group of 45 users in a 4th year B.E (Hons)/M.E. degree programme in Electrical & Electronic Engineering/ Electronic & Computer Engineering, in the National University of Ireland Galway. Results indicate a high level of user satisfaction, and effective independent learning and achievement.

The *RISC-V Online Tutor* complements RISC-V assembly program capture and simulation tools [7–9], and embedded RISC-V development platforms and tools [10], with formal online learning of RISC-V fundamentals: architecture, applications, assembly programming, with program execution and debug in hardware, and interfacing with memory-addressed peripherals. Course availability is timely, with the growing interest in the RISC-V open-source Instruction Set Architecture, the low number of available books on RISC-V [11–13], the call for training materials by the RISC-V International [14], and the increasing demand for practical online learning systems, particularly during the Covid-19 pandemic.

The structure of the paper is as follows. Section 2 presents the course structure. Section 3 illustrates the elements in the course browser. Section 4 summarises the reported vicilogic architecture, remote FPGA project prototype building process, course view and course lesson builder controls. Section 5 illustrates lesson examples. Section 6 highlights user experience, user opinion and analytics, and discusses the effectiveness of the course and platform. Section 7 concludes the paper and highlights future work.

Course contents: RISC-V Online Tutor: Architecture and Applications (duration, minutes)
Labs are included for all lessons, from lesson 5. Three **major** practical activities are **highlighted**.

+	1. Introduction: RISC-V Processor Architecture and Applications	90
	2. RV32I RISC-V instruction generator and viewer	
	3. RISC-V references	
	4. Tour of vicilogic online learning and prototyping platform	
+	5. **addi** register-immediate instructions, and ALU	120
+	6. **Tools tutorials:** assembly program, simulate, execute/debug	120
+	7. **add/sub** register-register instructions	40
+	8. Register-immediate instructions **addi, xori, ori, andi, slti, sltiu**	60
+	9. Register-register instructions **add, sub, xor, or, and, slt, sltu**	20
+	10. Constant-shift **srli, srri, ssai,** register-shift **srl, srr, sra**	35
+	11. **lui** load upper immediate instruction	15
+	12. Branch **beq, bne, blt, bge, bltu, bgeu** instructions, delay loop programs and Program Counter Control Unit (PCCU)	50
+	13. Program delay/shift x10 register left each sec **lui, add, addi, slli, bne**	25
+	14. Store/load to/from memory instructions **sw, sh, sb, lw, lh, lb, lhu, lbu**	60
+	15. **auipc** add upper immediate to PC instruction	20
+	16. **jal/jalr** jump and link instructions	25
+	17. Review of DEC instruction decoder component and EX Execute stages	30
+	18. **Optional lab:** RV32I VHDL model capture, simulation, synthesis	
+	19. Program functions and stack handling	40
+	20. Memory-mapped peripheral components (counter, in/out ports)	45
+	21. **Major lab:** Target game application demo and lab	
+	22. Pipelined RV32I RISC-V processor architecture	60
+	23. C program compilation to RISC-V assembly program	30

Fig. 2 Course contents, illustrating the instruction-by-instruction course structure, associated labs and typical lesson duration (excluding time for lab exercises)

2 Course Structure

Figure 2 illustrates the course structure. The *RISC-V Online Tutor* course duration is approximately 15 h (based on user feedback), excluding lab exercises. The course structure highlights the incremental learning and practice strategy, presenting each RV32I RISC-V instruction in turn (lessons 5–16), with dynamic and interactive images of instruction-specific hardware components and signals (in the browser 'view', with overlayed signal widgets). Users interact with the RISC-V hardware in course lessons, through interactive guided learning, knowledge checks, sandboxes and challenges, along with text-based course material and supporting videos. The course provides self-paced, structured training and practice and can be followed independently. The course integrates learning of RISC-V assembly program capture and simulation throughout, using Venus online assembly program capture and simulation tools [7]. Users execute downloadable course assembly program examples and capture/simulate their own lesson-specific program examples. Users then upload and execute program instruction machine code to a transparently allocated and configured remote RISC-V processor, and use the vicilogic debug/breakpointing tools to verify program operation.

If users have prior knowledge of digital design, hardware description language (HDL) capture, simulation and logic synthesis, e.g., on completion of vicilogic pre-requisite course *Digital Systems Design and FPGA Prototyping: Fundamentals, HDL and EDA Tools* [2], they can optionally complete lesson 18 laboratory to capture the HDL models for several RV32I RISC-V components, within a RV32I RTL project template, using Xilinx Vivado Electronic Design Automation (EDA). The lab provides a HDL testbench, enabling simulation and verification of the user's design, and synthesis of the complete processor RTL model, or its individual components.

A later course exercise (lesson 21) challenges users to design and develop an assembly program game application, executing on a remote RISC-V processor, at full clock speed, with a game application user interface, interacting with processor hardware signals. The game arena currently includes a 16 × 32-bit LED array, clk and rst signal widgets. Peripheral components (presented in lesson 20) include 16-bit in/out port interfaces and a 16-bit programmable counter which, for example, can be used as a random number generator in the game. Further hardware peripheral devices will be added, e.g., DAC, ADC, SPI and I2C components, and console interface.

Lesson 22 presents RV32I 5-stage pipelining, highlighting pipeline control and data signal behaviour, and illustrating pipeline hazard detection and handling. Lesson 23 presents example C programs, along with C to assembly language compiler toolsuite installation and execution. This provides a platform whereby example fundamental C program function execution on a RV32I RISC-V processor can be visualised, down to the processor component and signal level.

3 Course Browser Elements

This section presents the course browser and selected lesson examples. Lessons layout and logic is generated using the vicilogic course builder. Figure 1 lesson step example right hand pane is 'waiting for user' to click the clk widget once. Doing so toggles signal clk, executing the current instruction. The lesson logic polls specified processor signals to detect correct user action, and to gate automatic lesson progression.

Figure 1 'view' (left pane) includes the following RISC-V components and signals which support register-register and register-immediate instructions in the current lesson. Processor stages **IF**, **ID**, **EX**, **MEM** and **WB** are labelled along the bottom of the view. Active datapaths are highlighted in the view, using star widgets.

- Instruction Fetch (**IF**) stage
  - Instruction Memory (**IM**), with signal widgets in a dynamic window, centred on the current instruction memory byte address, highlighted by the * widget
  - Program Counter Control Unit (**PCCU**)
- Instruction Decode (**ID**) stage
  - Includes Register Bank (**RB**), displaying 16 (of the 32) RV32I registers x0–x15, and Instruction decoder (**DEC**)
- Execute stage (**EX**)
  - Arithmetic and Logic Unit (**ALU**)
  - **ALU** input signal B datapath select multiplexer (**selB**)

Figure 3 illustrates all RV32I RISC-V architecture components and signals, including memory (**MEM**) and writeback (**WB**) stage and branch components and signals which are not included in Fig. 1, for register-immediate or register-register instructions.

The 32-bit instruction and its fields are presented as LEDs along the top of the view. ALU input signals A, B and output ALUOut are also displayed as LEDs at the top of the view to aid bit-level presentation of the ALU arithmetic, logical and branch instructions. The right hand pane can include Action prompts, instructing the user to control input signal widgets. The 'view' (left hand of browser) includes a facility to force an instruction using signal **debugInstruction**(31:0), to over-ride the current Action instruction. The platform includes an Ask a Question facility in every lesson step (top right of the browser, highlighted in Fig. 1), activating a pop-up and enabling context-based queries (automatically emailed to the course tutor), or suggestions from users. In practice, this facility has been well used, and has significantly informed lesson enhancements. The course Contents link (top left of the browser, highlighted in Fig. 1) enables selection of any course lesson.

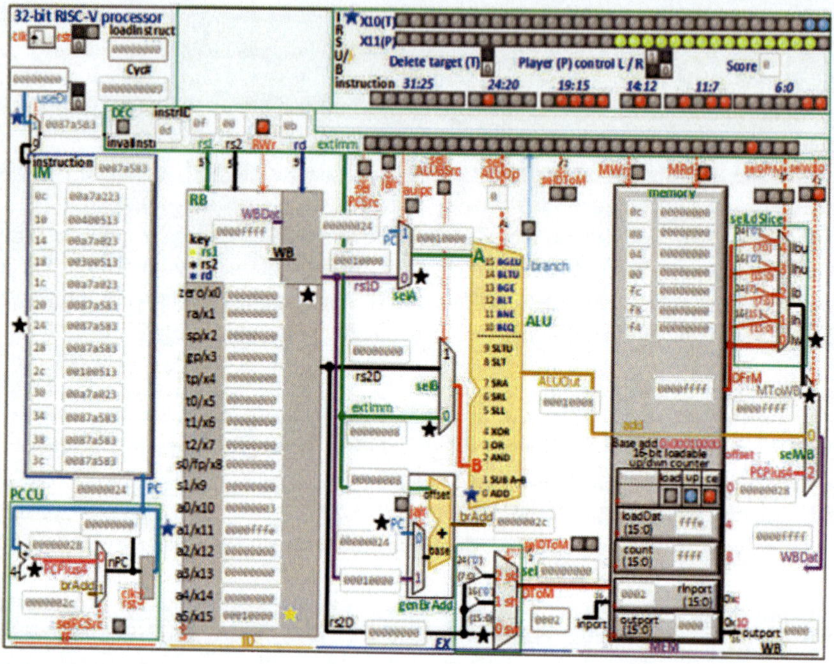

Fig. 3 Complete RV32I RISC-V architecture view, with all internal signals and overlayed widgets, presenting real-time state of signals in a remote RISC-V processor. Link to lesson [2]

4 Vicilogic Architecture, Remote FPGA Project Prototype Builder and Course View/Lesson Builder

This section summarises the reported vicilogic architecture, remote FPGA project prototype build process, and the course view and lesson builder [3].

Figure 4 illustrates the vicilogic architecture, including

- vicilogic server elements
- Xilinx PYNQ-Z1/-Z2 FPGA array in the cloud [6]

The PYNQ ARM-based controller handles communication between a user digital design (entity) (configured in hardware) and the vicilogic controller registry and client application. Client application requests to the vicilogic controller registry (Fig. 4(a)) include FPGA hardware configuration, FPGA entity input signal state change, and FPGA signal reads. When a controller comes online, it provides a list of its available entities, their types, and their address to the registry. Direct client-controller communication follows controller registry setup of the client-controller link. The server controller registry central resource management server transparently configures and allocates an entity to a requesting client session, based on available authorised resources.

The vicilogic architecture is scalable, i.e., the controller/entity architecture supports user interaction with

- vicilogic centralised FPGAs. 35 modules are currently available for use by any vicilogic user [2]. During the period Feb 2018–Jan 2021, 107,000 FPGA configurations have been performed by users and vicilogic platform developers.
- local ethernet-connected user FPGAs
- distributed FPGA clusters

The architecture also supports the addition of other FPGA/SoC device families.

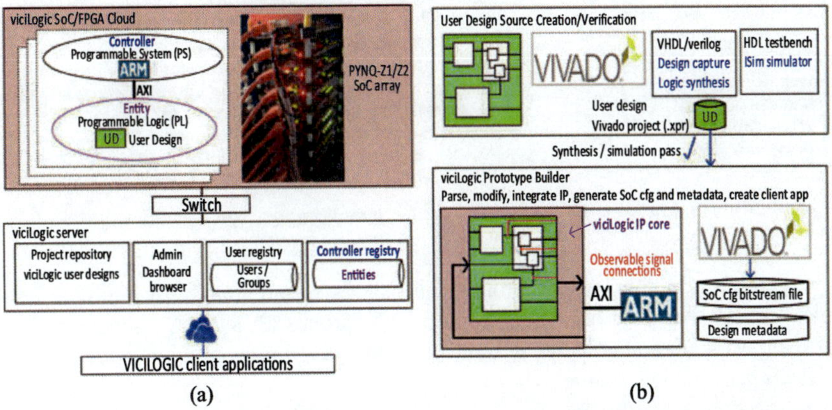

Fig. 4 **a** Vicilogic architecture **b** vicilogic prototype project build process [3]

Figure 4(b) illustrates the vicilogic prototype project build process [3], which integrates with the Xilinx Vivado industry-standard EDA toolsuite, to

(i) parse and modify the design HDL to enable probing of all signals in the design at the top level. This enables real-time signal controllability and observability of all FPGA signals from the course browser.
(ii) integrate the design HDL/IP with Xilinx Zynq ARM/AXI processor system, vicilogic resources (including BRAMs, clk generator), and generate/upload FPGA hardware configuration bitstream file and json metadata to the vicilogic project repository

The vicilogic course lesson builder defines the lesson content to appear in the browser. The course content pane (right side of browser) typically contains text, images and URLs. The course view pane (on left side) typically includes an interactive diagram, which connects to signals in a remote RISC-V processor. The view pane can also include embedded videos, text and images. The view creator selects a remote FPGA prototype project (configuration bitstream and related json metadata), and overlays signal widgets on the diagram view image. Widget types include switch, LED, hexadecimal text box, and stars (e.g., *, *) widgets (used to highlight active datapaths).

Figure 5 illustrates a selection of commonly used course builder commands, used to develop course lesson and course progress logic.

set_text	define the right hand pane lesson text (html)
set_left_text	add video/image/URL to left pane (no FPGA configuration)
set_view	allocate view (with FPGA configuration) to left hand pane
write	assign a state to an input signal
loop n end_loop	repeat a command sequence n times
start_batch send_batch	encapsulate and transfer a series of commands to the FPGA ARM processor for efficient batch execution
flash_icons stop_flash_icons	start/stop flashing defined signal widgets (coloured)
wait_for_sigs	suspend step progress until a signal enters a defined value
knowledge check	hide a selected signal on view, record user performance
allow _view_clicks	enable (sandbox/guided lesson) user control of input widget
stop_view_clicks	disable (captive lesson) user control of input signal widget

Fig. 5 Commonly used course lesson builder commands

Execution of a course step automatically allocates and configures a remote FPGA project, controls remote hardware input signals, probes all RISC-V processor signals and overlays signal widgets on interactive course diagrams.

Example vicilogic commands are

- write: apply a signal sequence to control an FPGA into any state
- wait_for_sigs: monitor specified signals to progress a course step only when the user controls the FPGA design into the required (specified) signal state.

Commands enable complex interaction with the 'view' and remote FPGA, including the development of guided lessons, where a user must follow a precise set of signal widget interactions to complete a lesson. The right pane provides confirmation and elaborates system operation as the lesson progresses.

5 RISC-V Lesson Examples

This section illustrates selected RISC-V lesson examples. Figure 5 presents an Arithmetic and Logic Unit (ALU) unit level lesson [2]. The user can control all ALU inputs implemented as a unit level module, and observe component and signal behaviour. The active datapath is highlighted by the *, currently executing an arithmetic shift right of 8 (B = 8) bits on data A = 0 × 80f73155. ALUOut result = 0xff80f731.

Figure 6 presents the complete RV32I RISC-V architecture view for a program application which executes peripheral write/read accesses [2]. All internal signals are included, with overlayed widgets. While the view may appear complex and busy at first sight, its elements have been introduced incrementally as the course progresses. Active data paths are highlighted with coloured star widgets. Instruction decoder (DEC) signals are included, mostly labelled in red. The view animates active signals and data for all RV32I instructions. The view includes MEM stage memory component with a hexadecimal data window, centred on the current memory byte address pointer.

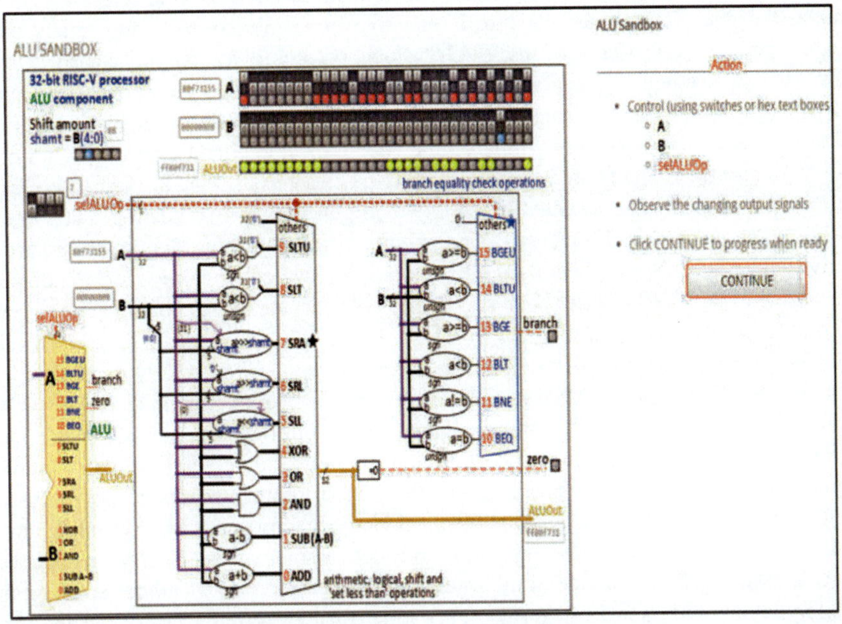

Fig. 6 RISC-V ALU unit level sandbox. Link to lesson [2]

6 User Experience, Opinion and Analytics

This section highlights user experience, user opinion and analytics, and discusses the effectiveness of the course and platform. The user group was a class of 45 4th year B.E (Hons)/M.E. degree programme in EEE/ECE, National University of Ireland Galway, independently following the *RISC-V Online Tutor* course during September–November 2020. Users worked from home throughout the period as a consequence of the Covid-19 pandemic. Figure 7(a) presents FPGA component usage records for the user group, during the period Friday 23/10–Thursday 12/11 2020 (weeks 5–7 of the 8 week RISC-V course). When a user enters a course lesson page, vicilogic automatically allocates an FPGA to the user and configures the FPGA as a remote RISC-V processor (or other digital logic component, depending on the lesson topic). Each colour highlights the use of a specific FPGA module in a course lesson. The size of each coloured column element represents the level of course step sessions which the FPGA has serviced, i.e., FPGA configuration and control by course lessons, sandboxes, challenges and user-loaded programs. Figures 7(b) and (c) highlight user A/B activity respectively, which contribute to the Fig. 7(a) chart.

The number of active sessions is in the 300 per day range on several days illustrating that the vicilogic architecture is effective in seamlessly managing many concurrent connections, user sessions, FPGA allocation, configuration and control, along with the course content provision. Figure 7 highlights the shared nature of the vicilogic FPGA array, with user sessions automatically and transparently accessing a range of FPGAs as they progress through the online course or upload and execute programs.

A significant assignment was due on 9th Nov (lesson 18, Fig. 2), with users capturing HDL models for a series of RISC-V components, and simulating and synthesising a complete RV32I RISC-V processor. Figure 7(a) highlights considerable activity on 8th–9th Nov, reflecting the level of use of interactive component sandboxes for RISC-V ALU, **WB** (write back stage) and EX (execute stage) components (lesson 17), to help their understanding.

The course philosophy is based on the use of the online *RISC-V Tutor* course as the primary learning mode, with users following the online course independently. An online Blackboard Virtual Learning Environment tutorial session was held each Tuesday and Friday. The combined duration of these online sessions was approximately one hour per week. The goals of the online tutoring sessions was to ensure regular contact with users, to provide feedback on questions raised on any course topic, and to ensure that practical activities and deliverables were clearly understood. Questions which arose outside of the weekly online sessions (often submitted using the vicilogic Ask-a-Question facility (Fig. 1) or email) were promptly responded to. Most users did not raise any question. The tutorial facility has been effective and is considered an important element in supporting inevitable questions which users raise, and in providing individual support which user may require. The incremental instruction-focused course lesson approach (lessons 5–16, Fig. 2) has worked well, and provides a useful instruction-based index.

User assignment submissions have been in the form of 3–5 min videos (groups of 2), describing program instruction operation, with reference to the RV32I RISC-V architecture instruction-specific views. Users have applied the vicilogic platform effectively themselves in describing selected RISC-V component behaviour, for their created programs, and for illustrating their level of understanding and achievement.

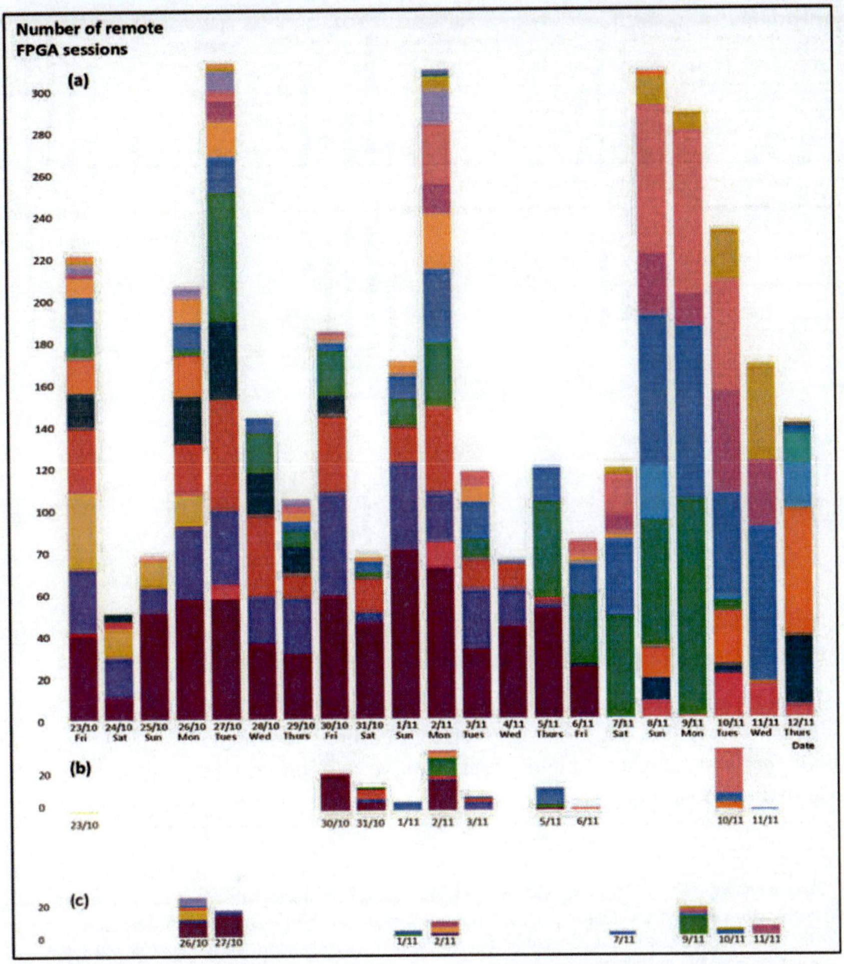

Fig. 7 **a** FPGA usage records for a class of 45 users, during period Friday 23/10–Thursday 12/11 2020. Each colour highlights the use of a specific FPGA module in a course lesson. The size of each coloured column element represents the level of course step sessions which the FPGA has serviced. **b** Activity of user A **c** Activity of user B

Figures 8(a)–(g) present user opinion captured using anonymous surveys at the end of lessons. User lesson ratings are in the range 7.9/10–8.3/10 (inset on graphs). User feedback indicates a positive opinion of the learning strategy, and acknowledges the visually-rich and interactive experience provided by the platform. Resulting high level of achievement by users in course labs has indicated effective independent learning.

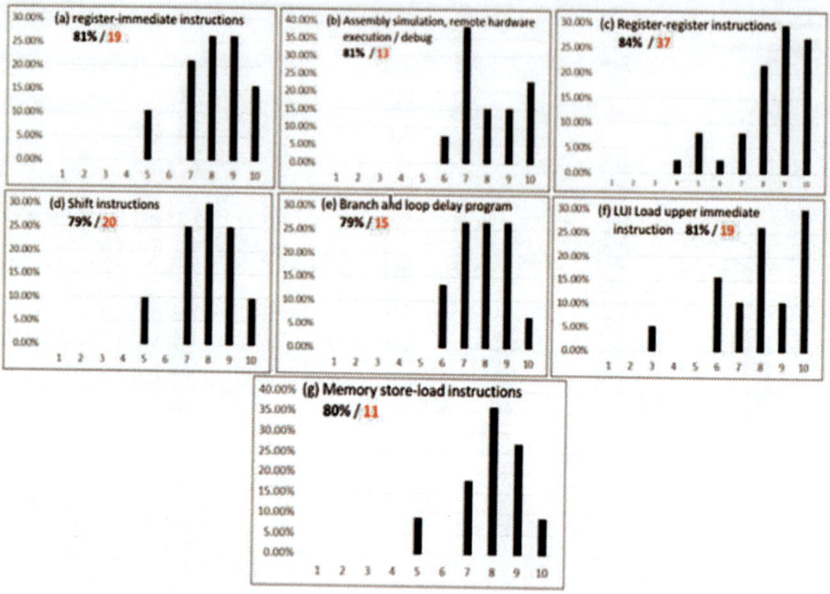

Fig. 8 User opinion on a range of course lessons. User lesson ratings are in the range 7.9/10–8.3/10 (inset in each chart), along with the number of respondents (red)

Figure 9 includes comments captured for question: "*What element(s) of the lessons did you find most useful?*", highlighting positive opinion on a range of aspects of the platform and pedagogy.

sandboxes, interactive diagrams and widgets, visualising internal signals, visualising registers,
combination of view and lesson text, video demos, breakdown of instruction step by step,
worked examples and step by step clocking through instructions, knowledge checks, challenges,
use of coloured widgets, labs and exercises, thorough coverage of each instruction,
star widgets indicating where to look (highlighting active datapath), good pace,
good flow between reading text and understanding what was happening on hardware,
signals being shown at the top of the screen (to aid bit-level activity),
being able to prove that I know this stuff, example assembly program descriptions and demos,
instruction generator/viewer is really useful.

Fig. 9 Comments captured for question: What element(s) of the lessons did you find most useful?

7 Conclusions and Future Work

The paper has presented the *RISC-V Online Tutor* course as a novel contribution to online self-paced formal training in RISC-V architecture, applications and assembly programming. The vicilogic platform transparently integrates diagrams and interactive signal widgets in the user browser, which interact with remote RISC-V hardware.

Lessons use a range of pedagogical techniques to provide a visually-rich, interactive, learn-by-doing experience. User experience, opinion and analytics indicate a high level of user satisfaction, and effective independent learning and achievement. The platform will be shared with the RISC-V user community. Opinion will be captured to inform future development. The scalable vicilogic architecture will extend to provide distributed FPGA networks, supporting locally-connected and distributed FPGA clusters. The UI will also be developed further. Selected sensor and controller peripherals will be integrated on remote FPGA modules to enable development of small training-level embedded systems and game applications, using the remote RISC-V hardware.

References

1. Codecademy. https://codecademy.com. Accessed 11 Jan 2021
2. Vicilogic course index and resources. https://www.vicilogic.com/static/ext/RISCV/REV2021/ReferenceLinks.pdf. Accessed 11 Jan 2021
3. Morgan F et al (2018) Vicilogic 2.0: online learning and prototyping of digital systems using PYNQ-Z1/-Z2 SoC. In: Proceedings of the International Symposium on Rapid System Prototyping (RSP2018), Torino, Italy, pp 76–82, October 2018
4. Skhiri R et al (2019) From FPGA to support cloud to cloud of FPGA: state of the art. Int J Reconfigurable Comput 2019, Article ID 8085461. https://doi.org/10.1155/2019/8085461
5. Vicilogic registration. https://www.vicilogic.com/register/. Accessed 11 Jan 2021
6. TUL PYNQ-Z2, Xilinx Zynq SoC. https://www.tul.com.tw/ProductsPYNQ-Z2.html. Accessed 11 Jan 2021
7. Venus RISC-V instruction set simulator. https://www.kvakil.me/venus/. Accessed 11 Jan 2020
8. Guillaume Savation, Emulsiv RISC-V Simulator. https://github.com/Guillaume-Savaton-ESEO/emulsiV Accessed 11 Jan 2021
9. Ripes, A graphical 5-stage RISC-V simulator for teaching purposes. https://github.com/mortbopet/Ripes. Accessed 11 Jan 2021
10. SiFive HiFive1 Boards. https://www.sifive.com/boards. Accessed 11 Jan 2021
11. Patterson D, Hennessy J (2017) Computer Organisation and Design: RISC-V Edition, The Hardware Software Interface, 1st edn. Morgan Kaufmann, Burlington
12. Patterson D, Waterman A (2017) The RISC-V Reader: An Open Architecture Atlas. 1st edn. Strawberry Canyon, Yavapai. ISBN 0999249118
13. Dos Reis AJ (2019) RISC-V Assembly Language. ISBN-13: 978-1088462003, Independently published
14. Learning RISC-V. RISC-V International. https://riscv.org/community/learn-about-risc-v/. Accessed 11 Jan 2021

FPGA Meets Breadboard: Integrating a Virtual Breadboard with Real FPGA Boards for Remote Access in Digital Design Courses

Shuowei Li[1(✉)], Heran Wang[1], Luis Rodriguez-Gil[2], Pablo Orduña[2], and Rania Hussein[1]

[1] University of Washington, Seattle, WA 98195, USA
{lis26,hw99,rhussein}@uw.edu
[2] LabsLand, Bilbao, Spain
{luis,pablo}@labsland.com

Abstract. With the transition to remote instruction, new modalities for conducting hands-on labs are needed. In particular, courses that entail major hardware components faced challenges in making the hardware available for students in a reliable and sustainable way. This paper presents our experience in implementing a virtual breadboard feature to interface with real Field Programmable Gate Arrays (FPGAs) boards located on the University of Washington's campus, where students can access the boards remotely to complete their lab assignments.

Keywords: Remote Laboratory · FPGA · Breadboard · Hardware Design

1 Introduction

Traditionally, universities teach Field Programmable Gate Arrays (FPGAs) courses using on-campus labs, which have lab benches, PC/workstation, and other supporting equipment. In order to complete assignments, students need to be physically present in the labs to interact with these pieces of equipment. However, this approach is not feasible during the COVID-19 pandemic. A new teaching strategy based on a global network of remote laboratories is appealing in this era.

Remote laboratory facilities have never been a new topic. It has been used in teaching physics, biology, and other subjects [1–3]. However, the majority of them are stand-alone systems that cannot be shared with different facilities. In the field of engineering education, VISIR designs a shareable system to provide seamless access for students from various universities [4]. It allows students to explore electronics, sound, and vibration or propagate radio signals through an online platform. Even though the previous works demonstrate how applicable the educational remote laboratory facilities are, the state-of-the-art applications lack the ability to teach the student how to use General Purpose Input Output (GPIO) ports to communicate with the FPGA boards.

In particular, FPGA has been a target of remote laboratories in the literature for many years, in different initiatives in different universities [9–12]. In [8], a distributed FPGA laboratory with multiple instances in multiple locations is described. In this

M. E. Auer et al. (eds.), *Online Engineering and Society 4.0*, Lecture Notes in Networks and Systems 298, https://doi.org/10.1007/978-3-030-82529-4_15

work, we added a new concept on top of [8]. This new concept intends to increase the scope of the lessons by providing an easy-to-use solution for students to work with the GPIO ports of the FPGA.

This paper introduces a feature that implements a virtual breadboard where students can use GPIO ports to interface with real FPGA boards that are accessed remotely through a global network called LabsLand [5, 8]. This remote FPGA laboratory has been used to teach a junior-level course in digital design at the University of Washington during the autumn quarter of 2020. It presents a pilot project that intends to serve the needs of the course post-COVID-19. This paper presents our experience in transforming a lab that traditionally used physical breadboards and FPGA boards to an online modality. We demonstrate how the lab used to be conducted pre-Covid-19 and how it has been restructured to fit online instruction.

2 Background

The laboratory experiment highlighted in this paper is one out of 5 other labs that students are expected to complete for the course. This lab experiment is a refresher of the finite state machines and introduces students to the onboard general-purpose input/output (GPIO). The students achieve this goal during the process of designing a parking lot occupancy counter, which consists of two pairs of photosensors monitoring the activity of cars. For educational purposes, students use switches to mimic the output signal from the sensors. The signals are also displayed on two LEDs to make the signals visible while introducing more circuit components to the laboratory. This gives students opportunities to practice implementing breadboard circuits and interfacing GPIOs, which are essential skills for hardware design.

Traditionally, in the on-campus laboratory environment, each student receives a breadboard assembly in conjunction with a DE1-SoC FPGA, which is shown in Fig. 1. Students will implement three switches and two LEDs on the breadboard for this laboratory experiment. An example of breadboard circuits is shown in Fig. 2.

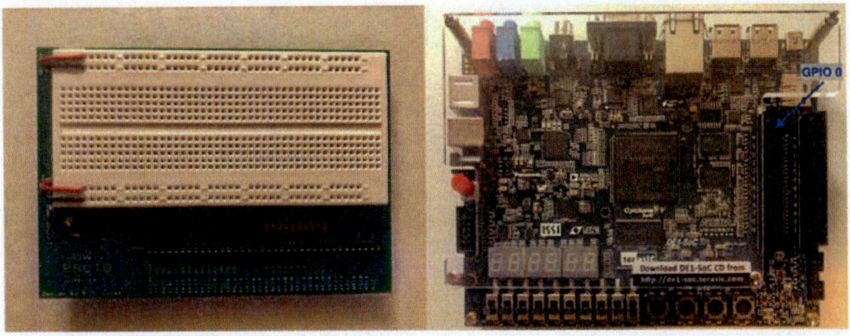

Fig. 1 The customized breadboard module (left) and the DE1-SoC FPGA (right). The breadboard module is connected to the FPGA board via the GPIO 0 header

Fig. 2 The breadboard circuits that are required by the laboratory experiment

However, the manufacture of these breadboard modules for large classes is time-consuming, and, most importantly, the physical laboratory kit is not accessible in the remote learning environment. Therefore, for this particular laboratory experiment, a remote laboratory solution that does not rely on the distribution of hardware is needed.

3 Technical Solution

3.1 User Perspective

To provide an uncompromising learning experience to students, we added a virtual breadboard interface to the remote laboratory facility LabsLand to simulate hands-on circuit building, providing students the opportunity to interface GPIOs in their SystemVerilog designs.

Design Decisions for the Breadboard User Interface. A breadboard configurator is added to the user interface switch to simulate the hands-on breadboarding experience. Students have access to several circuit components in the configurator, including three switches and two LEDs placed onto the virtual breadboard. We have a GPIO header that is internally connected to the DE1-SoC's JP1 in the backend of the system. Next to the breadboard's implementation area, we provide a pinout diagram for DE1-SoC's GPIO 0 (JP1) to help students understand the relationship between pin names and pin numbers and build the circuit. Figure 3 depicts the default state of the virtual breadboard.

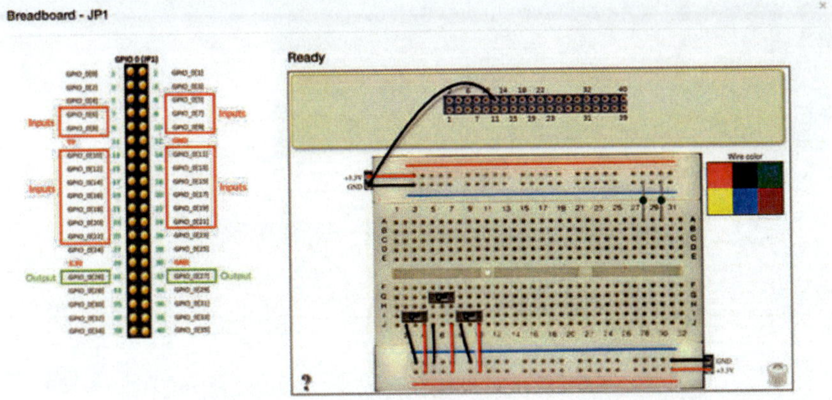

Fig. 3 The breadboard configurator on the LabsLand IDE

Students will connect each circuit component to an appropriate GPIO pin of their choice for this particular laboratory experiment. To start with, students need to figure out which pin to use by referring to the pinout diagram to the left. As the pin names (e.g., *GPIO_0[20]*) are deliberately not shown on the actual GPIO header, students are forced to use the relationship between pin names and pin numbers, which ensures that all the essential steps for working with the actual boards are taken on the web interface as well. Then, students click on a wire color in the wire selector to choose a wire. They can use their mouse as a pen for drawing wires, connecting the GPIO header and the breadboard. All the connections on the breadboard will be automatically saved to emulate the real-life experience.

Students can then proceed to the remote laboratory facility LabsLand code editor interface to continue implementing the constructed virtual breadboard circuit. Once they synthesize and upload the project to FPGA, the user interface shown in Fig. 4 will be presented to the students with all the previous changes made to the breadboard preserved. The students can toggle the switches and observe the LEDs' states along with other outputs on the DE1-SoC board, such as the HEX display, to verify their design. A tutorial specifically for the breadboard user interface is provided to the students to help them get familiar with the workflow.

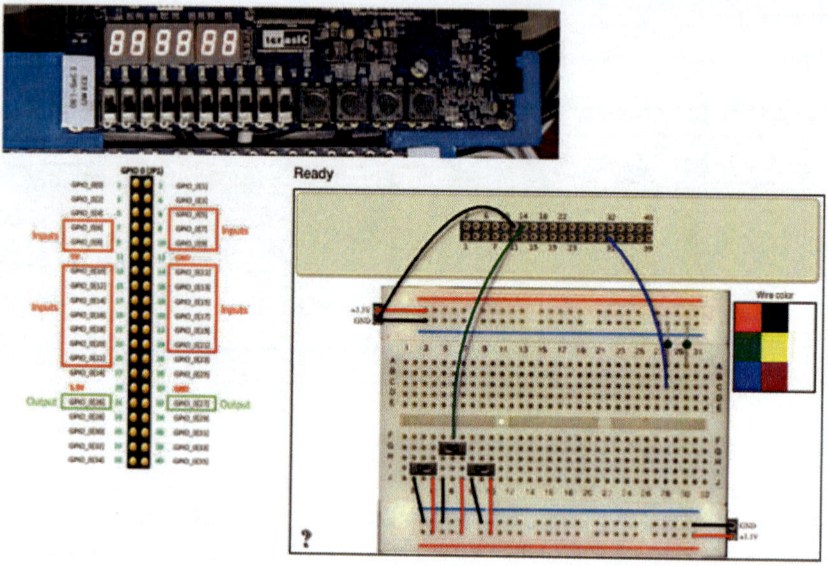

Fig. 4 The breadboard user interface for testing FPGA programs

This workflow provides several benefits over other possible implementations. Firstly, it closely emulates the process of building a breadboard circuit in a physical laboratory setting, as it allows students to build the circuit before or in parallel with writing code. Secondly, by always saving the state of the breadboard for each student, it simulates the irreversible and nonvolatile nature of working with physical hardware. Thirdly, while providing unlimited time for building breadboard circuits, it also reduces the test time each student spends on the limited number of the FPGA hardware, reducing queueing time and providing a better learning experience for more students.

3.2 Technical Perspective

For the breadboard user interface, the Open Source VISIR[1] client has been used as an external library, particularly the breadboard component. On top of the breadboard component, a JavaScript code has been added that checks if certain points of the user interface are connected to the GPIO's and the switches and LED's on the breadboard. The system automatically detects faulty connections to help students connect what they need to connect to the particular GPIO.

The wiring phase is typically done in the code editor [8]. Once the breadboard is wired, whenever students use the FPGA, they have the same circuit built before, and they can still modify it once they are connected to the FPGA itself. Internally, each change of the wires or each change of the switches state is translated to a call to the

[1] https://github.com/JohanZackrisson/visir_html5

server to send specific signals to the GPIO's. Once we detect any wire changing calls, the web browser will turn on and off the LEDs accordingly.

4 The FPGA with Breadboard Laboratory Experience at the UW

4.1 Contextualization

Due to the logistical difficulty of shipping physical DE1-SoC lab kits to students during the COVID-19 pandemic, all students in the Design Of Digital Circuits class at the UW in Autumn 2020 were asked to use the remote laboratory facility LabsLand along with the newly developed breadboard user interface. The remote laboratory does not require the shipping of any laboratory equipment and can be accessed anywhere and anytime. Before taking this class, all students are required to take a prerequisite junior-level Design of Digital Circuits and System course. Since the remote laboratory facilities just deployed for this offering. Students only have prior knowledge of using the physical DE1-SoC development board and breadboards.

4.2 Survey and Results

The students in Design Of Digital Circuits And Systems class were asked to complete an anonymous online survey to indicate their degree of agreement on certain statements based on a 5-point Likert scale [6], with one being "strongly disagree" and five beings "strongly agree." Out of 60 students enrolled in the course, 55 students completed the survey. The survey questions were designed to evaluate their experience using the remote platform with the newly developed breadboard user interface. The survey and the corresponding quantitative data for statistical analysis are given in Table 1.

Table 1 The online survey, questions, and quantitative data (N = 55)

#	Question	Min	Max	Mean	Std. Dev.	Var.
1	The pinout diagram shown on the Breadboard UI was helpful	2	5	4.02	0.89	0.8
2	Connecting wires on the breadboard UI was easy	1	5	3.35	1.25	1.56
3	Testing the breadboard circuit with my code was easy	1	5	3.55	1.1	1.22
4	The breadboard interface does not need improvement	1	5	2.95	1.18	1.39
5	The breadboard interface closely emulated the real physical breadboard	1	5	3.67	1.12	1.26
6	The Breadboard UI helped me understand how to use FPGA's GPIOs	1	5	3.64	1.08	1.16
7	I was less afraid of damaging the FPGA board than when I work with circuits in the traditional lab	1	5	3.58	1.46	2.14
8	The Breadboard UI worked out without any problems	1	5	3.38	1.19	1.43

The survey results indicate that the virtual breadboard served its purpose as an alternative to the physical breadboard. Furthermore, deliberate design decisions, such as the placement of the pinout diagram, helped students understand the topics being taught in the course. Moreover, while closely emulating the physical breadboard, the virtual breadboard is not hard to use, helping students become proficient in it quickly and thus saves them time on other tasks. The survey results also show the breadboard interface has room for future improvements.

5 Future Works

We plan to extend the UW remote laboratory to support the LTC connector and the ADC. The LTC connector on the DE1-SoC board contains the Serial Peripheral Interface (SPI) master, an I^2C, and a GPIO interface. The communication is enabled by the LTC connector and the hard processor system (HPS) system [7]. The real code will be passed to the ARM using the LTC connector. We are working on providing better support of ARM, for example, running bare-metal code.

We plan to support Analog-to-Digital Converter (ADC) and Digital-to-Analog Converter (DAC) on the DE1-SoC in the near future. There is two proposed usage of the ADC/DAC. Firstly, we can connect a single sensor to the FPGA, which captures the analog signal. Then we initialize the ADC to read some digital values based on the reading of the sensor. Secondly, we can connect ADC/DAC to another device, for example, the Raspberry Pi.

Combining ADC and LTC connectors, we are opening the doors for students to implement more exciting projects involving sensor readings. Additionally, we are analyzing options to improve the GPIO connectivity in the breadboard by providing virtual components, such as logic gates, to improve the power of the current solution.

6 Conclusion

During Autumn 2020, we taught Design of Digital Circuits and Systems at the University of Washington using the remote laboratory facility.[2] The remote laboratory facility is relying on international collaboration with other universities, including the ones described in [8]. We introduced a novel approach during this offering by using a virtual breadboard, which allows students to interact bi-directionally with the FPGA using LEDs and switches. Traditionally, FPGA remote laboratories only support using pre-wired switches and buttons, and LEDs seen on the camera. However, it does support wiring the GPIO freely. This contribution represents a first step of providing a hybrid simulated and real solution for teaching labs using GPIO in class, providing a certain degree of freedom to students compared to existing solutions. We have evaluated this approach by conducting an anonymous survey for the 60 students enrolled in the course who went through this pilot project. The results show that we were able to successfully transform a lab experiment that has been traditionally conducted in-person to an online modality. Our approach has room for improvement which our team plans to undertake in future work.

[2] https://weblab.ece.uw.edu

References

1. Brinson JR (2015) Learning outcome achievement in non-traditional (virtual and remote) versus traditional (hands-on) laboratories: a review of the empirical research. Comput Edu 87:218–237. https://doi.org/10.1016/j.compedu.2015.07.003
2. Jong TD et al (2013) Physical and virtual laboratories in science and engineering education. Science 340(6130):305–308. https://doi.org/10.1126/science.1230579
3. Feisel LD, Peterson GD, Arnas O, Carter L, Rosa A, Worek W (2002) Learning objectives for engineering education laboratories. In: 32nd annual frontiers in education, Boston, MA, USA, p F1D. https://doi.org/10.1109/FIE.2002.1158127
4. Gustavsson I, Zackrisson J, Håkansson L, Claesson I, Lagö TL (2007) The visir project–an open source software initiative for distributed online laboratories. In: REV conference 2007
5. Orduña P, Rodriguez-Gil L, Garcia-Zubia J, Angulo I, Hernandez U, Azcuenaga E (2018) Increasing the value of remote laboratory federations through an open sharing platform: LabsLand. In: Online engineering & internet of things. Springer, Cham, pp 859–873. https://doi.org/10.1007/978-3-319-64352-6_80.
6. Sullivan GM, Artino AR Jr (2013) Analyzing and interpreting data from Likert-type scales. J Grad Med Educ 5(4):541–542. https://doi.org/10.4300/JGME-5-4-18
7. DE1-Soc User Manual, 6th edn (2020)
8. Mayoz CA, da Silva Beraldo AL, Villar-Martinez A, Rodriguez-Gil L, de Souza Seron WFM, Orduña P (2020) FPGA remote laboratory: experience of a shared laboratory between UPNA and UNIFESP. In: 2020 XIV technologies applied to electronics teaching conference (TAEE), Porto, Portugal, pp 1–8. https://doi.org/10.1109/TAEE46915.2020.9163773
9. Morgan F et al (2011) Remote FPGA Lab with Interactive Control and Visualisation Interface. In: 2011 21st international conference on field programmable logic and applications, Chania, 2011, pp 496–499. https://doi.org/10.1109/FPL.2011.98
10. Mosquera F, Pottathuparambil R (2018) RESL: a web browser based remote embedded system laboratory. In: Proceedings of E-learn: world conference on E-learning in corporate, government, healthcare, and higher education 2018. Association for the Advancement of Computing in Education (AACE), San Diego, pp 1140–1150
11. Garcia-Zubia J, delpina DL, Orduna P (2006) Accessing WebLabs from cellular phones. In: IECON 2006 - 32nd annual conference on IEEE industrial electronics, Paris, pp 3779–3781. https://doi.org/10.1109/IECON.2006.347819
12. Andujar JM, Mejias A, Marquez MA (2011) Augmented reality for the improvement of remote laboratories: an augmented remote laboratory. IEEE Trans Educ 54(3):492–500. https://doi.org/10.1109/TE.2010.2085047

Non-traditional labs in the IoT and Industry 4.0 Domain

Smart Innovative Engineering for Smart Agriculture Modernization

Galyna Tabunshchyk[1](✉), Peter Arras[2], Karsten Henke[3], and Heinz-Dietrich Wuttke[3]

[1] National University "Zaporizhzhia Polytechnic", Zaporizhzhia, Ukraine
[2] KU Leuven, Campus De Nayer, Sint-Katelijne-Waver, Belgium
peter.arras@kuleuven.be
[3] Ilmenau University of Technology, Max-Planck-Ring 14, 98693 Ilmenau, Germany
{karsten.henke,dieter.wuttke}@tu-ilmenau.de

Abstract. Smart engineering, as it is taught in many engineering disciplines at universities becomes more and more important to the agricultural sector as well. In the paper there are considered such approaches as remote engineering, digital twins and online teaching tools, which could be effectively implemented for the tasks of Smart Agriculture. The authors discuss three examples of IoT-A related solutions in remote engineering to enrich the education in the agricultural field.

Keywords: Remote Engineering · IoT in Agriculture · GOLDi · Digital twins for electrical grid

1 Introduction

Population growth and increasing quality demand for agricultural products (necessary for high-quality export products adapted to EU-standards) press the sector to modernize. In EU 28 [1] there is a yearly "waste" of 20% of food production mostly in consumption. In medium- and high-income countries, food is wasted and lost mainly at later stages in the supply chain. Elsewhere in the world, the problem is harvesting numbers, due to bad monitoring, diseases, processes, with 60% of production only entering the food chain. Food waste and loss occur mainly at early stages of the food value chain and can be traced back to technical, financial and managerial constraints in harvesting techniques and secondary due to storage and cooling facilities.

In western and industrialized countries, there is a lot of automated agriculture and technological aids in the agricultural machinery. Many things happening there in on the measuring/sensoring/monitoring of soils, crops, harvests and the integration of these data in the farm-management system.

In less and not-industrialized farming countries, automation and exploitation of the data is coming both in a single revolution in farming in the near future.

M. E. Auer et al. (eds.), *Online Engineering and Society 4.0*, Lecture Notes in Networks and Systems 298, https://doi.org/10.1007/978-3-030-82529-4_16

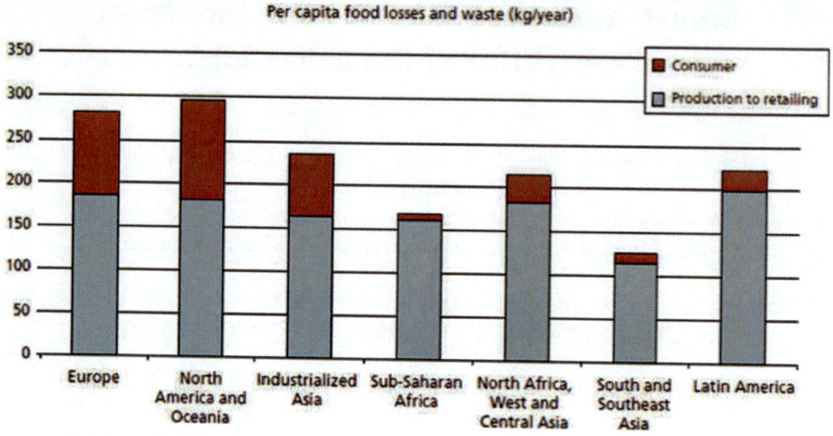

Fig. 1 Per capita food losses and waste, at consumption and pre-consumption stages, in different regions of the world [2]

Table 1 Estimated waste percentages for each commodity group in each step of the FSC for Europe incl. Russia

	Agricultural production	Postharvest handling and storage	Processing and packaging	Distribution: Supermarket Retail	Consumption
Cereals	2%	4%	0.5%, 10%	2%	25%
Roots and tubers	20%	9%	15%	7%	17%
Oilseeds and pulses	10%	1%	5%	1%	4%
Fruits and vegetables	20%	5%	2%	10%	19%
Meat	3.1%	0.7%	5%	4%	11%
Fish and seafood	9.4%	0.5%	6%	9%	11%
Milk	3.5%	0.5%	1.2%	0.5%	7%

Table 2. Estimated waste percentages for each commodity group in each step of the FSC for sub-Saharan Africa

	Agricultural Production	Postharvest handling and storage	Processing and packaging	Distribution	Consumption
Cereals	6%	8%	3.5%	2%	1%
Roots and tubers	14%	18%	15%	5%	2%
Oilseeds an pulses	12%	8%	8%	2%	1%
Fruits and vegetables	10%	9%	25%	17%	5%
Meat	15%	0.7%	5%	7%	2%
Fish and seafood	5.7%	6%	9%	15%	2%
Milk	6%	11%	0.1%	10%	0.1%

From the numbers in (Table 1 and 2) it is clear that there is a considerable loss during production, during handling and packaging in lesser developed regions. With the use of better techniques and monitoring, the numbers can be improved – and as such the losses can be cut.

A second big challcngc is to makc farming more sustainable, where the use of phosphates and nitrates need to be lowered and where lower emissions of GHG (greenhouse gasses) is high on the agenda. In the European Union, agriculture is responsible for 9% of the GHG-emissions.

It is forecasted that the use of Internet of Thing in Agriculture (IoT-A) could possibly increase the food production by 20%, so that agriculture would be feeding 10 billion of people globally [3]. IoT-A is helping to make smarter decisions, reduce costs and boost production. By increasing both quantity and quality of agricultural products, IoT-A is able to improve the intelligence of farms. Collection of massive amounts of information from soil mapping, crop fields, weather data, agricultural vehicles and usage of specialized sensors are among the many ways smart farming helps agriculture. In terms of environment issues, IoT-A-based smart farming can provide great benefits including more efficient water usage or optimization of the waste and a reduction of GHG-emissions.

It is the educational task of the universities to prepare their graduates to embrace and use these new technologies. For this an updated and workover of curricula is imposing itself to implement these IoT-A-technologies.

2 Examples of IoT-A Related Education

The number of IoT-A units is expected to increase through the years, from 30 million devices in 2015, to 75 million by 2020. From 2015 giants as Yahoo! Inc., Google, Alphabet Inc. ... invest in food- and farming-related start-ups [3] All these require highly prepared graduates in multidisciplinary topics and in IoT for Agriculture (Fig. 1 and 2) .

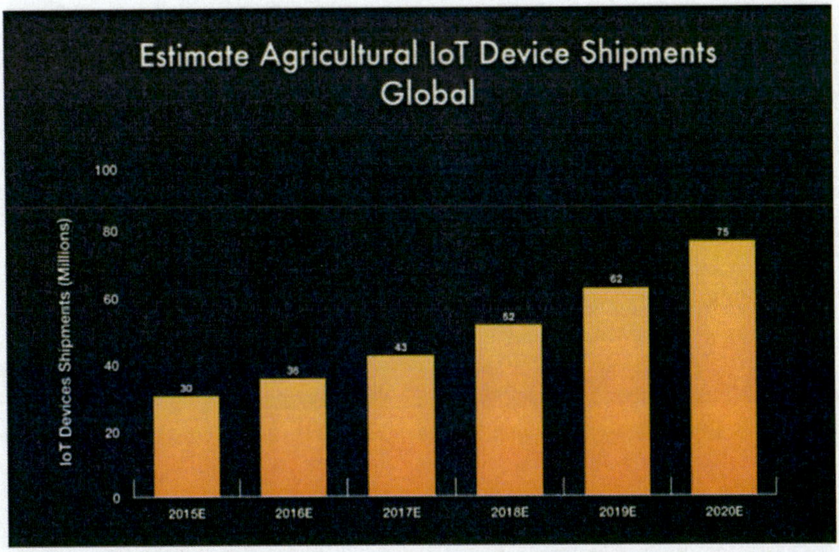

Fig. 2 Estimated growth of IoT devices in agriculture [4]

However, the gap between knowledge required in the field and skills of the agricultural specialists currently available at the labor market is increasing dramatically.

FAO-studies [3] clearly show the problems of the food-producing sector and pinpoint to two solutions: technological innovation and stimulating entrepreneurship in the rural regions.

Agriculture is a traditional sector, which suffers from an image of being old-fashioned and outdated.

This reflects on the traditional agrarian universities who have problems in attracting the necessary students. The traditional agrarian universities have a lack of knowledge and technical infrastructure to teach the use and implementation of IoT-A.

On the other hand, technical universities attract many students in study fields of software development and automation, which could serve the agricultural sector to go to the necessary innovations for the increasing demand in productivity to be able to produce food for the growing populations.

The authors describe an educational approach, which will give engineering graduate students in agriculture the necessary digital and technical skills to cope with the problems mentioned above. All (specialized) modules of courses are designed to serve students but also to be used as training materials for professionals from the agricultural sector so that they can update and upgrade their knowledge on new technologies.

Including modules on entrepreneurship in combination with the technical solutions offered is another innovative key factor in the approach for bridging the existing gap. Entrepreneurship was also mentioned as a solution for the food waste problem, mostly in the supply chain where elimination of intermediate level of wholesalers cause delays and storing cost in the throughput of agricultural products. Cutting intermediate levels will bring down the waste/loss of production.

A consortium of universities elaborated a strategy, which contains a 3 level educational approach: traditional course materials for classroom for use with Ba/MSc students; E-learning courses for students and the professionals (to integrate in SPOCS); Modular structured courses for upgrading/updating the knowledge of professionals (from industry).

Key modules, which aim to give students digital competences, include fields like green energy optimization for rural areas; modelling, simulation and control for agricultural and biological systems; smart design CAD-methods for agricultural machine design; hardware and software design for IoT and remote engineering, sensing and measuring systems for pets' detection and others.

The authors suggest using a remote experiment infrastructure based on remote laboratory GOLDi, ISRT system and CALM module.

The setup is explained with some examples of IoT-A related solutions.

Case 1 is about the setup of a set of interactive labs to train students in different aspects of IoT and embedded systems. The problem to solve is about the electrification and charging problems of electrified transport/robots.

Case 2 is about a platform of remote labs on the GOLDi platform, which students/professionals can use to train their skills in control logics and strategies.

Case 3 is about an E-learning setup – CALM – as an alternative for classroom teaching, integrating remote and virtual labs.

2.1 Case 1: Electrification of Agricultural Machinery and the Charging Problem

A growing population and the impact of climate change represent clear challenges for the agricultural sector.

The proliferation of electric motors and actuators in applications ranging from industrial processes and cars is indicative of the drive away from mechanical traction and actuation systems to electrically based systems. Large diesel vehicles are likely to remain in practical use for many years to come, however the optimum use of such vehicles has been to go as large as possible, which in itself leads to issues such as soil compaction and "brute force" delivery of fertilizers, herbicides and pesticides. The migration from monolithic, fossil-fuel-based agricultural platforms to fleets of smaller electric powered robotic platforms offers the possibility of much lower emissions with locally generated power. Examples of robotic platforms in use are drones (for monitoring, for spraying herbicides) and robots for automatic feeding, milking, cleaning stables in dairy production.

In recent years an increase in the use of agricultural land for solar photovoltaic, wind turbines and anaerobic digestion plants has been seen in the EU [5]. Therefore, the potential for dual use of not only the land, but also the electricity generated, is of interest to the agricultural robotics community. Many agricultural implements are driven directly from the prime mover (often a tractor) via a mechanical linkage. By using electric drives, the efficiency can be much higher and the whole system made safer as a result. One of the most common sources of injury and death on the farm is the mechanical linkages in large farm machinery. Therefore, there are potentially major health and safety benefits to electrification and automation of farm equipment.

One advantage of modern robotics is their ability to be built using low-cost, lightweight and smart components [3]. Using collaborative and cooperative behavior in a fleet of robots further provides the opportunity to spread tasks over multiple platforms and thereby reduce the damage caused by heavy conventional agricultural platforms on the soil or existing crops. The robotic fleet can also take advantage of multiple data sources to calibrate the task, reduce waste and focus on areas of greatest need, potentially reducing fertilizer costs and environmental impact [6].

Electric vehicles/robot charging systems are needed in order to correctly manage upcoming smart agriculture. An electric vehicle charging system is a complex system, not only from the technology side but even more from the connectivity part (towards the grid for the energy source and for the payment of energy and services) [7] (Fig. 3).

From the energy input side such charging station should have a choice from net-powered, wind-powered and solar energy. To optimize the commercial side of the charging system it is proposed to extend it with an energy management system which – on the basis of cost optimization – will autonomous decide what energy source will be used for charging [8, 9].

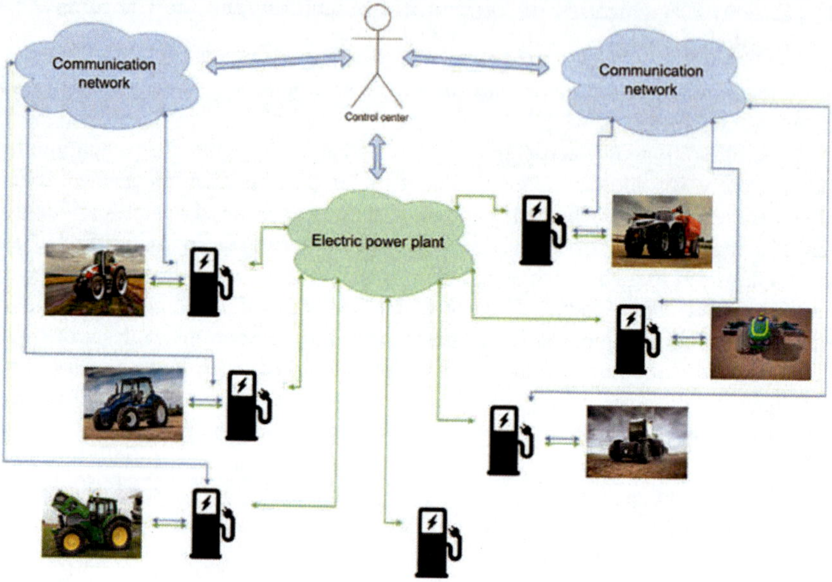

Fig. 3 Cyber-physical nature of an EV charging system in the agricultural sector

The setup is integrated in the ISRT (Interactive platform for Embedded Software Development Study) [10]. The ISRT-server is a platform for remote laboratories, in order to train students in IoT-tasks. It is ideal for the scalability of the charging station case. A ready-made black-box charging station module is available in the ISRT so that students can work on the communication and client layer. Within this work there was developed an extension for an energy management system, which could improve student's knowledge in the development of the decision making systems, solving optimization tasks and used as simulator for developing digital skills in the managing smart grid.

2.2 Case 2. Remote Engineering for the IoT System Design

The GOLDi laboratory flexibly realizes variety of combinations of virtual and remote experiments (Fig. 4). Thus it is possible to carry out all experiments either completely virtually or on real devices or in a combination of both. In the experiments, the students have the task of designing a control algorithm (CA) for a certain control object (CO) running on certain control-units (CU) [11, 12].

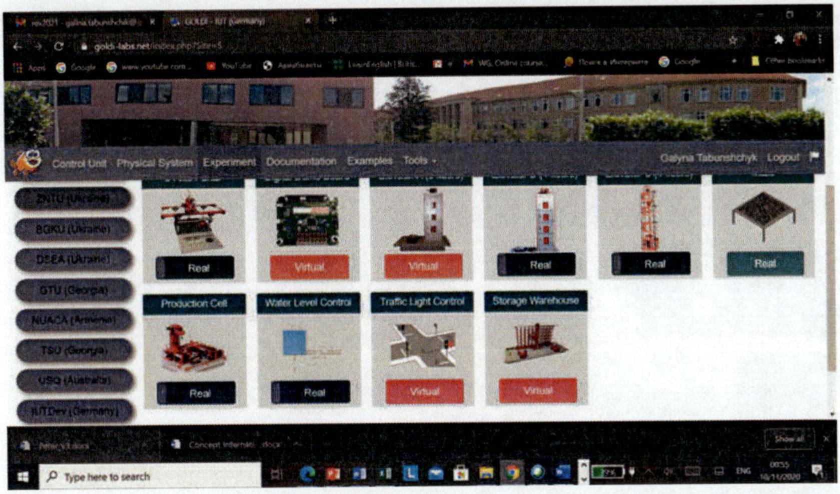

Fig. 4 Selection of COs and CUs

Extensions to the remote lab GOLDi in IoT applications for Smart Agriculture provide an e-learning environment with SPOCs (Small Private Online courses) [13] in connection with the Moodle learning management system (LMS).

Modelling, simulation and control for agricultural and biological systems are possible subjects for experiments with the extended GOLDi-lab. That way, automation and remote engineering for agricultural hardware and software solutions in the IoT-A domain (e.g. digital electronics, digital system design, and software design for microcontrollers) can be taught in a more practical way. A cluster of SPOCs enables a set of remote and virtual experiments.

2.3 Case 3: CALM (Computer Aided Learning Module)

We put forward a model of blended learning to appeal to the interests of present-day students and to stimulate their self-study. As an example we have an hypertext supported learning environment - CALM - which explains the theoretical basis of material sciences, provides students will all documents concerning laboratory sessions and assignments, shows examples of phenomena, and links to literature to on the subject (Fig. 5).

The theory itself is supported with real physical lab sessions, virtual laboratories to let students experiment on the phenomena in an extensive way, and remote laboratories used by students to check the theory with real data and to give the feeling of connection to reality.

The hypertext-supported material is used during the classroom teaching lessons on material sciences and afterwards by students for self-study. It contains all materials, slides etcetera used in the lessons. As such, the eLearning part enables students to study the course in one clearly defined manner: the look and feel of classroom teaching is the same as the look and feel of the self-study part. This increases efficiency of the self-learning.

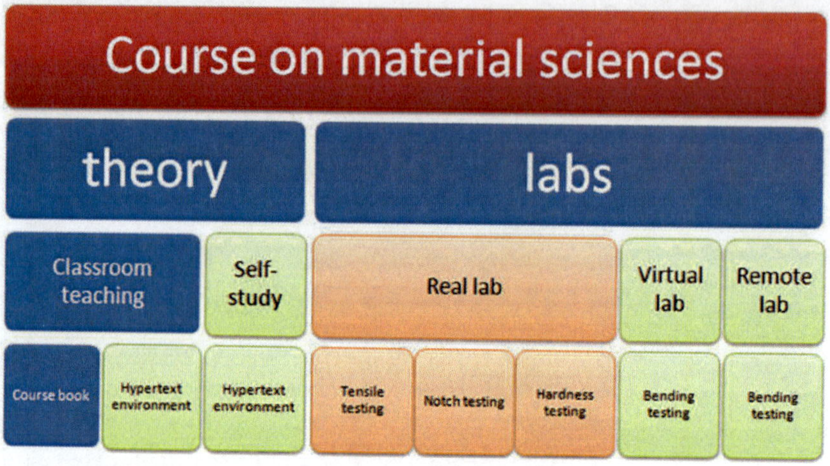

Fig. 5 Structure of the course on material sciences integrated in the CALM

3 Conclusions/Recommendations/Summary

HEIs in the agrarian sector need an upgrade to incorporate the new and digitalized technologies (IoT, automation) in their curricula to make them more attractive and to innovate the sector to be able to cope with the increasing demand of productivity.

A close collaboration between technical and agricultural HEI's is embedded and input/co-creation of content with and from the agricultural sector ensure the relevance of the setup.

The paper provides innovative ideas in teaching future agricultural experts.

Smart and electronic technologies will improve the early stages of the food value chain and thus cutting the losses. On the field monitoring (e.g. with drones to control growth and diseases) for immediate (and local) solutions, monitoring crops and harvesting, improve traceability and food safety with IoT-A solutions are some the aspects.

The implementation of remote laboratories and digital twins allow to improve digital skills of the agricultural experts.

In the paper, authors describe a digital infrastructure and suggest educational scenarios.

References

1. Estimates of European food waste levels, Stockholm, 31 March 2016. 80p. ISBN 978-91-88319-01-2. https://www.researchgate.net/publication/301216380_Estimates_of_European_food_waste_levels
2. Gustavsson J, Cederberg C, Sonesson U (2011) Global food losses and food waste, International congress "Save FOOD!", Interpack2011, Düsseldorf, Germany. ISBN 978-92-5-107205-9

3. Global Perspectives Studies. http://www.fao.org/global-perspectives-studies/en/
4. Kuprenko V (2019) IoT in Agriculture. Why it is a future of connected farming world. https://theiotmagazine.com/iot-in-agriculture-why-it-is-a-future-of-connected-farming-world-70b64936627c
5. Energy balance sheets. https://ec.europa.eu/eurostat/documents/3217494/10077623/KS-EN-19-001-EN-N.pdf/59b44e6f-ff33-488b-a85f-9c4f60703afc?t=1566379390000
6. EU Robotics AISBL (2014) Robotics 2020 multi-annual roadmap for robotics in Europe, call 1 ict23 – horizon 2020. https://www.eu-robotics.net/cms/upload/downloads/ppp-documents/Multi-Annual_Roadmap2020_ICT24_Rev_B_full.pdf
7. Granda F et al (2017) Spatial characterization of radio propagation channel in urban vehicle-to-infrastructure environments to support WSNs deployment. Sensors 17:1313. https://doi.org/10.3390/s17061313
8. Dey KC, Rayamajhi A, Chowdhury M, Bhavsar P, Martin J (2016) Vehicle-to-vehicle (V2V) and vehicle-to-infrastructure (V2I) communication in a heterogeneous wireless network – performance evaluation. Elsevier. https://doi.org/10.1016/J.TRC.2016.03.008
9. Tuttle DP, Fares RL, Baldick R, Webber ME (2013) Plug-In Vehicle to Home (V2H) duration and power output capability. In: 2013 IEEE transportation electrification conference and expo (ITEC), Detroit, MI, pp 1–7. https://doi.org/10.1109/ITEC.2013.6574527
10. Tabunshchyk G, Van Merode D, Arras P, Henke K, Okhmak V (2018) Interactive platform for Embedded Software Development Study. In: Auer M, Zhutin D (eds) Online engineering and internet of things. Springer International Publishing, pp 315–321. https://doi.org/10.1009/978-3-319-64352-6_30
11. Wuttke HD, Henke K, Hutschenreuter R (2020) Digital twins in remote labs. In: Auer M, Ram BK (eds) Cyber-physical Systems and Digital Twins. REV2019 2019. Lecture Notes in Networks and Systems, vol 80. Springer, Cham. https://doi.org/10.1007/978-3-030-23162-0_26
12. Henke K, Wuttke HD, Nau J (2020) EIFEL—a new approach for digital education. In: Uskov V, Howlett R, Jain L (eds) Smart Education and e-Learning 2020. Smart Innovation, Systems and Technologies, vol 188. Springer, Singapore. https://doi.org/10.1007/978-981-15-5584-8_2
13. Fox A (2013) From MOOCs to SPOCs. Proc ACM 56(12):38–40

Internet of Postal Things - Current Issues of Scientific Research in the University Environment

Maria Rostasova(✉), Tatiana Corejova, and Martin Kincl

University of Zilina, Univerzitna 8215/1, 010 01 Zilina, Slovakia
Maria.Rostasova@fpedas.uniza.sk

Abstract. The issue of the Internet of Postal Things has come to the forefront of scientific research at the university, where one of the study programs of bachelor's and engineering studies is focused on the training of managers and experts in the field of postal technologies and services. The area of the use of the Internet of Postal Things has not yet been thoroughly examined on a global scale, although in professional circles we already find indications of concrete results of the use of the Internet of Things in the conditions of the postal sector. For this reason, in this paper we present an approach to understanding this issue and finding the correct interpretation of the term "Internet of Postal Things". As an educational institution, we also needed to map the current perception and understanding of this issue in the segment of our students, but also those who do not complete university studies, but use postal services and are interested in encountering information and communication technologies, such as in the form of the Internet of Postal Things.

Keywords: Post operation · Internet of Postal Things · Perception of the Internet of Postal Things by Customers

1 Context

The issue of the Internet of Postal Things has come to the forefront of scientific research at the university, where one of the study programs of bachelor's and engineering studies is focused on the training of managers and experts in the field of postal technologies and services. This area has not yet been thoroughly researched on a global scale, although in professional circles we already find indications of concrete results of the use of the Internet of Things in the conditions of the postal sector. It is already unequivocally confirmed in professional circles that the Internet of Postal Things represents a near future. At present, the postal market around the world is characterized by strong pressures to increase the efficiency of postal processes and services, to reduce the share of human labor, to the prosperity of postal companies, to improve the services of postal clients and to increase the security of postal employees, who are the main actors in the so-called "last mile", that is, the delivery of consignments to the addressees.

M. E. Auer et al. (eds.), *Online Engineering and Society 4.0*, Lecture Notes in Networks and Systems 298, https://doi.org/10.1007/978-3-030-82529-4_17

The Internet of Postal Things is understood as a postal infrastructure equipped with low-cost sensors that allow data to be collected and communicated based on the Internet of Things. Postal services as such work with large amounts of data, and this technology helps them in areas such as efficiency, customer service, and helps develop new services and business models, enabling process optimization and automation. It provides effective solutions in the interconnection of the postal network and several opportunities for new, smarter applications, especially in the field of transport and logistics, but also in the management of buildings and state services. IoPT responds to the increased need of customers to provide modern postal products and services. At the same time, the postal company, through partnerships with suppliers, local governments, and customers, is experimenting with new business concepts, such as "connected mailbox" or others. The Internet of Postal Things talks about creating an open platform for data that is stored, managed, shared and analyzed. A strong privacy policy is also in place. However, the postal company must invest not only in technology, but also in security, privacy and the problems associated with the organization of relocation. IoPT collects, organizes, and efficiently uses large amounts of data that have already been pre-processed by existing information systems. The postal physical infrastructure and the digital world of information are thus combined into one shared network. The interconnection of the IoPT ecosystem with other infrastructures (houses, cities) expands the possibilities of using IoPT. However, implementing IoPT is not easy and will not happen overnight. It must be implemented gradually, in small steps, according to a precise plan and well-chosen priorities (e.g., first the operational efficiency of postal activities and only then the introduction of new business models). These aspects are key and lead to the successful implementation of IoPT in the practice of postal companies. Postal companies should not overlook this innovative technology and should take it as an opportunity to strengthen and expand their business in the market [1, 2].

The Internet of Postal Things could also have a positive side effect on other sectors that do not belong to the postal sector. Several postal companies already use different types of sensors to solve operational problems, such as predictive fleet maintenance. Innovative solutions are already working experimentally in areas other than postal operations, such as the use of handheld terminals (small-scale mobile electronic devices), data sources provided to customers or the use of sensors on postal courses (cars) to collect environmental data from which they could benefit. For example, local governments [3].

IoPT technology could be based on three functions of the postal service and the characteristics of the postal network:

- **Extensive infrastructure.** The providers of postal services require an extensive infrastructure of stationary and mobile facilities, which cover the areas in question and create a seemingly endless number of connections daily. Many of these objects do not currently communicate or collect data with each other. The density and reach of this network provide IoPT with almost unlimited potential for its application. This could create an information network that complements the physical network. The traditional infrastructure provides an opportunity to replace it with new low-cost sensors. Postal companies plan to continue modernizing existing buildings to make their infrastructure more sustainable.

- **Experience with data collection and analysis.** Postal companies have demonstrated experience in using large data sets, creating data-based processes and products, and experimenting with sensors. Postal processing centers manage a large amount of generated data. As they sort the millions of mails that go through the mail system, they collect large amounts of data during each day. Postal companies accept the opportunity to do business in the field of information provision and have therefore begun to build Internet-based information technology architectures. They use these to provide analyzes that streamline the managerial decision-making process and increase added value for the customer. In the provision of postal services, some IoPT applications are already used abroad to meet the strategic goals of raising the profile of postal businesses and optimizing the processes that are used (for example, using RFID chips to measure the quality standards of postal services in international mail).
- **Customer information requirements.** Postal customers are increasingly demanding more and better data on the relocation process of their items, which can ultimately be of great help to IoPT. For example, customers who have requested parcel services expect information about tracking their parcel shipment, the expected delivery date, more options in terms of the place of delivery and a simplified return procedure in case of a complaint. Same day delivery as the shipment was shipped and dynamic real-time redirection have the potential to increase as customers shop intensively online. Postal companies are already responding to this trend and meeting customer requirements. The use of sensors on the vehicle, mailbox, portable handheld device or the mail item itself enables these customer requirements to be met through effective monitoring of the relocation process [3, 4].

The maximum challenges in the use of IoPT are sent all over the world mainly by the national operator in the USA. U.S. Postal Service (hereinafter USPS) is already the creator of many innovative solutions. Through IoPT, it inserts sensors and other data collection technologies into physical objects and infrastructures. The postal network infrastructure allows the use of a large amount of data to which the USPS has access. He has developed several IoPT applications that have a high potential for use in postal operations.

The first area of use of the IoPT are **transport and logistics applications for monitoring the system of condition and performance of vehicles throughout the value chain**. The goal of these applications is, for example, to reduce fuel costs, to reduce the need for manual human intervention, such as in maintaining and optimizing the way personnel, systems, and other assets work together. The most important applications in this area are fuel consumption management, load optimization of vehicles providing postal courses, predictive vehicle maintenance and efficient and dynamic real-time vehicle route planning [5, 6].

The second area of use of IoPT are **intelligent postal buildings**, in which systems are implemented, the main task of which will be the effective management of energy consumption, security and protection of buildings and systems to reduce the cost of building maintenance.

The third area is represented by **innovative and better-quality letter and parcel services**, which include applications to ensure the reliability and security of the provided postal services. Sensors generating the necessary data, which increase the added value for senders and recipients of postal items, enable the creation of new postal services (for example, the application of the so-called smart mailbox).

The last area of use of IoPT is represented by **e-government services**, respectively **services of local self-government** towards the population, focusing on applications that use IoT technologies to take advantage of the presence of postal vehicles and carriers in the same territory at the same time [7].

2 Purpose

Such focused scientific research was motivated by the great interest in finding answers to research questions, what is the current rate of acceptance of the use of the Internet of Postal Things by postal operators and, of course, what is the rate of acceptance of the use of the Internet of Postal Things by postal customers. Although at first glance it may seem that every progress and use of the most modern information and communication technologies is received with enthusiasm in specific companies, the view of experts at the university was not clear. There have been regular discussions about whether the problems and threats of using the Internet of Things in general will not outweigh the positive aspects and opportunities that this modern technology brings in practice. The task of the scientific research was to state and subsequently verify hypotheses, the confirmation or non-confirmation of which will determine the degree of acceptance of the use of the Internet of Postal Things in a selected customer segment. There have also been regular discussions on university grounds about whether postal operators are literally "flocking" to the use of the Internet of Postal Things, whether they want to innovate processes in the postal operation in this direction.

As far as the Slovak Republic is concerned, it can be stated that the Internet of Things is significantly in the center of attention, especially at the university, in educational and research institutions, among researchers who follow the latest global trends. As for the other companies, they are showing this phenomenon, there is not much interest. However, over time, interest in this technology is expected to expand among other companies that have decided to apply and invest in IoT technology. There is a growing awareness in professional circles that the time is right to start experimenting with this technology in such a way as to achieve greater efficiency and competitiveness throughout society. The bibliometric analysis of this concept of the Web of Science database testifies to the extent of interest in publishing professional articles on the Internet of Things published worldwide [8]. Next Figs. 1 and 2 shows a map of all the keywords and 25 keywords in the documents that include the term Internet of Things in their title [14].

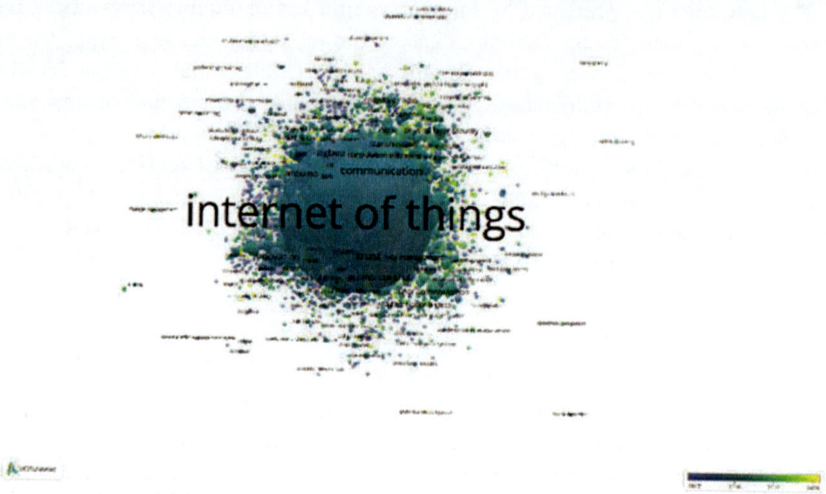

Fig. 1 Map of all the keywords in the documents from WoS that include the term Internet of Things in their title [14]

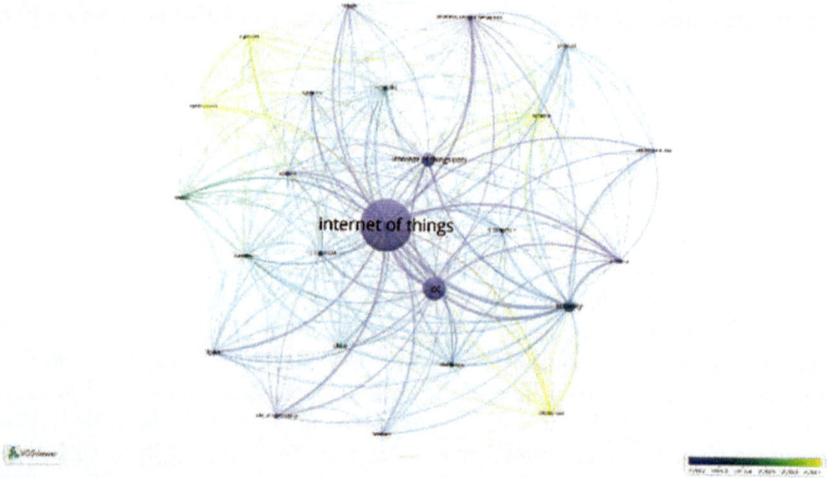

Fig. 2 Map of all the keywords in the documents from WoS that include the term Internet of Things in their title [14]

The growth of IoT-related technologies and software solutions in 2019 was faster than the expected growth predicted from 2018. The estimated number of active devices in 2019 using IoT technology was approximately 9.5 billion. This number of connected IoT devices does not include laptops, smartphones, tablets or similar devices. By 2025, the total number of connected IoT devices is expected to reach 28 billion, with new low-energy LPWAN and 5G broadband connections [9]. The following table (Table 1) shows the most important events, advances and technologies related to the Internet of Things in 2019.

Table 1 A selection of Internet of Things technologies in 2019 according to a survey by Gartner's industry [9, 10]

Technology	Description
IoT security	• there is no single common protocol that connects the world of the Internet of Things together; this opens the way for hackers who can gain access to sensitive user data
IoT analytics	• continuous data flows from connected devices can be archived, sorted and analysed using BigData and used for meaningful decision making • analytical data will help predict certain problems, understand customer behaviour, respond to negative product experiences
IoT device management	• the need to monitor and manage IoT devices over long distances • with the increasing number of connected devices and their placement over long distances, it will be necessary to use systems for their continuous monitoring
Short-range IoT networks with low power consumption	• a large proportion of IoT installations will consist of low-energy, short-range networks
Long Range Low Energy Networks (LPWAN)	• they will use real-life applications in densely populated localities, such as cities, construction sites, energy networks, etc. • help get a wide range of "things" under one roof that would not otherwise be connected via RFID, M2M or similar protocols
IoT processors	• sensors and nodes form the power of IoT processing • powerful small IoT processors can be involved in two-way communication—depending on the complexity of the processors, their level of security, power consumption, design, etc. will also change
IoT operating systems	• operating systems such as Mac OS, Android and Windows are not designed for IoT devices (mainly due to high memory requirements) • IoT devices require an independent operating system that can be easily scaled, controlled, and also updated • they need an operating system with specific languages and communication protocols for seamless data exchange • development of IoT-ready operating systems
Event processing	• traditional database systems are not enough to manage them—powerful platforms for data flow distribution are needed to streamline and process these events
IoT platforms	• these platforms provide compatibility for the entire IoT ecosystem, allowing companies to connect to enterprise systems
Internet of Things standards and ecosystems	• there are currently no clear standards under which all IoT devices are governed by a common protocol; these standards emerge when manufacturers come together to create IoT devices that can work and communicate with each other without problems • similarly, new ecosystems will be connected, such as interconnected households, interconnected production, interconnected healthcare, interconnected cities, which will make the world a huge network of interconnected things

Figure 3 presents the areas of investment in IoT in 2019 in the Slovak Republic.

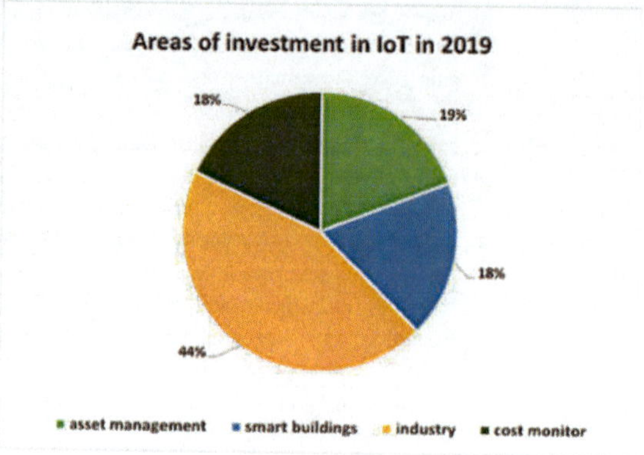

Fig. 3 Areas of Investment in IoT in 2019 [11]

Table 2 provides an overview of the Internet of Things networks in Slovakia.

Table 2 Networks of Internet of Things in Slovakia (as of 2019) [12]

Operator	Technology	Coverage
Antik	LoRaWAN	more than 25 cities throughout Slovakia, including Bratislava and Kosice
O2	NB-IoT	n
Orange	LoRaWAN	Bratislava, Zilina, Banska Bystrica, Ziar nad Hronom, Sala, Kosice, Zlate Moravce and Velky Saris
SimpleCell	Sigfox	87% of the territory and 92% of the population of Slovakia
Slovak Telekom	NB-IoT	75% of the territory and 88.5% of the population of Slovakia
Slovanet	LoRaWAN	57 localities, including all regional cities and more than half of district cities

Scientists and other IoT experts have clearly identified that IoT is in the maturity phase and will become a full-fledged business tool in the coming years. IoT is evolving at a rapid pace, and experts and lay people alike cannot underestimate its development in the years to come. Investors may experience some market fluctuations, but the market for IoT stocks is set to grow. Among the companies worth investing in in the coming period, they have the largest revenue streams from the IoT space, some of which are more involved in hardware, while others are exclusively involved in the creation of IoT software platforms and services. These include, according to Robots. net: Microsoft, IBM, Amazon, Apple, Alphabet, Intel. ATM, AT&T and Verizon, Sirra Wireless, Marvell Technology [13, 15, 16].

3 Approach

The purpose of this research was to clearly outline the change or addition to the content of education of future experts in the management of postal traffic at our university. The fulfillment of the objectives of the research activities was carrying out secondary and primary research, with the aim of verifying the established research assumptions. These research assumptions resulted from the setting of relevant research objectives. As the discussion of experts at the university campus also concerned the justification and correctness of the use of the term "Internet of Postal Things", one of the results of the research was a linguistic analysis of the term in question. Subsequently, a primary research of a qualitative nature was carried out, aimed at determining the level of use of the Internet of Things in the environment of the national postal operator in the Slovak Republic, which is Slovenska posta (Slovak post). Quantitative primary research was focused on the acceptance of the Internet of Postal Things in a selected segment of customers (Generation Y), who we assumed were strongly inclined to use new technologies and support any innovation that simplifies and improves their own lives and work. These research activities were preceded by the compilation of a framework (brief) SWOT analysis of the use of the Internet of Things, where the knowledge base is already shifted further, as is the case with the use of the Internet of Postal Things. An association analysis of the obtained data related to the research assumptions in question was used to express the final statements (created research assumptions).

To fulfill one of the goals, which was to reveal the interest in solving the problem of the Internet of Postal Things (hereinafter IoPT), a bibliometric analysis was processed in the Web of Science database (Fig. 4).

Fig. 4 Bibliometric analysis of term Internet of Postal Things in WoS database [14]

Another goal—**to determine the degree of acceptance of the use of IoPT** in a selected segment of customers was fulfilled through the implementation of primary research of a quantitative nature (Table 3 and Table 4).

Table 3 Statistics of all collected primary research responses [15]

Number of all answers	776
Age minimum Age maximum Age average Age median Age modus	16 55 23,59 22 21
Number of answers of people with permanent residence in the Slovak Republic and aged 18 to 30 years	668
Number of women Number of men	483 185
Total number of forms with neutral answers (%) Number of completed forms after deduction of neutral answers	40,42% 398

The following Table 4 contains an evaluation of the merits of the questions that were part of the processed questionnaire.

Testing of Research Assumptions

As part of the solution, an association analysis of the obtained data related to the set research assumptions was performed [15].

Research assumption 1: **The selected customer segment of the postal company (Slovak citizens from 18 to 30 years) accepts 80% of the offer to use IoPT in the postal sector** has been **confirmed**. By averaging the individual answers to the questions concerning the acceptance rate of the IoPT service, the acceptance result was 79.64%.

Research Assumption 2: **The acceptance rate of the IoPT service depends on the (current) rate of customer use of mail services** was tested using a chi-square test and correlation coefficients, confirming only a weak statistical dependence. The research assumption was **not confirmed.**

Research Assumption 3: **The fact that a respondent has ever heard of IoPT is related to a personality type in relation to his knowledge of information and communication technologies** was also tested using a chi-square test and correlation analysis. No statistical dependence was found, and the research assumption was **not confirmed** either.

Table 4 Statistics on merits of all collected primary research responses [15]

<table>
<tr><th>Evaluated aspect</th><th>Evaluation</th></tr>
<tr><td>Level of awareness of the concept of "Internet of Postal Things"</td><td>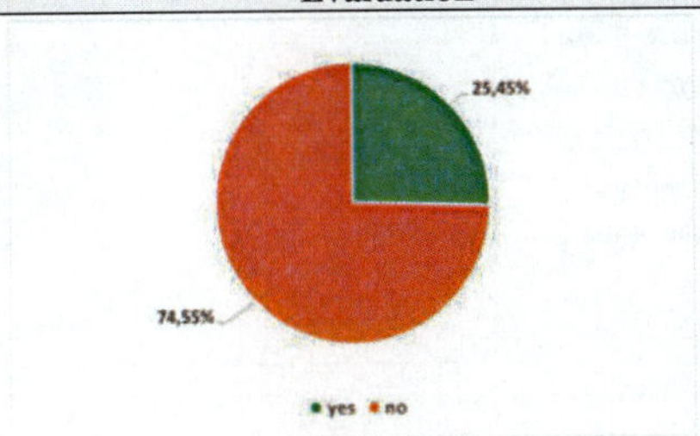
</td></tr>
<tr><td>The degree of acceptance of the use of the Internet of Postal Things in applications in transport and logistics</td><td>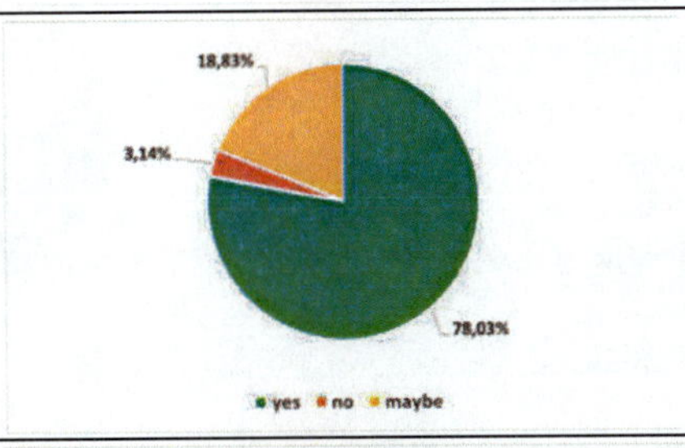
</td></tr>
<tr><td>The degree of acceptance of the use of the Internet of Postal Things in intelligent postal buildings</td><td>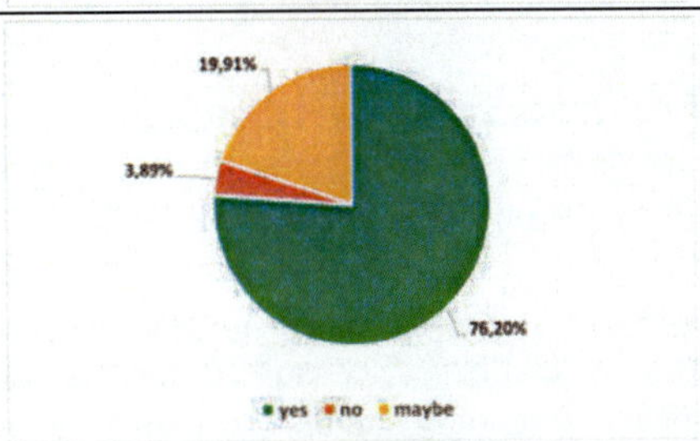
</td></tr>
<tr><td>The degree of acceptance of the use of the Internet of Postal Things in extended letter and parcel services</td><td>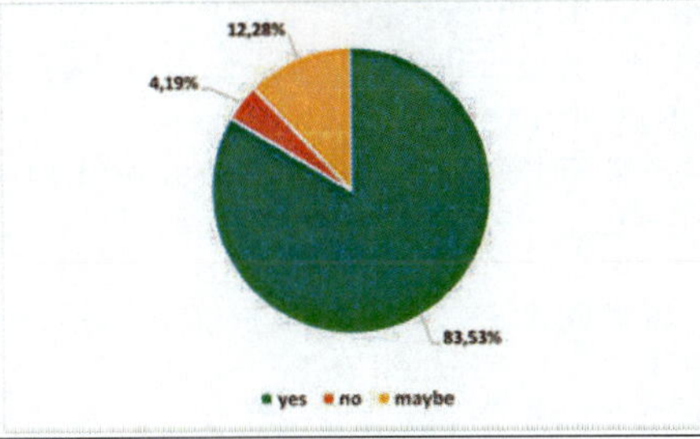
</td></tr>
</table>

The rate of acceptance of the use of the Internet of Postal Things in local services	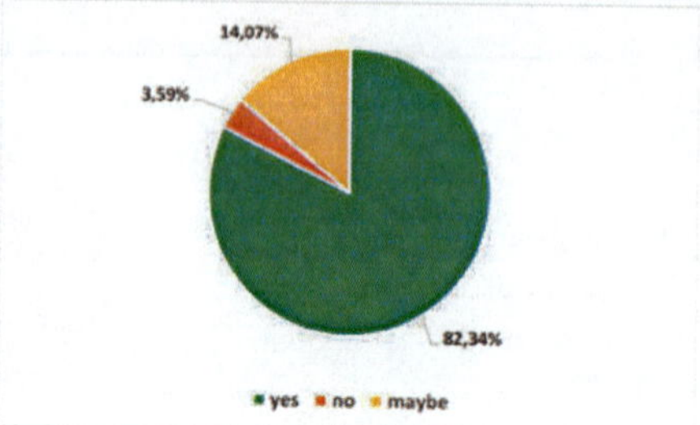
The overall rate of acceptance of the use of the Internet of Postal Things in Slovakia	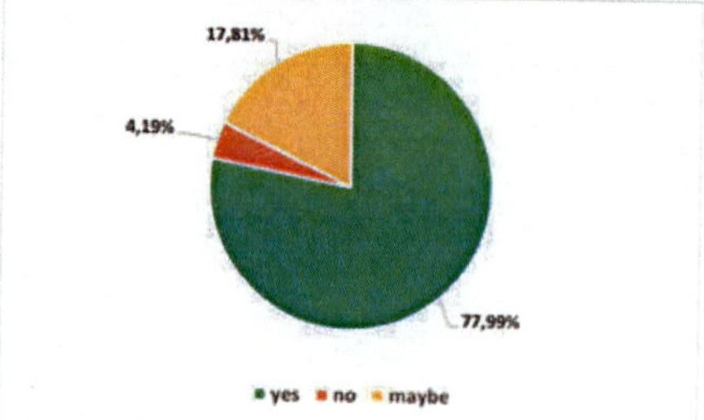

4 Actual or Anticipated Outcomes

Regarding the results of the linguistic analysis, it was possible to compile a table of the frequency of occurrence and frequency of the term across all sources entering it. Based on its evaluation, the correctness and justification of the use of the term "Internet of Postal Things" was confirmed, because in the vast majority of cases, foreign authors, resp. Practitioners commonly use the term "Internet of Postal Things" and its meaning and content are clearly defined.

The results of qualitative research (conducted by a structured interview with competent representatives of the company) are related to obtaining the idea of the Slovak national postal operator on the issue of the Internet of postal matters. However, the provider of the universal postal service in Slovakia knows about this issue only from professional journals and from the contributions of experts at the worldwide conference POSTEXPO, which has been dealing with this topic for several years in professional lectures presented at this conference. The Slovak postal operator considers the Internet of Postal Things to be an "interesting concept" that can bring benefits not only to customers but also to postal operators. With the introduction of the Internet of Postal Things, it expects to improve its position in the market of postal operators, which is characterized by a highly competitive environment. It also expects significant cost savings. From the information that Slovak post has, it appears to it that these are "cheap and easy to implement technologies". It is therefore apparently a contradictory reality, which is characterized by the fact that the Internet of Postal Things has not yet been implemented at all in this postal operator. Nor can they estimate the time frame in which they would consider introducing this technology. They see the reason for this thinking that it is necessary to wait for the use of the Internet of Postal Things in the world to prove successful.

The results of quantitative research focused on a selected customer segment of the postal company (Slovak citizens aged 18 to 30) pointed to a relatively high acceptance of the offer to use IoPT in the postal sector. Regarding the dependence of the rate of use of the new technology - the Internet of Postal Affairs - on the current rate of use of postal services by this segment of customers, it showed only a very weak statistical dependence. The research confirmed the recommendation for postal companies that they should gradually implement postal internet technologies into the offer of postal services, because the monitored segment of customers already and now and in the future expects and positively receives innovations of this type.

5 Conclusions/Recommendations/Summary

The results of research in this area have significantly highlighted the Internet of Things as a postal infrastructure equipped with low-cost sensors that allow data to be collected and communicated on the Internet of Things. The provision of postal services and the provision of postal processes is associated with the processing of large amounts of data. New technologies help achieve cost efficiency, customer service flexibility, help develop new postal services and business models, and enable the optimization and automation of processes in mail relocation. It provides an optimal layout of the postal network and more opportunities for new, smarter applications, especially in the relocation process, but also in building management and public utility services, in which the post office falls under the critical infrastructure of the state. The Internet of Postal Things is responding to the increased demands of customers for more accessible, faster, and more secure postal services. At the same time, through partnerships with suppliers, local governments and customers, the postal service is experimenting with new business concepts. The Internet of Postal Things allows you to create an open platform for data that is stored, managed, shared, and analyzed. It also helps protect personal information. However, postal companies will have to invest in technology, but also in security, privacy and the problems associated with providing postal companies. The postal physical infrastructure and the digital world of information would thus be merged into one shared network. Services. Interconnecting the postal ecosystem with other infrastructures would further expand the possibilities of using the postal Internet. Implementing the Internet of Postal Things will not be easy and will not happen overnight. It will have to be implemented in gradual steps and according to correctly chosen priorities.

Acknowledgements. The work was created in connection with the scientific research project VEGA 1/0011/21 Research on the interactions among new emerging technologies, the performance of enterprises and industries based on network technology infrastructure, the application of new business models and the institutional regulatory, environmental, and social environment.

References

1. Postal News, Postal Information, Postal Employees and more. http://www.postal-reporter.com/blog/postal-service-explores-concept-of-internet-of-postal-things-to-bring-data-collection-to-next-level. Accessed 15 Jan 2020
2. Marsh B, Piscioneri P (2015) The Internet of postal things. In: International Conference on CTS. IEEE, Arlington
3. Internet of Postal Things. https://www.uspsoig.gov/sites/default/files/document-library-files/2015/rarc-wp-15-013_0.pdf. Accessed 07 Jan 2020
4. Internet věcí. http://i2ot.eu/internet-of-things. Accessed 10 Jan 2020
5. Pojmy poštového trhu. http://www.postovesluzby.sk/postovy-trh/pojmy-postoveho-trhu. Accessed 10 Jan 2020
6. Madlenakova L, Fanova J (2018) Hodnotenie informácií súvisiacich s monitorovaním pohybu poštových zásielok. Pošta, telekomunikácie a elektronický obchod, pp 27–36
7. The Internet of Everything for Cities. http://www.cisco.com/c/dam/en_us/about/ac79/docs/ps/motm/IoE-Smart-City_PoV.pdf. Accessed 15 Jan 2020
8. Internet of Things in 2020: Roadmap for the future. https://docbox.etsi.org/erm/Open/CERP%2020080609-10/Internet-of-Things_in_2020_EC-EPoSS_Workshop_Report_2008_v1-1.pdf. Accessed 07 Jan 2020
9. IoT 2019 in Review: The 10 Most Relevant IoT Developments of the Year (2020). https://iot-analytics.com/iot-2019-in-review. Accessed 17 Jan 2020
10. Top 10 Internet of Things Technologies for 2019. https://www.techjeny.org/internet-things-iot-technologies. Accessed 18 Jan 2020
11. Powerful IoT Examples in 2019. https://www.softwaretestinghelp.com/best-iot-examples. Accessed 18 Jan 2020
12. Slovenskí operátori zabrali, siete pre Internet vecí výrazne narástli. https://www.etrend.sk/trend-archiv/rok-2019/cislo-12/operatori-zabrali-siete-pre-internet-veci-vyrazne-narastli.html . Accessed 18 Jan 2020
13. What is Industry 4.0?. https://blog.thewholebraingroup.com/what-is-industry-4.0. Accessed 15 Jan 2020
14. Kincl M, Rostasova M (2020) VOSviewer software – web of Science database
15. Lelek L (2020) Poštový trh s využitím internet vecí. University of Zilina, Zilina
16. Industry 4.0 Challenges and solutions for the digital transformation and use of exponential technologies. https://www2.deloitte.com/content/dam/Deloitte/ch/Documents/manufacturing/ch-en-manufacturing-industry-4-0-24102014.pdf. Accessed 15 Jan 2020

Physically Realistic Simulation of Mechanical Assembly Operations in a Virtual Reality Training Environment

Florian Dyck[1(✉)], Marc Pilates[2], Lisa Masjutin[3], and Jörg Stöcklein[1]

[1] Fraunhofer Institute for Mechatronic Systems Design IEM, Paderborn, Germany
florian.dyck@iem.fraunhofer.de
[2] Hamm-Lippstadt University of Applied Sciences, Hamm, Germany
[3] Bielefeld University, Bielefeld, Germany

Abstract. Virtual technologies are becoming more and more important in the manufacturing industry. In particular when it comes to the education of employees and production work, cognitive assistance systems offer a high potential for enhancing efficiency. The use of Virtual Reality (VR) for an assembly simulation allows for cost-effective training of employees. In such assembly simulations, however, it is essential to make the training content as realistic as possible in order to avoid unacceptable training effects. This research paper presents a technical solution of a virtual training environment that allows a physically realistic simulation of mechanical assembly operations. As part of this project a VR application was developed in which the trainee learns the assembly processes for the production of a complex mechanical component. To provide the most realistic learning experience possible, various functionalities have been implemented. A methodology has been developed which physically correctly reproduces the kinematic sequences of the components in the respective assembly processes. In addition, virtual tools such as a hammer or a torque wrench are provided to the trainee, which can be operated in a natural manner. Furthermore, interactive 3D visualizations are used to cognitively support the trainee during the assembly tasks. To validate the developed VR application, an evaluation with trainees was conducted, which showed that the VR application is more satisfying and easier to use compared to paper-based instructions. Therefore, this technical solution offers a scalable alternative for companies and schools to teach mechanical assembly tasks.

Keywords: Virtual reality · Assembly training · Realistic interaction

1 Introduction

The increasing complexity of products and assembly systems in the manufacturing industry requires appropriate training of the employees who have to operate these systems. Such training is already provided in vocational schools to prepare students for assembly work. In most cases, the teachers use paper instructions, which might include sectional drawings and textual assembly instructions. However, this procedure highly

M. E. Auer et al. (eds.), *Online Engineering and Society 4.0*, Lecture Notes in Networks and Systems 298, https://doi.org/10.1007/978-3-030-82529-4_18

demands the cognitive skills of the learner as he or she has to transfer the abstract information to real three-dimensional operations [1]. An additional or alternative approach is training on-the-job, whereby the assembly tasks can be practiced on a real assembly system. This involves training employees in running production lines. The consequences can be a decrease in productivity or an increase in error-proneness, which in both cases results in costs. Alternatively, real duplicates of the assembly systems with real components can be used for training purposes only. However, this approach is associated with additional costs and space requirements. Especially for vocational schools, real components and tools would have to be available in large quantities in order to be able to teach multiple students simultaneously.

The use of Virtual Reality (VR) in this context has the potential to address the challenges described above. With this technology, physical structures and operations can be simulated virtually, thus eliminating the need for real systems and components. This has the potential to save costs since components and tools cannot get damaged in the virtual environment. At the same time, operator safety can be increased because there is no risk of injury. Furthermore, VR offers the possibility to place interactive visualizations freely in space and thus provide information without being restricted to physical presentation devices or surfaces. However, the VR adaptation of physical operations, structures and tools must be designed very realistically. Otherwise, a negative effect can occur, and false learning results might be obtained when the learner is transferred to the real assembly environment.

This paper examines a training scenario of a commercial-technically oriented vocational college dealing with education in the age of digital transformation (Education 4.0). One of the training subjects is mechanical engineering, in which the challenges for Industry 4.0 are taught. Students are introduced to the fundamentals of mechanical engineering in theory and practice. The apprentices receive complex commissioning, maintenance and changeover jobs in various learning situations. A concrete application is the assembly process of a mechanical chuck. Since the physical components are not available to the college for this application, the learners had to learn the contents on the basis of drawings. Therefore, a VR training environment for this assembly scenario shall be realized and examined regarding the learning outcomes.

The remainder of this paper is structured as follows. Section 2 presents the related work in research on virtual assembly. Section 3 deals with the implementation of the virtual training environment and its features. The procedure and the results of the accomplished evaluation are explained in Sect. 4. Finally, a conclusion is given and plans for future work are outlined in Sect. 5.

2 Related Work

The idea of using virtual technologies for the simulation of assembly processes has existed for many years. In 1997 Jayaram et al. demonstrated the feasibility of creating valued design information using VR tools with the Virtual Assembly Design Environment (VADE) [2]. VADE enables engineers to create product and process information by performing assembly tasks virtually. It can be used to evaluate tolerance issues, select optimal component sequencing and generate assembly and disassembly process plans. However, the topic has attracted renewed attention only as a result of technological advances in recent years. The developments of Oculus and HTC in the VR consumer area contributed to making VR systems more powerful and affordable. Therefore, also in the field of VR assembly simulation, more and more applications in industry and research have been documented in recent years.

A VR system for assembly validation using hand tracking for interaction is presented by Jayasekera et al. [3]. The authors developed a mechanism for interaction between planar faces of components for virtual assembly. On this basis, further algorithms can be used to validate and extract planar surface geometries. Furthermore, a constraint management system was implemented, which monitors the allowed degrees of freedom of the components.

Studies of Abidi et al. have revealed, that immersive Virtual Reality training can result in improving the time to complete assembly tasks [4]. Furthermore, the authors reported about a better transfer of training when participants completed assembly operations on fabricated parts and the virtual training resulted into more user satisfaction for performing the assembly tasks.

Al-Ahmari et al. demonstrated a virtual manufacturing assembly simulation system (VMASS) [5]. The authors' system is intended to bridge the gap between assembly design and planning and assembly training, implementation and execution. In order to provide a tool that enables training operators and letting them acquire and improve their assembly skills, this work integrates concepts and studies from training simulations. These concepts include the evaluation of assembly training effectiveness and transfer of training. The authors extended the system following the objective to make the assembly training easy and realistic through the usage of multi-modal input and output [6]. A study was conducted in order to evaluate the systems effectiveness and functionality. One main finding was that the multi-modal feedback mechanism has a positive influence on the performance of the users. Furthermore, the participants trained by VMASS committed fewer errors and took lesser time in actual product assembly when compared against the participants from the traditional training group.

Dwivedi et al. also focused their research on the use of virtual assembly as a training tool [7]. The authors investigated the influence of immersion as well as the use of text and images as support during the execution of manual assembly tasks. Their findings showed that immersive VR glasses lead to a more effective training compared to training on a 2D monitor, since training tasks could be performed faster and more accurately in a virtual environment.

As shown, the use of VR in the training of assembly tasks can have a positive effect on aspects such as productivity, error-proneness or learning transfer. However, little attention has been paid to the effects of realistic component interaction and physically correct process mapping on learning success. This could be particularly interesting, since realistic component interactions are key to a successful immersive virtual assembly experience [8].

3 VR Training Environment

Starting from the initial situation described in Sect. 1, the main goals of this research work were the realization of the given training scenario in a VR application and its evaluation. This application reproduces the assembly process of a chuck virtually and intuitively. It offers the learners the opportunity to better understand the function and structure of the assembly. To achieve the overall goal of improving the learning outcomes several key requirements were defined for the application. A particularly relevant requirement is a **realistic simulation of the assembly tasks**. It is important that the assembly processes are depicted as realistically as possible and that joining processes, such as screwing and locking, are made tangible through haptic and acoustic feedback. The necessity of using virtual tools increases plausibility and fidelity of reproduction. A critical aspect is the correct representation of the kinematic motion sequences of the components during the assembly. In the virtual assembly simulations known so far, this aspect is often simplified by snapping to the final position and orientation. For the realized VR application a solution was aimed at that comes closer to reality. Another requirement is to **perform the VR training without the need for external guidance**. So, on the one hand the application needs to be usable intuitively. On the other hand, the user should be guided within the application to perform the assembly simulation without previous knowledge. All this was realized exemplarily in the VR training environment and made available to the learners. Through the implementation of several difficulty levels, the learner should on the one hand learn the assembly process with the help of detailed instructions. On the other hand the learner should carry out what he or she has learned independently without the help of aids.

3.1 Setup and System Overview

To realize a virtual training environment, a hardware solution is required in addition to the software to be developed. The **Oculus Quest** was selected for this purpose. The wireless usage and low acquisition costs are the main advantages of this VR device. Thus, a parallel usage by several persons at any distributed places is possible. As input for the software implementation, an assembly plan was required in addition to the 3D models of the mechanical chuck and the tools. With the help of this plan the logic about the component behavior could be developed. A virtual training environment with a typical workbench was initially created using the game engine Unity. The environment is designed in a clean and straightforward manner so that the learner can concentrate on the assembly (see Fig. 1).

All necessary components and tools for the assembly of the mechanical chuck are available on the workbench. In order to be able to guide even a learner without any previous knowledge through the application, a virtual clipboard was developed, which can be placed freely in space. During the virtual assembly, the clipboard shows the current assembly task to be carried out. The board updates itself automatically with the corresponding progress. In addition, the clipboard serves both as a menu guide and to explain the controls of the application. As a visual support for the assembly tasks, an interactive hologram is shown behind the assembly table.

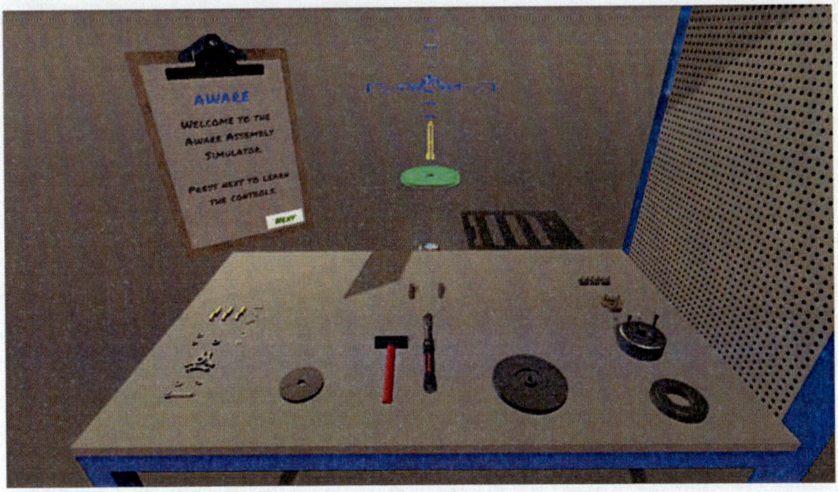

Fig. 1 Overview of the virtual assembly environment

In the following chapters, the basic interaction functionalities and visual guidance features are explained.

3.2 Operation of Assembly Tasks

A large part of the development effort was spent on the kinematically correct representation of the assembly processes. For this purpose, a methodology was developed that allows the exact motion sequence of the individual parts during the assembly to be depicted. This methodology includes a specific collision detection and partially restricts the learner's movements in order to correctly execute the kinematic workflow. The principle of operation is explained below with the aid of Fig. 2, using the example of a washer to be mounted on a screw.

First, the operator is able to grab any component or tool and move it freely in space. In this state, the parts collide with each other in a physically correct way based on Unity's physics engine. However, joining processes cannot be realistically simulated with this engine because the collision behavior is not precise enough for complex CAD data. Therefore, the application switches to constraint-based operation when a joining process is to be performed. Hence, this methodology is a hybrid approach of physics and constraint-based assembly simulation. Triggers for the switching are *start colliders* (see Fig. 2, left), which detect parts being within the collider box. In the example shown in the illustration, a washer is to be attached to the screw by an axial movement. When the washer is moved into the *start collider*, the system checks if the part has to be mounted on the screw and if its orientation is correct. If these conditions are met, the washer is fixed to the blue marked axis of the screw and can then only be moved in the direction of this axis and rotated around this axis (see Fig. 2, right). Simultaneously the *start collider* is switched off and a *guide collider* (see Fig. 2, right) is switched on. As long as the operator holds the part inside the *guide collider*, it remains fixed on the axis.

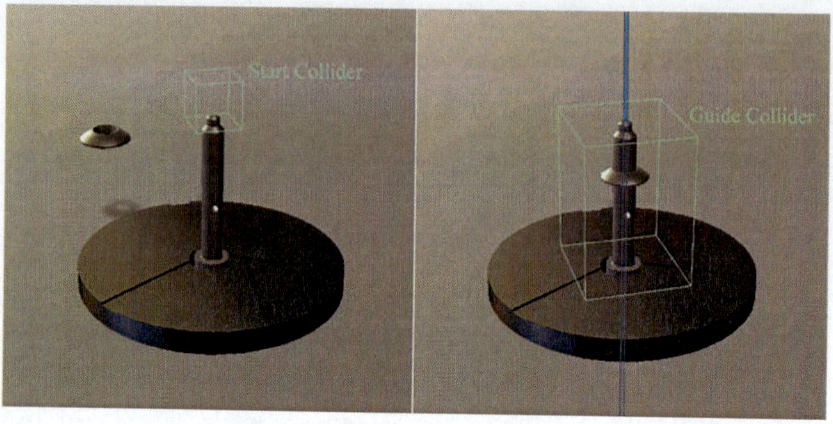

Fig. 2 Collider concept for kinematic sequence of components. Left: *Start collider* as trigger for constraint-based operation. Right: *Guide collider* and axis as constraints for the mounting

This ensures that the washer stays on the screw and can only be released from its fixation by moving it sufficiently upwards. If the user guides the component to its target position at the bottom of the screw, the system detects this with the help of tolerance parameters and the user can release the component. This completes the joining process and the *guide collider* is switched off. Simultaneously the *start collider* is reactivated so that further components can be mounted on the screw.

Interaction with the Hammer. In the assembly process it is necessary to guide several metal pins into the corresponding drilling to fix different components. For the real assembly a hammer is used. The hammer needs to be available in the virtual training environment as well. In a first approach, a collider was added to a virtual hammer head and a pin, which moves the pin partially into the drilling when these two intersect. However, a simple collider in the virtual hammer head and the metal pin was not sufficient to represent the process realistically. The pin could be hit from the side or even from below, which does not correspond to the expected use. To avoid this, two colliders were attached to the striking surfaces of the pin, a thin *collider A* and a wider and flatter *collider B* (see Fig. 3, left). To strike the pin into the component, the collider in the hammer head must first touch the *collider A*, then the *collider B*. This can only be done when the hammer hits the intended side of the pin. Blows from below or from the side trigger the *collider B* first, then the *collider A* and the pin is not pushed into the drilling. A successful strike is conveyed both visually and with an appropriate tone and vibration of the controller.

Interaction with the Torque Wrench. In a later stage of the assembly process, it is necessary to insert screws of different sizes into the corresponding threaded hole of the component in order to fix individual parts. A torque wrench is used for these assembly processes. Therefore, a virtual representation of this tool including its functionality was developed for virtual assembly. The screw should stick to the torque wrench when it is placed in the corresponding position. For this purpose, the model of the screw is placed

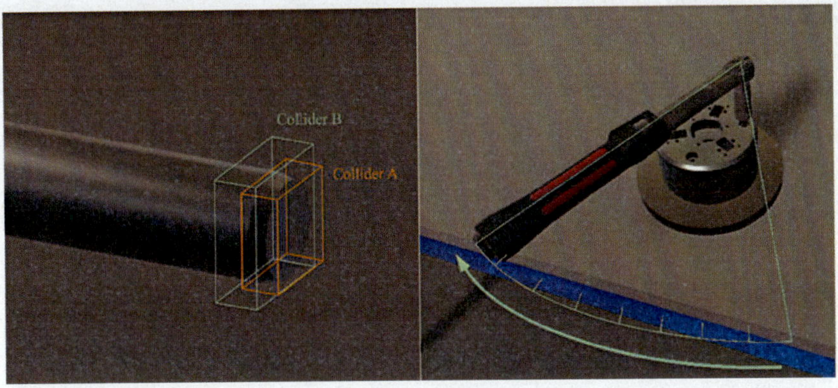

Fig. 3 Collider arrangement on a pin (left) and the operation of a torque wrench (right)

on the torque wrench when two colliders of the objects overlap and align according to the straight line between the screw and the user's hand. In order to turn the attached screw into the component the learner rotates the torque wrench clockwise (see Fig. 3, right), while rotation in the opposite direction causes an idle state and no rotation of the screw. For this functionality, it must be determined in which direction the user is currently turning the torque wrench. Therefore, the angle of rotation axis of the last time frame is compared with the angle of the current time frame and a momentary direction of rotation can be deduced. The screw rotates in synchronization with the torque wrench and, when turned clockwise, is gradually displaced into the component so that the impression of screwing is created. When turning counterclockwise, the characteristic "clacking" is played as a sound to give the user the impression of idling. The vibration function of the controller is also used here to give the user the haptic feedback intended when operating a real torque wrench. Once the user opens the hand by releasing the trigger on the controller or moves the torque wrench far enough away from the screw, the turning process is stopped, and the screw is released from the torque wrench.

Since a torque wrench does not work in reality without an attachment that belongs to the screw, this attachment was created for each screw type. The attachments are attached to the head of the torque wrench. Without the corresponding attachment, a screw cannot be placed on the torque wrench.

3.3 Visual Guidance

With the aim of a good learning success, the learner should be able to conduct and understand the individual assembly processes and steps in an intuitive way. Written instructions for the assembly processes were already available. However, it is not useful in VR to provide the user with the instructions via text only. The option of having the textual instructions read aloud was discarded because of concerns about reduced concentration and thus reduced learning success. Furthermore, Virtual Reality offers far more possibilities that can enhance the training environment. Therefore, a

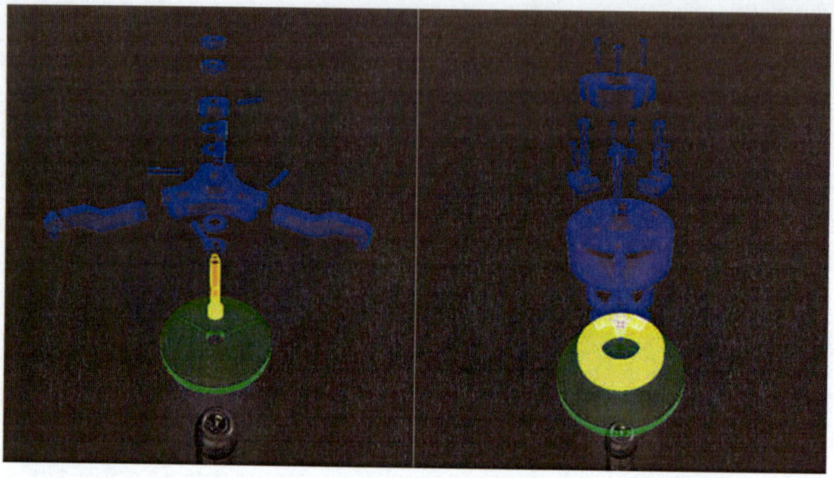

Fig. 4 Holograms indicating different stages of the assembly process

three-dimensional hologram is implemented in virtual space (see Fig. 4), which the user can circle around and take a closer look in order to understand the process graphically and spatially. The hologram consists of semi-transparent models of the individual components of the assembly process. The components are located at their final relation to each other, but in an exploded view, so that each component floats freely in space and is clearly visible to the user. Each component in the hologram has the status "*not yet mounted*", "*to be mounted next*" or "*completely mounted"*. These states are visualized with a blue, yellow and green color of the components, with blue being the default color meaning "*not yet mounted*". When a part is successfully assembled, the corresponding part in the hologram changes its color to green and the part to be assembled next to yellow. This is done in a logical grouping of individual parts. For example all screws of one kind which are to be mounted next are highlighted in yellow at the same time, instead of each one individually one after another. This is possible, because the order in which grouped screws are mounted does not matter.

In order to remain clear and not to overburden the learner, the entire assembly process is not shown at once, but staggered at certain points. This means when one stage of the assembly process is reached, the hologram changes its appearance and shows the next stage. If the support of the hologram is not necessary or not desired, the learner can deactivate it.

4 Evaluation

To validate the VR training environment the following research question was derived: *Which additional value is provided by giving students the opportunity to use the VR application?* Altogether three hypotheses have been formulated for this: *When using the VR training simulator, satisfaction is higher, ease of use is easier and the gain in knowledge is higher.* These hypotheses are to be proved by means of questionnaires. The procedure and results are described below.

4.1 Design and Procedure

The experimental evaluation was conducted in a vocational school in Germany. The participants were 20 male industrial mechanic trainees. The age ranged from 17 to 27 years (M = 18.70, SD = 2.27). The evaluation was structured as follows. The students were informed that participation is voluntary, and no personal data is collected, so that all information given is processed anonymously. The teacher explained the procedure and randomly assigned the students into one of two groups of equal size. Both groups received paper-based assembly instructions for the mechanical chuck. The documents included a sectional and exterior view and a parts list of the chuck. Their task was to become familiar with the mechanical component and to create a disassembly instruction. Ten of the students, who were randomly selected for the experimental group, additionally had the opportunity to use the VR application to assemble the chuck. Both groups had two hours to complete the task. After two hours, an evaluation questionnaire was handed out to every student.

4.2 Questionnaire

The questionnaire consisted of twelve items regarding satisfaction, utility, knowledge, ease of use and an overall rating. Items measuring satisfaction, utility and knowledge were based on the Questionnaire for Professional Training Evaluation (Q4TE) [9]. The Q4TE is based on the four-level framework of training evaluation [10]. According to this framework, the four levels of training evaluation are reaction, learning, behavior and organizational results. The first level, the reaction, is multidimensional and consists of the affective response to the training and the perceived utility. The second level refers to the knowledge required in a training. Only the first two levels were included in this evaluation, since the short-term effects of the VR application were to be examined. Original items were adapted on the given training scenario and specified for the two groups. Items measuring ease of use were adapted from the Technology Acceptance Model (TAM) [11]. TAM postulates that the perceived ease of use is an important factor that influences the behavioral intention to use a technology. Items were translated into German by Mlekus et al. 2020 [12]. Overall rating was measured with a self-constructed item. The item was "Overall, I liked the learning with the VR glasses" for the experimental group and "Overall, I liked the learning with the assembly instructions" for the control group. All items were rated on a scale from 1 (I do not agree with this statement at all) to 5 (I strongly agree with this statement). The actual learning performance was measured subjectively and objectively. The subjective learning performance was measured with a self-constructed item as follows: "I am now able to describe the disassembly of the chuck". The rating of the objective performance was based on their disassembly instructions. A model solution was handed out to the students, so they could check their results and rate their performance on a scale from 1 to 5 based on the number of mistakes. The teacher additionally verified their ratings.

Table 1 Descriptive statistics and correlations for dependent variables

Variable	*M*	*SD*	1	2	3	4	5	6	7
1. Satisfaction	3.45	1.16	—						
2. Utility	2.95	1.05	0.44	—					
3. Knowledge	3.02	1.01	0.54*	0.57**	—				
4. Ease of use	3.11	0.96	0.71**	0.25	0.12	—			
5. Overall rating	3.70	0.98	0.80**	0.55*	0.49*	0.61**	—		
6.Subj. performance	3.70	1.03	0.63**	0.47*	0.57**	0.57**	0.58**	—	
7.Obj. performance	3.15	0.88	0.45*	0.38	0.56**	0.24	0.42	0.64**	—

Note. $N = 20$. *M* and *SD* represent mean and standard deviation, respectively *indicates the correlation is significant at the 0.05 level, **indicates the correlation is significant at the 0.01 level (two-tailed).

4.3 Results

Subjective and objective learning outcomes of vocational students who either used paper-based instructions and the VR application or only paper-based instructions for learning were analyzed. Table 1 provides the means and Pearson correlation coefficients of the variables measured in the experiment. There is a moderate positive correlation between satisfaction and objective performance ($r = 0.45$), meaning that high values in satisfaction are rather associated with a high objective performance in the data. This correlation is statistically significant ($p < 0.05$).

To analyze the effect of the VR application on learning outcomes, further statistical analyses were conducted. As two independent groups were investigated – the control group and the VR group – t-tests with independent sample data were chosen to show whether a difference between the groups is random or statistically significant. Results are provided in Table 2.

Table 2 Effects of use of the VR application on dependent variables

Variable	Control		VR		*t*(18)	*p*	Cohen's *d*
	M	*SD*	*M*	*SD*			
Satisfaction	2.74	0.89	4.15	0.97	3.357	0.004	1.501
Utility	2.70	1.14	3.20	0,95	1.069	0.299	0.478
Knowledge	2.70	0.89	3.35	1.06	1.490	0.154	0.666
Ease of use	2.68	0.87	3.55	0.89	2.230	0.039	0.997
Overall rating	3.30	0.95	4.10	0.88	1.960	0.066	0.839
Subjective performance	3.40	1.27	4.00	0.67	1.327	0.201	0.593
Objective performance	3.00	1.05	3.30	0.68	0.758	0.458	0.339

Note. Control = group with only paper-based instructions ($N = 10$), VR = group with paper-based instructions and additional use of the Virtual Reality application ($N = 10$). *M* and *SD* represent mean and standard deviation, respectively. The results of t-tests are shown comparing the means between the two groups. Cohen's *d* depicts the standardized difference between two means

As shown in Table 2, students in the VR group are significantly more satisfied than the students in the control group. The p-value for this difference is $p = 0.004$, meaning the probability for no difference between the groups is 0.4%. The effect size is $d = 1.501$, indicating a very large effect [13]. Moreover, a large effect on perceived ease of use ($d = 0.997$; $p = 0.039$) was found. Participants rated the VR application to be significantly easier to use than paper-based instructions.

Significant effects on perceived utility, perceived gain in knowledge, overall rating, and subjective as well as objective performance could not be found.

5 Conclusion

This paper presents a technical solution for a virtual training environment for mechanical assembly tasks. The solution addresses the challenges that arise when training for new employees or new assembly systems is required. A learner is brought to a virtual assembly station with the help of VR glasses and is supported by various features in the learning process of assembling a mechanical chuck. One feature to be highlighted is the kinematically correct representation of the respective assembly processes, which was realized by a special methodology. This should help to present the virtual processes as realistically as possible. In order to be able to assess the influence on the learning success, an evaluation was carried out. The main findings are that the VR application is more satisfying and easier to use compared to paper-based instructions. Thus, two of three hypotheses were confirmed. A positive influence of virtual assembly on gain in knowledge could not be proved. This can be explained by the small scope of the assembly tasks and is to be examined in more detail with a larger assembly scenario as future work. However, this technical solution offers a scalable alternative for companies and schools to teach mechanical assembly tasks. It can replace traditional learning methods and save costs while improving learning outcomes. Looking ahead, the integration of further assembly processes and tools into the virtual learning environment is aimed at in order to be able to represent a broader spectrum of mechanical learning scenarios.

Acknowledgements. This research and development work is based on "AWARE", a project of the "it's OWL" Leading-Edge Cluster, partially funded by the German Federal Ministry of Education and Research (BMBF).

References

1. Porter ME, Heppelmann JE (2017) A manager's guide to augmented reality – why every organization needs an augmented reality strategy. Harv Bus Rev 95(6):45–57
2. Jayaram S, Connacher HI, Lyons KW (1997) Virtual assembly using virtual reality techniques. Comput Aid Des 29(8):575–584
3. Jayasekera RDMD, Xu X (2019) Assembly validation in virtual reality—a demonstrative case. Int J Adv Manufac Technol 105:3579–3592
4. Abidi MH, Ahmad A, El-Tamimi AM, Al-Ahmari AM (2012) Development and evaluation of a virtual assembly trainer. Proc Hum Factors Ergon Soc Ann Meet 56(1):2560–2564

5. Al-Ahmari AM, Abidi MH, Ahmad A, Darmoul S (2016) Development of a virtual manufacturing assembly simulation system. Adv Mech Eng 8(3): 1–13
6. Abidi MH, Al-Ahmari A, Ahmad A, Ameen W, Alkhalefah H (2019) Assessment of virtual reality-based manufacturing assembly training system. Int J Adv Manufact Technol 105:3743–3759
7. Dwivedi P, Cline D, Joe C, Etemadpour R (2018) Manual assembly training in virtual environments. In: 18th International Conference on Advanced Learning Technologies (ICALT), pp 395–399. IEEE, Mumbai
8. Seth A, Vance JM, Oliver JH (2011) Virtual reality for assembly methods prototyping: a review. Virt Real 15(1):5–20
9. Grohmann A, Kauffeld S (2013) Evaluating training programs: development and correlates of the questionnaire for professional training evaluation. Int J Train Dev 17(2):135–155
10. Kirkpatrick D, Kirkpatrick J (2006) Evaluating Training Programs, 3rd edn. Berrett-Koehler Publishers, San Francisco
11. Venkatesh V (2000) Determinants of perceived ease of use: integrating control, intrinsic motivation, and emotion into the technology acceptance model. Inf Syst Res 11(4):342–365
12. Mlekus L, Bentler D, Paruzel A, Kato-Beiderwieden AL, Maier GW (2020) How to raise technology acceptance: user experience characteristics as technology-inherent determinants. Gruppe. Interaktion. Organisation. Zeitschrift für Angewandte Organisationspsychologie (GIO) 51(3):273–283
13. Sawilowsky SS (2009) New effect size rules of thumb. J Mod Appl Stat Meth 8(2):597–599

Challenges and Solutions to Integrate Remote Laboratories in a Cross-University Network

Hadi Adineh[1(✉)], Matteo Galli[2], Birte Heinemann[3], Nils Höhner[4], Davide Mezzogori[2], Matthias Ehlenz[5], and Dieter Uckelmann[1]

[1] Hochschule für Technik Stuttgart, Schellingstraße 24, 70174 Stuttgart, Germany
hadi.adineh@hft-stuttgart.de
[2] Department of Engineering and Architecture, University of Parma, Parco Aree Delle Scienze, 181/A, 43124 Parma, Italy
[3] Learning Technologies, RWTH Aachen University, Ahornstr. 55, 52704 Aachen, Germany
[4] University of Koblenz-Landau, Koblenz, Germany
[5] RWTH Aachen University, Aachen, Germany

Abstract. Location-independent networking of laboratory infrastructures is opening new possibilities for teaching and learning. In order to make full use of the possibilities it makes sense to form associations and use real, partly digitized and fully digital laboratories shared across locations. The connection of different laboratories with different equipment and different conditions confronts us with technical challenges. This article presents challenges, which are faced in an international project, as well as possible technical solutions and the way to our decision.

Keywords: Remote laboratories · Engineering education · AMQP · MQTT · VPN · Digital transformation

1 Introduction

The Internet of Tings and its implementations are bringing about rapid and massive changes in many areas due to the increasing shift of value creation from the physical to the digital world [1]. This process is not without consequences for the design of degree programmes in the engineering sciences, as they are required to take into account the emerging information technology content from the IoT spectrum in their curricula. Engineering students should have the opportunity to get to know the diverse areas of digitalization related to the IoT, to experience them in a practical way and to help shape them [2, 3]. Hands-on laboratories have been used to train students correspondingly. High acquisition, maintenance and support costs of university laboratories are limiting the utilization of these labs, though. Cross-university collaborations are a good way of networking industry-related, physical and digital laboratory environments and making them available to a larger group of users. Additionally, remote access to digitalized laboratories is overcoming the need of physical presence for students and researchers.

M. E. Auer et al. (eds.), *Online Engineering and Society 4.0*, Lecture Notes in Networks and Systems 298, https://doi.org/10.1007/978-3-030-82529-4_19

Digitization thus enables new forms of location-independent networking of laboratory infrastructures. In the project Open Digital Lab for You (DigiLab4U[1]) we explore the challenges, possibilities, opportunities and benefits of networked laboratories. One main part of the project is the digitization of real laboratories and to link them with virtual components. We research benefits of the combination of real laboratories and their digital or partly digital counterparts. The location-independent access and distance learning possibilities are promising, as they can improve individual and self-regulated learning. Among other things, the software architecture that is currently being developed should enable not only the connection of laboratories, which are already part of the project, but should provide a possibility to include further laboratories as easy as possible. The software consists of various components and interfaces that are intended to form a holistic infrastructure to support networking, the learning process, remote working with the labs and analytics.

One part of the software is the Laboratory Management System (LabMS)[2] which should be deployed inside laboratories to manage lab devices and the entire process of experiment. It also should be compliant and interact with the Learning Management System (LMS), in our case Moodle. It is the core service in our platform by providing the interface for most stakeholders. Therefore, we need to develop a solution, which is able to facilitate interactions between Moodle and the LabMS. In this paper, we will present our roadmap to the choice of technology and our argumentation in the decision-making process.

2 Related Works

Numerous works describe the realization of virtual and remote laboratories. According to Heradio et al. [4], for example, simulations of experiments are becoming even more popular, especially in STEM applications. Indeed, the offering of virtual and remote laboratories is expanding in the recent years. Moreover, innovative new types of laboratories, such as serious games and virtual reality labs, are becoming increasingly more common. Cross-institutional networks need to deal with organizational and technical challenges [5].

To better understand current implementations of remote laboratories and networks, a review of the literature from 2000 has been done, in order to find a description of architecture to serve as benchmarks of the proposed architecture. We focused on those that are still ongoing and are sharing most of our goals, namely: LiLa [6], UniLabs [7], NetLab [8] and Go-Lab [9].

The architecture of LiLa consists of four main components: a Web server, hosting the LMS, a database for the storage of experiment data, and a database for the management of the booking system; the implementation relies on a SCORM package without proper integration within its network.

[1] https://digilab4u.com/

[2] in the following referred to as LabMS

NetLab focuses on laboratory experiments based on microcontrollers, in which each lab exposes a webcam for video streaming; without the use of a dedicated LMS, tutorials, procedures, user information reservations are exposed through a web server. The Go-Lab architecture is composed of an Inquiry Learning Space, which contains course models and learning materials. A specific component is dedicated for user authentication, and Learning Analytics tools are used to monitor student learning progress. Finally, a range of web services abstract lab device details.

Unilabs is a network of 15 remote and virtual laboratories of different Spanish universities. Applications, remote experiment dashboard, as well as learning materials provided on the central LMS. With respect to other notable works, Sáenz et al. [10] use web-sockets to exchange data in JSON format between client and server. Finally, Cornetta et al. [11] propose their solution to enhance existing FabLabs, making them remotely accessible: this further underlines the appeal shown for remote laboratory networks.

None of the mentioned labs focussed on IoT-applications. Addressing this specific market segment distinguishes DigiLab4U from other approaches and will benefit the envisioned community building process. Because of this IoT-focus, typical lab-interfaces such as LabVIEW are not available. Instead connections to specific measurement and handling systems need to be provided.

While some technical documentation on other lab-networks are available, their considerations prior to choosing their preferred solution are difficult to find. Therefore, we have chosen to publish our considerations, so that other researchers can reflect on these.

3 Overview of Our Platform

As shown in Fig. 1, the digital learning environment consists of three logical components. The Learning Management System (**LMS**), the Laboratory Management System (**LabMS**) and the actual lab hardware which is controlled by an Experiment Runner (**ER**). These three parts can be located at different locations and on different servers. The network contains one LMS server, 1…n LabMS instances depending on the number of laboratories and 1…n ERs within the LabMS based on the number of available experiments.

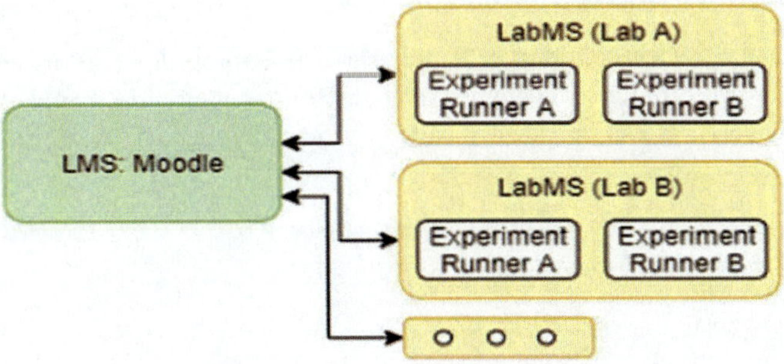

Fig. 1 Learning management system and laboratory management systems

The main interface for users such as lecturers and learners is the learning management system “Moodle” [12] which resides on a central server. This server is used to display the user interfaces of remote laboratories in course contexts and allows enrolled users to perform remote experiments at certain time slots. Additionally, learning material and experiment results are provided.

The actual administration of the remote experiments available within a laboratory is done by a specifically developed LabMS. It receives and validates experiment parameters and commands from the user interface, which resides on the LMS server.

The actual laboratory hardware is controlled by an application (Experiment Runner), which can control specific hardware components via interfaces.

To successfully conduct a remote experiment, the three components have to be connected via network. The user input on the LMS server has to be forwarded to the corresponding LabMS server which in turn controls the actual hardware. To realize these connections different technologies are feasible, which are described in the following chapters and are compared with each other using the Digilab4U architecture as an example.

4 Technical Considerations

Considering the system overview in Sect. 2, multiple services are collaborating with each other in compliance with the concept of Service Oriented Architectures [13]. It supports multiple laboratories as well as being flexible for accepting additional services (e.g. Learning Analytics).

To be compliant to open web standards we chose to implement the interface as a REST API [8]. This ensures simplicity in service development and service collaborations. However, services with REST APIs should be directly called by the service user. For example, in our case, Moodle must be able to call LabMS functionalities through REST APIs directly. DigiLab4U internal services are deployed in the cloud, so their IP addresses could be called from other services. The consequences if a laboratory is not able to provide direct access to its PC which is running LabMS is critical. Following, we mainly discuss different scenarios which could be investigated in this manner.

4.1 LabMS with a Static IP Address

If the laboratory owner is able to provide a static IP address for LabMS and open the institutions firewall for inbound connections, the following approach applies. As Fig. 2 reveals, LabMS APIs can be directly called by other DigiLab4U services.

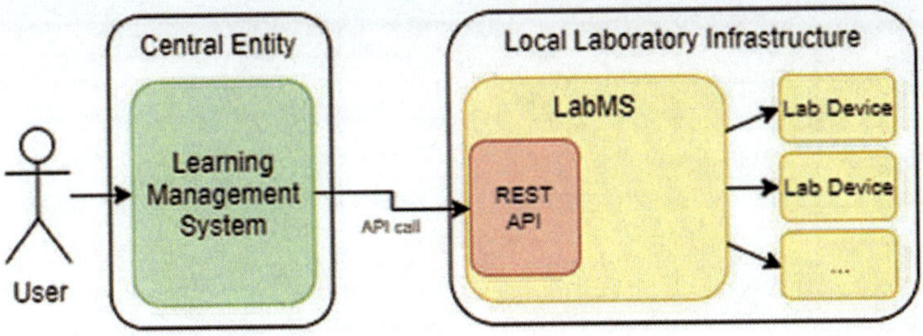

Fig. 2 LabMS and LMS connection, when LabMS equipped with a static IP address

4.2 LabMS Without a Static IP Address

Network infrastructures vary between different universities. There are various (security) strategies IT departments deploy, which bring different challenges to connecting via static IP-addresses. Some departments prefer to dynamically address clients, possibly due to a lack of IP subnets of sufficient size for permanent assignment, others employ network address translation (NAT) and use the special IP ranges for private subnets internally, which are not globally unique. At some institution, security concerns lead to restrictive firewall policies which prohibit the exposure of hosts and enforce this by appropriate firewall configuration. Whichever it is, all of those reasons prevent our central entity (Moodle) from initiating a direct connection to the APIs of the LabMS instance. In this condition, we need to think about implementing a third-party server to facilitate such communications. This server should be accessible by all LabMSs and Moodle servers. There are possible solutions on different layers of the TCP/IP stack of which the most common have been scrutinized and will be examined in this contribution. A VPN server could be one solution usually employed on the network layer. A second solution can be based on messaging brokers like MQTT broker as an application layer solution. Each of them has its pros and cons, which are explained in more detail in the next sections.

Virtual Private Networks

A Virtual Private Networks (VPN), is a private network constructed within the public network such as the Internet. In our VPN-based solution, LabMS instances and Moodle should reside in a DigiLab4U VPN. Within the VPN, servers are able to communicate directly with each other, even without static IP address.

The VPN connections are handled by a VPN server, which will create a virtual network of laboratories, in which the Moodle server can communicate to the LabMS instances, see Fig. 3. The Moodle server acts as a bridge between the VPN and the user, who is not part of the VPN network.

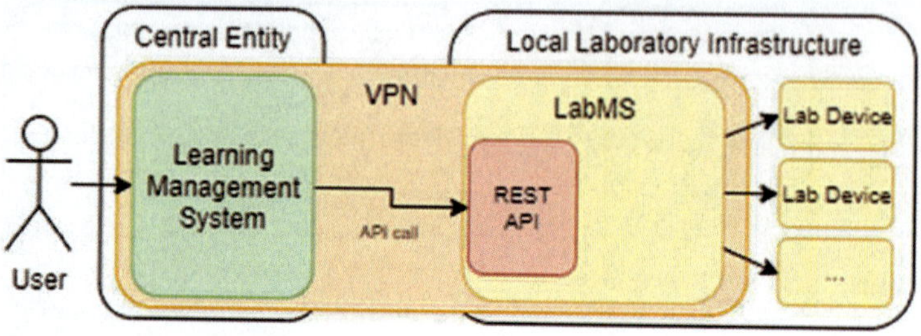

Fig. 3 LabMS and LMS connection, when LabMS cannot provide a static IP address and is connected to DigiLab4U cloud via VPN

By connecting the LabMS and Moodle through VPN, Moodle is able to call LabMS REST APIs directly. But this solution raises another problem. For each experiment, there is a User Interface (UI) which is written in HTML and JavaScript which calls the LabMS API. When a user is accessing the UI, HTML and JavaScript codes are executed on his local PC, which is not part of DigiLab4U-VPN. Thereby, restricting the access to the LabMS API, which is just accessible from the Moodle Server. To solve this problem, another layer has to be provided in the Moodle host, for example, using a wrapper around LabMS. This wrapper will be added to Moodle as a module and can call LabMS REST-APIs inside the Moodle host. In the following, JavaScript codes just need to connect to the Moodle server (without needing the LabMS real IP address).

A big challenge is to create a harmonic network without interfering with the labs' and institutes' internal infrastructure due to the heterogeneous approaches. A complex split tunnel configuration will be required as well as a dedicated subnet of sufficient size. One other main issue of this solution is security. As the number of connected laboratories increase, this VPN could potentially share the resources of each institute to the other ones. Therefore, securing such a network would be a big issue. For that reason, in some institutes VPN connections will not be allowed which rules out this approach.

Messaging Brokers and Web Sockets

Besides the discarded solution with the VPN approach, we have thought about another possibility, where we use a Messaging Broker for communication which would avoid initiating "inbound" connections to the LabMS instance. Within this approach, we have considered several possibilities, which we will now present.

MQTT: MQTT [11] is an open connectivity transport standard. It generally works on top of TCP within the application layer which leads to a simple reliable stream [2]. We implemented MQTT in LabMS to facilitate communications between LabMS and lab devices. To be compliant with the publish-subscribe pattern, MQTT protocol could hide every related entity from each other. Each MQTT client can act as publisher, subscriber or both. They just need to be able to connect to the MQTT broker, not to each other. This fits our requirements well. So, the first solution which we considered was implementing the MQTT broker in between the Moodle and LabMS instances as shown in Fig. 4.

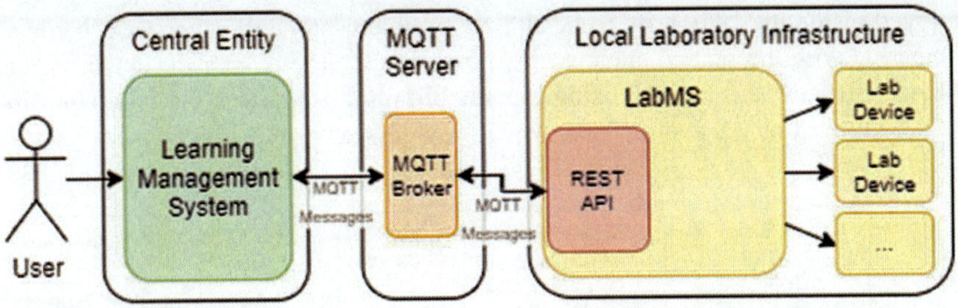

Fig. 4 LabMS and LMS connection, when LabMS cannot provide static IP address and a MQTT would be implemented as messaging protocol between Moodle and LabMS instances

While providing a promising perspective, this approach does not suit the initial requirements of providing a REST API connection to Moodle. So, this solution was dropped as well.

As a second attempt, we made some changes in previous architecture so that we could provide a MQTT client in each lab (see Fig. 5.). This client is responsible to get commands from a central LabMS entity and do related tasks on lab devices on one hand, and on the other hand, get devices states and command results from them to send to the LabMS.

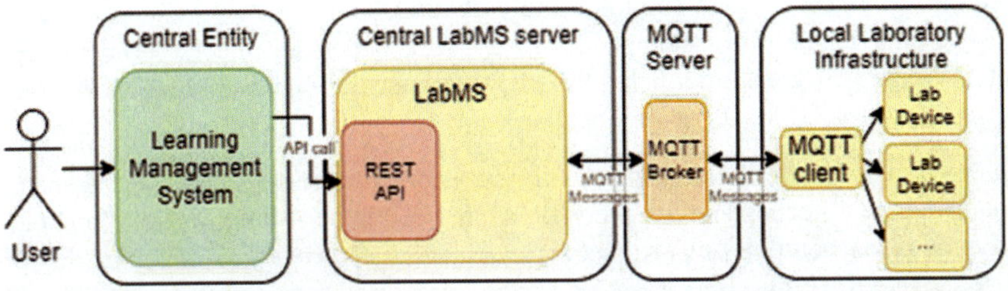

Fig. 5 LabMS and LMS connection, when LabMS cannot provide static IP address and a MQTT would be implemented as messaging protocol between Moodle and LabMS instances

However, the following considerations prevented us from choosing this solution. First of all, LabMS is responsible for managing lab devices, and performing experiments. Long physical distance between LabMS and real lab hardware devices could reduce the security and reaction speeds due to high latency in harmful situations.

As Fig. 6 reveals, in a third suggested architecture, LabMS would be split into two parts. One part, REST-API Server, provides the REST API to the Moodle and should be deployed in the server which is equipped with a static, accessible IP address. This server component could be deployed as a cloud service (i.e., AWS Elastic Cloud, Google Cloud Platform, MS Azure etc.). The second part provides the other responsibilities of the LabMS. We named this part “Experiment Server” which should be deployed inside the laboratory to be able to communicate with experiment hardware. Comparing with the second attempt, this proximity (and removing the single instance as the bottleneck and potential single point of failure) increases the communication speed between LabMS and devices, while providing more reliability.

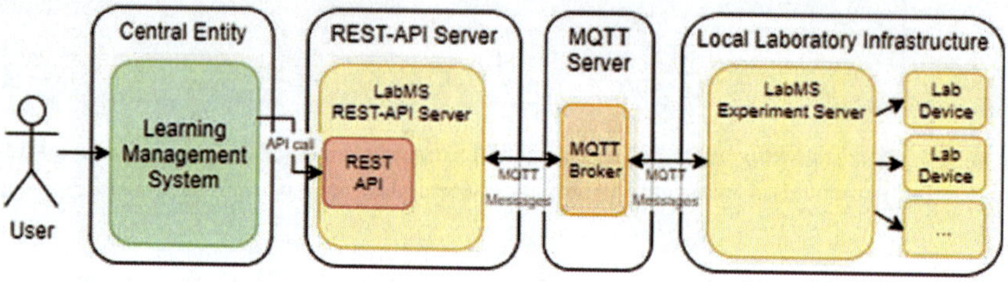

Fig. 6 LabMS and LMS connection, when LabMS cannot provide static IP address and a MQTT is implemented as messaging protocol between REST-API server of LabMS and its Experiment Server

Concerning the connection between Moodle and LabMS, the second and third attempts are equivalent to the first scenario: indeed, the lab owner institute is required to provide a server with a static IP. However, this server does not need to be the laboratory server and is not required to reside in the lab. It only needs to be accessible from the lab and be able to communicate outbound.

Regarding to the third attempt, the REST-API server and the Experiment server communicate with each other using the MQTT protocol, which is able to work around the lack of static IP for the experiment server.

Alternative Approach, Web Sockets: Till now, laboratory servers could be reached by Moodle server independent of their IP addresses. MQTT helps us by facilitating communication between devices and remote servers. However, in our case, this protocol shows problems with service integration scenarios (integrating Experiment Servers and REST API servers). Especially, it does not guaranty message orders. Therefore, we need higher-level message transportation protocols. Looking for a full-duplex, bidirectional communication channel Web Sockets look desirable. Web Sockets operate through a single socket over web [15]. As it could be seen from Fig. 7, The REST API server part of the LabMS, also responsible for acting as Web Socket server. Then, the Experiment Server part can connect to it and enable communication with each other.

Web Sockets work well if we mainly need message transportation between separated parts of LabMS. Due to LabMS requirements, REST API should be its main interface. While REST-API is based on request-reply pattern, normally Web Sockets natively does not match this pattern, further action would be required to adapt and overcome potential pitfalls and harmonize both strategies.

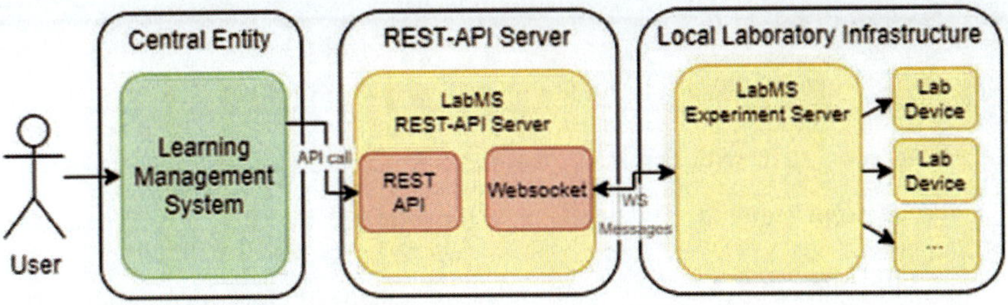

Fig. 7 LabMS and LMS connection, when LabMS cannot provide static IP address and a web socket would be implemented between REST-API server of LabMS and its experiment server

Advanced Message Queuing Protocol (AMQP): AMQP is an open standard application layer for protocol-oriented message-oriented middleware service [16]. This protocol is based on the publish-subscribe pattern. Looking for a solution to support both publish-subscribe and request-reply pattern, we found RabbitMQ (an open-source message broker software). It supports Remote Procedure Call as well as AMQP. Figure 8 shows the new architecture with RabbitMQ.

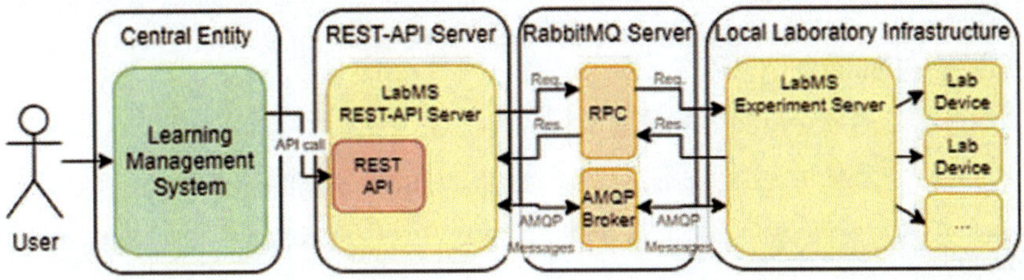

Fig. 8 LabMS and LMS connection, when LabMS cannot provide static IP address and a AMQP would be implemented as messaging protocol between REST-API server of LabMS and its experiment server

By combining RPC and AMQP, laboratories integrate into DigiLab4U without requiring a separate static IP address.

5 Comparisons and Discussion

The presented architectures will be compared according to the following criteria:

- the requirement to have a static IP,
- required effort by developers with further expansion in mind,
- security and safety considerations
- as well as organizational aspects.

5.1 The requirement to have a static IP

This criterion investigates if LMS component require a static IP with a matching firewall configuration to allow users to reach it from the Internet. Three scenarios for the LabMS ecosystem with respect to the static IP feature have been looked at:

- Static IP is not required: This is the case for solutions 2 (VPN) and 3 (first MQTT attempt).
- Static IP required for each laboratory instance: This is the case for solution 1 (static IP in the laboratory).
- Static IP required for a laboratory cluster: This is the case for all other solutions (4, 5, 6 and 7). In this case, it is necessary that the static IP is assigned to the machine that is hosting the REST API server. This machine can then be shared by several laboratories.

5.2 Required Effort by Developers

The solutions involve the use of specific communication technologies. The technologies presented can be implemented in different languages including the most popular (Java, Python,.NET, …). The need to have corresponding skills represents a critical element as it requires learning the skill itself and gaining experience in its application.

The development of UI components is shared between all solutions and is therefore not reported in Table 1.

Table. 1 Comparison of each solution related to skill requirements

Required Competencies Solution	REST API	MQTT	WebSocket	RabbitMQ
LabMS with Static IP	YES	NO	NO	NO
VPN	YES	NO	NO	NO
First MQTT Solution	NO	YES	NO	NO
Second MQTT Solution	YES	YES	NO	NO
Third MQTT Solution	YES	YES	NO	NO
Web Socket	YES	NO	YES	NO
AMQP	YES	NO	NO	YES

5.3 Security and Safety Notes

The static IP solution requires all communications to be based on the HTTP protocol. Security is achieved by forcing communications on secure layers such as TLS or SSL using the HTTPS protocol. This protocol is widely used and documented.

The VPN solution natively offers a high degree of protection to the external environment thanks to the encryption of all communications. In this specific solution the focus must be directed to the shared resources inside the virtual network and to the definitions of control policies.

Security issues for MQTT, Web Socket and AMQP are currently being investigated and are part of future developments [17, 18].

5.4 Organizational Aspects

At an organizational level, the following are the main criticalities which need thorough consideration:

- Bureaucratic or procedural complexity in obtaining a static IP enable access to the laboratory equipment through the Internet.
- Bureaucratic or procedural complexity to include one or more machines within a VPN.
- Funding difficulties for purchasing one or more machines in the "Cloud" to serve as REST API server(s). The recurring cost (monthly, annual, pay-per-use) of these solutions is difficult to reconcile with projects for which a budget must be defined in advance.
- At the organizational level, choosing a solution for which the institution does not have staff with required skills involves the need to invest resources for training staff or the need to refer to external consulting entities.

5.5 Further Discussions

In order to be able to choose, the most suitable solution whenever a new remote laboratoriy needs to be implemented, an effective and efficient decision-making process must be provided.

An initial decision-making method to identify feasible solutions quickly is based on an elimination process targeting at a solution which is applicable in a wide range of heterogeneous infrastructures. Therefore, the criteria defined above can be applied. At each step the number of implementable solutions should be reduced. If more than one solution remains satisfying all those baseline requirements, these can be evaluated according to a cost analysis in terms of implementation time and budget.

The DigiLab4U team faced a decision-making process starting from discarding the VPN based solution as it is not compatible with the policies of every institution. Solution 3 was discarded because of the formal and pre-set requirement to use REST API in the communication with accessory systems. The first application layer solution was debated by some institutes as a fullback because of the difficulties in obtaining static IPs but in the end, it was chosen as a common ground applicable in every environment contributing in the current funding phase. The further supposed extension described as the discarded solutions 4, 5, 6 and 7 have been dropped for concerns regarding sustainability. An initial setup and testing phase for the server instances might have been conducted on existing hardware, but does not scale beyond the first funding phase, major investments in further labs side would have been required. For that same reason a solution involving cloud plans with recurring payments has been considered to diverge from the idealistic vision of longevity of this project beyond governmental funding.

6 Lessons Learned

Choosing the best method, the one which suits most requirements, was a challenge. The first and foremost goal was to find common ground which would fit into the vastly heterogeneous landscape of IT ecosystems in (educational) institutions. To recap, excluding VPN-based solution for practical reasons as well as security concerns, other ones require the lab owners to be able to provide public access to their server with a static IP.

Our premonition was, that the first solution is most straightforward, but the danger of losing potential partner laboratories due to formal restrictions denying them access to a static IP has been too big of a risk to take. Besides the already defined requirement to maintain extensibility, we also discussed concerns about privacy and security.

Meanwhile, all DigiLab4U consortium members decided to start out with the first solution and provide a static IP address for LabMS instance in their laboratory. Due to some security concerns and restrictions, the VPN solution was discarded as currently not applicable.

Due to IT infrastructure policies at some universities, some laboratories are not able to provide a static IP addresses. In such situations, solutions based on messaging brokers will be a good choice and is currently under closer evaluation for implementation. The integration of multiple components of LabMS in combination with AMQP protocol and Remote Procedure Call feature of RabbitMQ is our preferred solution for such institutes. A modern, modular approach will ensure extensibility and might support a drop-in replacement if technology changes and better solutions appear. Currently, there are no considerable downsides in this approach apart from a slightly increased effort in implementation. Comparing to MQTT, AMQP supports advanced queuing solution, authentication and other high-level features. Moreover, the RPC feature of RabbitMQ enables us to directly implement a request-reply pattern on remote applications which is not publicly accessible via a static IP.

7 Future Work

Since the initial research has been done and the decision has been made, the next step is to complete the implementation of the software and all interfaces as well as the evaluation in practical application. In addition to the decision for the public accessible Static IP approach for the currently connected lab environments, the consortium partners in charge of the conceptual design and realization of software components is already working on the integration of MQTT and evaluating RabbitMQ as AMQP approach into their laboratory as a prototype for further extensibility of the ecosystem.

As soon as these integrated concepts are fully implemented, this proof of concept will help to integrate further laboratories, even beyond the current consortium, and test the scalability of the infrastructure. The roadmap is already set and partner labs for this phase have been chosen. Last but not least, sustainability is one of the top priorities: Beyond the expansion of our network, we will further investigate on security of protocols and interfaces, stability and reliability as well as clean, modular integration in learning management systems like Moodle.

Acknowledgements. Thc projcct on which this paper is based was funded by the Federal Ministry of Education and Research (BMBF), Germany under the funding code 16DHB2112. The responsibility for the content of this publication lies with the authors.

References

1. Heyde M, Auth G, Hartmann A, Erfurth C (2017) Hochschulentwicklung im Kontext der Digitalisierung – Bestandsaufnahme, Perspektiven, Thesen. In: Eibl M, Gaedke M, (Eds.), Informatik 2017. Gesellschaft für Informatik, Bonn, pp 1757–1772. https://doi.org/10.18420/in2017_175
2. Kortuem G, Bandara AK, Smith N, Richards M, Petre M (2013) Educating the Internet-of-Things generation. Computer 46(2):53–61. https://doi.org/10.1109/MC.2012.390
3. Terkowsky C, Jahnke I, Pleul C, May D, Jungmann T, Tekkaya EA (2013) PeTEX@Work: designing CSCL@Work for online engineering education. In: Goggins SP, Jahnke I, Wulf V (eds) Computer-supported collaborative learning at the workplace. CSCL@Work (Computer-supported collaborative learning series). Springer, New York, pp 269–292
4. Heradio R, De La Torre L, Galan D, Cabrerizo FJ, Herrera-Viedma E, Dormido S (2016) Virtual and remote labs in education: a bibliometric analysis. Comput Educ 98:14–38. https://doi.org/10.1016/j.compedu.2016.03.010
5. Orduña P, et al (2015) An Extensible architecture for the integration of remote and virtual laboratories in public learning tools. IEEE Revista Iberoamericana de Tecnologias del Aprendizaje 10(4):223–233. https://doi.org/10.1109/RITA.2015.2486338
6. Richter T, Tetour Y, Boehringer D (2011) Library of labs - a european project on the dissemination of remote experiments and virtual laboratories. In: Proceedings - 2011 IEEE international symposium on multimedia, ISM 2011, pp 543–548 IEEEhttps://doi.org/10.1109/ISM.2011.96
7. Sáenz J, Chacón J, De La Torre L, Visioli A, Dormido S (2015) Open and low-cost virtual and remote labs on control engineering. IEEE Access, 3:805–814 https://doi.org/10.1109/ACCESS.2015.2442613 (2015)
8. Husain Siddiqui M, Purohit V, Mane S (2016) Embedded web server based netlab for remote access. In: Proceedings of the international conference on inventive computation technologies, ICICT 2016, pp 1–5. IEEE https://doi.org/10.1109/INVENTIVE.2016.7830192
9. Govaerts S, et al (2013) Towards an online lab portal for inquiry-based stem learning at school. In: International Conference on Web-Based Learning, pp 244–253. Springer, Berlin, Heidelberg
10. Sáenz J, de la Torre L, Chacón J, Dormido S (2019) A new architecture for the design of virtual/remote labs: the coupled drives system as a case of study In: 2019 24th IEEE international conference on emerging technologies and factory automation (ETFA), pp 769 775) IEEE
11. Cornetta G, Touhafi A, Togou MA, Muntean GM (2019) Fabrication-as-a-service: a web-based solution for STEM education using Internet of Things. IEEE Internet Things J
12. Moodle Homepage, https://moodle.org/. Accessed 16 Nov 2020
13. Erl T (2008) SOA. Principles of service design, Prentice-Hall service-oriented computing series. Prentice Hall, Upper Saddle River, NJ, London
14. Liu X, Zhang T, Hu N, Zhang P, Zhang Y (2020) The method of Internet of Things access and network communication based on MQTT. Comput Commun 153:169–176

15. Pimentel V, Nickerson BG (2012) Communicating and displaying real-time data with websocket. IEEE Internet Comput 16(4):45–53
16. Fernandes JL, Lopes IC, Rodrigues JJ, Ullah S (2013) Performance evaluation of RESTful web services and AMQP protocol. In: 2013 Fifth international conference on ubiquitous and future networks (ICUFN), pp 810–815. IEEE
17. Andy S, Rahardjo B, Hanindhito B (2017) Attack scenarios and security analysis of MQTT communication protocol in IoT system. In: 2017 4th international conference on electrical engineering, computer science and informatics (EECSI), pp 1–6 IEEE
18. Erkkilä JP (2012) Websocket security analysis, Aalto University School of Science, 2–3

Remotely Teaching Engineering During a Pandemic

Patrick Kane(✉)

Cypress Semiconductor, An Infineon Technologies Company, San Jose, USA
patrick.kane@infineon.com

Abstract. The sudden onset of the COVID-19 pandemic in early 2020 caused schools to close their campuses and switch to online classes for the Spring term with very little notice. Many institutions continued to hold online courses for the balance of 2020. This mixed methods study is a snapshot of what technologies, methods, and course delivery tools were employed to teach engineering courses online in Spring 2020. The study included 40 professors and lecturers from 35 institutions and 25 countries. A 27-question survey regarding how the educators coped with the sudden change from in-person to online teaching was created and uploaded to Survey Monkey. Of particular interest were hands-on engineering lab courses and the equipment students used to remotely complete the lab experiments. The results indicate that Zoom and Moodle were the most popular course delivery tools among those that participated in the study. Many institutions used more than one course delivery technology. Arduino "shield compatible" (not actual Arduino) development kits from several manufacturers were the most popular type of hardware used in hands-on engineering lab courses.

Keywords: COVID · Teaching · Engineering · Online

1 Introduction

The sudden onset of the COVID-19 virus caused a sea change at educational institutions world-wide. Starting in March and April of 2020, virtually all academic institutions prohibited on-campus activities and courses were moved online with very little time to prepare for the change. (The Chinese Ministry of Education closed campuses in China on January 20, 2020 however, none of the participants in the study was teaching at a university in China) [1]. While many subjects lend themselves to an online format, hands-on engineering laboratories is not one of them [2]. Yet, hands-on laboratories are an essential component in awarding technical degrees [3]. This study surveyed 40 educators as to their experiences teaching engineering courses online in the Spring of 2020, during this unprecedented pandemic event. Special attention was focused on the challenges of teaching hands-on engineering lab courses online.

The study was purposely kept to a simple survey questionnaire. One goal of this study was to obtain educator's perceptions of their online teaching experiences as the Spring 2020 school terms were ending. Another goal of this study was to discover from the participant's perspective what worked and what did not work. Pertinent to the study was how technologies developed for the classroom were adapted to online use. The

M. E. Auer et al. (eds.), *Online Engineering and Society 4.0*, Lecture Notes in Networks and Systems 298, https://doi.org/10.1007/978-3-030-82529-4_20

focal point of the study was what hardware was used in hands-on labs and how effective it was in the online environment.

The author assumes that there will be many more in-depth studies of the COVID-19 pandemic and how it affected engineering education. This study is meant to be a snapshot in time (Spring 2020) while the memory was still fresh in the participant's minds. As such, this study should be viewed as a precursor to future research.

2 Methodology

A 27-question survey was created and posted to Survey Monkey. Invitations were sent out to 200 potential participants during June 2020. Out of this number of potential participants, 40 responded and completed the survey. The survey collector was opened on June 16, 2020 and closed on September 18, 2020.

This mixed-methods study took the following approach to analyzing and presenting the data. It combines qualitative instructor perspectives of the success and impressions of their experience along with quantitative data concerning the instructors, disciplines, the tools used, the courses taught, and other key metrics of concern. The questionnaire data was segregated into qualitative and quantitative chunks. The data was then grouped together where it made the most logical sense and analyzed (Fig. 1).

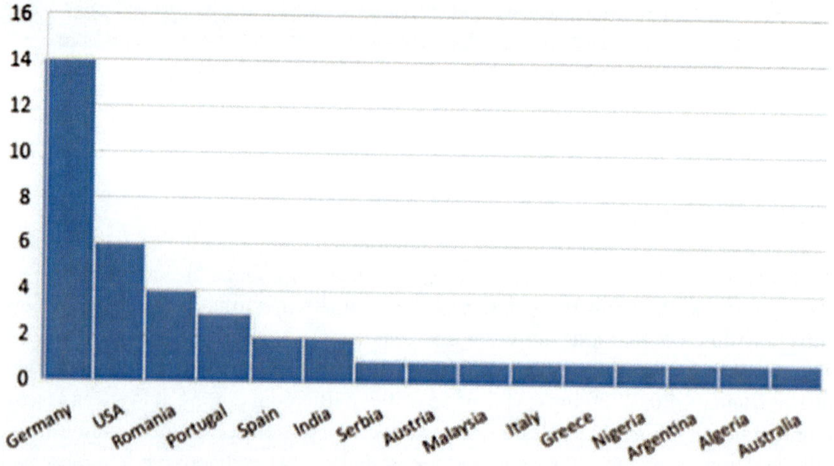

Fig. 1 Number of survey participants by country

3 The Data

3.1 Online Teaching Experience and Total Years of Teaching Experience

The years of teaching experience among the participants ranged from 3 to 44 years. Of the 40 respondents to this question, almost half (19) had previous experience teaching online. Four of the respondents reported that they had been using the flipped classroom or blended learning styles of teaching, onc participant had cxpericnce teaching a Massive Open Online Course (MOOC), and the other 14 respondents reported experience with a fully online course. Almost all participants with ten or more years of teaching experience had taught some form of online course. One interesting data point is that with two exceptions, those with less than ten years teaching experience had not taught an online course.

3.2 Number of Courses Taught, Types, and Course Levels

The course levels taught spanned the sequence from first year to graduate level. Notably, graduate courses – which lend themselves to teaching online as they typically have no lab sections [4], were in the plurality with 21 out of the total of 86 total courses reported on by the respondents. The highest number of courses taught by a single participant was 9.

Eighteen of the participants reported that they had taught hands-on lab courses in the Spring 2020 school term. These types of courses usually require some type of hardware for the students to work with. Even a rudimentary DC circuits course typically requires resistors, capacitors, inductors, a Digital Multi Meter (DMM), and a power source to complete lab exercises, unless they are using simulation tools as two participants reported.

3.3 Teaching Technologies Used

The number of different technologies and tools that were used to deliver the courses was 27. The tools used ranged from the most current technology to simple email. Zoom was used by 20 of the participants, while Moodle came in second with 14 participants using this tool. It is notable that Zoom charges fees for its tool while Moodle is an open-source tool. Other notable technologies used were a variety of Google-based tools such as Google Meet and Google Classroom as well as BigBlueButton. Several institutions reported using more than one technology in a single course. For example, one school used recordings for the lectures, Zoom for office hours, Moodle for quizzes and all lecture materials, and OpenMeeting for live meetings. This same institution later replaced OpenMeeting with Zoom. Zoom offered several tutorials on their Web Site (still available as of December 2020), including "Tips & Tricks: Teachers Educating on Zoom" and "Zoom for Online Learning updates." Another school started with Skype and then switched to Black Board Collaborate. The course delivery tools used are summarized in Fig. 2.

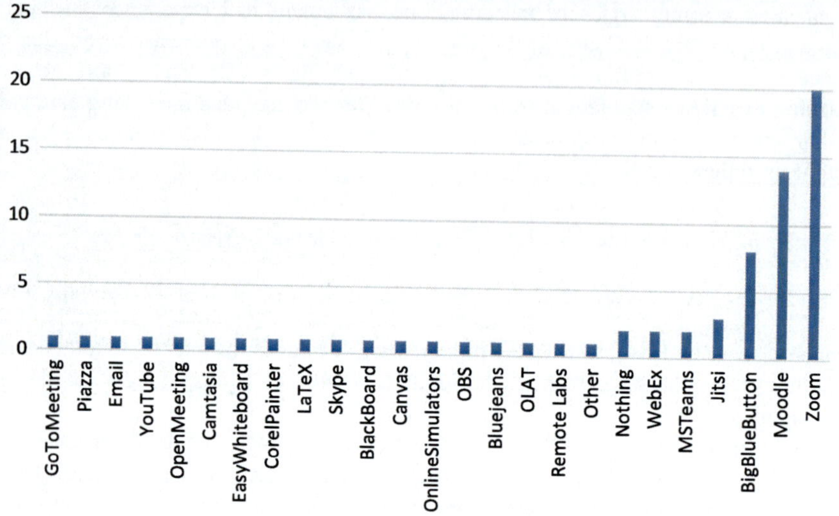

Fig. 2 Course delivery technology used

3.4 Number of Students in Class, Student Location, and Internet Access

The number of students per class ranged from 8 to 600. The person who reported 600 students was teaching a graduate level lecture course. It was not clear how many sections were contained in the course reporting 600 students, but the participant had been teaching online prior to the pandemic, and therefore was somewhat ready to take all of the sections online.

Given the circumstances surrounding the timing of the pandemic crisis, many foreign students were in their home countries (many for Lunar New Year, or Spring Break), and did not (or were unable to) return to the country in which the institution they were attending was located, but attended classes from their home. In some cases, this resulted in time zone issues with students having to attend classes very early in the morning or late at night. The participants reported that 62.5% of their students were not in the same country in which the institution is located during the Spring 2020 term. While teaching virtual hands-on workshops March through July 2020, the author learned that many students, while not on campus were still in the vicinity as conversely, travel restrictions would not allow them to travel home to their native countries.

The participants reported that most of their students did not have issues with Internet access. However, a substantial number (35%) did have issues. Those who had issues reported that the issues were mostly due to slow or unreliable internet. One student did not have a PC and used a smartphone for four weeks until able to obtain a PC and was able to catch up with the rest of the class. Another student using a Mac had an issue running some of the required software (Fig. 3).

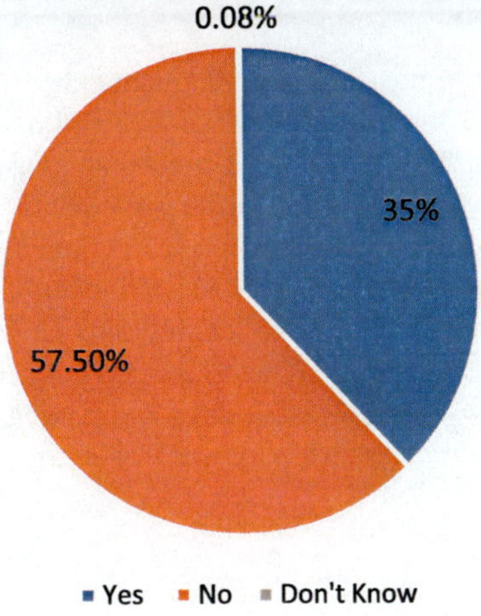

Fig. 3 Computer or internet access issues

3.5 Hardware Used

There was a broad selection of hardware used in online engineering labs according to the participants. This ranged from "nothing in this short turnaround time" to fairly sophisticated systems such as the NI ELVIS. While there were three remote labs, the majority of hardware used was in the student's hands. The university would typically ship a lab kit to students for their use during the course. The most frequently reported test equipment used was a Digital Multi-Meter (DMM). FPGA kits from Xilinx, Lattice, and Intel were used. The majority of the hardware used were "Arduino Shield Compatible" MCU kits from several manufacturers. It is interesting to note that only one respondent reported using an actual Arduino board. The responses are reported in Table 1.

Table 1 Hardware used in online engineering lab courses during pandemic

Manufacturer/Hardware	Responses
Cypress - PSoC 4/5, FM4, FreeSoC (SparkFun)	10
Breadboard, DMM, Resistors, Capacitors, Diodes, 7-Seg display, TFT, Joystick	7
Simulation only (MATLAB, LTSpice)	5
Remote Labs	3
Microchip - Atmel ARM 7DT, Arduino	2
STMicro - Nucleo (Mbed)	2
Xilinx - Zybo Z7	2
Texas Instruments - LaunchPad, TIVA	2
Digilent/NI - Analog Discovery,Explorer,Pmods/myDAQ,myRIO,Elvis	2
Laptop, Phone, Tablet	2
Nothing - too short of time to prepare/ no budget	2
Intel/Altera - DE10	1
Lattice - FPGA Dev kit	1
Raspberry Pi	1

3.6 Online Preparation

Responses regarding how much time the participants had to prepare to teach online spanned the gamut from "not prepared at all" to "I record my lectures for years now (as supplemental material for students). I could immediately use those recordings for the current semester." This correlates with those participants who had previous online teaching experience versus those who did not. Institutionally, 17 of the 40 institutions did not have any online courses prior to the pandemic and therefore, had to decide on course delivery tools on very short notice. Schools on the quarter system had more time to prepare as the Winter quarter was ending as the pandemic began to impact education and there was a natural break between Winter and Spring quarters. Schools on the semester system had to make changes mid-course and cope with the situation as best they could.

3.7 Industry Support

When asked about industry support, the top two responses were remote internships and providing thesis and capstone topics. This was followed closely by providing state of the art technologies, lab materials, and development kits so that students are able to learn "today's technologies and not yesterdays." Several respondents pointed out that having low priced or free tools available was preferred.

As to support provided by industry in Spring 2020, 27 out of 40 participants reported that they received no industry support. Those respondents who did receive support reported that they received kit donations (from Cypress and Digilent /NI) and free licenses (from MATLAB). While two participants reported that they did not ask for any support from industry, one respondent reported "In fact, concerned parties made it more challenging (Fig. 4)."

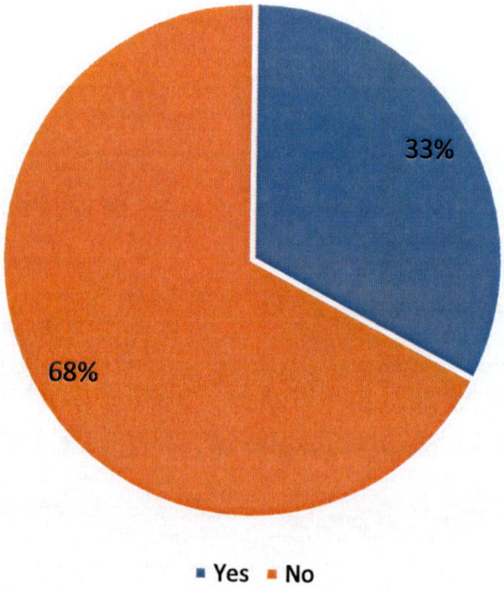

Fig. 4 Support from industry during spring 2020 pandemic

3.8 Pedagogical Differences

Several respondents reported that the lack of physical proximity was an issue. When asked about pedagogical issues, one respondent wrote "I think the main challenge is losing the physical proximity with the students. I think it is only now that we fully understand how important physical proximity is." Another comment related to proximity explained that it was "trickier to figure out how to help them debug hardware remotely."

Professors not using a tool with video capabilities reported that they thought that weaker students who are "afraid to ask for help" were disadvantaged because the professor could not see body language that might indicate they were having trouble with the lab. Even those using tools with video capabilities mentioned lack of body language as a challenge. One respondent wrote "direct personal contact is irreplaceable" and another wrote "personal contact and face to face discussion cannot be replaced by an online system." The data analysis suggests that proximity is an important factor.

Other pedagogical issues included that it was easier to manage the class, reticent students would ask a question in the chat, and the fact that it takes more effort to motivate the students. The instructor cannot see the student's progress in the lab unless the student is sharing the screen. Several participants mentioned that going forward, they will pursue a combination of online and in class courseware.

3.9 Unexpected Benefits

The participants reported a multitude of unexpected benefits One of the benefits mentioned several times was the fact that travel time was essentially nil for both professors and students. This was seen by some as more important for the working student who did not have to expend the time or expense of traveling to and from campus. Other unexpected benefits mentioned were: able to teach while traveling, easier to organize time (for both educators and students), improved quality of work and improved participation, asynchronous learning via videos which can be viewed many times, and students became more independent because they had to take initiative to set up their own home labs and make time for studying.

One participant stated that after their online experience in Spring 2020 their institution was going to create an online degree program. Several participants mentioned that they did not expect the level of efficacy that resulted from creating videos that could be viewed many times. One participant wrote that once the video was created, fundamentals did not need to be explained over and over again. Students were directed to view the video and they could ask questions later if they did not understand something. Other participants reported that they were able to assign more challenging and rewarding homework online. In this same manner, another participant responded that their teaching was more practically oriented and that students had more to measure and more circuits to build. These students had to create and upload a video to demonstrate and explain how they solved an assigned task.

Eleven participants did not think there were any unexpected benefits and one participant wrote that they had to reduce the content by 30% and thought that weaker students were suffering. This participant also reported that there was no infrastructure to teach online at the beginning of the pandemic and that they "had to fight to get the infrastructure up and running."

3.10 Challenges

There were many challenges reported by the survey participants. However, the one challenge that stood out was "electricity." This was from a professor in a developing country. Those of us who live in developed countries often forget the challenges of not having everyday conveniences people take for granted such as reliable electricity when teaching an online course. Beyond this, several participants wrote about "the blurred frontier between work and family life" and "having a quiet environment for synchronous classes at home." This is a challenge not only for educators, but has also been echoed by remote workers world-wide since the pandemic forced people out of the workplace and into the home for those fortunate enough to be able to work remotely.

Another major challenge was having enough time to create online lessons. A common sentiment was that it takes much more time to prepare lessons for an online environment. One participant responded that they were working over 70 hours per week between preparing lessons and teaching. There are no last-minute changes that one can make just before or even during an online class [4]. Maintaining student participation and attention was mentioned as another big challenge. Not knowing how the students were participating and understanding the material due to the lack of body language feedback was frequently mentioned as an issue. It is difficult to know what inactive students are doing and if they are engaged. Keeping both teachers and students motivated after eight hours of Zoom classes was an issue pointed out by several participants. Finally, specifically in lab classes, simply redesigning the projects for a new environment and creating meaningful exams with little time to do so was mentioned as a challenge by several participants.

3.11 Ethics and Integrity

Ethics and integrity refer to knowing that the person getting the credit is that same person who is doing the work and taking the test. In academic terms, authentication is a means of ensuring that the student receiving the grade is the person doing the work [5]. This issue has yet to be resolved as two important issues in online courses remain: authentication and identification [6].

According to the participants, the most common way that this was accomplished was for students to present their student ID on camera before giving a presentation or taking a test. Other solutions mentioned were to have online proctors, LMS login based on password authentication, and in some cases, not to have any test at all. In the latter case the requirement was to demonstrate completion of the labs with photos, lab reports, code, and videos. Another solution was to administer written tests with a video discussion regarding elements of the test to follow.

While most participants had solutions or partial solutions to the issues of ethics and integrity, some participants wrote that ethics and integrity were not issues. One participant wrote that "I trust my students!" However, another participant emphatically wrote that this issue was not resolved. The issue of authentication and identity is one of the most difficult issues to address, not just in online engineering courses, but in online courses at large. The data suggests that the issue is controversial as well.

4 Discussion

Many of the issues described by the participants are issues that would occur in any online course. Pedagogical differences - primarily not having person to person proximity to the students as an example. The issue of authentication and identification in online courses is also not new due to the pandemic. However, some of the corner cases such as not having reliable electricity, or not even having a PC and having to take the course on a smartphone, were unforeseen as the courses at these institutions would never have been put online except for the pandemic.

There is also an issue of the students learning preferences. Today's engineering students are primarily in the cohort known as Generation Z [7]. Research has shown that Generation Z students have an attention span of eight seconds as they are multitasking across five screens. By contrast the previous generation – the Millennials – have a 12 s attention span while multi-tasking across three screens. This may indicate that an unexpected benefit of the current situation is that the students were learning and consuming information in the way that best suits them given their demographics and the fact that Generation Z are the first true "Digital Natives" [8].

4.1 What Worked

Data analysis suggests that the underlying technology worked. Commercial virtual meeting tools such as Zoom, Webex, and various tools from Google and others were successfully adapted and used to conduct the courses online. Some institutions struggled with the technology, but that was due more to the urgency of getting it up and running quickly rather than the technology itself.

Creating videos was mentioned by several participants as an unexpected benefit. Several participants discovered that by creating a video, they only had to teach a fundamental topic once (on a video), and then they could dive deeper into the topic during the class time. One participant opined that the COVID-19 situation sped up the adoption of digitalization that was already happening.

4.2 What Didn't Work

With the exception of the participants who had already been teaching online, the other participants had little or no time to prepare for the online courses. Some institutions had no online delivery method and had to adopt tools such as Zoom and WebEx with little or no time for faculty and student training on the tools. Keeping track of inactive students and knowing if they were engaged in the course was a challenge. Creating meaningful exams, authentication, and identification were also issues at some institutions. Based on the data, students who had Internet issues were in developing countries, or the institution itself was in a developing country.

5 Conclusion

While the participants did an admirable job dealing with the unprecedented move to teach courses online due to the sudden onset of the COVID-19 pandemic, there are still issues that need to be resolved and questions to be answered. In fact, there are more questions than answers. Many of these questions cannot be answered immediately. For example, did the pandemic have any effect on retention rates in engineering as a major? Did the students learn as much online as they would have in a traditional classroom? How many courses will remain fully or partially online when conditions are once again conducive to in-person courses?

What this study provides is a snapshot in time with incomplete results. It was not meant to be comprehensive, but rather considered urgent to gather data while memories were still fresh given the nature of the situation. The answers to the questions uncovered by the study will be left for future work. If this pandemic had occurred even five years ago and most certainly ten years ago, academic institutions may have had to just close their doors and hold no classes whatsoever for the bulk of 2020. The technology and the internet bandwidth simply would not have been sufficient. Students would have lost a year of their academic lives. Thankfully, this did not occur.

References

1. Ning A, Corcoran B (2020) "How China's schools are getting through COVID-19" *EdSurge* April 20 2020. https://www.edsurge.com/news/2020-04-20-how-china-s-schools-are-getting-through-covid-19#:~:text=China's%20Ministry%20of%20Education%20(MOE,Tutoring%20centers%20and%20daycares%20closed.
2. Hsiung SC, Deal WF (2013) Distance learning: Teaching hands-on skills at a distance. Technol Eng Teach, 72(5):36–41. http://www.iteea.org/Publications/TTTarticleindex15.htm
3. Gustavsson I, Nilsson K, Zackrisson J, Garcia-Zubia J, Hernandez-Jayo U, Nafalski A, Nedic Z, Gol O, Machotka J, Pettersson M et al (2009) On objectives of instructional laboratories, individual assessment, and use of collaborative remote laboratories. IEEE Trans Learn Technol 2:263–274
4. Kane P (2018) Exploring the experiences of faculty teaching an online hands-on engineering laboratory course. Doctoral dissertation. University of Phoenix 2018.
5. Gearhart D (2010) The issues related to student authentication in distance education. Int J Technoethics (IJT) 1(1):60–69. https://doi.org/10.4018/jte.2010081005
6. Wiklund M, Mozelius P, Westin T, Norberg N (2016) Biometric belt and braces for authentication in distance education. Paper presented at the european conference on e-learning, Prague, Czech Republic, 2016. Retrieved from: https://search-proquest-com.contentproxy.phoenix.edu/docview/1860070766?accountid=134061
7. Cilliers EJ (2017) The challenge of teaching generation Z. PEOPLE: Int J Soc Sci 3(1), January 2017
8. Granados N (2017) Gen-Z media consumption: it's a lifestyle, not just entertainment. Forbes Magazine, June 2017. https://www.forbes.com. Accessed Jan 2020

Forecasting and Monitoring Smart Buildings with the Internet of Things, Digital Twins and Blockchain

Xichun Yang[1](✉), Ananda Maiti[1], Jinhao Jiang[1], and Alexander Kist[2]

[1] Discipline of ICT, University of Tasmania, Newnham, Australia
{xichuny, jinhao.jiang}@utas.edu.au
[2] School of Mechanical and Electrical Engineering, University of Southern Queensland, Toowoomba, Australia
kist@ieee.org

Abstract. Smart buildings have grown in numbers and capabilities worldwide. Smart buildings rely on the sensor networks and Internet of Things (IoT), which allows the development and optimization performance through various methods, including efficient energy management solutions and predictive technologies for future decisions. Smart buildings generate data in real-time, and human-based choices are often taken in real-time. IoT provides connectivity for the various sensors. Smart buildings must ensure the proper operation with minimum time and energy with regards to dynamic human demands and performance in real-time. Blockchain technology can ensure data integrity and regulated autonomous flow of information between users who have a varying degree of trust. Therefore, the combination of Blockchain and the IoT can achieve data integrity between IoT devices installed in the smart buildings where multiple humans are involved and have different non-overlapping expectations from the building's environment and response. In this paper, a set of predictive technologies are analyzed, and their common parameters are shown to affect the possible prediction and visualization of the IoT data for users. A digital twin based on Blockchain is proposed where individual users owing some sensors can use and visualize processed data based on rules set by the owners in the Blockchain.

Keywords: Internet of Things · Blockchain · Digital twin · Predictive analysis · Smart building

1 Introduction

Smart buildings have become a popular trend worldwide. Mordor Intelligence forecasted that the market of the smart building would grow from 2020 to 2025 with a 23% annual growth rate [1]. Smart building applications optimize diverse factors including management efficiency, energy consumption and comfort of people.

Internet of Things (IoT) sensors in the smart building can monitor parameters such as temperature, humidity, and personnel activities. This allows managing microenvironments in buildings in real-time. Other sensors, such as the PIR sensor can monitor the real-time occupancy in the smart building. Based on the collected data, the IoT

M. E. Auer et al. (eds.), *Online Engineering and Society 4.0*, Lecture Notes in Networks and Systems 298, https://doi.org/10.1007/978-3-030-82529-4_21

system can optimize the space configuration and maximize the use of reasonable space, which can achieve the purpose of optimizing the efficiency of the building management. Smart buildings often rely on a network of IoT sensors, which allows the development and optimization of smart buildings through multiple layers:

- collecting raw data from sensors
- process the raw data to see past events or possible future.
- Identify methods to improve the performance of the building, including the human elements. This includes giving responsibilities to people to perform tasks and associated authorizations to specific people to do them, e.g. asking someone to manage a room's sensors [2–4].

Blockchain can support these objectives. Blockchain is a peer-to-peer implementation for record-keeping which provides the IoT networks with transparency. Blockchain also enables autonomous exchange of information while maintaining the privacy of the owner of the information.

A blockchain stores information in a series of linear blocks of fixed size. Each block contains the hash of the previous block contents. The entire Blockchain is available in multiple nodes. This way, the Blockchain can guarantee the reliability of the source of the data as well. Blockchain has been used in many application areas for these characteristics, such as supply chain [5, 6], asset management [7] and crypto currency [8, 9]. Blockchains can be permissioned (where the member of the Blockchain have a level of trust between them) or permission less (the members are unknown to each other). In the context of this paper, only the permissioned Blockchain, e.g. Hyperledger Fabric, is considered [10].

The key contribution of this paper is an analysis of different time series forecasting/prediction methods that can potentially be run in a blockchain. It is shown that while prediction and related processed information is possible, it depends on many *parameters* that can be unique to different users. So, a digital twin-based methodology for integrating and managing user preferences maintaining their privacy is proposed as well.

The remainder of the paper is organized as follows: Sect. 2 discusses the related work on the prediction algorithms, digital twin and Blockchain. Section 3 presents the proposed digital twin based processed and raw data visualization mechanism with Blockchain. Section 4 presents the analysis results of various prediction algorithms and discussion on the means of integrating the prediction into the Blockchain.

2 Related Work in Smart Buildings

In this section, the various factors that determine the quality and quantity of forecasts based on time series data is discussed. IoT sensors in smart buildings primarily generate time-series data. The basic strategy of using digital twins with IoT for easy visualization of data is also discussed.

2.1 Smart Building

A significant benefit of deploying smart building is to achieve real-time monitoring. It enables users to understand the real-time state of buildings through intuitive visual data analysis icons and optimize and adjust system operating strategies based on application conditions. It is essential to consider how to maximize the efficiency of limited smart sensors in buildings. The sensors can monitor and report various environmental problems, including fire alarms, office air quality, detection of hazardous chemicals and structural integrity of buildings [11–15].

2.2 Prediction and Anomaly detection Mechanisms for IoT Data

The objective of anomaly detection is to find abnormal activities in the available dataset. The statistics-based and machine learning-based techniques are commonly used anomaly detection processing methods. The statistical method is mainly to make assumptions about the distribution of data and find out the "anomaly" defined under the assumptions. The time series prediction models are typically statistics-based anomaly detection method. The statistics-based techniques are suitable for low- dimensional data and have good robustness, but they heavy dependence on assumptions. The anomaly detection technique has been applied in different aspects in real life, including bank frauds, medical service, network intrusion detection, traffic safety and industrial equipment failures [21–23]. All these methods can be used with a blockchain. They take raw sensor data from IoT systems and then process them to cure data.

An effective machine learning field for prediction on time series data is called time series forecasting. Time series means the data collected at constant time intervals. Time series forecasting figures out the regular patterns in the past time series data and predict future events. Thus, the sequence of events is one of the main elements for time series forecasting. Time series forecasting analysis has various models and five of them will be compared in this research, including Autoregression (AR), Autoregressive Moving Average (ARMA), Autoregressive Integrated Moving Average (ARIMA), Simple Exponential Smoothing (SES) and Vector Autoregression (VAR) [16–19].

In addition to the above five-time series prediction models, the Holt-Winters algorithm is used in this research for time series forecasting as well. The Holt-Winters algorithm is based on a simple exponential smoothing technique [20]. The simple exponential smoothing follows two principals. The first is forecast of the future, using the current level to predict the point at the next moment. The second principle is to estimate the current level, using the average of the current observation value and forecast value. The Holt-Winters algorithm is the promotion of simple exponential smoothing in trend and seasonality. It can describe trends and cycles and can quickly adapt to changes in trends and cycles. However, Holt-Winters only uses the forward sequence information without smoothing, so it is more sensitive to anomalies. Anomaly detection is fundamental to establish the training dataset.

2.3 Smart Building and Blockchain

The combination of Blockchain and the IoT can achieve data security between IoT devices installed in the smart buildings. Blockchain can help solve equipment warranty issues in smart buildings. The data of the Blockchain is distributed in many computers, and the establishment of device *smart contracts* ensures the uniqueness and correctness of the results.

Blockchain is opening up new IoT features because it allows exchange of value and digital assets between participants automatically without the need for intermediaries. Value can be approval in the form of services, products, or smart contracts. The Blockchain creates a trusted chain of events, transactions, assets and key project details. The integration of email, project management systems, and accounting systems can create verified records for all transactions on the Blockchain and ensure that data will not be lost.

Within Blockchain a *smart contract* is program that is run in a distributed manner among the peer nodes of the Blockchain. This guarantees that the program that is running known to all and is processing the right data in the right manner to generate a processed output. The processed output in this context is the predicted sensor in the near future.

The *smart contract* plays a key role in the interaction between the digital twin and blockchain network. Only if a smart contract is installed and instantiated, is it possible for the digital twin program to get the latest data from the blockchain network. The smart contract is normally installed in one of the nodes in the blockchain network, and each node is inside a peer. Besides, multiple smart contracts can be created under each blockchain channel.

Energy consumption billing in various offices in a smart building is a difficult task. The application of Blockchain can support billing in a separate building, replacing the cumbersome manual billing method. Blockchain can also help solve equipment warranty issues in smart buildings [24, 25]. The data of the Blockchain is distributed on millions of computers, and the establishment of device smart contracts ensures the uniqueness and correctness of the results. Suppliers will not be confused due to equipment failures, and the responsible body will be clearer. The combination of blockchain and smart buildings is getting closer.

2.4 Digital Twins

Digital twins are virtual representations of physical entities that have become in the field of smart buildings. A virtual representation of a building is driven by a simulation process relaying on building models and sensor readings. The mapping in the virtual space reflects the life cycle of the physical building. The smart building projects rely on a combination of data warehouse and AI technologies. Digital twins in the smart building field make use of 3D model computing to assist the design of the building. IoT devices that collect building data. Those devices include temperature sensors, light sensors, PIR sensors. The collected data forms part of the digital twin which reflects the dynamics of the physical environment in the building. Based on the real-time data collection and tracking, data can be manipulated through AI algorithms and data mining technologies. Figure 1 shows thc Digital Twin in Smart Building's logical structure [26].

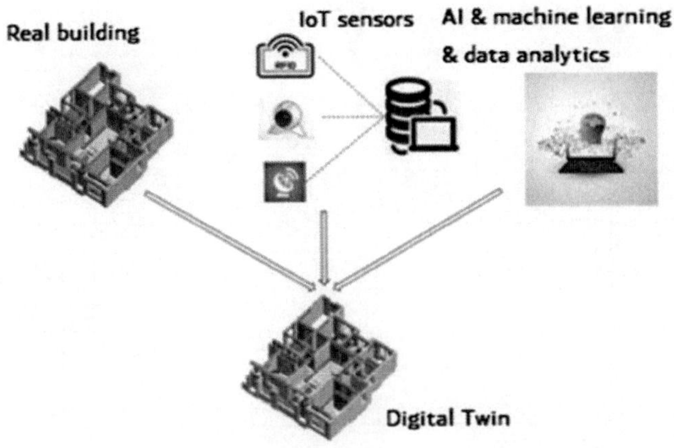

Fig. 1 Digital twin in smart building's logical structure

3 Proposed Digital Twin based Visualization Architecture

The data collected by the IoT sensors must rely on a management mechanism for authorization and processing of the data. The aim of the IoT system in smart building is to predict future data points and identify anomalies. A digital twin-based visualization architecture is proposed here to implement this analysis. To increase trust in the smart building management system, the use of Blockchain is propose as it is both non-tamperable and decentralized. Trust created by blockchain technology can provide strong security guarantees for the data that is used by the system. The main focus of this paper is to combine the characteristics of Blockchain and IoT to propose an effective smart building management system supporting predictive forecasting and real-time monitoring. The prosed architecture is presented in Fig. 2 and described in the following sections.

3.1 Systems Settings and Data Collection Process

Thc core of thc smart building system is integration and informatization, and this smart building system has three layers of functions including.

- data collection layer,
- data processing layer,
- digital twin layer,

The data collection layer is the foundation of the system because all operations depend on sensor and actuator data. A comprehensive understanding of the physical environment in the building is the prerequisite of data collection and the digital twins.

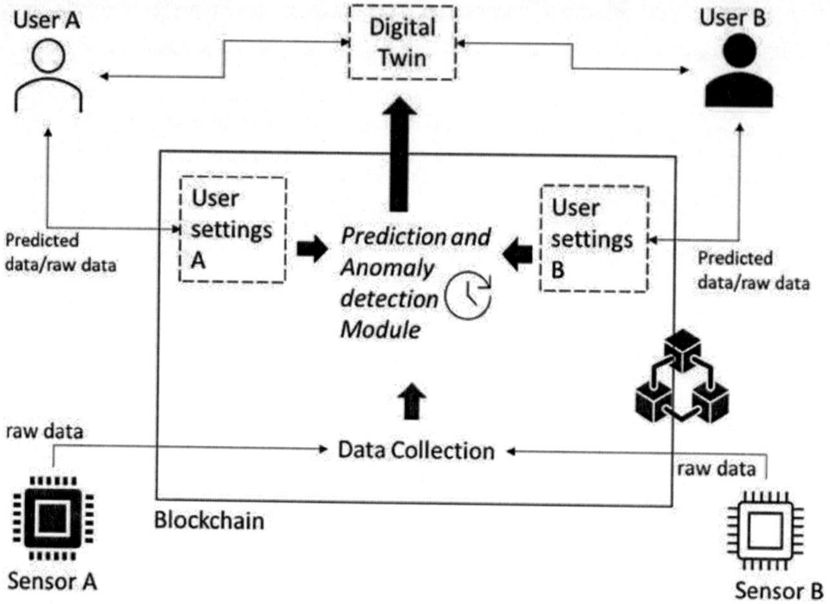

Fig. 2 The proposed system architecture

As an example of the system described in this paper, light sensors, temperature sensors, and PIR sensors are deployed in different rooms of a building to provide real-time and reliable data sources. These sensors collect data every few minutes as defined by the user, for example, every 5 min. Over time a considerable amount of historical data is accrued in the database in the Blockchain.

The *data processing* layer *or prediction and anomaly detection* layer is the place where IoT data are analyzed and actions are created. The change in the physical environment in each room creates changes in sensor reading which is stored in blockchains. Within the blockchain several data pre-processing techniques can implemented in smart contracts: *data cleansing* and *data redundancy* algorithms. This layer also takes various inputs based on the users' settings which are stored in the Blockchain. Each user has isolated settings, and these settings are used by the smart contract discreetly. The data processing layer can convert the raw data from the data collection layer to predicted data. Predicted data consists of the following:

- forecasts i.e. possible future time series of the sensors data, based on historical data
- anomalies within the forecasting based on the historical data
- patterns of the past data

For the prediction algorithms, the proposed system makes use of *lightweight* machine learning algorithms with respect to their space/time complexity. Users can see only the raw sensor data if they have authorization and the processed or predicted data. This is managed by Blockchain.

The *digital twin* layer is used to visualize the information based on the outputs of the data processing layer. The digital twin layer can also provide key calculation results and advisories. The 3D model gives information about the actual smart building status so that users can have deeper insights into what should be done to optimize the whole building. The users can store individual settings through this interface.

3.2 Blockchain Based IoT System

Figure 2 indicates the proposed Blockchain based IoT system in this research. There are different users in the smart building environment and different types of sensors will collect raw data. Prediction and anomaly detection module could make assumptions based in the raw data. Data is stored reliably in the Blockchain under the *smart contract*. The Blockchain based prediction mechanism can have two approach towards the human demands for information from the IoT system:

- it could manage requests as much as possible and
- if not possible then the system will notify the related user.

The digital twin in the smart building makes use of 3D model to assist the design of the smart building. Different sensors are connected with the blockchain network through network channels, smart contracts, and nodes. These sensors are temperature sensors, light sensors, and PIR sensors. At definite intervals, sensors send a value and timestamp to the blockchain network. Then the digital twin can retrieve the data (raw or processed) and generate the digital twin customized for a specific user. In this way, the digital twin uses the data generated from IoT sensors.

Blockchain is a decentralized technology and is mainly used to avoid or resolve disputes. As every data is stored and encrypted in multiple nodes, there is no way for any peer or members to manipulate the data. Also, if the peer fails in providing the required data in time, that can be recorded as well, and this will be indisputable among the members.

4 Results

This section presents some results on the proposed architecture. First the different parameters required to the prediction/anomaly detection mechanism is discussed. Then a digital twin mechanism if presented which uses the Holt-Winter algorithm and then the application in the Blockchain is discussed.

4.1 Rate of Data Collection vs the Accuracy of Prediction

Sensors in the IoT system can collect data at specific rates. Depending upon how much of that data is used for predictions, different levels of accuracy can be achieved. Some predication algorithms can be time consuming.

A test was done to see the accuracy of the different candidate prediction models and algorithms that can be run from Blockchain. A data set consisting of a temperature time series has been included into each model and detects the variance of differences between the actual and the future predictions. The actual temperature data is collected every minute. Data set is trained on 7000 data points from 8000 data points available. Prediction is done for the next 100 (i.e. 7001 to 7100) data points keeping the time interval the same.

In order to compare the difference between predicted temperature data and real temperature data, the accuracy of the six different prediction models is calculated in three steps. The first step is to use the absolute value to compare the difference between predicted temperature and real temperature. Then the variance will be calculated based on the temperature prediction. The third is to calculate the accuracy of the different time series models based on the variance and the maximum and minimum prediction results.

The results of the above experiment show that the ARIMA and the Holt-Winters model has the highest accuracy i.e., the least amount of variance between the actual and predicted values. Compared with findings of previous research, it is confirmed that the ARIMA model include the advantages of AR and ARMA model. ARIMA model has a simple structure and high efficiency of prediction. Moreover, the nonstationary time series could only be analyzed by the ARIMA model. Therefore, the ARIMA model has been chosen as the method of prediction during the reconfiguration process of the IoT system in the smart building. Alternative algorithms may be chosen such SES as well depending on the actual sensor and nature of data (Table 1).

Table 1 The performance of time series prediction models

Time Series Prediction Models	The variance of the difference between the actual and predicted temperature	Simulation Execution Time
AR	0.07%	2.38 s
ARMA	4.035%	31.21 s
ARIMA	0.068%	181.39 s
SES	0.086%	2.56 s
VAR	0.229%	98.25 s
Holt-winters	0.064%	134.45 s

4.2 Anomaly Detection Rate

The following two tables show the number of anomalies detected by two different sensors using ARIMA. For this, the same data set was considered, but the amount of data was altered for each prediction. The *time gap* considered between the sensor data was changed from every five to every minute. With the change of the interval, there is a higher probability to encounter more anomalies in the smart building IoT system.

As shown in Table 2 (Sensor A), if the proposed system is running every 5 min, there are 4 anomalies detected. When the time gap changes to every 1 min, the number of anomalies has increased to 54 anomalies. The same trend can be seen for Sensor B. The number of anomalies has increased from 14 to 44 when the interval changed from every five to one minute.

Table 2 Temperature sensor anomaly detection overview with different time gap with ARIMA

Time Gap	Anomalies in Sensor A	Anomalies in Sensor B
Every 5 min	4	14
Every 4 min	6	17
Every 3 min	8	21
Every 2 min	16	34
Every 1 min	54	44

There are several parameters in the above discussions that determine the quality of the predictions and information viewed by the users. Such parameters are based on individual preferences and needs. Depending on what the user requires, these parameters can be set accordingly:

- Time Gap for determine the prediction
- Sensor selection
- The future timeline i.e. how far the prediction should go
- Prediction model

4.3 The Digital Twin

A digital twin of a building was created as shown in Fig. 3. There are a number of sensors:

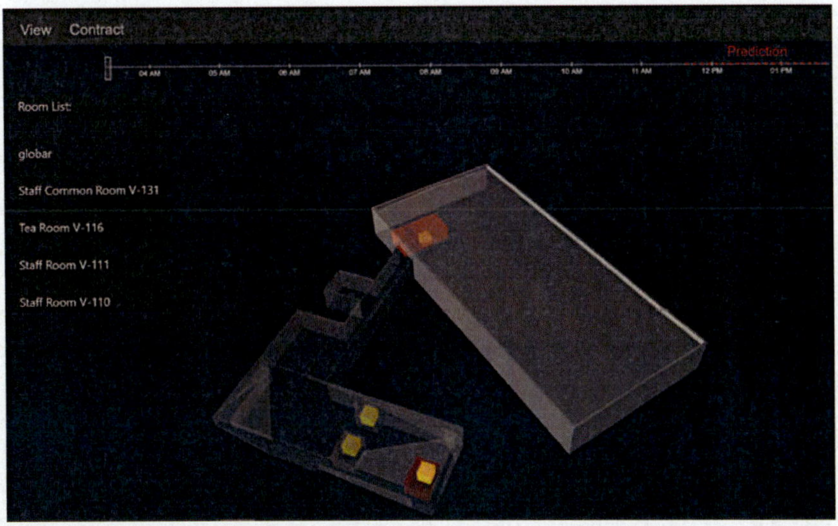

Fig. 3 Web-based digital twin interface

- Two common room sensors collecting human movement (with PIR sensor) and temperature
- Two sensors in individual rooms, each owned by private individuals.

The digital twin has the following components:

- An observation panel (left side) to observe individual room data if the viewer has permission to view the data. The common room data is available to all. The private sensor' raw data may be restricted the owner. There is also global view to see a bird's eye view of the building.
- A time bar at the top for visualization according to time. The figure below shows a prediction time of up to 2 h using the raw data from the sensors. The Holt-Winters algorithm is sued for this prediction, but others can be used as well.
- A 3D model of the object. The concerned object can be given different color patterns which may be changed according to the shifting in the time bar at the top. This allows the viewer to see the change of the past and the future prediction of the areas in the building.
- A contract or settings panel to change the user setting if they need to.

4.4 Integration in Blockchain

The above programs and equations can be implemented using NodeJS, python or Java in *smart contracts* within Blockchain. Several parameters regarding the above can be set for individual users in the blockchain system:

- The parameters to decide the prediction e.g. how far they want to predict and how much accuracy is desired based on the prediction model.
- The parameters to determine data collection strategies from the sensors, e.g. time gap.

Blockchain can guarantee the privacy of each user, their data and their settings in the Blockchain and yet be able to share the information with others for predictions. The digital twin built on top of the blockchain system can be used to visualize the data in a realistic manner.

The smart contracts can ensure the data provide by each user is collected timely and reliably. They can also ensure the data provided for predictions is also genuine and the prediction is valid. The Blockchain can then return the prediction to another user, but not the underlying raw data. But as blockchain smart contracts are run by Blockchain, the smart contract guarantees that the provider of the raw data is following the rules. Each user can see.

- their own information as a whole in raw format
- raw data from other if permitted through the Blockchain
- predicted information including future possibilities and the anomalies

This may include raw data from other users with permissions set in the smart contracts of the Blockchain. The Blockchain can be run by multiple peers responsible for management of the building and the owners:

– Personnel responsible for individual building management
– Authorities responsible for managing multiple buildings within a campus or organization
– Individual users (only to a certain extent proportionate to their stake in the data collected from the sensors they are handling in the system)
– External authorities for maintaining regulations.

5 Conclusions

In recent years, people have paid more attention on the development of smart buildings. In terms of building management, it is essential to forecast and monitor smart buildings efficiently. In this paper, we have combined the characteristics of Blockchain and the IoT network to establish a digital twin-based visualization architecture, which could achieve the predictive forecasting and real-time monitoring of smart buildings. The ARIMA model and Holt-Winters has the highest accuracy in the forecasting performance. Users can view the information and determine the quality of the predictions based on several parameters to satisfy their own need. The 3D model provided by the digital twin gives information about the actual smart building status so that users can have deeper insights into what should be done to optimize the whole building. The privacy of each user can be guaranteed by the Blockchain, including their data and their settings. The integrated Blockchain also helps to achieve information interaction of different users. The proposed IoT system can be used further to involve actuation to change the status of smart buildings and raise the management and maintenance of smart buildings to a higher level.

References

1. Smart building market - growth, trends, forecasts (2020 - 2025). Mordor Intelligence (2019)
2. Ghayvat H, Mukhopadhyay S, Gui X, Suryadevara N (2015) WSN- and IOT-based smart homes and their extension to smart buildings. Sensors (Basel) 15:10350–10379
3. Minoli D, Sohraby K, Occhiogrosso B (2017) IoT considerations, requirements, and architectures for smart buildings—energy optimization and next-generation building management systems. IEEE Internet Things J 4:269–283
4. Jin X, Wang G, Song Y, Sun C (2018) Smart building energy management based on network occupancy sensing. J Int Council Electr Eng 8:30–36
5. Jangirala S, Das AK, Vasilakos AV (2019) Designing secure lightweight blockchain-enabled RFID-based authentication protocol for supply chains in 5G mobile edge computing environment. IEEE Trans Ind Inform
6. Banerjee A (2018) Blockchain technology: supply chain insights from ERP. In: Advances in computers, vol 111, pp 69–98. Elsevier
7. Verma D, Desai N, Preece A, Taylor I (2017) A block chain based architecture for asset management in coalition operations. In: Ground/Air Multisensor Interoperability, Integration, and Networking for Persistent ISR VIII, pp. 101900Y. International Society for Optics and Photonics
8. Underwood S (2016) Blockchain beyond bitcoin. ACM New York, NY, USA

9. Crosby M, Pattanayak P, Verma S, Kalyanaraman V (2016) Blockchain technology: beyond bitcoin. Appl Innov 2:71
10. Cachin C (2016) Architecture of the hyperledger blockchain fabric. In: Workshop on distributed cryptocurrencies and consensus ledgers
11. Wei X, Zhang X, Geng L (2019) Smart building temperature monitoring system using matplotlib. J Comput Sci Inf Technol 7
12. Jia M, Komeily A, Wang Y, Srinivasan RS (2019) Adopting Internet of Things for the development of smart buildings: a review of enabling technologies and applications. Autom Constr 101:111–126
13. Xu W, Zhang J, Kim JY, Huang W, Kanhere SS, Jha SK, Hu W (2019) The design, implementation, and deployment of a smart lighting system for smart buildings. IEEE Internet Things J 6:7266–7281
14. Qolomany B, Al-Fuqaha A, Gupta A, Benhaddou D, Alwajidi S, Qadir J, Fong AC (2019) Leveraging machine learning and big data for smart buildings: a comprehensive survey. IEEE Access 7:90316–90356
15. Hyman B, Alisha Z, Gordon S (2019) secure controls for smart cities; applications in intelligent transportation systems and smart buildings
16. Xu W, Peng H, Zeng X, Zhou F, Tian X, Peng X (2019) Deep belief network-based AR model for nonlinear time series forecasting. Appl Soft Comput 77:605–621
17. Baptista M, Sankararaman S, de Medeiros IP, Nascimento C Jr, Prendinger H, Henriques EM (2018) Forecasting fault events for predictive maintenance using data-driven techniques and ARMA modeling. Comput Ind Eng 115:41–53
18. Zhang X, Yu Y, Xiong F, Luo L (2020) Prediction of daily blood sampling room visits based on ARIMA and SES model. In: Computational and mathematical methods in medicine 2020
19. Luo L, Luo L, Zhang X, He X (2017) Hospital daily outpatient visits forecasting using a combinatorial model based on ARIMA and SES models. BMC Health Serv Res 17:469
20. Tanyavutti A, Tanlamai U (2018) ARIMAX versus holt winter methods: the case of blood demand prediction in Thailand. Int J Enviro Sci Educ 13:519–525
21. Ahmed M, Mahmood AN, Islam MR (2016) A survey of anomaly detection techniques in financial domain. Future Gen Comput Syst 55:278–288
22. Wang L, Jones R (2020) Data analytics for network intrusion detection. J Cyber Secur Technol 4:106–123
23. Weller-Fahy DJ, Borghetti BJ, Sodemann AA (2014) A survey of distance and similarity measures used within network intrusion anomaly detection. IEEE Commun Surv Tutor 17:70–91
24. Xu Q, He Z, Li Z, Xiao M, Goh RSM, Li Y (2020) An effective blockchain-based, decentralized application for smart building system management. Real-Time Data Analytics for Large Scale Sensor Data, pp 157–181. Elsevier
25. Singh SK, Rathore S, Park JH (2020) Blockiotintelligence: a blockchain-enabled intelligent IoT architecture with artificial intelligence. Future Gen Comput Syst 110:721–743
26. Altun C, Tavli B, Yanikomeroglu H (2019) Liberalization of digital twins of IoT-enabled home appliances via blockchains and absolute ownership rights. IEEE Commun Mag 57:65–71

IIoT Laboratory Model for Remote Control System Applications

Marjan Golob(✉), Božidar Bratina, Milan Rotovnik, and Nenad Muškinja

Faculty of Electrical Engineering and Computer Science, University of Maribor, Koroška cesta 46, 2000 Maribor, Slovenia
marjan.golob@um.si

Abstract. This paper describes how the Industrial Internet of Things (IIoT) technology can be used to build an efficient laboratory learning model of a remote and distributed control system. We describe the development of simple Internet of Things (IoT) devices for industrial application of online process measurements. Our experiences are presented with implementation of the IIoT data concentrator based on different hardware platforms. The data are transferred to the cloud system, where the application of fault detection and analysis was realized and visualized by free web technologies. A Fault Detection and Isolation (FDI) application was developed for a laboratory model of connected hydraulic tanks, by implementing the data driven Principal Component Analysis (PCA) method with partial models for isolation purposes.

Keywords: Industrial Internet of Things · Distributed control systems · Remote laboratories · Control education

1 Introduction

The term Internet of Things was proposed by Kelvin Ashton in 1999 to describe a system where the Internet is connected to the physical world via ubiquitous sensors [1]. He gave the name to demonstrate the idea that objects could be connected to the network manually. Industrial Control Systems are generally developed using Distributed Control System (DCS), Programmable Logic Control (PLC), Supervisory Control and Data Acquisition (SCADA) system, Remote Terminal Units (RTU), control servers, Human–Machine Interface (HMI), and many other industry specific devices. This integration between legacy industrial information systems and IoT is known as the Industrial Internet of Things (IIoT). The IIoT has been implemented in a variety of industries, like Health Care, Logistics, Avionics, Waste Management, the military, smart city and many other fields under the auspices of Industry 4.0 initiatives [2]. IIoT applications connect machines, people, processes, Cyber-Physical Systems, and refers to the interconnected sensors, actuators, instruments, and other devices with computers to support the manufacturing industry. Future engineers will be facing technology complexity challenges. As stated in the study [3], one of the required competences for Industry 4.0 is domain knowledge, which includes methodologies, languages and tools that are important when dealing with the requirements of Industry

M. E. Auer et al. (eds.), *Online Engineering and Society 4.0*, Lecture Notes in Networks and Systems 298, https://doi.org/10.1007/978-3-030-82529-4_22

4.0 projects` implementation. The new paradigms, such as Cyber Physical Systems, Cloud-based Manufacturing, Smart Manufacturing, and others, are becoming important [4, 5]. For engineers, a deep understanding of interrelations between the electrical, mechanical and computer components will be a vital ability. For example, Industry 4.0 devices can contain complex systems of IIoT devices, cloud-computing based analytics, management complexity associated with networked systems, soft-ware, configuration, compliance, and security. For engineers, a deep understanding of interrelations between the electrical, mechanical and computer components of digitalized and intelligently managed production processes will be a vital ability.

A concept of a laboratory model for IIoT is proposed in this paper. Details are presented of the various stages of development of an introductory laboratory model based on IoT. It starts with tackling the challenges and designing laboratory experiments with limited resources available. A project-based learning approach is adopted to develop this laboratory through students.

2 The Laboratory Model of a Remote and Distributed Control System

In the past ten years we have built an efficient Laboratory Model of the Remote and Distributed Control System (LMRDCS). The main objective of the laboratory model presented in Fig. 1 is to perform research, and for teaching our students how to design and implement a remote and/or distributed control system based on modern communication and web technologies. We use this laboratory model for performing practical experiments as a part of courses such as Modeling of Processes, Industrial Automation, and Process Control Systems. The model can also be used in distance learning, as web-based teaching and learning with performing remote experiments. The lecturer can access the experiment from a distant lecture room, as well as can use the experiment in practical exercises, where, for instance, students can handle the process control system themselves to see the process responses at certain disturbances, set point changes, and controller parameter tuning. We have described our pedagogical experiences with LMRDCS in papers [6, 7].

Several laboratory processes, such as the distillation process, electric furnace, hydraulic system, air-conditioning system, heat exchanger and building automation model, which are automated locally, are connected and supervised with a distributed control system. Different communication technologies and Standard Protocols are employed for realization of this industrial communication network. The industrial Standard Protocols PROFIBUS, CAN, LON and TCP/IP are used with wire (Ethernet) and wireless (GSM, GPRS) communication technologies. The model is suitable for study, implementation and testing of industrial communication networks such as:

- Wide Area Network (WAN) – an industrial network that covers a broad area and uses public communication links.
- Industrial WLAN and VPNs – wireless industrial networks based on GSM/GPRS technologies and remote-control applications with point-to-point communication technologies.

- Remote Terminal Units (RTU), Industrial Gateways and Data Concentrators, and other incoming communications technologies, which are specific to the implementation of Internet 4.0.

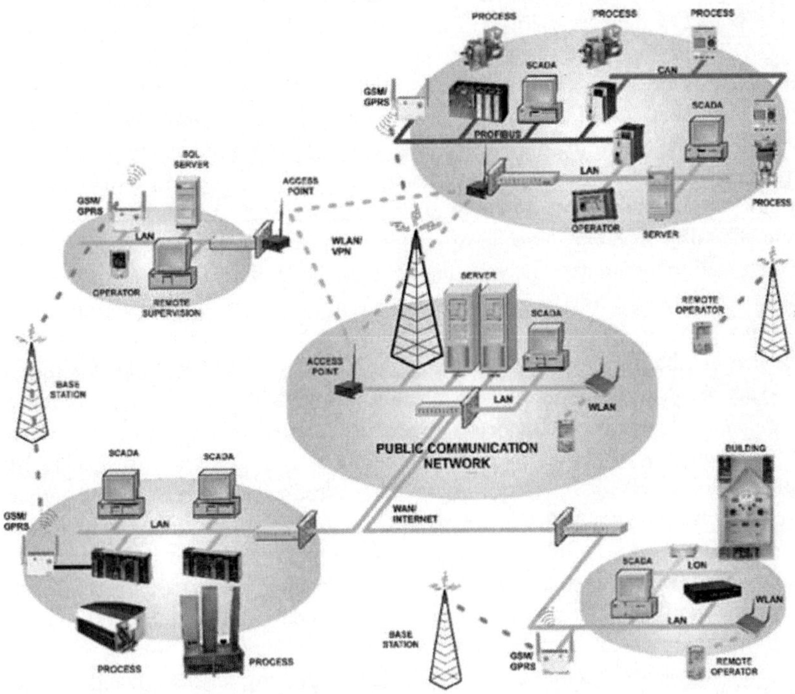

Fig. 1 The Laboratory Model of Remote and Distributed Control System (LMRDCS) realized in the Laboratory for Process Automation, FERI, University of Maribor

The existing web and remote laboratory platform received a positive feedback from students and Professors of the undergraduate and postgraduate studies in the fields of Electrical Engineering, Mechatronics Chemical Engineering at local and foreign institutions. However, with the development of newer communication technologies, an improved platform was designed towards the Industry 4.0 concept.

3 Development of Modules for Connecting Laboratory Processes with IIoT Technology

Nevertheless, factories today represent a relatively closed environment, where communication takes place mostly at the local level, and not necessarily with the outside world via the Internet. Communication between control systems and devices (controllers, sensors, actuators) takes place via wires. IoT concepts are approaching reality with the development of wireless technologies that will enable cost-effective and fast communication in an interfering industrial environment. For years, measurements from remote locations have been transmitted to a central monitoring system via VHF or GPRS modems. Technological design and unique features minimize abuse. When implementing IIoT it will be necessary to consider whether the advantages are sufficient given the potential dangers (viruses, malware) posed by the connectivity of internal systems to the Internet. One of the motives for the introduction of IIoT devices in the industry is the price. Compared to expensive industrial equipment and closed industry Standards, the use of affordable IIoT devices in relatively clean industrial environments makes sense today. Small instruments can be developed quickly and integrated into the existing industrial environment, and include data on process operation (process variable measurements), machine or line condition (operating hours, number of ejections per line), etc. Existing machines can be upgraded with sensors and algorithms for work cycle optimization, efficiency analysis, fault and congestion detection, etc. Data can be processed locally at the line level or within the company to optimize internal systems (edge computing) or sent higher (to the cloud) for business analysis. The range of services that can be developed and offered to industry is extensive, with different and vast amounts of data. Some companies already allow their customers to monitor and optimize devices based on a connected device in the cloud.

The following is a laboratory implementation of IoT, designed to capture measurements at the lowest level of the industrial environment (current loops) and send them to the cloud. The controller in the SCADA system is not presented in the existing measuring association. A closed management system can represent a (financial) limitation of the upgrade and the more comprehensive application. Data processing is shown in the fault detection application on the model of hydraulically connected vessels. From the point of view of security, data on measurements are not recorded in parallel. Still, the needs of application measurements are only processed, and the results are transmitted to the operator in real-time. Thus, only the meta-model process is stored in the cloud. The application of fault detection and isolation is performed using the Principal Component Analysis (PCA) method and takes place in real-time. The results and concluding thoughts are given at the end.

3.1 IIoT Modules for Current Loop Measurements

In Industry 4.0, the IIoT represents an essential part of industrial devices' connectivity, so some global manufacturers are already enabling the integration of IIoT devices with the existing industry protocols, ProfiNet, OPC UA. However, connectivity to existing equipment is not always possible, either due to limitations in the control system, equipment obsolescence, or an unwanted process interruption. It is possible for long-term use to install additional industrial equipment during maintenance work, or to use the existing one by installing independent IoT devices. We tested the latter idea of connecting to existing current loops (4–20 mA) a measuring system that sends data from the measured object to the cloud with various development boards and microcontrollers. The measurement of the physical quantity (current, voltage) on the microcontroller's side is performed via its own A/D converter or an external measuring system. Figure 2 shows the additional resistor's location in the current loop of the process. For example, in the Figure, the resistance value is 150 (Ω). The input of the microcontroller A/D converter is connected in parallel to the resistor and measures the voltage drop (normalized by the resistor) in the range between 0.6 V and 3 V.

The obtained measurement data are sent to the cloud and are equipped with information about the serial number of the measuring device, the measurement time, the process number, the measurement number and the measurement value. In this way, we can locate exactly where particular measurement data belong, and such a record allows measurements of several different quantities in different processes. We can also process data from various devices with one application, where we choose to display the data according to the process number.

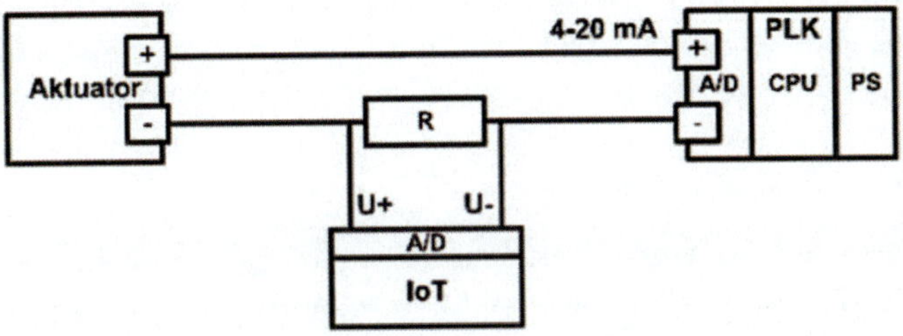

Fig. 2 IoT connection circuit to an existing current loop

The described measurement data recording system allows us to use various microcontroller systems for measurement, connecting to the Internet, wired, or wirelessly. The microcontroller must compile a unique URL in the code, consisting of the server's IP address (cloud) and measurement data. The cloud can represent a data server, IoT device, smartphone, etc. For further processing, the script reads the URL data and writes it to a temporary location. Figure 3 shows the various commercial microcontrollers with which the data acquisition was tested.

(a) (b)

Fig. 3 Some microcontroller development modules for IoT development needs

Figure 3(a) shows the various ESP modules which are the so-called cheap IoT starter kits. In combination with the Arduino module, they can be programmed in the

Arduino IDE programming environment, where the program code is written in the C programming language. ESP8266 can also be programmed in the Python programming language, which may otherwise be limited by the support for the I/O functionality offered by these systems. The board also has a WiFi module for sending data wirelessly to the cloud in addition to the controller. This one also has a 10bit A/D converter. Figure 4(b) shows the MCB1700 development board with an ARM Cortex M3 processor that acts as a hub and is wired to the web. A hub is a base unit to which remote units are wired or connected wirelessly. Remote units measure/collect data and send them to a hub, which processes them appropriately and sends them to the cloud. Due to various limitations, remote devices cannot communicate directly with the cloud, so they use intermediate points to transfer data to the cloud. Due to their affordability, the described modules are readily usable for IoT purposes, even in combination with less demanding development systems such as Arduino and Raspberry.

3.2 Laboratory Model of the IIoT

For testing purposes, we created a demonstration model for testing telemetry devices and the IIoT system for control and industrial automation applications. The model's goal was to demonstrate the usability and test the developed building blocks and the software's functionality. The physical appearance of the model is shown in Fig. 4.

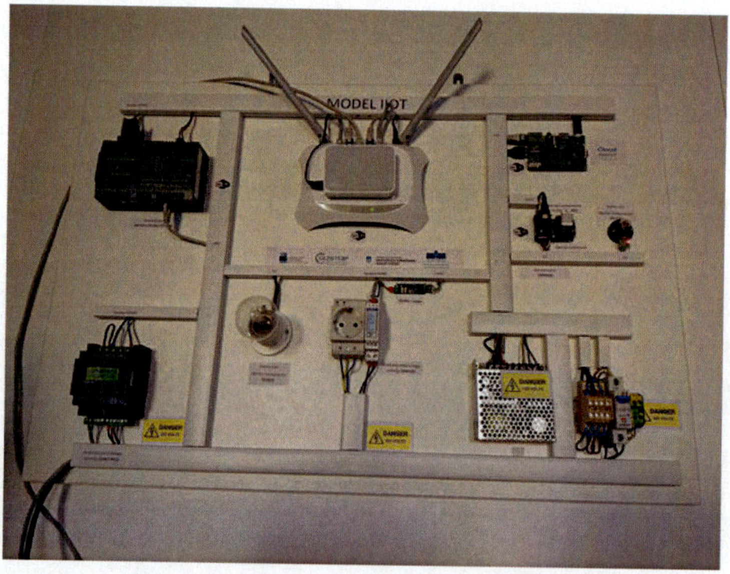

Fig. 4 The Laboratory Model of remote and IIoT

The model allows remote access to laboratory processes, the implementation of exercises and different experiments, and technologies. We have developed and tested the:

- SIEMENS SIMATIC IOT 2040 [8],
- Raspberry Pi,
- GEMALTO EHS6T [9].

Developed communication agents enable:

- Collection and transfer of process data from edge locations to a central process historian that is suitable for use in "Big Data" systems or cloud services.
- Collection of data from measuring devices (energy system) connected to the MBUS or MODBUS.
- Sending data to the historian data center using different communication technologies with minimal bandwidth usage.
- Storing data and forwarding mechanism in case the data center is unavailable.
- Secure data transmission.

The system is open, flexible, and the price affordable. It enables the implementation of different communication protocols. The processed data can also be visualized and displayed in a web environment. We prepared several applications, where students used an experimental laboratory model as part of project-based learning. Some are described below.

Application for Capturing Process Data Based on SIEMENS SIMATIC IOT2040. We have prepared a laboratory assignment for students, where they have to perform a remote reading of the industrial process's energy consumption. Wiring of an experimental laboratory model realized with SIEMENS SIMATIC IOT2040 is presented in Fig. 5.

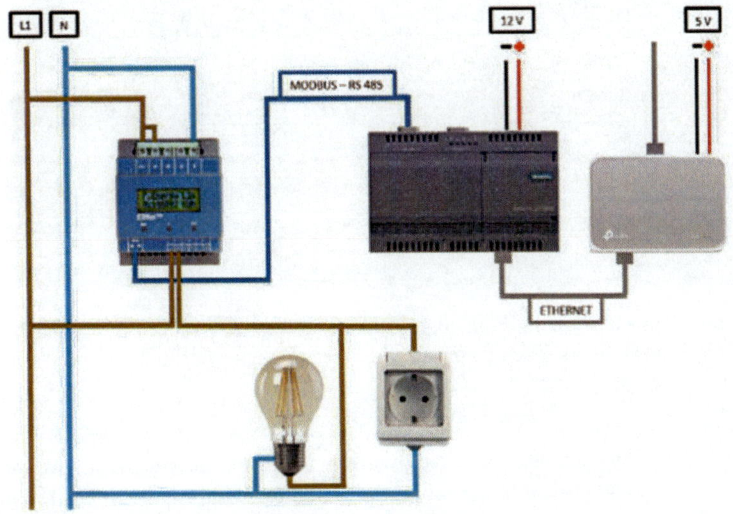

Fig. 5 Schematic diagram of the binding of an experimental laboratory model realized with SIEMENS SIMATIC IOT2040

In testing the application for measuring energy consumption with the ELNET PICO 5/60 energy multi-meter, the measured electrical quantities are transferred with the MODBUS RS485 protocol to the SIEMENS SIMATIC IOT2040 data concentrator. The data concentrator is on a local wireless Ethernet network, and sends data to the cloud. In our example, the cloud server is implemented on the Raspberry Pi platform. Software for storing measurements and visualization of results can be realized in various ways, from simple solutions using the MQTT protocol and Node-RED program to display measured values to complex programs with the database, REST, JSON JavaScript software technologies.

Data capture application based on Raspberry Pi and WiFi interfaces ESP 8266. To implement the data concentrator, we opted for the Raspberry Pi 3 microcomputer. We first set up the Raspbian operating system. The Raspberry Pi open-source platform is available at https://www.raspberrypi.org/. Students downloaded the following software: Apache2 webserver, PHP, MySQL database, and other utilities (phpMyAdmin, PHP MySQL Connector). Next, they connected the sensor, actuators, and other devices with WiFi interfaces ESP 8266. This module was expanded with the WEMOS D1 mini pro module. An example of a simple application is presented in Fig. 6.

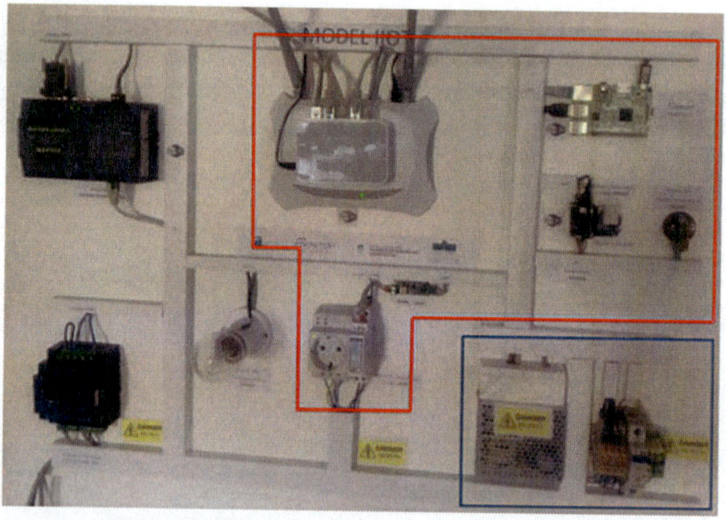

Fig. 6 An experimental laboratory model's connection was realized with a Raspberry Pi data concentrator. Three devices (energy consumption meter, temperature and humidity sensor, and light bulb actuator) are marked with a red frame

GEMALTO EHS6T-Based Data Acquisition Module. Instead of a Raspberry Pi, students can use another data hub, such as the Gemalto EHS6T USB. It is a reliable and straightforward communication unit that enables machine-to-device (M2M) device-to-device communication via wireless technologies. Thus, with such a device, it is possible to establish a connection quickly, reliably, and cheaply with various industrial applications. The device has an integrated truncated Java ME (Java™ Micro Edition) software environment for embedded systems. The design enables connection via a 3G and lower wireless mobile network. The device also has a built-in General-Purpose Input/Output interface (GPIO) for communication with external sensors and other devices.

4 Examples of Student Application and Test Results

In the following, we will describe the results of using the developed IoT laboratory model in the case of a student's practical exercise of using the PCA model in three related hydraulic tanks.

The hydraulically connected tanks process consists of three tanks of different dimensions, where two parallel tanks are placed above the third. A separate PID level control is performed in the upper tanks with a frequency-controlled pump and liquid level sensors. A Graphic User Interface for online operation of the laboratory model is presented in Fig. 7.

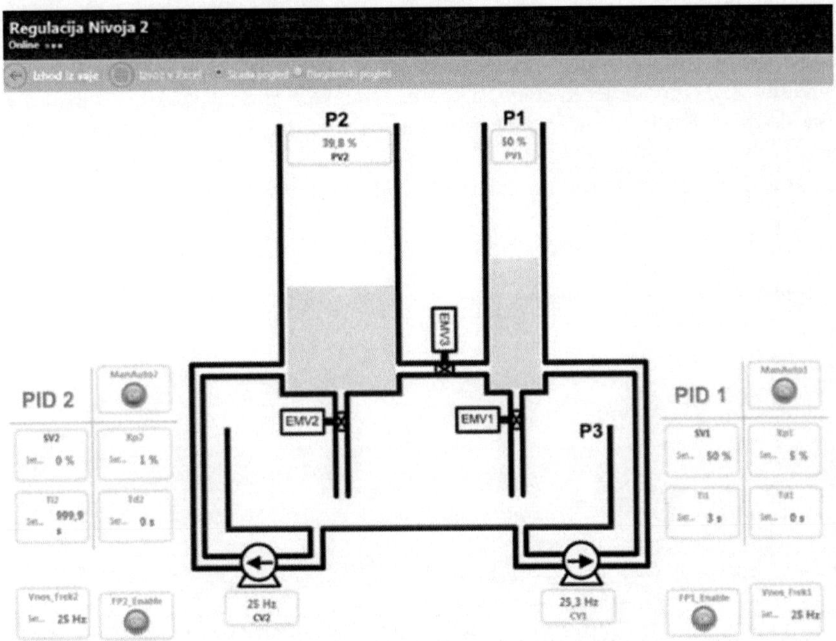

Fig. 7 Graphic User Interface for online operation of the three tanks laboratory model

The critical measurements for developing the Principal Component Analysis (PCA) [10] model are measurements of the controllers' output level and value (pump frequency). A valve connects the vessels. In the event of a fault in one vessel, this is reflected in the other vessel. In addition to the general PCA model, which describes the error-free operation, we also developed the partial PCA models needed to isolate individual faults. The faults we modeled are of the additive type, namely, the level sensor's movement in both tanks and the pipe’s clogging, or pump failure. Measurements of process quantities were sent via current loop IoT modules wired and wirelessly to the cloud, where they were first normalized to the desired range. Before that, it was necessary to prepare an environment for storage and processing within the SQL database and tables. Necessary information in the database includes:

- device number and process number,
- constants for calculation of normalization and scaling,
- measurement concerning physical quantities,
- constants for data processing for the PCA, and
- PCA model number stored in the appropriate matrix.

Measurement data are processed in the cloud and stored temporarily only for the PCA method. Developed data models of the process in different states (normal operation, actuator faults, etc.) are written in the form of *P* matrices, results, or the T^2-hits, and are saved in Tables for graphical display. Processing real-time measurements with all PCA models, combined with the appropriate logic (determining whether the values are within the allowable range), irregularities of the process operation are displayed in graphs. The developed graphical interface enabled the display of the type of error and the current plotting of the T^2 measure for individual PCA models. Due to system limitations (AD converter resolution, noise, small error amplitude), errors on the closed-loop PCA sensors were not detected successfully. On the contrary, clogging faults of the inlet and pumps were detected successfully. With the method of partial models and appropriately set threshold values, the errors were isolated unambiguously, as can be seen from Fig. 8.

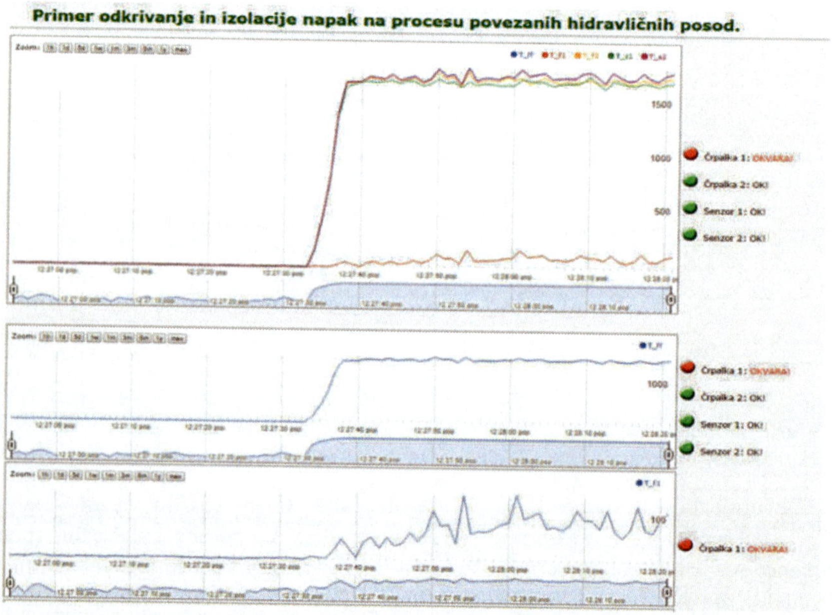

Fig. 8 Graphic User Interface for online operation of the three tanks laboratory model

5 Conclusions

Challenges and future perspectives in education for the future applications and demands of Industry 4.0 are discussed in this paper, together with IIoT control and industrial automation system experiments. The integration of various affordable micro-controllers enables the development of embedded devices in the form of IoT or IIoT technologies. Connectivity between devices is becoming comfortable, and developing web technologies, real-time visualization, and data processing is possible, either in the cloud or locally. We present an example of developing a measurement system in the form of individual IoT devices that enable the transmission of data to higher systems in the cloud.

Capture and processing with the PCA method and plotting via a web interface for drawing graphs are shown with the implementation of the error detection and isolation application. Implementing a simple IoT system for fault detection on a model of hydraulically connected tanks in real-time has proven successful, but only for significant deviations in process performance, such as pump failures. The partial faulty PCA process model used shows the deviation between the process operation's normal and faulty state quickly.

However, this model is also used for teaching our students how to design and implement remote and/or distributed control systems based on modern communication and web technologies. We use this platform for performing practical experiments as a part of courses such as Control System Components, Industrial Automation Process Modeling and Identification, and Process Control Systems. The model can also be used in distance learning, as web-based teaching and learning with performing remote experiments. The lecturer can access the experiment from a distant lecture room, as well as can use the experiment in practical exercises, where, for instance, students can handle the process control system themselves to see the process responses at certain disturbances, set point changes, and controller parameter tuning.

Acknowledgements. The presented research work was financed partly in the scope of the Competence Centre for Advanced Control Technologies and Program GOSTOP, that are financed partly by the Republic of Slovenia, Ministry of Higher Education, Science and Technology and the European Union (EU) - European Regional Development Fund.

References

1. Kramp T, van Kranenburg R, Lange S (2013) Introduction to the Internet of Things. In: Bassi A, Bauer M, Fiedler M, Kramp T, van Kranenburg R, Lange S, Meissner S (eds) Enabling Things to Talk: Designing IoT solutions with the IoT Architectural Reference Model. Springer, Heidelberg, pp 1–10
2. Kalunga J, Tembo S, Phiri J (2020) industrial internet of things common concepts, prospects and software requirements. Int J Internet Things 9(1):1–11
3. Erol S, Jäger A, Hold P, Ott K, Sihn W (2016) Tangible Industry 40: a scenario–based approach to learning for the future of production. Procedia CIRP 54:13–18
4. Lee J, Bagheri B, Kao H-A (2015) A cyber-physical systems architecture for industry 4.0–based manufacturing systems. Manufacturing Letters 3:18–23

5. Thames L, Schaefer D (2016) Software-defined cloud manufacturing for industry 4.0. Procedia CIRP 52:12–17
6. Golob M, Bratina B (2013) Web-based monitoring and control of industrial processes used for control education. In: Proceedings of 10th IFAC symposium on advances in control education, pp 162–167, Sheffield, UK
7. Golob M, Bratina B (2018) Web-based control and process automation education and industry. Int J Eng Educ 34(4):119–1211
8. Siemens. https://mall.industry.siemens.com/mall/en/WW/Catalog/Products/10321262. Accessed 15 Nov 2020
9. Gemalto. https://www.thalesgroup.com/sites/default/files/gemalto/M2M_ehs6t_hardware.pdf. Accessed 15 Nov 2020
10. Jolliffe IT (2002) Principal Component Analysis, 2nd edn. Springer, New York

Quality Assurance for Remote-Lab Systems by New Reporting Tool

Elmissaoui Taoufik[1(✉)], Charradi Sahbi[2], and Selmi Wafik[2]

[1] University of Carthage, SUP'COM,
INNOV'COM Lab && University of Kairouan, Kairouan, Tunisia
[2] University of Kairouan, Kairouan, Tunisia

Abstract. With the development of information and communication Technology (ICT), the concept of a remote-lab system has become inevitable. This system encourages students to follow remotely courses and practical works. With this solution of the remote system, students can manipulate real laboratory equipment. The insurance quality for this education system becomes a necessity especially online practical works. Our goals in this paper is to develop a tool that enable us to have a detailed report about the student and teacher's satisfaction concerning online practical works. This process was achieved through questionnaires to actors using remote-lab system like teachers and students. The feedback was able to enhance our platform and the practical works scenario. The application developed in this work provide some interaction with the user in real time. In fact, users can manipulate the system parameters and have a dynamic results and also select the output format. The invention of the new system "Industry 4.0", the RL is implemented for majority of the education disciplines.

Keywords: e-Learning · Remote-lab · Quality insurance · Reporting tool

1 Introduction

The development of the ICT in the world provides for scientist to concept and realize a remote-lab system (RL). Before COVID19, the RL has been developed in order to make the totally transfer from the learning face-to -face to e-learning. But, nowadays, the RL solution became a necessity for education. In fact, the e-learning system it is the better solution to solve the outbreak caused by the pandemic. With the development of learning mythology and the invention of the new system.

"Industry 4.0", the RL can be implemented for the majority of the education disciplines.

Multiple solutions and architectures are developed to realize the RL system in the literature. In the last decade, multiple remote-lab have been created and tested with success [9, 10]. For example, LABS-LAND create and tested an FPGA laboratory with success [9].

The RL able the universities to ensure both course and practical work using e-learning system. The same, university can economic the material cost. In fact, the university can buy one materiel for all establishment. Each institute can remotely use the materiel.

M. E. Auer et al. (eds.), *Online Engineering and Society 4.0*, Lecture Notes in Networks and Systems 298, https://doi.org/10.1007/978-3-030-82529-4_23

The materiel for practical works can mounted in same building in order to facilitates the maintenance and enhance the security. The quality insurance it is a very interested process. In fact, this activity must be in the beginning, during and after each online pedagogical activity (courses, practical work). The quality of the e-learning does not have a precise parameter's it's national, reginal and domain depend [1]. There are multiple methods able us to ensure quality in the literature. In fact, Lorenzo has fixed the five pillars of quality in the online education that include learning effectiveness, student satisfaction, faculty satisfaction, cost effectiveness and access [2]. The Analytic hierarchy process (AHP) has been implemented to evaluated quality of the e-learning system in [3–8]. In fact, each author uses the AHP to analyze some factors like the study of “Chinese attitudes “by realized by Hang et al. in [4]. In our paper we focus on the quality insurance of the use of RL in the e-learning system. In fact, we must ensure the satisfaction of the RL users like students and teachers in order to enhance our system and the practical work scenarios.

In this paper, we present the RL system architecture in the first section. The second section will be dedicated to describe our proposed tool to ensure the quality of the RL system. The results of our application will be discussed in the third section. Finally, we present the conclusion, and our futures works related the RL system enhancement.

2 Remote-Lab System

The remote laboratories can be insured by two methods: remotely simulator and remotely hardware manipulation in real time. In fact, the second kind of laboratory enables student to remotely use and control the laboratory equipment's and has the same benefits of someone inside the laboratory.

The RL present multiple advantages such as; enable university to provides a full online degree, let the person that has the specific need to do at home as well as their practice work, provide a solution for university to stay active in the major force that induce the outbreak like COVID-19, Inundation, war… The principal actor of RL are students and teachers. Our RL application enable students to connect with their connection parameter's to remotely do their practice work. For the same students can submitted results and interpretations to their teachers.

The teacher has a most important roles in our system. In fact, each teacher should prepare the necessary for the remote practice work like the needed equipment's, the practice work goals, the needed results and practice work scenarios which ‘is very delicate activity. Our remote lab doesn't provide the possibility to be used by multiple students in the same time. To solve this constraint, each registered student must reserve in advance the time of their practice work. Student can access to our RL everywhere and in the reserved time. The architecture of our Remote lab is presented by the figure Fig. 1. Our laboratory is operational 24 h 7/7. In fact, the student has the possibility to realize their work in the weekend. The same, students can manipulate the hardware lab equipment's and do their work directly with the real equipment's. The webcam installed in our laboratory enable students to see in real time the lab equipment's and can remotely interact and interpret the results. The lab equipment's must provide the possibility to be manipulated by Laptop or by any other module that contain an operator system like raspberry Pi. The switched module enables the users to choose the practice work and move to another practice work.

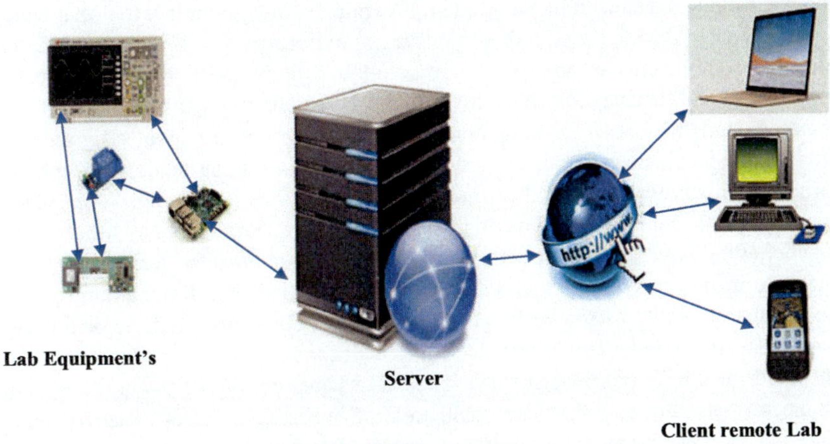

Fig. 1 Remote lab architecture

The figure Fig. 2 shows our real remote laboratory that is created in the frame of EOLES (530,466-TEMPUS-1–2012-1-FR-TEMPUS-JPCR) project under TEMPUS program with our partner's.

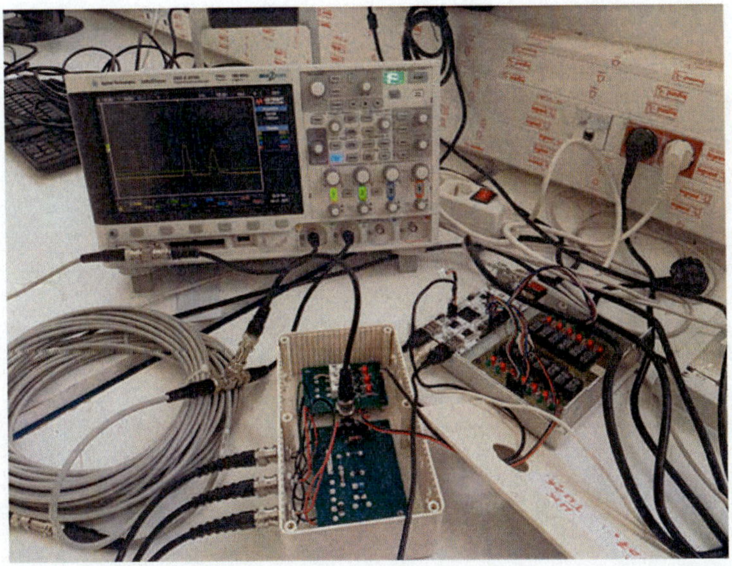

Fig. 2 Real remote lab developed by EOLES team

3 Reporting Solution Tools MSRT

The incorporation of remote lab into educational system will play a vital role to problem solving towards enhancing quality learning. The system is a tool that enables the collection of detailed report about the student and teacher's satisfaction concerning online practical works remotely. The remote-lab exploit questionnaires to generate reports that will enable decisions for the enhancement of infrastructures and also increased student access to equipment, a wider range of possible assignments or activities, and enhanced opportunities for collaboration among students. A practical work scenarios is likely to result in more and/or more varied learning. When labs are accessed online, students can (potentially) be engaged in learning at any time and from anywhere they have Internet access, as opposed to hands-on activities that rely on the times that institutions are open and staff members are available. Remote-lab opens the door to activities that may take longer than a typical class meeting time or multi-part assignments that require students to use equipment for several short periods over a long time, both of which pose logistical barriers in a hands-on environment. This system enhances opportunities for student collaboration when labs are accessed online, by removing the same-time, same place constraints posed by traditional work groups.

The first step of our solution consists of preparing the necessary questionnaire for students and teachers. The opinion of these two actors is very important to optimize our practical work and our remote laboratory infrastructure. The questionnaire depends on the practical work and the e-learning platform, also the addition of questionnaire to our e-learning platform and finally, we install our application (plugins) in our platform and generate the report. The development of a metric application for our RL, we need a flexible tool to extract relevant data from questionnaires. Initially, we identified the best solution been the use of Moodle plugin that can be integrated directly to Moodle platform. We develop our own reporting tool "The Moodle survey reporting tool" (MSRT) is a standalone application developed to create reports based in survey created in Moodle platform. The tool can be beneficial on several levels:

1. Easy to use and simplicity of the tool,
2. Ergonomic interface,
3. Rapid report generation,
4. Modular tool,
5. without need for extra skills,
6. No need of external tools.

After answering the questionnaire by trainees, the teacher can do the following steps:

1. Backup the questionnaire.
2. Directly import the generated backup file into the reporting tool.
3. Customize the report structure.
4. Generate the PDF report.

Our application provides for the user an interactive service. In fact, the user can focus his work in one question or questions group. The same, we can also write some interpretations. The MSRT enable user to put the necessary information and results in the report.

4 Results and Discussion

In order to test our MSRT, we have addressed our questionnaire for 14 participants by the Moodle platform. We import the questionnaire results from the Moodle survey page. We create our report structure with drug and drop from the element list. Our MSRT application present multiple functionality such as: Header of report, Footer of report, Titles of report, Free texts inside the report, Question Result, Number of Participants, Question comment, Grouped Questions, Average Question, Graphs, Insert Page break and Insert Line Break. The figure Fig. 3 provide the welcome interface of the MSRT.

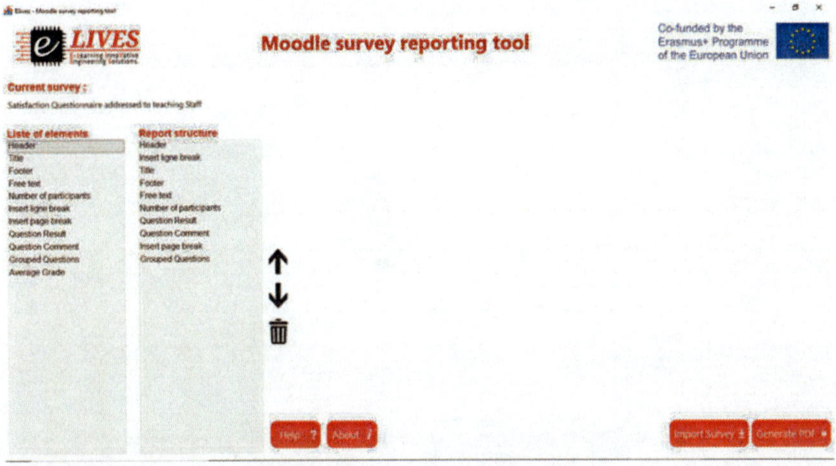

Fig. 3 MSRT welcome interface

The figure Fig. 4 shows the generated report by our application. The figure Fig. 5 illustrate the histogram and the table generate by our tool. This document contains the necessary results that enables the remote lab actor to take the best decision to optimize the remote practical work. The MSRT present a metric tool that enable us to make the diagnostic of our remote lab system. This application can be considered as a decision support system for pedagogical response (teacher, dean …).

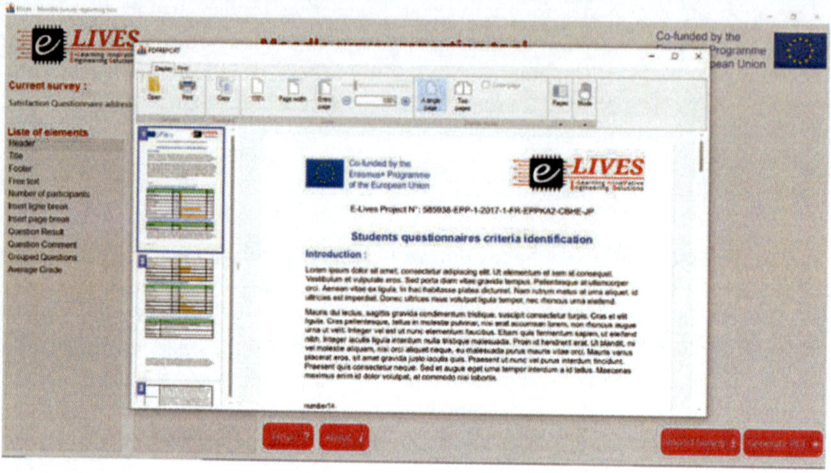

Fig. 4 MSRT generated PDF report

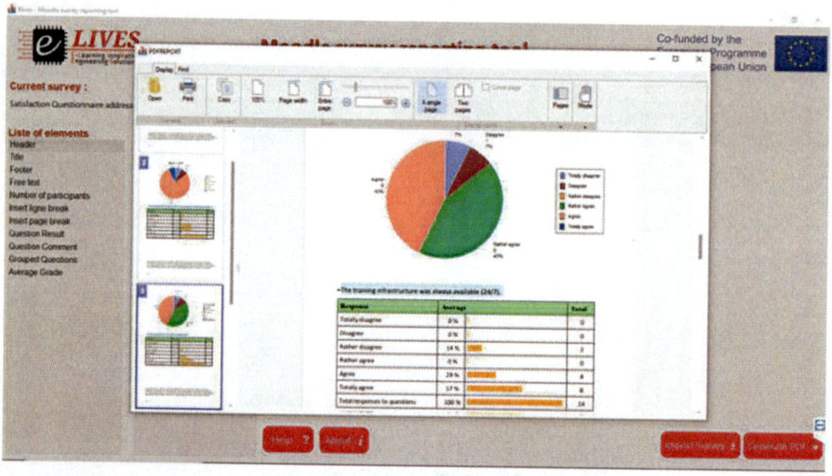

Fig. 5 Example of results generated by MSRT

5 Conclusion and Future Work

The questionnaire measures the functional and didactical aspects of the online learning, as how the teacher experiences it [11]. This questionnaire is used to provide the course and practical work, from the students' point of view. Our tool MSRT allow us to exploit questionnaires response in order to produce the report. The output file of our application provides the necessary data's that enable us to take the best decision in order to enhance our remote laboratory.

Our application is operational only with Moodle platform. In our future work, we will enhance our application to be valuable with e-learning system and especially the remote laboratories. The application will be well optimized for it to give a standard result. For the input data, we will find a solution that can make our MSRT to get information from another standard platform. In fact, we can prepare a help file that enables user to present their input data in order to be adapted to our solution.

Acknowledgements. This work is supported by university of Kairouan and by the e-LIVES project e-LIVES : E-learning innovative engineering solutions, 585938-EPP-1-2017-1REPPKA2-CBHE-JP funded by the European Commission under agreement number 20172891/001-001

References

1. Pawlowski, JM (2003) The European Quality Observatory (EQO): structuring quality approaches for e-learning. In: Proceedings 3rd IEEE international conference on advanced technologies, IEEE, Athens, Greece
2. Lorenzo G, Moore JC (2002) Report to the Nation: Five Pillars of Quality Online Education, The Sloan Consortium, Needham, MA. http://www.sloan-c.org/
3. Naveed QN, Qureshi MRNM, Shaikh A, (TCSE, IEEE), Alsayed AO, Sanober S, Mohiuddin K (2019) Evaluating and ranking cloud-based e-learning critical success factors (CSFs) using combinatorial approach, IEEE Access, vol 7, IEEE, 23 October 2019
4. Zhang L, Wen H, Li D, Fu Z, Cui S (2010) E-learning adoption intention and its key influence factors based on innovation adoption theory. Math Comput Model 51(11):1428–1432
5. Sharma R, Banati H, Bedi P (2011) Incorporating social opinion in content selection for an e-learning course. In: Proceedings of the 6th international conference computing science education, pp 1027–1032, August 2011
6. Gupta V, Chauhan DS, Dutta K (2013) Incremental development & revolutions of E-learning software systems in education sector: a case study approach. Hum-Centric Comput Inf Sci 3(1): 8
7. Jeong H-Y, Yeo S-S (2014) The quality model for e-learning system with multimedia contents: a pairwise comparison approach. Multimedia Tools Appl 73(2):887–900
8. Kassim SSM, Salleh M, Zainal A, Husin ARC (2017) Risk tolerance and trust issues in cloud-based E-learning. In: Proceedings of the 2nd international conference on internet things data cloud computing, pp 73 March 2017
9. Mayoz CA, da Silva Beraldo AL, Villar-Martinez A, Rodriguez-Gil L, de Souza Seron WFM, Orduna P (2020) FPGA remote laboratory: experience of a shared laboratory between UPNA and UNIFESP. In: 2020 XIV technologies applied to electronics teaching conference (TAEE), IEEE, 8–10 July 2020
10. Bose R (2013) 'Virtual labs project: a paradigm shift in Internet-based remote experimentation.' IEEE Access 1:718–725. https://doi.org/10.1109/ACCESS.2013.2286202
11. E-LIVES best practices guides. https://e-lives.eu/?page_id=10241

Batch Experiment: A Fruitful Way of Combining Hands-On Laboratory and E-Learning

Eleonora Bottani, Davide Reverberi(✉), Giovanni Romagnoli, Maria Ustenko, and Andrea Volpi

University of Parma, Parma, Italy
{Eleonora.bottani,davide.reverberi, giovanni.romagnoli,maria.ustenko, andrea.volpi}@unipr.it

Abstract. In this study, the authors present the utilization of laboratory-based Learning at the University of Parma and its integration with DigiLab4U (Open Digital Laboratory for You). At the University of Parma, the course of AUTO-ID in production and logistics from the very beginning has exploited the lab-based learning to convey practical skills gained in RFID (Radio Frequency IDentification) Laboratory to the students through a hands-on laboratory. However, due to the COVID-19 pandemic and consequent measures taken by the Italian government, the very same method could not be used anymore during the spring semester of the year 2020. To solve the issue a batch experiment method has been adopted, where the parameters were set by students and the experiment was carried out by a researcher. During the experiments, the researchers gather raw data and send it back to the students. Having these data, the students were able to perform the data analysis. To give the students the possibility to better understand the environment of the laboratory, pictures, and videos of the experiment performance have been supplied too. To complete the batch experiment, a tutorial has also been provided with the details of the software set up and the components used. The total remotization of RFID Laboratory is scheduled for late 2021.

Keywords: Lab-based learning · Auto-ID · Batch experiment

1 Introduction

Since October 2018, when the DigiLab4U project (http://digilab4u.com) has seen the light, the five universities of the consortium (University of Parma, HFT Stuttgart, IWM Koblenz-Landau, BIBA Bremen Institute for Production and Logistics, and RWTH Aachen University) have started to work on the experiments' scenarios. The mission of the project is to provide a common platform of hybrid education and research in non-traditional laboratories for the Internet of Things technologies in cross-institutional cooperation. In Parma, the collaboration involves the course of "AUTO-ID in production and logistics", which is addressed to the students of master's degree in management engineering. The course takes place in a spring term and consists of face-to-

M. E. Auer et al. (eds.), *Online Engineering and Society 4.0*, Lecture Notes in Networks and Systems 298, https://doi.org/10.1007/978-3-030-82529-4_24

face lectures and practical experience, during which, originally, the students used to go to the RFID Laboratory of the University of Parma to run practical tests. Through the collaboration with DigiLab4U, the RFID Laboratory started organizing itself towards its automation stage to become remotely accessible in order to enable the students to perform experiments without a physical presence. The lab is planned to be fully available by remote starting from the year 2021. In the meantime, teaching should have continued commonly. However, due to the COVID-19 pandemic crisis, the presence of students was strictly prohibited. To cope with the situation and give the possibility to learners to continue gaining practical experience, a proper solution has been found. The University of Parma provided the online theoretical lessons for the students and together with DigiLab4U elaborated a batch experiment method. In "batch experiment" students fully specify the experimental conditions in advance and submit the specification as a request. The experiment is then performed according to student's specifications [1] and raw data are shared with them for further analysis.

The goal of this work is to report an application of the batch experiment methodology, which might enable students to remotely perform lab activities without utilizing a remote lab. In this way, the benefits of lab learning might be collected by students even without physical interaction with the lab. To prove the applicability of our method we report a case study carried out at the University of Parma during the year 2020.

The remainder of the paper is organized as follows. In Sect. 2 the authors will provide a literature review on the batch experiment utilization, the hands-on and remote laboratories, the utilization of documents, videos, and pictures as a tutorial for the students, and the modification had in lab-based teaching due to the COVID-19 pandemic. In Sect. 3 the structure of the AUTO-ID course will be presented together with the experiments that were originally performed in the RFID Laboratory. Section 4 will follow with the arise of COVID-19, the design, the implementation, and the deployment of batch experiment methodology. Sections 5 and 6 will close the paper with the discussion of the results and the future steps.

2 Literature Review

The paragraph discusses the importance of laboratories and hands-on experiments in STEM (Science, Technology, Engineering, and Math) disciplines. Hands-on experience helps to obtain applicable knowledge that students will use in practical situations. Hence, the lab is essentials in university education [2]. Teachers and researchers use hands-on laboratories to run the didactical experiments. The laboratory's goal is to give the students the technical skills as well as the understanding of the process in the "real" environment. However, usually traditional labs have high costs as they require many essentials elements to be provided: (i) safety, (ii) large area spaces to provide observability, (iii) availability of the teacher. Ergo, many universities have started to use remote laboratories. In the literature, experts share different opinions about remote and hands-on labs. Remote laboratories have various advantages as (i) reusability, (ii) interoperability, (iii) convergence, and (iv) the possibility that the students use them independently [1]. Utilizing a remote lab may reduce expenses and save students time, as they can access to the experiments anywhere and anytime via internet. Moreover, it enhances the students' autonomy in understanding the experiment and its limitations. On the other hand, remote laboratories hardly support the development of collaborative learning skills. Indeed, teamwork is important during the laboratory's activities as well as the possibility to "watch" the crucial real experience. Thus, there is a need of maintaining a "blended-learning approach" which is a balance between two types of laboratories [3]. Remote laboratories are successfully used in universities to teach different topics. For example, in measurement engineering, the learners study the mechanical properties of materials where the remote experiment helps students to develop skills in a physical laboratory [4]. At the University of Tennessee, a remote chemistry lab is used. Here the students use web access to manage the laboratory stations. Separated into teams they control liquids flow and its physical properties. Students insert settings according to the objectives of their learning goals, then collect and elaborate data [5]. Some remote laboratories provide a data repository to collect experiments' data. For example, in iLab, students control the lab instruments by setting parameters. iLab is a network of remote laboratories, thus is possible to share resources between each lab. To do so, it provides data capability storage and other generic services [6]. Lab2Go repository focuses on creating a common framework to gather and analyze data of any laboratory in the world based on web technologies [7]. Lila is an online repository based on a meta-infrastructure, from which it is possible to download the resources from its web portal [8]. Another relevant lab-network is Labicom that is used for hosting and sharing online laboratories. It has three types of data: static client content, real-time data, and live streaming video [9]. The next lab analyzed is RemLabNet. It collects information under controlled values and measures data results to experiment with the same set up that can be reproduced offline. The system creates storage of all information and experimental data that are registered by users for later use. Moreover, the data warehouse is a part of a system storage and system central united data analysis for remote laboratories [10]. The last network analyzed is SCY (Science Created by You). SCY is a learning environment based on platforms where students learn about science topics resolving a socio-scientific problem. Students can save work in repository service and also, they can invite other students to work with them [11]. Due to the COVID-19 pandemic, universities were forced to search for new ways to teach practical courses. To continue the laboratory's activities Dept. of Engineering at the University of Parma, developed batch experiments. A batch experiment is an experiment where different settings are chosen in advance, then used to collect raw data. [12]. In this type of experiment students do not need to be present "physically" in a lab, however, can use data collected from real experiments previously. Being at home the students decide the parameters to be set, while a technician goes into the laboratory to run the experiment using the parameters requested. After that, a technician will send the raw data to the students. In this way, the students work on real raw data, analyze it, and reach the finals goals. For example, the University of Michigan uses the batch lab method. The university has created a remote chemistry laboratory for the student that should attend all subject in the laboratory. The teacher made use of a textbook with a web platform to select chapters and videos. Students, instead, used simulation software to create experiments and results, and at the end, they generated a video describing the experimental procedure [13].

3 AUTO-ID in Production and Logistics

It is strongly believed that a combination of theoretical and practical activities helps to assimilate studied cases [14]. The "AUTO-ID in production and logistics" course has always been taught as a sequence of face-to-face theoretical lectures and practical experience in the RFID Laboratory. The goal of the course is to teach the basic AUTO-ID technologies notions, with a focus on RFID, in the logistics and production systems. The theoretical knowledge provided by the course leads the students to acquire the skills needed to design an AUTO-ID solution in different application contexts. To enhance these skills, a practical experience completes the course. Here, the students are divided in groups of three or four people, and one scenario out of two is assigned to them. In the first scenario, the students must determine and analyze an RSSI (Received Signal Strength Indicator) curve of products having different RF natures (i.e. RF-friendly; -reflecting; -absorbing). The second scenario objective is to optimize the parameters of an RFID system to achieve the highest percentage of reads of wanted tags and the lowest of unwanted ones. Both scenarios are dived in different phases:

- **lab experience**, where students go to the RFID Laboratory, to set up the system with the supervision of an expert, run the experiment, and collect the data.
- **data analysis**, where the collaborative learning aspects, that represent a fundamental part of the engineering students' education [15], are used the most. The students discuss the future steps to be taken and then proceed with the data elaborations.
- **report creations and discussion**. In this final part, the students present the outcomes of the data analysis and discuss their conclusions with the professor.

Both scenarios are designed with the scope to introduce the learners to an industrial-like environment, allowing them to understand how to behave within a real production system, and how the acquired knowledge can be applied in practice. The RFID Laboratory consists of a circular structure equipped with an industrial roller, a belt conveyor, and three non-automatized roller conveyors that work as unload buffers. Moreover, two Near Field (NF) and two Far Field (FF) are installed on one side of the belt conveyor. To trigger the antennas' activation, a photoelectric cell is installed. All these five components are connected to a reader. The function of the reader is to link the antennas and the trigger to the reader's software installed on a PC. The software is used to select the antennas, set parameters (i.e. reading time, power level, trigger event), and collect the data from the antennas. To this end, different boxes are located on the conveyor and circulate on it, with the photoelectric cell triggering the readings of the antennas, and the reader setting the parameters and storing the data on the PC. To have an environment even closer to a real one, sixteen tags have been attached on a wall next to the antennas to simulate the bias noise (i.e. the presence of unwanted tags) findable in an industrial background, and some errors have been added to the RFID reads, as more external tags or extra number of cycles.

3.1 Scenario 1 "RSSI Curve Creation"

During the theoretical lessons, the students are taught about the theoretical RSSI curve trend of an RFID tag moving at a constant speed in an electromagnetic field [16]. The RSSI value is an indicator of the power present in a received radio signal and its theoretical value can be easily calculated. However, in an industrial facility, the RSSI curve is influenced by numerous factors, including the frequency and the surrounding environment. With this experiment, the students can elaborate an RSSI curve offline, which resembles the real-time measurement. Thus, the students reach the overall learning objective which is - compare the theoretical estimated curve with the practical one calculated during the data analysis.

Learning Environment

In the Design of Experiment (DOE) (Table 1) for the RSSI curve creation, there are three boxes containing the same product, running on the belt conveyor ring. Every time the three boxes complete a lap of the ring, a cycle is computed. The boxes run five cycles for each power level value, that equal to: 18 dBm, 21 dBm, 24 dBm, 27 dBm, and 30 dBm. This procedure is repeated for the products "A", "B" and "C" (i.e. RF-friendly, RF-reflecting, RF-absorbing), and two types of antennas (FF and NF).

Table 1 RSSI curve creation DOE

Products	RF friendly	RF absorbing			RF reflecting
Power (dBm)	18	21	24	27	30
№ of cycles	5				
Antennas configuration	NF		FF		

During each cycle, once a box passes in front of the photoelectric cell, the reader receives a signal and triggers the activation of the antennas selected. Their active time is defined as a parameter of the reader. In the current experiment, the reading time was set at three seconds. Each of the nine boxes is labeled with a single Dogbone Paper Face Monza R6 tag. The three products inside the packages were: (i) oil - RF-friendly, (ii) canned tomato sauce - RF-reflecting, and (iii) bottled water - RF-absorbing.

Experiment Execution

Before starting the experiment, the students were divided into groups of three to four people to enhance collaboration skills, fundamental in the engineering field. Thus, the students place the boxes on the circular belt conveyor with three meters between them. The PC is connected to the reader via Ethernet, and through the reader's software the parameters are set. When a package passes in front of the photoelectric cell, the reader activates the antennas; the trigger time is t_0. The reader switches off the antennas at $t_f = t_o + 3$ s. During these three seconds, the antennas stimulates the tag, collect data from them, and send the data through the reader to the PC. Here, once the five cycles are over, a.csv file is created. This file stores as many text strings as the number of reads. In turn, each string reports: Electronic Product Code (EPC) value, Time Stamp, Run Number, RSSI, Reader, Channel, Power, and Antenna. Once the data collection process is complete, the students start the data analysis. Here, the data are organized and the RSSI curve is calculated in function of the time stamp. At this point a well understanding of the industrial environment is paramount - the students should be well oriented in the environment they worked in to interpret the obtained data correctly.

Final Results
Once the data analysis will be made on all the products, power levels, and antennas configurations, the students will have 30 different configurations (three products for five power levels, and two antennas type) that have to be compared; the related outcomes have to be presented. To do so, the difference between transmitted and received power, the meaning and the factors that influence the RSSI curve, the reason for their impact, and the deviations of the real curve from the theoretical one, must be clear.

3.2 Scenario 2 "Reading Optimization"

One of the main challenges in the RFID area, and especially in an industrial environment, is the configuration of the parameters that optimize the readings. During the lectures, the students are introduced to this topic with the calculation of α- and β-types errors. α-type error, or type-I error, denotes the percentage of tags that must be read but were not read during the reading time. It can be calculated with the formula (1):

$$\alpha - type\ \ error = no\,of\,expected\,tags\,not\,read/total\,no\,of\,expected\,tags \qquad (1)$$

On the other hand, β-type error, or type-II error, indicates the percentage of unexpected tags that have been read on the total number of unexpected tags that must not be read. Its formula is:

$$\beta - type\ error = no\,of\,unexpected\,tags\,not\,read/total\,no\,of\,unexpected\,tags \qquad (2)$$

In our scenario, the total number of unwanted tags was 32, divided in 16 ghost tags attached on a wall near the belt conveyor, and 8 tags attached on each one of the two boxes not analyzed. The minimization of these two errors has to be achieved with an appropriate configuration of the input parameters (i.e. power level, type of antennas, and antennas orientation), which is the students' final objective.

Learning Environment
In this scenario, only one product (i.e. aluminum cans of tomato sauce) is present. Three boxes containing this product are placed on the belt conveyor ring, and, as in the previous experiment, a cycle is computed anytime a complete round of the ring is concluded. The DOE (Table 2) reports the configuration of the experiment: five cycles per power level, from 30 to 18 dBm with a decrease of 1 dBm every five cycles. This procedure is replicated for four different antennas configurations: (i) NF parallel to the belt conveyor (scenario id: NF); (ii) FF parallel to the belt conveyor (FF-//); (iii) FF oriented 45° in motion direction (FF + 45°); (iv) FF oriented 45° opposite to motion direction (FF–45°). Each box is labeled with eight Dogbone Paper Face Monza R6 tags, two on each side face,

one oriented vertically, and another one horizontally [1, 15]. As for scenario 1 “RSSI Curve creation”, the reading time was set at three seconds.

Table 2 Reading optimization DOE

Product	Tomato sauce (3 boxes)												
Power (dBm)	18	19	20	21	22	23	24	25	26	27	28	29	30
№ of cycles	5												
Antennas configuration	NF				FF-//			FF + 45°			FF-45°		

Experiment Execution
As previously, in this scenario the students are divided into groups of three to four people. The boxes are placed on the conveyor ring with a sufficient distance between them to avoid reading overlay. Time zero (t0) is taken as the moment when a single box activates the photoelectric cell, by passing in front of it, and consequently, the reader receives the signal and switches on the antennas. Starting from t0, the antennas are set in reading mode for three seconds (tf = t0 + 3 s). Here, the antennas transmit data to the reader, and then to the reader’s software installed on the PC. The software allows the data exporting in.csv format as for the RSSI Curve Creation scenario, displaying the same data. With 52.csv files (the combination of 13 power levels and four antennas configurations), the students start the data analysis. The learning goal of this scenario is to define the combination of power level and antennas configuration that minimizes both α- and β-type errors. To do so, the students must detect the box that activates the photoelectric cell. This box is labeled with eight tags that are expected to be read, while all the remaining tags read in the same run are to be considered as “noise” (unexpected tags). A tag is considered as “read” when it is scanned at least once by one of the two antennas. This analysis has to be repeated for all three boxes, resulting in three calculations of the α- and β-type errors. It is also particularly important for the students to identify all the tags and correctly map them to box A, B, C, or ghost reads.

Final Results
The theoretical knowledge provided before the experiment will introduce the students to how the reads will change according to the different types of antennas or with a different antennas’ orientation. Having both errors calculated for each box, power level, and antennas configuration, the students will be able to define the most effective combination of factors to minimize them.

3.3 RFID Laboratory Evolution

RFID Laboratory at the department of engineering and architecture at the University of Parma has been active since the year 2006 and became an experimentation station for RFID applications locally and internationally. Since the year 2018 RFID Laboratory has been added to the consortium of DigiLab4U project focused on providing a hybrid education platform. Thus, since then laboratory upgrade has taken place, which embraces various areas of the environment, i.e. hardware-software connectivity, learning environments, teaching scenarios. According to thc plan, full remotization will be achieved by the end of the year 2021. In the meantime, the RFID Laboratory should continue functioning normally. However, the spreading of the COVID-19 pandemic has forced the Italian government to adopt critical countermeasures that made it impossible to perform the experiment on-site. As the RFID Laboratory is not remotely accessible yet, as proper connectivity hardware is missing as well as the software layer to manage it, as a middle step, a batch experiment method has been implemented. A batch experiment is an unforeseen solution to go up with an emergency situation. It will be taken out of the curriculum once the world pandemic situation comes back to normal. In the meantime, the laboratory upgrade is continuing as it was previously.

4 Batch Experiment

Ever since the pandemic has started the Italian authorities had to ensure students' school continuity to grant the students the possibility not to lose the 2019/2020 school year. Owing to the remote laboratory at the University of Parma being fully activated only next year, some specific techniques and actions were undertaken to continue long-distance lab education. The batch experiment is one of these techniques. The essence of the batch experiments was to give the students the possibility to define the parameters, thus enabling the DigiLab4U staff to go to the laboratory, run the existing scenarios according to the parameters set, and provide the resulting raw data to the students for subsequent analysis. The main challenges were: (i) to execute an experiment having absolutely new conditions; and (ii) to figure out how to remotely guide the students through the batch experiment. To this end the batch experiment was performed in three steps: (i) gathering parameters, (ii) collecting data, and (iii) analyzing data.

4.1 Data Gathering

The first step requires the definition of the parameters by the students. After the gathering of the settings, the presence of DigiLab4U and the University of Parma staff in the lab is required to run the two scenarios. In case of a batch experiment, two full professors at the University of Parma and two members of DigiLab4U team were present in the RFID Laboratory to collect raw data. Preparation of the tests includes customization of the software, adjustment of the antennas, reader, and a photoelectric cell, placing the last one on a specific place depending on the configuration of each scenario. Additionally, one of the participants was asked to prepare the examined samples (see Tables 1 and 2) as well as to attach the required amount of tags on the boxes with those samples inside, and ghost tags on the wall near the antennas (as in Fig. 1). The experiment was executed following exactly the same methodology as described in the previous sections. In both scenarios, anytime a box passed in front of the photoelectric cell, the reading round was triggered, and the readers received values of readings from both antennas. Due to the synchronization with the photoelectric cell, the reading rounds were not 'automatically' numbered by the reader; rather, this must be done by students, keeping in mind that the reads of each reading round were clustered in a time interval of max 3 s. A cycle, each of which counts 3 reading rounds, is defined by a box moving along the whole length of the conveyor, thus completing its tour. However, we noted that some reading intervals may last longer than the expected 3 s. This fact aims at simulating possible manual errors while performing the experiments. Students should therefore try to identify those errors and correct them. This correction may not have a significant influence on the overall result of this experiment, but it may be very educational to introduce this inaccuracy in the analysis.

4.2 Preparing Materials

After running the tests, significant attention was given to the design of the learning materials. It was quite a challenge to form the studying sheets in such a way to (i) keep up the motivation of learners, even if they were not able to go to the lab personally; (ii) maintain their curiosity; and (iii) familiarize with how a semi-industrial environment works. Detailed learning sheets were developed to explain to the students the experimental setups, the tasks, and assignments, and recap the theoretical concepts. Additionally, the designed learning materials contained: (i) the motivations for the adoption of the batch experiment; (ii) the theoretical notes linked to each scenario; (iii) the preparation of the experiment; (iv) the materials and methods applied; and (v) an example of expected results. Videos, and other digital tools are certainly useful supplements for increasing classroom learning as well as equity, compensating for real-life obstacles to laboratory education. To create a more realistic picture of the laboratory, several detailed videos were therefore recorded. Those videos explain the structure of the laboratory and all possible experiment configurations to allow asynchronous instruction.

Fig. 1 Experiment configuration

4.3 Batch experiment completion

As a second step, the material described above was uploaded on a common learning platform (Moodle) that was successfully adopted for assignments, logistics, and distributing course materials at the University of Parma. Consequently, students were serendipitously partially prepared for transitioning to remote learning. The collected RFID raw data was provided in.csv file format and was uploaded in that platform together with the learning materials. Students were separated into groups of two to three people and were advised to carefully study the materials before starting the experimental phase of the course. After this, each student worked on his/her own inside every single group, solving the given tasks. The students were free to choose by themselves how to elaborate the raw data and which data analyzing tool to use. As it is believed that teaching presence is a foundational concept of online education, and particularly the batch method, so the students were continuously supported by the tutor, who was available for a chat and/or videoconferencing to offer help and guidance. After finishing analyzing data, students were expected to create a report, according to the objectives of the scenario, and discuss it with the professor during the exam for the final evaluation.

5 Discussion

The approach presented in this paper enabled the possibility for the students to have a semi-practical experience. In that way, they have been able to apply the theoretical knowledge learned and face the difficulties of a real environment. This effort maintained the very same course format as the previous academic years, even with the use of different methods (i.e. e-learning, recorded videos, informative material, batch experiment). Moreover, from a didactical point of view, the authors: (i) analyzed the critical point of the learning materials when provided online without face-to-face lectures; (ii) introduced the tutor and analyzed the interactions of students with her; and (iii) adapted the data analysis to the workload required. Until now, lab-based learning has always been held with hands-on laboratory and the same should have been happened for the AUTO-ID course during the year 2020. However, due to the COVID-19 pandemic, a batch experiment approach was adopted. Such a way to resolve the occurred situation not only helped to continue lab-based education but also served as a starting point to prepare a blueprint for the future lab remotization. The authors suggest that utilizing a batch experiment may significantly reduce potential problems with the integration of lab scenarios, as those scenarios, together with supplementary materials, have been designed, tested, and improved to fit the online learning platform. The adoption of such a method is also a possible example of how to confront occurred circumstances in a short period and continue teaching. Using a batch experiment method promoted an opportunity to introduce the students to an industrial-like environment, and its peculiar characteristics, in a virtual way, while still complying with the COVID-19 crisis. Such an approach is obviously a temporary solution to go up against the pandemic situation, however, can be furtherly used as an emergency measure in case of hardware/software malfunctions or to comfort an individual request of a student or a teacher. In Table 3 can be saw the characteristics of the different type of laboratory.

From a practical point of view, the students were evaluated by completing an online exam and succeed quite well. After the exam students participated in the evaluation of the new method. The interview structure was developed by an expert in evaluation, also a member of DigiLab4U, and contained both general questions and some particular information about the execution of the experiment. Students reported positive feedback. By acknowledging that the students' suggestions to improve the experimental design were valid and worth pursuing, we raised the chances that they will remember what they learned and be able to transfer it to novel problems in the future.

Table 3 Different type of laboratory characteristics

	Hands-on laboratory	Batch experiment	Remote laboratory	Virtual laboratory
Environment interaction	Real environment	Videos, pictures, files	Webcam, videos, AR/VR[a]	Virtual / Simulated
Parameters customization	Full	Limited	Full	Partial/Full
Data gathered	Real	Real	Real	Simulated
Availability	On booking	Always	On booking	Always

[a]Augmented Reality/Virtual Reality

6 Future steps

We are now working on making the lab available for remote access in the nearest future, with a planned hardware upgrade of the lab by the end of 2020, as the first step of its complete remote automation expected by 2021. Towards this goal, our future step will be the upgrade of the software layer linked to the laboratory hardware. By following those steps, and improving the existing scenarios, the AUTO-ID course will be accessible from all over the world, trough the DigiLab4U Learning Management System, and will be enriched by new technologies and alternative teaching methods.

Acknowledgements. This research was funded by the German Ministry of Education and Research BMBF, grant numbers 16DHB2112 (HFT Stuttgart), 16DHB2116 (University of Parma), and 16DHB2115 (IWM Koblenz), and was developed within the project DigiLab4U (https://digilab4u.com/). The responsibility for the content of this publication lies with the authors.

References

1. Romagnoli G, Esposito G, Rizzi A, Zammori F, Bertolini M, Uckelmann D (2020) Lab networks in engineering education: a proposed structure for organizing information. Int J Online Biomed Eng 16(05):41–70
2. Gravier C, Fayolle J, Bayard B, Ates M, Lardon J (2008) State of the art about remote laboratories paradigms - foundations of ongoing mutations. Int J Online Eng 4(1):19–25
3. Restivo MT, Mendes J, Lopes AM, Silva CM, Chouzal F (2009) A remote laboratory in engineering measurement. IEEE Trans Ind Electron 56(12):4836–4843
4. Henry J (2002) Using the modern chemical engineering laboratory at a distance. In: ASEE annual conference proceedings, pp 841–849
5. Marincean S, Scribner SL (2020) Remote organic chemistry laboratories at university of michigan-dearborn. J Chem Educ 97(9):3074–3078
6. Gitelman DR (2002) ILAB: a program for postexperimental eye movement analysis. Behav Res Methods Instruments Comput 34(4):605–612
7. Richter T, Tetour Y, Boehringer D (2011) Library of labs - a European project on the dissemination of remote experiments and virtual laboratories. In: Proceedings - 2011 IEEE international multimedia, ISM 2011, pp 543–548

8. Titov I (2013) Labicom.nct - Thc on-linc laboratories platform. In: IEEE global engineering education conference, pp 1137–1140
9. Titov I, Glotov A, Andrey Y, Petrov V (2016) Labicom labs: remote and virtual solid-state laser lab, RF & microwave amplifier remote and virtual lab: interactive demonstration of Labicom labs in winter 2016. In: Proceedings 2016 13th international conference on remote engineering and virtual instrumentation, REV 2016, pp 336–338
10. Friedrich M et al (2011) Early results of experiments with responsive open learning environments. J Univers Comput Sci 17(3):451–471
11. de Jong T et al (2012) Using scenarios to design complex technology-enhanced learning environments. Educ Technol Res Dev 60(5):883–901
12. Zutin DG, Auer M (2014) A simple LabVIEW based framework to facilitate the deployment of iLab batch lab servers. In: Proceedings 2014 11th international conference on remote engineering and virtual instrumentation, REV 2014, pp 328–331
13. Hardison JL, DeLong K, Bailey PH, Harward VJ (2008) Deploying interactive remote labs using the iLab shared architecture. In: Proceedings - frontiers in education conference, FIE, pp 18–23
14. Feisel LD, Rosa AJ (2005) The role of the laboratory in undergraduate engineering education. J Eng Educ 94(1):121–130
15. Burghardt M, Ferdinand P, Pfeiffer A., Reverberi D, Romagnoli G (2020) Integration of new technologies and alternative methods in laboratory-based scenarios. In: 17th international conference on remote engineering and virtual instrumentation, REV 2020, pp 297–317
16. Pfeiffer A, Lukarov V, Romagnoli G, Uckelmann D, Schroeder U (2020) Experiential learning in labs and multimodal learning analytics. In: Ifenthaler D, Gibson D (eds) Adoption of data analytics in higher education learning and teaching. Springer, Cham, pp 349–373. https://doi.org/10.1007/978-3-030-47392-1_18

Towards the Operationalization of Trust Relationships in Networked Organizations

Valentin Kammerlohr[1,2](✉), David Paradice[2], Jannicke Baalsrud Hauge[3,4], and Heiko Duin[4]

[1] Hochschule für Technik Stuttgart, Stuttgart, Germany
Valentin.Kammerlohr@hft-stuttgart.de
[2] Auburn University College of Business, Auburn, USA
[3] KTH-Royal Institute of Technology, Södertälje, Sweden
[4] BIBA – Bremer Institut für Produktion Und Logistik GmbH, Bremen, Germany

Abstract. The sharing economy is a widespread concept of commonly used resources either among individuals or across organizations putting efforts into jointly utilizing existing resources. Within digital networked organizations, trust behavior needs to be supported by mechanisms to establish and sustain the collaborative network. This paper presents methods for visualizing and scoring collaborative behavior as steps towards operationalizing the trust relationship within networked organizations. Using a case study involving a sharing digital laboratory, we investigate how transparency influences collaborative work and how it affects digital collaborations as a whole. We apply Social Network Analysis (SNA) and the Supply Chain Operations Reference (SCOR) model to visualize and score trust behavior within a case study for digital laboratories. The result of the trust behavior will be extended to an overarching sharing economy business model that promotes the network effect for networked organizations.

Keywords: Trust · Sharing economy · Network analysis · Collaborative behavior · Digital laboratory

1 Introduction

New concepts for sharing goods and services between individuals and companies are emerging worldwide. Well-known examples such as Share Now and Airbnb are seizing new business areas with innovative business models. However, shared resources in academic settings, especially across the boundaries of institutions, remain scarce. Addressing this scarcity requires an understanding of sharing behavior so that it may be promoted. The term sharing means *"[...] the act and process of distributing what is ours to others for their use and/or the act and process of receiving or taking something from others for our use"* [7].

Sharing is an alternative to private ownership, as two or more people can enjoy the benefits or share the costs of using something [7]. Sharing can include physical assets but also more abstract things like knowledge, responsibility, or power at agreed

M. E. Auer et al. (eds.), *Online Engineering and Society 4.0*, Lecture Notes in Networks and Systems 298, https://doi.org/10.1007/978-3-030-82529-4_25

conditions. The development of Web 2.0 allows (compared to Web 1.0) a bi-directional provision of information, which is where digital sharing began (e.g., chat rooms, web sites, online encyclopedias, file sharing) that occurs today [8]. Intangible things such as information, opinions, images, and ideas are easier to share, especially when they are in digital form, and we do not lose them by sharing [7].

Within digital marketplace business models such as the multi-sided platform model, resources are exchanged between buyers and suppliers [1]. More users (buyer & supplier) increase the value of the marketplace, the so-called network effect [1]. The more resources are shared by suppliers the more buyers enter the marketplace and vice versa (cross-side network effect). The greatest challenges for the network effect are the strengthening of user trust, the further development of marketplace offers, and the acquisition of new user groups [6, 14]. This paper focuses on the first challenge – strengthening of user trust within the sharing networked organization.

Sharing is a traditional social practice carried out between trusted persons such as family or friends [6]. In comparison to physical markets, a digital marketplace can provide a transparent evaluation system for the quality and reliability of the stakeholders [26]. Many digital marketplaces rely on a user-centered evaluation system, in which the suppliers are evaluated by the buyer or vice versa, usually by a rating score (e.g. on a scale of 1 to 5) [26]. Williamson [32] argues that trust appears as a consequence of the absence of opportunistic behavior when there are no penalties for such behavior. This means that resources can be contracted out (outsourced) to the networked organization if the necessary trust in the organization is present and it makes sense from an economic point of view. A networked organization is a group of companies or units that act like a single larger entity and use social mechanisms for coordination and control [16]. The units that make up a networked organization are usually legally independent units that gather their potential from each other [10]. Consequently, it is necessary to strengthen user trust within the networked organization in order to maximize the network effect of the business model.

In academic settings, digital laboratories (labs) are a commonly shared resource within networked organizations [18]. Thus, viewing the operation of a digital lab for educational purposes with a sharing economy perspective exposes benefits from sharing knowledge, infrastructure, and facilities. Labs provide students with practical experience and professional techniques and practices [19]. Lab work is a costly activity for any educational or training institution because labs often require specific and expensive equipment and infrastructure as well as trained and qualified staff [4, 18, 19]. In contrast, the utilization of highly specialized labs is often low and access is limited to specific user groups [19], usually those working or studying in the institution to which the lab belongs. The use of digital labs reduces access barriers, enabling global access to shared lab resources.

Through the transformation of the traditional business model of educational labs to a sharing economy, it is necessary to strengthen the users' trust in the networked organization. In an earlier study we have identified variables and factors influencing trust behavior, and a model of trust behavior for networked organizations has been developed [4]. With this paper, we aim to tie in with the research of Baalsrud Hauge et al. [4] by further investigating the trust relationships in networked organizations in order to operationalize the sharing conditions. We present methods for visualizing and scoring collaborative behavior in networked organizations. Therefore, measures and methods for visualizing the interaction of participants should make the network transparent leading to an opportunity to improve the network effect. The research methodology utilizes a mixed method approach: a literature review and participatory action research [11], where the participants act as active researchers in their interaction of a case study for lab sharing. Initially, the research topic is specified, and the research questions (RQ) are developed.

- RQ1: Which methods are suitable for visualization and scoring of trust behavior in collaborative networks?
- RQ2: How can a cognitive model of trust be operationalized to support the network effect?

The rest of this paper is structured as follows. A literature review (LR) [29] in the field of methods for visualizing collaborative behavior was conducted with a three step approach: planning the review, conducting the review and reporting and dissemination of results. These three steps were followed by a participatory research [11] investigating a case study for lab sharing as a real-world situation from which to discern trust scoring attributes within collaborative networks. Further investigation of the case study in a co-creation workshop within the user target group examines findings on the operationalization of the cognitive trust model.

2 Visualizing and Scoring Collaborative Trust Behavior

One of the biggest challenges for the sustainability of a sharing economy is to establish trust [6, 17]. Trust is a widely used term and there are many different definitions. Trust is still a research topic in different fields, and so far no common conceptual understanding of trust has been established [16]. For the use in shared networked organization for digital labs, six trust characteristics apply as analyzed by Baalsrud Hauge et al. [4], 1) virtual enterprises [16], 2) role-based trust [24], 3) inter-personal and inter-organizational trust [3], 4) institutional trust [5], 5) static and dynamic trust factors [12] and 6) the elements of service relationships including personal trust [21].

Mayer et al. [23] defined trust as the willingness of one party to be vulnerable to the actions of another party, based on the expectation that the other party will perform a particular action important to the trustor, regardless of the ability to monitor or control that other party. The relationship of trust exists between a trustor and trustee. The trustor is the actor who desires to assume trust, to rely on someone. Conversely, the trustee is the actor who is to be trusted, to be relied upon [12]. When a trust relation exists, service will be provided to an appropriate standard and expectation [17]. Behrendt et al. [6] describe building a reputation that a buyer and supplier acquires by evaluating peers in repeated transactions as way to establish trust in a digital marketplace. By building a peer evaluated reputation, we have investigated in out earlier study five variables and factors influencing trust behavior for networked organizations [4]: *"(1) states (conditions) of shareable assets in regard to capacity, presence and/or (idle time), capability; (2) previous experience in the sharing of same resource;*

(3) restrictions and compensation; (4) level of behavioural congruence of actors participating in the sharing; (5) regulatory issues and dispute resolution".

2.1 Visualizing Collaborative Behavior

A tabular framework was used to collect and structure information about several visualization methods. The following information was considered: name of method, application areas, description, expected results, application details, time required, tool support, strengths, weaknesses, literature, creator, and date. Mind Mapping (MM) has been identified as good for representing structure but has no direct support of visualizing and/or structuring of collaborative networks [9]. BPMN 2.0 (Business Process Model and Notation 2.0) also provides a graphical representation. It is widely used but the generation of meaningful models is difficult. Social Network Analysis (SNA) is used for the analysis and measuring of structural properties of networks of independent dyadic relationships [25]. In SNA the relationship between the centralities of all nodes can reveal much about the overall network structure. However, network visualizations are often frustrating when network graphs become dense. Collaborative Network Relationship Analysis (CNRA) attempts to quantify relations between organizations engaged in a collaborative networked organization [15]. CNRA allows one to analyze the structural and functional dimensions of connections, but like SNA it is a snapshot of the collaborative situation and is not dynamic. Dynamic Collaborative Network Relationship Analysis (DCNRA) is a dynamic extension of CRNA that supports different types of nodes representing different entity classes [13]. Unfortunately, the models tend to get overly complex and there is no specific software available to create DCNRA models.

2.2 Scoring Collaborative Behavior

Various trust models [5, 12, 16, 31] appear in the literature. For example, Daudi [12] created a supply chain model that evaluates in three phases the propensity, action, and task of a trustee by scoring the expectation and observation of a task. According to Baalsrud Hauge et al. [4], this model could be applicable to the sharing of digital labs, but logistic specific factors should be adapted.

For sharing digital labs, model attributes should score the relationship between actors within the networked organization. To some extent the underlying ideas of creating shared labs infrastructures owned by different organizations shows the same characteristics as those of virtual (networked) organizations and dealing with similar challenges [22, 28], both in terms of what driving the motivations (a common goal) as well as in terms of technical challenges as described in Tagarev et al. [28]. A main challenge is still, as discussed in previous chapter, how to achieve and measure organizational trust. Also here we find some reference work in the field of virtual organizations in the formation phase, e.g. by Seifert [27]. However, a main difference between virtual organization and shared labs network is the organizational structure at universities and their payment models. Therefore, we argue that although the shared labs network has the characteristics of a virtual organization during project phase, it will have more in common (from its characteristics) with supply chain networks after the project phase. For those, we already have several KPIs that can be used for measuring important trust factors like organizational reliability, responsiveness and flexibilities at different organizational levels which is based on different ways of producing the product.

Learning from supply chain networks, these attributes could be a score to measure processes of a supply chain operation. The Supply Chain Operations Reference (SCOR) model is a process reference model [2]. A more detailed discussion can be found in Huan et al. [20]. Even though this model is developed for supply chains, the performance attributes as well as the metrics are applicable in different sectors [12]. We assume that they may also be applied in the context of inter-connected labs, since most of the activities could be seen as a supply of services where a lab supplies a purchaser (a different university or department). In this context, the Perfect Order Fulfilment would tell the purchaser to what extent a lab or other service provider fulfils an order, whereas the cash-to-cycle time will indicate the time a lab waits from, e.g., paying a lab assistant until the purchaser has paid the invoice. A very interesting metric for labs, with their high cost for infrastructure and equipment, is the return on working capital, which in a well-managed lab with shared access would be achieved much faster than a single institution would contribute to this metric.

SCOR is one of the most used reference models within supply chain management and widely used for management operations. There are other reference models that also focus on interoperability in supply chain [30], but SCOR includes the typical supply chain activities such as Plan, Source, Make, Deliver, and Return processes, and users of SCOR believe that a complete supply chain can be fully mapped using it. Zhou et al. [33] have performed an empirical validation of the SCOR model where the impact of the different processes on each other has been studied, and the conclusion is that representing the supply chain by the SCOR model adds value towards the processes. SCOR is a layered model, each level offering more specific details on a process. The first level deals with the process types, level two deals with the process categories, and level three is the process element level. However, for each level, different Key Performance Indicators help to score company performance. These specific strategic metrics can be measured across the supply chain, which makes it possible to monitor and improve the overall performance.

3 Case Study DigiLab4U – Network for Sharing Labs

The cross-institutional research project DigiLab4U intends to offer a digitized lab environment that enables an education marketplace for digital lab facilities. The goal is to develop an integrated, hybrid learning and research environment consisting of a large variety of lab technologies for a digital offering that can be used by any kind of students from bachelor to doctoral students. The DigiLab4U learning marketplace consists on the one hand of digital labs from several suppliers and on the other hand of users who access the labs. The cooperation between universities, research institutions and industry allows pooling the supplier resources so that users, like learners and researchers, have access to a larger variety of digital courses supported by different digital labs. An essential step in this research project is to increase the number of cross-side participants (suppliers and buyers) to create a network effect that enhances the value of the platform [1, 14]. A major starting point for sustainable development of the marketplace is to strengthen trust between suppliers, buyers, and marketplace providers [1].

3.1 Social Network Analysis to Visualize Trust Behavior

In this section, the visualization models from Sect. 2.1 are first evaluated and then the SNA is further analyzed to visualize the case study network. A multiple criteria decision analysis was applied to the methods and approaches to visualize collaborative behavior through 9 criteria to decide for the case study. A panel of experts from the case study consisting of five persons defined the criteria and measured each approach described in Sect. 2 on a scale from 1 (bad) to 5 (very good), as presented in Table 1. CNRA scores full points in the first criteria (#1) because it can represent multiple aspects of collaboration compared to the other models. DCNRA supports different types of entity classes and can thus, compared to the other models, cover several types of collaborative entities (#2). Visualization of people and collaborative relationships (#3, #4) are all covered by the models in various forms, however, there is a strong distinction in dynamic modeling (#5). With the exception of DCNRA, the models are not suitable to cover a dynamic environment. In the case of operationalization for digital labs, a continuous establishment of trust behavior is being attempted. This means that either a) events are created that restart the model or b) that the model has to be returned to DCNRA. The widely used SNA offers a different standard and methods for the analysis of the network (#6), which is also related to criteria #9, as this model is supported by various software solutions. Except for CNRA and DCNRA, the remaining models are easy to learn and use (#8, #9).

Table 1 Analysis of methods and approaches to visualize collaborative behavior

Criteria	MM	BPMN 2.0	SNA	CNRA	DCNRA
#1 Different aspects of collaboration	3	2	2	5	4
#2 Explicit modelling of collaborative entities	1	2	3	1	5
#3 Explicit visualization of persons	2	5	5	2	5
#4 Explicit visualization of collaborative relationships	2	3	4	5	5
#5 Modelling of dynamics	1	2	1	1	4
#6 Methods for analysis of model	1	2	5	1	1
#7 Easy to learn	4	4	4	2	1
#8 Easy to use	4	3	4	2	1
#9 Availability of modelling software	1	5	5	3	1
Average	2.11	3.11	3.66	2.44	3.00

Based on Table 1, SNA is best suited for the purpose of modelling cooperative behavior. However, the model has its difficulties in a dynamic environment (compared to DCNRA) regarding what should be reflected within the operationalization of the model. Following, a more detailed description of SNA is given.

SNA has especially proven to be useful in identifying the structure of a community, the strong and weak connections in a community, and in providing information on the development of the community itself. Also, the identification of trusted/influential participants and leading participants by SNA can lead to a faster way of intervening or supporting the community. SNA is used as an orientation towards the understanding of human behavior that focuses on the importance of social relations. A simplified visualization of the DigiLab4U network applying SNA is given in Fig. 1. The three colors (orange, green and purple) of the nodes represent the different organizations that share labs, like HFT Stuttgart, BIBA Bremen or University of Parma. The individual nodes represent users within an organization that share labs in the marketplace. The strength of the edges symbolizes the strength of the relationship of trust between organizations. The example shows that if Ms. Apfel ever leaves the organization, the network will become unstable for further lab sharing between the networked organizations, which has a direct impact on the network effect. This is just a snapshot and changes dynamically based on the sharing events. The following section investigates how to measure the strength of the edges between the different nodes.

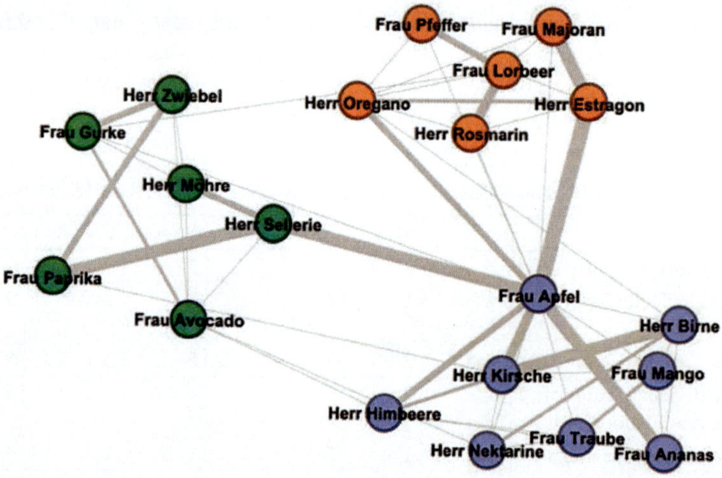

Fig. 1 Social network analysis to visualize collaborative behavior within DigiLab4U

3.2 Supply Chain Operations Reference Model to Score Trust Behavior

Within a co-creation workshop of the DigiLab4U project, a brainstorming with target users was conducted with the aim of identifying important indicators that influence the cognitive model of trust. The aspects collected during the workshop were of both qualitative and quantitative nature. The challenge is to operationalize these indicators for the inclusion in the cognitive model of trust.

To operationalize the personal and organizational indicators of the workshop, the indicators were used to identify operational performance factors grouped by the five SCOR level 1 performance attributes, as shown in Table 2. As a result, the workshop findings could be applied, but did not cover all SCOR metrics. For example, the relative number of digital lectures in labs (load factor) or redundancy of lab infrastructure could be allocated to reliability. Within agility, several level 1 metrics could be assigned, such as Upside Supply Chain Flexibility/Adaptability to the percentage of commercial companies in the network and number of cooperation agreements with well-known institutions. The costs could be allocated to the value of lab equipment in all labs and lab network turnover. Asset Management Efficiency, on the other hand, was for the workshop participant more difficult to allocate. There were no direct indications in the workshop, which does not mean that this dimension is not applicable.

Based on the results of the workshop, further performance characteristics and additional dimensions should be defined for networks dealing with education in addition to the operational metrics from the SCOR model. For example, assessments are needed for the teaching and learning performance in the lab, the value of the research pursued in the lab, the perception of the lab in the marketplace, the amount of innovation that originates in the lab, the level of cooperation that occurs in the lab, and the overall commercial success of the lab are some of the areas that come to mind.

Table 2 Operationalization of the workshop indicators with the SCOR level 1 metrics

Level 1 metrics	Performance indicators
Performance attribute reliability	
Perfect order fulfilment • The percentage of orders delivered, delivery performance to customer commit date, documentation accuracy, and perfect condition	• Percentage of commercial companies in the network • Level of PR activities Relative amount of created research data • Density of (social) media connections among involved persons • Relative number of digital lectures in labs (load factor) • Redundancy of lab infrastructure
Performance attribute responsiveness	
Order fulfilment cycle time • The duration of source cycle time, make cycle time, delivery cycle time, and delivery retail cycle time	• Redundancy of lab infrastructure
Performance attribute agility	
Upside supply chain flexibility • The company's ability to act flexible against the number of days needed to meet unplanned demand	• Percentage of commercial companies in the network • Number of cooperation agreements with famous institutions

(*continued*)

Table 2 (*continued*)

Level 1 metrics	Performance indicators
Upside supply chain adaptability • The company's ability to adapt a maximum percentage of the number of items that can be sent if there is increased demand	Percentage of commercial companies in the network • Number of cooperation agreements with famous institutions
Downside supply chain adaptability • The company's ability to adapt in terms of percentage reduction in the number of orders to be delivered in the next few days	• * (Rate of unused lab hours)
Overall value at risk • The value of the risk borne by the company as a whole	• Risk of equipment failure • Risk of personnel unavailability
Performance attribute costs	
Supply Chain Management Cost • The costs of overall supply chain management (cost to plan, cost to source, cost to make, cost to deliver, cost to return, and mitigation cost)	• Value of lab equipment in all labs • Lab network turnover
Cost of goods sold • The sale main costs that consist of direct labor cost, direct material cost, and indirect cost related to production	• * (Amount of costs like labor, material, indirect that is related to lab usage)
Performance Attribute Asset Management Efficiency (Assets)	
Cash-to-cash cycle time • The company cycle time to get money after buying raw material	• * • (Cash-to-cash Cycle Time) • (Open bills)
Return on supply chain fixed assets • The company return in fixed assets such as tools or technology information that is used in production activities	• * • (Return on fixed assets)
Return on working capital • The company return from the financial capital of work that is used and also its profit	• * • (Return on working capital)

* It is obvious that not all areas of the Level 1 metrics are covered by indicators derived from the workshop results. Additional performance indicators need to be defined

4 Discussion

This work shows how a combination of SNA analysis of the existing networked organization and the five performance attributes in the SCOR-model can be used as an approach to score trust relationship. Corresponding to the SCOR level 1 metrics a set of KPIs that are more related to shared resources and the trust in networked organizations have been presented. The aim is to use these to detect weaknesses within the network. As an example, a KPI relevant for the performance attribute 'reliability' is the number of shared media contributions (social media, publications, lab descriptions etc.). These can be measured, but first by combining these with an SNA analysis that shows if the authors, the kind of interactions, the frequency of interactions is changing over time, we are able to retrieve more detailed insights of how the network evolves and develop over time. Then we can develop suitable strategies that strengthen the sustainability of the network.

The presented cognitive model of trust intends to visualize the current trust behavior in a shared environment to strengthening the networked organization. Since the current relation is dynamic through the effect of external events, operationalization is of central importance. Therefore, events or transactions should be used to operationalize the model. An event is a situation that is characterized through dynamic or change. A given state is changed by an event that leads to a transition from one state to another. In a sharing environment, events affect the current state of the trust relationship within the networked organization. Conditions for individuals and organizations change through events such as employee turnover, employee iteration, new hires, employee promotion, skill development, position change, or interactions such as resource sharing, meetings, or events and affect the current trust relation. Therefore, the developed cognitive model of trust is characterized by events, which allows an operationalization and a dynamic integration into the SNA.

5 Conclusion and Outlook

The paper explored two questions: how trust can be measured and visualized, and how the trust relationship can be operationalized to support the network effect of a business model such as a multi-sided platform. From the literature, six trust characteristics were identified for shared networked organization that can be used when interacting with stakeholder: virtual enterprises, role-based trust, interpersonal and inter-organizational trust, institutional trust, static and dynamic trust factors, and the elements of service relationships including personal trust. However, these depend on the conditions (states) of the shared assets, previous experience with resource sharing, restrictions and compensation, the degree of congruence of the behavior of the actors involved in sharing, and regulatory and dispute resolution issues. To measure these factors, the Supply Chain Operations Reference Model (SCOR) was evaluated in a co-creation workshop as an approach to score trust relationship within a cognitive model of trust. It has been shown that the logistic-specific performance attributes of the SCOR model are applicable, but the metrics needs additional attributes for education, digital labs, and sharing networks. In a participatory research of a case study on lab sharing, nine criteria for collaborative behavior are examined in a multiple criteria decision analysis to select from five visualization models. Social Network Analysis (SNA) is intended to be a suitable method for visualizing collaborative behavior in shared networked organizations. Limitations are the missing dynamics and the one-dimensional visualization of structural and functional dimensions of connections.

The paper contributes towards the operationalization of trust relationship within networked organizations by defining a cognitive trust model consisting of trust factors for networked organizations, a trust model that reflects the relationship between actors, measures to score trust interaction based on events, and a visualizing method for assessing vulnerabilities of a networked organization.

Future work includes the establishment of a sustainable business model for the sharing economy, which will be successful because of an increasing network effect. Trust is a key component in this process, therefore, an integration of the cognitive trust model could reveal changes in the network and possible improvements for the business model.

Acknowledgements. The project on which this paper is based was funded by the Federal Ministry of Education and Research (BMBF), Germany under the funding code 16DHB2112. The responsibility for the content of this publication lies with the authors.

References

1. Abdelkafi N, Raasch C, Roth A, Srinivasan R (2019) Multi-sided platforms. Elec Markets 29 (4):553–559. https://doi.org/10.1007/s12525-019-00385-4
2. APICS (2017) Supply Chain Operations Reference (SCOR) model. https://www.apics.org/docs/default-source/scc-non-research/apicsscc_scor_quick_reference_guide.pdf. Accessed 11 Jan 2021
3. Ashnai B, Henneberg SC, Naudé P, Francescucci A (2016) Inter-personal and inter-organizational trust in business relationships. Ind Mark Manag 52:128–139. https://doi.org/10.1016/j.indmarman.2015.05.020
4. Baalsrud Hauge J, Kammerlohr V, Göbl B, Duin H (2019) Influence of trust factors on shared laboratory resources in a distributed environment. PRO-VE 2019:624–634. https://doi.org/10.1007/978-3-030-28464-0_55
5. Bachmann R, Inkpen AC (2011) Understanding institutional-based trust building processes in inter-organizational relationships. Organ Stud 32(2):281–301. https://doi.org/10.1177/0170840610397477
6. Behrendt S, Henseling C, Scholl G (eds) (2019) Digitale Kultur des Teilens. Mit Sharing nachhaltiger Wirtschaften. Springer Fachmedien, Wiesbaden
7. Belk R (2007) Why not share rather than own? Ann Am Acad Pol Soc Sci 611(1):126–140. https://doi.org/10.1177/0002716206298483
8. Belk R (2014) You are what you can access: Sharing and collaborative consumption online. J Bus Res 67(8):1595–1600. https://doi.org/10.1016/j.jbusres.2013.10.001
9. Buzan T (2010) Use your head. Think your way to success. BBC Active, Harlow
10. Camarinha-Matos LM, Afsarmanesh H, Ollus M (2008) Methods and Tools for Collaborative Networked Organizations, 1st edn. Springer, Boston
11. Checkland P, Holwell S (1998) Action research: its nature and validity. Syst. Pract. Action Res. 11(1):9–21. https://doi.org/10.1023/A:1022908820784
12. Daudi M (2018) Trust in sharing resources in logistics collaboration. Dissertation, Universität Bremen
13. Duin H, Eschenbächer J, Thoben K-D (2011) Modelling dynamics in collaboration: an extension to the collaborative network relationship analysis. PRO-VE 2011:287–294
14. Eisenmann TR, Parker G, van Alstyne MW (2006) Strategies for two-sided markets. Harvard Bus Rev 2006:1–12

15. Eschenbächer J, Seifert M, Thoben K D (2009) Managing distributed innovation processes in virtual organizations by applying the collaborative network relationship analysis. PRO-VE 2009:13–22
16. Fladnitzer M, Grabner-Kräuter S (2006) Vertrauen als Erfolgsfaktor virtueller Unternehmen. Grundlagen, Rahmenbedingungen und Maßnahmen zur Vertrauensbildung, 1st edn. Deutscher Universitäts-Verlag, Wiesbaden
17. Goudin P (ed) The cost of non-europe in the sharing economy. economic, social and legal challenges and opportunities. European Parliamentary Research Service, Brussels (2016)
18. Haertel T, Terkowsky C, Dany S, Heix S (eds) Hochschullehre & Industrie 4.0. Herausforderungen - Lösungen - Perspektiven. W. Bertelsmann Verlag, Bielefeld
19. Heradio R, de La Torre L, Galan D, Cabrerizo FJ, Herrera-Viedma E, Dormido S (2016) Virtual and remote labs in education: a bibliometric analysis. Comput Educ 98:14–38. https://doi.org/10.1016/j.compedu.2016.03.010
20. Huan SH, Sheoran SK, Wang G (2004) A review and analysis of supply chain operations reference (SCOR) model. Supply Chain Manag (1):23–29. https://doi.org/10.1108/13598540410517557
21. Johnson D, Grayson K (2005) Cognitive and affective trust in service relationships. J Bus Res 58(4):500–507. https://doi.org/10.1016/S0148-2963(03)00140-1
22. Katzy BR (1998) Design and implementation of virtual organizations. IEEE Comput Soc 1998:142–151. https://doi.org/10.1109/HICSS.1998.655269
23. Mayer RC, Davis JH, Schoorman FD (1995) An integrative model of organizational trust. Acad Manag Rev 20(3):709. https://doi.org/10.2307/258792
24. Müthel M (2006) Erfolgreiche Teamarbeit in deutsch-chinesischen Projekten. Dissertation Universität Lüneburg, 2005, Wiesbaden
25. Otte E, Rousseau R (2002) Social network analysis: a powerful strategy, also for the information sciences. J Inf Sci 28(6):441–453. https://doi.org/10.1177/016555150202800601
26. Schallmo D, Rusnjak A, Anzengruber J, Werani T, Jünger M (eds) Digitale Transformation von Geschäftsmodellen. Grundlagen, Instrumente und Best Practices. Springer Fachmedien, Wiesbaden
27. Seifert M (2009) Collaboration formation in virtual organisations by applying prospective performance measurement. Bremer Schriften zur integrierten Produkt- und Prozessentwicklung, vol 63, Aachen
28. Tagarev T, Yanakiev Y (2020) Business models of collaborative networked organisations. DESSERT 2020:431–438
29. Tranfield D, Denyer D, Smart P (2003) Towards a methodology for developing evidence-informed management knowledge by means of systematic review. Br J Manag 14(3):207–222. https://doi.org/10.1111/1467-8551.00375
30. van Oosterhout MP, Veenstra A, Meijer M, Popal N, van den Berg J (2007) Visibility platforms for enhancing supply chain security. In: The international symposium on maritime safety, security and environmental protection
31. Walber M (2000) Konstruktivistische Erkenntnistheorie als Legitimation für den Einsatz von multimedialen computerbasierten Lernumgebungen in der beruflichen Bildung. Diplomarbeit, Universität Bielefeld
32. Williamson OE (1987) The economic institutions of capitalism. Firms, markets, relational contracting. Free Press, New York
33. Zhou H, Benton WC, Schilling DA, Milligan GW (2011) Supply chain integration and the SCOR model. J Bus Logist 32(4):332–344. https://doi.org/10.1111/j.0000-0000.2011.01029.x

Real-Time Fluid Simulation for an Interactive Plate Column Virtual Lab

Mario Wolf(✉), Pascalis Trentsios, and Detlef Gerhard

Digital Engineering, Ruhr University Bochum, Bochum, Germany
{mario.wolf,pascalis.trentsios,detlef.gerhard}@rub.de

Abstract. Fractionating columns are used for most of all technical distillation processes. The fundamental understanding of the principles of these critical components is often only provided in theory or in small-scaled laboratory setups.

The concept and application presented in this paper aim at educators who wish to offer their students the possibility for individual experimentation in the context of process engineering and more specifically plate columns. A dynamic, immersive virtual lab may solve the problem of not representing real-life dimensions and allows to show of the direct impact of internal structural changes not only in theory or graphs, but with real-time fluid and particle simulations.

For this reason, the relevant areas of a plate column, such as the infeed area, the column head, the condenser, the column sump and vaporizer are 3D visualized and backed up by McCabe–Thiele diagrams to show the consequences of changes in the design of the column. The virtual lab application was compiled as an executable Windows application with multiple views in 3D and 2D containing dynamic fluid simulations, which were then optimized for performance.

Keywords: Virtual lab · Process engineering · Real-time fluid simulation

1 Introduction

Fractionating columns are used for about 95% of all technical distillation processes in the USA [2], while the distillation itself accounts for about 40% of the total energy consumption in a typical chemical plant [4]. The fundamental understanding of the principles of these critical components is often only provided in theory or in small-scaled laboratory setups at universities. The classic educational methods neither transport the real-life dimensions of fractionating columns, nor the accompanying quantities of feed and product flows. In this context, not only do physical dimensions matter, but the implications of the engineering of the inner workings and their consequences for the performance of the fractionating column.

With this motivation in mind, use cases in the domain of fractionating columns are highly interesting subjects for the creation of a real-time fluid simulation and to assess technical limitations for virtual lab use in process engineering.

M. E. Auer et al. (eds.), *Online Engineering and Society 4.0*, Lecture Notes in Networks and Systems 298, https://doi.org/10.1007/978-3-030-82529-4_26

2 Related and Previous Work

Getting practical hands-on experience in the field of thermodynamics is more often than not bound to real laboratories.

It is crucial to visualize the thermofluidic procedures happening on the inside of a column during a distillation process for educational purposes. To accommodate that fact, special columns are constructed with an outer shell made from glass [6] as shown in Fig. 1 (right side). However, those columns are specially fabricated and only feasible for laboratory and educational purposes. Additionally, acquisition, maintenance and operation cost must be considered. Also, these columns are stationary like any other big, industrial machinery, and thus limit the availability for students when not at campus. Scaled-down fractionating columns like the one shown in Fig. 1 (left side) still require a well-appointed laboratory to be operated for educational use, because a base set of equipment is required for every fractionating column-based distillation process, independent from its size.

Fig. 1 Different types of glass columns. Figures from: Fisher Scientific (Austria) GmbH (left), C3 Prozess- und Analysentechnik GmbH (right)

Besides real labs, virtual labs are widely used for educational purposes in various engineering domains [3]. They allow visualization of complex engineering problems in a configurable, expandable, and low-cost manner [1]. Virtual Labs have the potential to enhance the overall educational quality and can be used as an addition to an already existing educational curriculum [7].

Virtual labs can be tweaked to fulfill specially designed or niche applications, without the inconvenience of operating a large piece of industrial equipment for a single use every semester.

Virtual labs can provide the possibility to gather close to real life data [5]. In the field of process engineering, virtual labs use mathematic algorithms and methods to approximate the behavior of real laboratory equipment, or, more precisely, the underlaying thermodynamic principles. An example for that is the web-based virtual lab “LaVIP” presented by Roman-Sanchez et al. that uses a MATLAB backend to simulate realistic behavior [8].

Fluid simulations are widely used in the animation, gaming, and overall entertainment industry, and can visually approximate the behavior of real fluids [11]. Various approaches for fluid simulations exist, for example particle based, height field and cubic Eulerian grid simulations [9]. Special mathematical and computer scientific methods enable real-time fluid simulation [10]. In this context, real-time is defined as the simulation being calculated for each frame rendered in the 3D engine. This implies that the simulations should be efficient enough to be rendered with a minimum of 30 fps and preferable higher, depending on the targeted medium.

The authors have previously worked on multiple projects regarding virtual labs and remote labs [12] as well as mixed reality applications and possible approaches to start developing such virtual experiences [13]. The development of the virtual lab presented in this paper benefited from the experiences gained in the previous projects.

3 Concept

The approach to the concept at hand was not to develop a physically 100% correct fluid simulation, but an approximation that can, when used in conjunction with further learning material, improve the learning experience by offering real-time, interactive changes to the virtual laboratory and showing the broader view of consequences of these changes.

One can think of that like the current use of physical glass columns in education scenarios. The only difference, besides the non-physicality of a virtual lab, is that the virtual lab, or more precisely the virtual fractionating column, suffers no limitations in its ability to quickly change operating modes or inner workings.

To achieve this goal, the approach was to replicate physical properties of real distillation processes and a selection of the relevant equipment used in these processes.

The identified process parameters and equipment then lead to the conceptual design and finally the creation of a visually appealing version of a virtual laboratory, which mimics the real-life distillation process with enough recognizability, so that the theory learned in regular lessons works as a basis for virtual experimentation.

Due to this focus, thc application makes use of the widely used McCabe–Thiele method and diagrams on the process engineering side of things together with real time fluid and particle simulation, to visualize and clarify the workings of the virtual components.

The virtual lab should be easy to access and expandable for future upgrades and offer a variety of common visualization examples for education purposes in process engineering. As the approach was planned by a team of process engineers, mechanical engineers, and computer scientists, the concept focuses on the process engineering basics first, then shows the basics of 3D modeling the desired scene and finally discusses the real-time visualization of the different flows in the system from a performance standpoint.

Regarding the overall design, Fig. 2 shows the mockup of the planned GUI, which focusses on the visualization as the center point (no. 3), diagrams and further textual information are provided on the right-hand side (no. 2) and the options and scene menu is located on the left-hand side (no. 1). The idea is to switch viewpoints based on the selected balance area and providing fitting information regarding the current context.

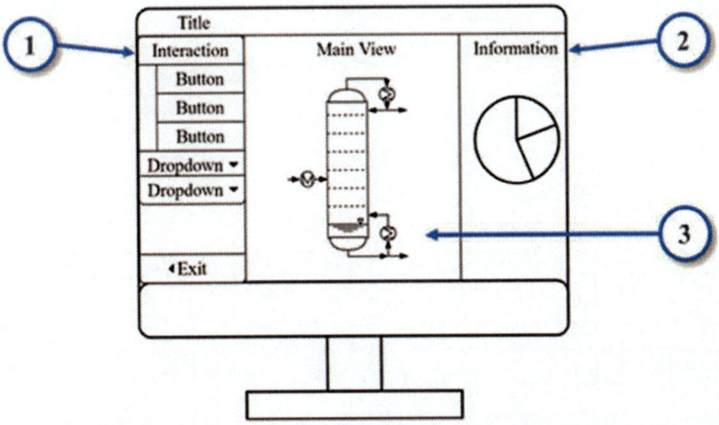

Fig. 2 Mock-Up for the rough GUI for the virtual lab

Process Engineering Basics

The main learning objective is to offer an introduction to rectification columns using the McCabe–Thiele method and diagrams to determine to amount of plates/separation stages for separating a liquid two-phase mixture.

First, we need to establish an understanding for relevant steps in learning experiences around the McCabe–Thiele method and how the theory is reflected in visual representations. For that, the whole of the plate column should be subdivided in several focus areas based on their phase equilibrium with individual characteristics, like their discrete operating status and process parameters.

Relevant areas of a rectification column are the column as a whole (Fig. 3 , *area I*), the infeed area (Fig. 3, *area V*), the column head (Fig. 3 , *area III*), the condenser (Fig. 3, *area II*) and the column sump and vaporizer (Fig. 3, *area IV*). Each area should be explained and visualized in detail in the application to show of the consequences of changes in the design of the column.

Modeling the Virtual Scene

The first test of feasibility was to model both a perforated plate and a bubble cap plate and import them in the Unity 3D engine in a tessellated 3D format (.fbx) after a slight polygon reduction. With that out of the way, the test for the behavior of the fluid simulation in conjunction with the given 3D geometries was conducted. As the principle of rectification columns relies on both fluid and vapor interaction, the next test was of the particle simulation to represent the rising steam in the plate column.

The first prototypes each consisted of a single plate with a simulated fluid emitter on top to ensure the correct function regarding the behavior of the fluid in a minimal working example.

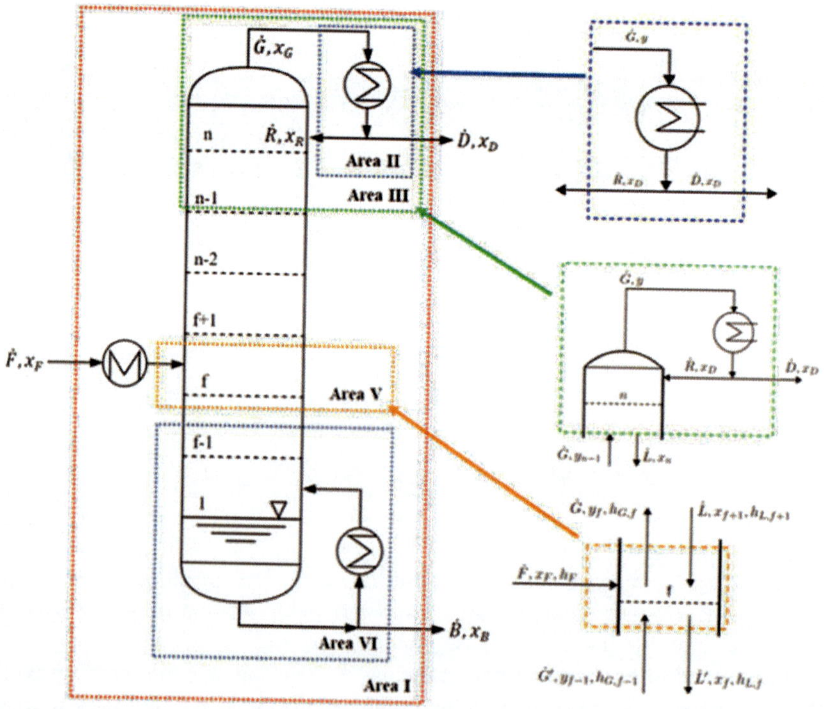

Fig. 3. Relevant balance areas of a rectification column with some exemplary highlighted balance areas.

Figure 4 shows the theoretical function of two perforated plates in a rectification column (left) and the first fluid simulation test with added particle effects for the visualization of vapor.

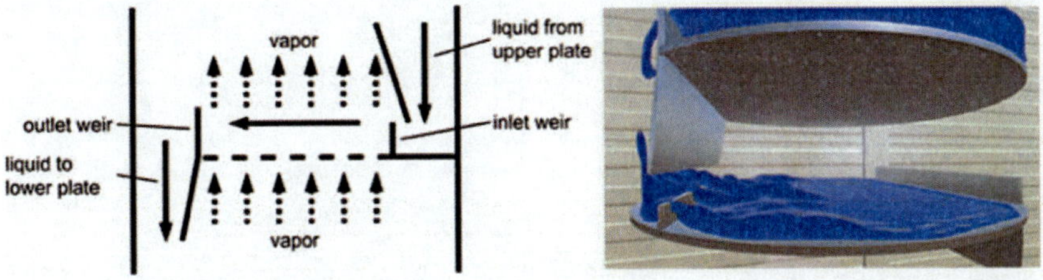

Fig. 4 Perforated plate concept (left) and prototype (right)

The next prototype tested the geometrically more complex bubble cap plate. Figure 5 shows the theoretical function of two bubble cap plates in a rectification column (left) and the first fluid simulation test with added particle effects for the visualization of vapor.

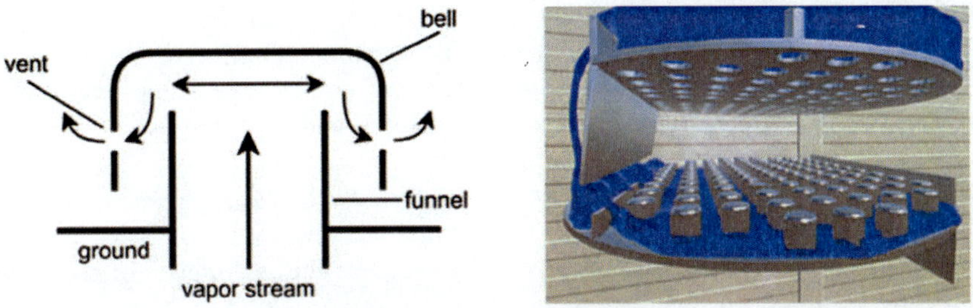

Fig. 5 Bubble cap plate concept (left) and prototype (right)

After making sure the individual components of the rectification column, meaning in this case the different types of plates, worked sufficiently, the outer shell of the column, the auxiliary hardware and piping were designed in a CAD software and imported into a surrounding level representing a laboratory, shown in Fig. 6.

Fig. 6 Unity editor view, virtual laboratory with rectification column and inactive fluid simulation (left), wireframe of the 3D-models (right)

Designing the User Experience

Based on the created virtual laboratory scene, the next step is to create the user experience through dialogs and animation. To recreate and educate about the five identified areas of interest (cp. Fig. 3) of the plate column, the application should guide the user through several different steps to familiarize with both the use of the application as well as the process engineering topics.

Figure 7 shows the planned process of the first use of the application. After the initial start of the application, the user is given an overview over the general GUI and the balance areas present in the application, as well as the importance of each component in the process of distillation.

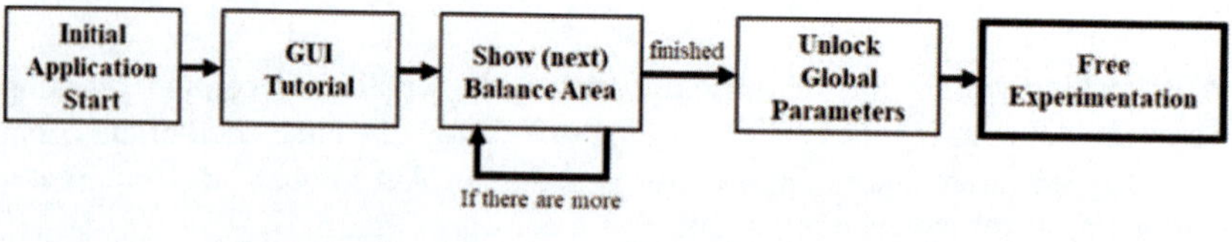

Fig. 7 Process of using the application for the first time

Additionally, the built-in tutorials show which parameters can be adjusted to change the simulation for the particular part of the plate column.

After a deep dive into the individual components, the information about global parameters concerning the plate column are presented and unlocked for customization, after which the user is free to experiment with the virtual plate column lab.

4 Implementation

The concept presented in this paper aims at educators who wish to offer their students the possibility for individual experimentation in the context of process engineering and more specifically plate columns. A dynamic, immersive virtual lab may solve the problem of not representing real-life dimensions and showing of the direct impact of internal structural changes not only in theory or graphs, but with real-time fluid and particle simulation.

The application's goal in further development is to offer gamification-like progress tracking as self-control mechanism for students and to guide them through an introductory course about plate columns, plate design and different engineering considerations regarding product purity in a distillation process via a learning management system. The current implementation already provides activity completion metrics, which should be the only metric shared with the supervising educator, as the individual experimentation is solely aimed at the student's curiosity.

The virtual lab itself was developed using several laboratory 3D assets to make it visually appealing and a fluid simulation asset (Obi Fluid by Virtual Method Studio) as well as the Unity 3D engine's own particle system to approximate the real-life thermo-fluid dynamic processes of the distillation process. The initial starting screen of the *Virtual Plate Column Lab* application is depicted in Fig. 8.

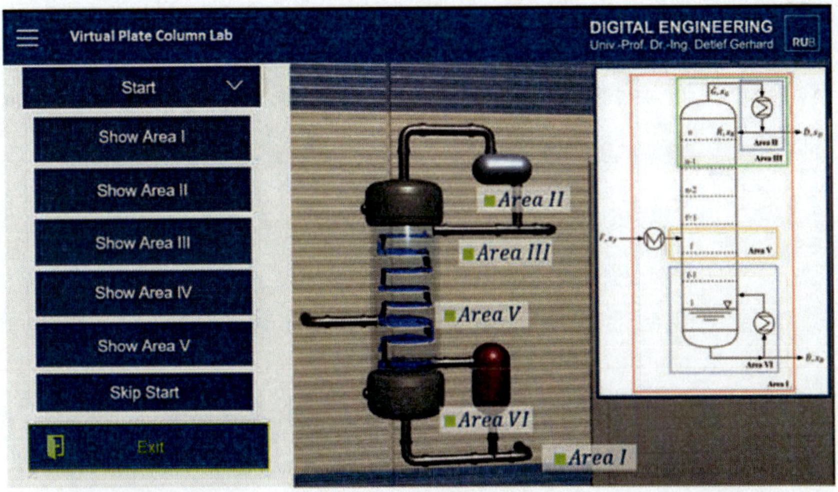

Fig. 8 Starting screen of the 'Virtual Plate Column Lab' application

The *Virtual Plate Column Lab* application was compiled as an executable Windows application, with preparations for using RESTful web services or LTI interfaces for interacting with learning management systems.

In a real plate column, the vapor passes through the trays, enclosing liquid particles and dragging them upwards. This kind of behavior is a) playing out on a very, very small scale and is b) very hard to approximate numerically in real-time. The authors tested the interaction between particles of both vapor and liquid but found that the performance was not sufficient for the intended use as a desktop application and even less so when considering future virtual reality use, which requires stable framerates above 90 frames per second.

So, as expected, compromises must be made to ensure both interoperability and adequate performance for a real-time application.

One of these compromises is that only a single fluid is simulated in real-time and not a mixture of two separate fluids with different boiling points. Instead, an approximation of the general visual impression is achieved by using the fluid simulation exclusively to render a single uniform fluid with a randomized life span for the individual liquid particles and Unity's particle effects to visualize the evaporation on the plates and in the sump. The result is a seemingly physically correct acting fluid with reducing volume over time for the impression of vaporization.

Figure 9 shows how the fluid simulation works and real-time simulation is achieved. To reduce complexity, only a finite number of physical particles (around 2000 on a mid-tier gaming PC) are calculated with a given set of physical properties. These particles are called "spherical imposters" as they are way bigger than real fluid particles, but therefore more efficient to simulate. The result can be seen in Fig. 9, left side in both 3D view (top) and 2D view (bottom). Of course, this is not the intended visual appearance. Using a special shader, which is provided with the fluid simulation

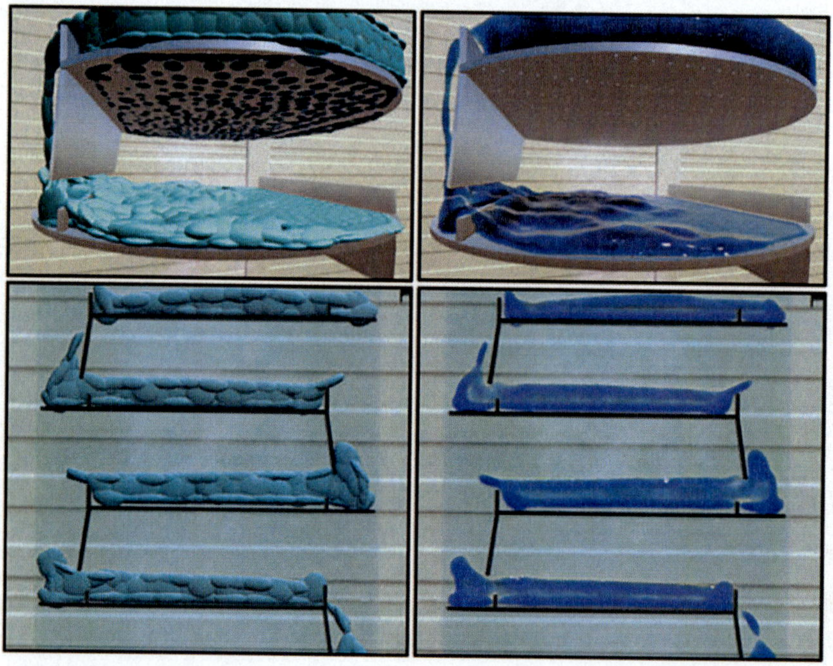

Fig. 9 Spherical imposters in 3D view (top left) and 2D view (bottom left), same scene with fluid shader in 3D view (top right) and 2D mode (bottom left)

asset, the "spherical imposters" are rendered into an approximation of a fluid with a varying density, as seen in the 3D view (Fig. 9, right side, top) and even more pronounced in the 2D view (Fig. 9, right side, bottom). The (background) calculation of "spherical imposters" and the rendering into fluid via the shader enables the visualization of a flow through the plate column.

The effect is shown in a screenshot in Fig. 10, where the lab scene focusses on balance area V (cp. concept in Fig. 3) with McCabe–Thiele diagram and the fluid and vapor simulation.

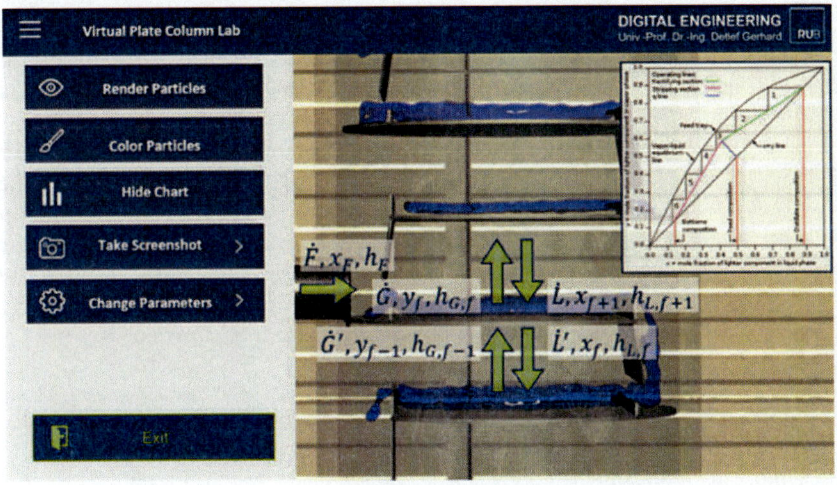

Fig. 10 Detailed view of the balance area V

The authors created the 2D views of the virtual lab to mimic the figures found in common thermodynamic and fluid dynamic textbooks. The obvious advantage of the 2D version is that is utilizes less resources for calculation than an equivalent 3D version. The visualization of the feed area is shown in Fig. 11. It is up to the user which kind of visualization is preferred, as the application offers both.

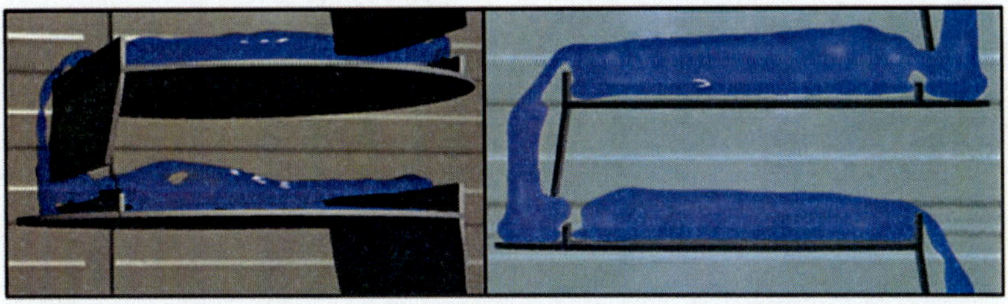

Fig. 11 Difference between 3D view (left) and 2D view (right) in use in a detail view

The downside of this choice between 3D and 2D visualization is that the performance strongly relies on the hardware of the device running the application, resulting in stuttering simulation if the 3D mode is engaged on a weak machine. A virtual lab with high end visuals and very detailed simulations would not be recommend, since it could only run on high-end devices such as gaming PCs.

To achieve a solid balance between visual fidelity and performance, various performance improvements methods were applied. For example, the main view of the full fractionating column is visualized with a 2D fluid simulation instead of a more complex 3D simulation. Additionally, every relevant area of the fractionating column can be viewed in a detailed, but separate 3D scene. In those cases, the view focuses the simulation only on the concerning area and thus offers a more detailed 3D simulation without performance wasted on other areas that are not in the field of view at the time. Static objects, their textures and the ambient light are baked, which means they are pre-calculated, to reserve performace for the fluid simulation.

All mentioned visualization technologies existed prior to the implementation of the scenario at hand. However, to overcome the technical difficulties of bringing them together in a coherent use case requires a considerable amount of experience in several domains. Running a full real-time fluid simulation in a 3D (game) engine causes massive performance problems, as these code-library-based simulations are often CPU-based. This hinders such a simulation to run smoothly on systems with lesser computing power like mobile devices or applications in a web-based context.

5 Experimental Implementation for Future Work

The authors also implemented a version where the vapor physically interacts with the fluid, as is the case in a real plate column. For this more realistic approach, the vapor was not simulated with unity's particle system, but instead with the same tool that was used for the fluid simulation. Due to the physical simulation of both vapor and fluid particles with different physical properties, they can interact with each other.

This simulation is shown in Fig. 12 with increasing time steps from step *a* to step *e*. In this example, the physical bodies of the particles are rendered to visualize the interaction between the blue fluid particles and the white vapor particles. This method allows to approximate the real process happening in plate column, where the vapor passes through the active tray, carrying liquid with it.

The implemented simulation currently works in a small scale with around 600 vapor particles and only a few liquid particles. It is not feasible to simulate the whole column with this approach with the current implementation, as a real time simulation would need a minimum of 30 fps for desktop use.

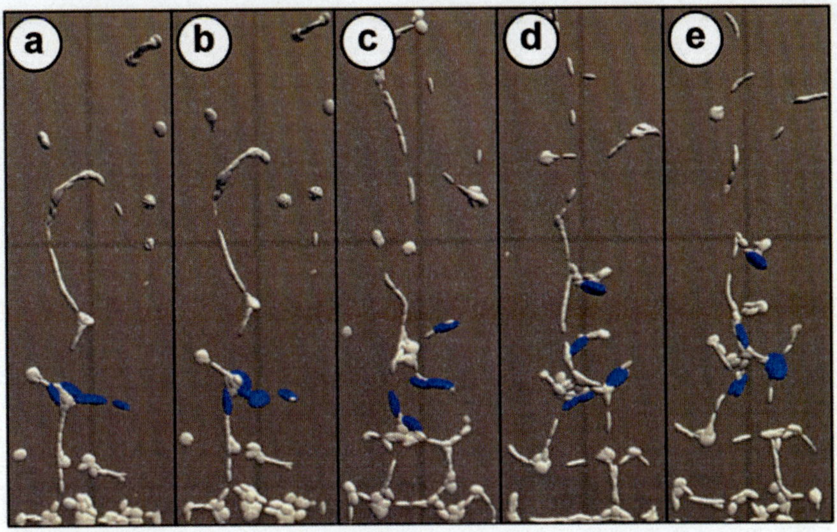

Fig. 12 Timeline of vapor particles (white) floating upwards with liquid particles (blue) getting dragged up by them

The next logical step, after fixing the performance issues mentioned above, is to actually reverse the initial decision to only emit a singular fluid as a simulation basis and instead emit two, let them mix and then let the mixture flow into the feed of the plate column. The motivational aspect of this is, besides depicting reality more precisely, that in the virtual setup we can alter the distinct substances not only in regard of their simulated physical properties, but also in color. As seen in Fig. 12, it is possible to change the color to visually distinguish the fluids and vapor to illustrate the processes within the plate column. Alternatively, the color of a substance can be changed over time, to show the continuing mixing of "old and new" material within the plate column during the distillation process.

Regarding the learning management platform integration, the Unity-based application is ready to send and receive RESTful web service calls, as well as to use the LTI interface of the Moodle LMS.

In a wholistic learning experience, the tools provided by the Moodle platform can be leveraged to offer chat tools, discussion forums, quizzes, test, and of course further learning material around the theory of distillation processes. Due to the already existing gamification and progress tracking tools, the learning platform can guide the students in their navigation between web-based learning materials and local 3D applications.

6 Conclusion

The authors showed that it is possible to visualize the approximation of the inner workings of a plate column and the consequences of changes in their layout by combining efficient visualization methods. This is done to allow real-time interaction and parameter change, rather than numerical precision.

Even though a web-based version of the application was planned, it could not be implemented in a satisfying manner. The main reason for that is the inability of JavaScript-driven applications running in a web browser to use multi-threading, which is needed for the fluid simulation.

The author's plan is to connect with educators in both academia and industry to enhance the experience and gather student and educator feedback. It will be critical to further investigate what is actually needed to convey realistic impressions as part of a virtual experience to enhance current teaching methods.

Acknowledgements. The authors would like to thank the undergraduate students Lasse Christian Bömkes, Jan Niklas Gößling and Ioannis M. Nikolaou Bennour who were part of the development team for this virtual laboratory.

References

1. Ahuja G, Gupta A, Wardhan H, Choppella V (2015) Assessing the impact of virtual labs: a case study with the lab on advanced VLSI. In: IEEE 15th international conference on advanced learning technologies (ICALT). IEEE, pp 290–292
2. Al-Haj Ibrahim H (2014) Design of fractionation columns. In: Bennett K (ed) MATLAB applications for the practical engineer. InTech
3. Cherner YE, et al (2019) Implementation of interactive and adjustable cloud-based e-learning tools for 21st century engineering education: challenges and prospects. In: IEEE world conference on engineering education (EDUNINE), pp 1–6
4. Felder RM, Rousseau RW (2005) Elementary principles of chemical processes. With integrated media and study tools, 3 edn
5. Gastaldi R, Campardo G (2020) Electronic experiences in a virtual lab. Springer, Cham
6. Górecki H, Duda JT, Byrski W, Janik K (1997) Laboratory distillation column for education on process control. IFAC Proc Vol 30(12):267–272
7. Keddi D, Frerich S (2021) Enhancing engineering education by virtual laboratories. In: Auer ME, May D (eds) Cross reality and data science in engineering, vol 1231. Springer, Cham, pp 359–365. https://doi.org/10.1007/978-3-030-52575-0_30
8. Roman-Sanchez DI, Lopez-Arenas T, Sales-Cruz M (2018) An interactive virtual laboratory for process engineering education. In: Computer aided chemical engineering, pp 1707–1708
9. Schmidt M, Savoye Y (2015) Fluid simulation. In: Lee N (ed) Encyclopedia of computer graphics and games. Springer, Cham, pp 1–5
10. Wan F, Yin Y, Zhang S (2018) 3D parallel multigrid methods for real-time fluid simulation. 3D Res 9(2)
11. Zadick J, Kenwright B, Mitchell K (2016) Integrating real-time fluid simulation with a voxel engine. Comput Game J 5(1–2):55–64
12. Trentsios P, Wolf M, Frerich S (2020) Remote lab meets virtual reality – enabling immersive access to high tech laboratories from afar. Procedia Manuf 43:25–31
13. Trentsios P, Wolf M, Gerhard D (2021) Comparing virtual reality SDK potentials for engineering education. In: Auer ME, May D (eds) Cross reality and data science in engineering, vol 1231. Springer, Cham, pp 375–392. https://doi.org/10.1007/978-3-030-52575-0_32

Design of a Mixed Reality Game for Exploring How IoT Technologies Can Support the Decision Making Process

Jannicke Baalsrud Hauge[1,2(✉)], Prabahan Basu[1,3], Fatima Sundus[1,3], Anindya Chowdhury[1,3], and Artem Schurig[2]

[1] BIBA – Bremer Institut Für Produktion und Logistik GmbH, Bremen, Germany
baa@biba.uni-bremen.de
[2] KTH-Royal Institute of Technology, Södertälje, Sweden
[3] Universität Bremen, Bremen, Germany

Abstract. There is a trend in industries towards adoption of new technologies coming from Industry 4.0 to improve logistics operations. This has also led to new requirements on engineering education. This paper presents work in progress on a mixed reality game that is intended to be used for the education on engineers and engineering students in the blended lab environments. The game itself consist of a physical and a digital component, and the focus in this article is to describe the design and prototypical implementation of the physical part and how that can be operated in a remote lab environment.

Keywords: Educational serious games · Mixed reality · IoT · Logistics operation · Engineering experimental education

1 Introduction

The relevance of understanding the concept of Industry 4.0 and the benefits and barriers of a potential usage in logistics and manufacturing operations is increasing [1–4] performance. The industry 4.0 concept that allows physical objects to connect to virtual spaces anywhere at any time is likely to improve user experience and improved process performance [5–8]. However, a successful implementation and operation requires proper training and awareness raising for all involved stakeholders as well as future employees, such as students [4–6]. Gamebased learning as method has shown potential as an educational tool for supporting the process competence development among engineering students [9–12], even though there is a need for making existing games more pervasive and adaptable to the different user needs [9]. This work-in-progress article aims to present how an existing digital educational serious game on CPS [10] can be extended with a physical component to increase the immersiveness with real-time, real world user feedback during game play, taking the experience with such mixed environments into account [2, 3, 12]. The focus in this work-in-progress article is therefore to investigate the role of sensors in a mixed reality (MR) environment consisting of a virtual and a physical component on the immersiveness and the learning process from a design perspective.

M. E. Auer et al. (eds.), *Online Engineering and Society 4.0*, Lecture Notes in Networks and Systems 298, https://doi.org/10.1007/978-3-030-82529-4_27

2 Methodology

We used action-based research for the conceptual design of the game. The design is a demanding operation, given the complexity of the resulting system and the difficulties in taking into account the different viewpoints involved. There are several different frameworks that are proven suitable for designing games [9, 11, 13–17], all with different advantages and drawbacks as well as also developed to fit different needs [9, 10, 12]. This work is an extension of the work described in [10], and thus we have selected the same game design framework mostly using the frameworks described in [11, 17, 18].

3 The Overall Game Design

This section briefly first introduces the Activity Theory-based Model of Serious Games (ATMSG) [17], before we use the model for designing the MR game. This model supports a systematic and comprehensive representation of educational serious games, showing the ways about how the game elements are associated to each other throughout the game, and how these elements help to the achieve the pedagogical goals.

3.1 Learning Scenario

The presented work will be a part of a course and we have designed the following description of the learning scenario (Table 1).

Table 1 Learning scenario technology supported decision-making

Course title	Decision making technology implementation
Possibility of distance learning	Yes, as a part of the shared lab infrastructure or through access to the games and/or simulations and digital twins
Abstract	The module will be used in both strategic and operative decision making. The main objective is the participant shall get a deeper understanding on how different technologies can support the seamless and transparent information flow and material handling in a warehouse, but also on factors that influence the decision making process, because they are not changeable At the operative level, it will be a decision making process related to daily operations The KPI used are those from first and second level SCOR model
Goals and objectives of the course in terms of competences and skills	Apply methods for strategic and operative decision making for technology introduction (S) and usage (O) Applying knowledge & methods on how to analyse risks and opportunities related to technologies in a warehouse in a supply chain Assessment Design layouts of WMS Understanding and Assessing of in-stocking and out-stocking technologies Understanding on how to implement automated material movement in an existing warehouse area Safety issues related to HCI

The corresponding game uses an open narrative, where the players get a set of tasks related to typically operational processes in a warehouse, like in- and out stocking, ordering, picking and placing etc. In the current gameplay, the players are first asked to get familiar with the logistics operations and then to evaluate how different IoT technologies can support the operational process in terms of process quality, efficiency and as well as cost and time reduction versus the implementation and operational costs of the technologies. The narrative opens up for difference in the goals for individual players- i.e. can also be used for assessing privacy and ethical issues. It is possible to solely focus on the technical implementation, be more specific related to how to select what type of sensors depending on what we are interested (humidity, tilt of forklift, temperature, real-time tracking etc.) depending what type of goods we have or more focusing on the flow of material.

3.2 ATMSG Application for Serious Games Analysis

In order to design a mixed reality game scenario that can allow the students to reach the above listed intended learning outcome, we need to redesign the existing computer game and to focus more in the game play of the haptics in the game. As described above, a main reason for choosing ATMSG, is that we used that in the previous design, but secondary also because it allows us to use the same model for analyzing and developing a game. The ATMSG model (Fig. 1) has the purpose in supporting the analysis and design of serious games for both experts in serious games, and non-experts who are engaged in serious games related projects, such as application domain experts.

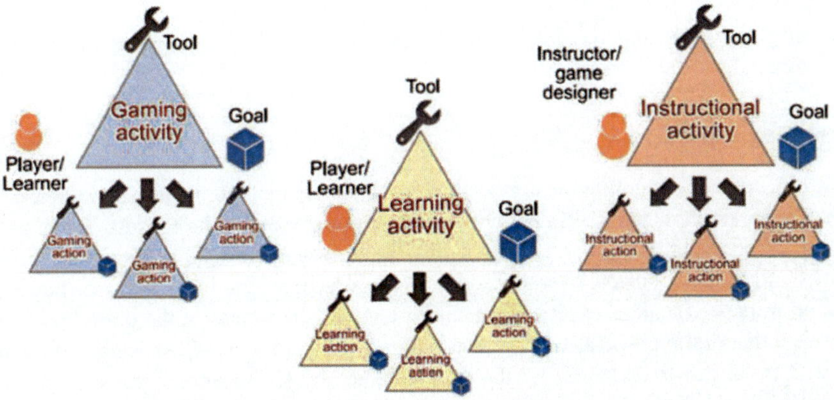

Fig. 1 Overview ATMSG methodology (based on [17])

The ATMSG four-step approach (Fig. 2) for SGs analysis directs in order to help analysing a game to have a more concrete understanding about how learning should take place in the game. These steps directs from a high-level understanding of the activities to the concrete components that execute those activities. It helps recognizes game components with the help of the taxonomy of SG components.

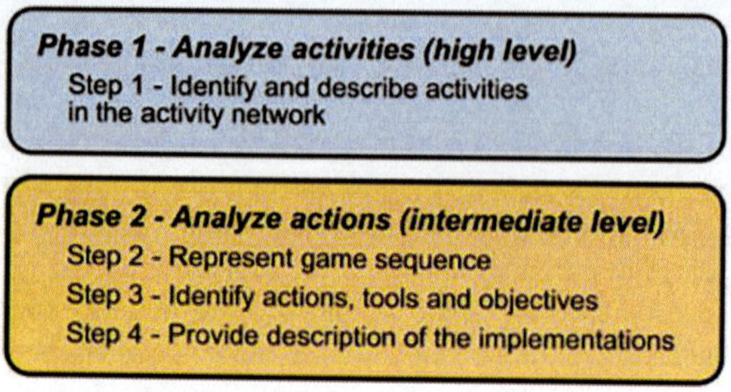

Fig. 2 (fig. 6 [17]) The four-step approach for applying the Activity Theory-based Model of Serious Games (ATMSG) to the analysis of educational serious games

Applying this approach (Fig. 1, Fig. 2) it is possible to construct a connection between learning activities with game playing activities as well as connect these with the instructional activities. We have applied this analysis in the game described above and in more detailed in [10]. The evaluation of the virtual game play showed that the learning objectives were met, but that some of the players had difficulties in players in assessing the differences between similar technologies for specific problems only based on what happens in the virtual game play. Therefore, based on previous work [2, 3, 12] we assume that the introduction of data from a physical environment can help in overcoming this and make the game play experience more pervasive.

Next section therefore outlines the first thoughts of which sensors to implement in the physical environment (i.e. a miniature warehouse in the lab) and how these can be used for a better learning experience and for aligning the game play activities with the learning activities and the instructional activities in order to improve the learning outcome.

3.3 Physical Part

The objective of extending the digital game with physical components is to increase the players possibilities to experiment with different components (like sensors, actuators or the tracking technologies). The player may then investigate what impact small differences might have on the overall results in a more practical way, and still have the opportunity to visualize, manipulate and re-play the game play (in the digital part). To develop that in the physical world, we use a model forklift and add sensors and motors to it. The forklift is operated and the data collected by the sensors is transferred back. The basic concepts of the physical model are discussed here. The sensors used, the reason for selection of the sensors, how the sensors can assist the use of the forklift will be discussed in detail. This is just the basic idea of how the sensors will be implemented on the forklift, more sensors might be added later to enhance its performance. The sensors used for this purpose are:1. Ultrasonic sensors, 2. Vibration sensors, 3. Touch sensor. Different sensors have different purposes and are installed in the forklift to assist the performance of the forklift. These are described below.

Ultrasonic sensors (also known as transceivers when they both send and receive, but more generally called transducers) work on a principle similar to radar or sonar, which evaluate attributes of a target by interpreting the echoes from radio or sound waves respectively. Ultrasonic sensors generate high frequency sound waves and evaluate the echo, which is received back by the sensor. Sensors calculate the time interval between sending the signal and receiving the echo to determine the distance to an object. This will be used on all four sides of the forklift to determine the distance of the forklift from any surrounding object. The more area the ultrasonic sensors cover, the more accurate the measurement will be. The main purpose of installing the sensor is to create an alert for the forklift driver of nearby surrounding objects.

Vibration sensor, whose internal structure is like a metal ball that is fixed in a special spring as a pole, around it is the other pole. When the vibration gets to an extent, the two poles are connected so as to judge the shock occurs. Vibration sensor outputs digital signal. The Vibration sensor is used for vibration detection. This sensor has two contact pins. When an external force is acted upon either by movement or vibration, the sensor's two contact pins are closed and contact is made between the two pins. When the force is removed the sensor terminals return back to open contacts. The vibration sensor can be installed on the forks of the forklift to measure any kind of vibration. If the given vibration limit is exceeded, an alert system can be instilled to let the driver of the forklift be alerted of certain vibrations or irregularities. The vibration sensors can be installed on the forks of the forklift, where most of the vibrations will occur due to lifting of materials. If the materials are more than the given limit, the sensor can create an alert. A vibration sensor can also be installed at the rear end of the forklift for extra safety measures.

Touch sensors work similar to a switch. When they are subjected to touch, pressure or force they get activated and act as a closed switch. When the pressure or contact is removed, they act as an open switch. Capacitive touch sensor contains two parallel conductors with an insulator between them. These conductor plates act as a capacitor with a capacitance value C0. When these conductor plates, they act as a conductive object and typically there is an uncertain increase in the capacitance. A capacitance measuring circuit continuously measures the capacitance C0 of the sensor. When this circuit detects a change in capacitance it generates a signal. The resistive touch sensors calculate the pressure applied on the surface to sense the touch. These sensors contain two conductive films coated with indium tin oxide, which is a good conductor of electricity, separated by a very small distance. Across the surface of the films, a constant voltage is applied. When pressure is applied to the top film, it touches the bottom film. This generates a voltage drop which is detected by a controller circuit and signal is generated thereby detecting the touch.

In this scenario the sensors can be used to activate certain switches or can be installed on the forks to automatically the fork movement after the materials to be lifted comes in contact with the touchpads of the sensors. The touch sensor can also be placed under the seat of the forklift, which will signal if a driver is present on the seat. If not, the forklift will not start. Other sensors can also be incorporated in the model. The infrared sensors can be installed along with ultrasonic sensors. Although they will serve the same purpose, the output of the sensors will be different and will give the user a different perspective on the model. The infrared sensors can also be installed on the sides or on the four corners of the forklift to maximize its coverage area. Weight sensors can be installed under the forklift and a weight limit can be set to avoid carriage of excess weight on the forklift (Fig. 3).

Fig. 3 Showing the integration of the different sensors in the physical devices

In the above image, dots of three different colors are visible. The different colors signify different sensors. The blue dots signify the ultrasound sensors, the red dots signify the touch sensors and the green dot signifies the vibration sensor. The dots are the position in which the sensors will be positioned. The sensors will be programmed using an Arduino and if the Arduino is used for setting up the model, a portable battery will be installed along with it. That would require more sensors and accuracy in measurement. Human supervision will always be necessary.

4 Discussion and Conclusion

This paper present current work on how to extend an existing digital game with physical components, in order to build a mixed reality environment in which players can experience and learn about technology assessment in complex systems. In our case the scenario is a warehouse environment and the technologies to be assessed are related to the concept of Industry4.0. The objective of the extension is to overcome a challenge in the learning process that was observed in the digital game [11] and to take advantage of what we have learned from other work [3, 9, 12, 18]. In order to support the re-design, we have used the ATMSG framework [17] and the findings of the LM-GM narratives [11] and the components from the pervasive game design framework [13, 18]. The existing game already provides a learning environment on technology assessment in warehouses, but with some limitations. The presented design is at conceptual level. A physical prototype that will allow testing if this new design increase the player's immersiveness is expected to be ready in June 2021. If successful it is expected that the different scenarios and different in game challenges will engage the players, and that the new game play, will allow higher personalization to players' needs and competence level, providing different challenges. With the option to choose different sensors in the game, a user can learn the different advantages and disadvantages of different sensors in different scenarios and tasks. We think this can overcome the limitation in a fully digital game, since while performing a task, a user can use different sensors one by one or together to understand the usefulness of sensors for a specific operation and situation. To maximize the effect of these sensors the option to use one at a time will help the use understand the effect it makes in the overall scenario. The players can also perform the tasks without the sensors. In two separate modes, they can perform a set of tasks both by using the sensors and without the assistance of the sensors. This will improve their understanding of the sensors as well as giving them an idea about their skills. The players are supported in the game play with the tutorials and videos, but the classes is supposed to be used in a blended setting with an experienced facilitator.

Acknowledgements. The project on which this paper is based was funded by the Federal Ministry of Education and Research (BMBF), Germany under the funding code 16DHB2112, European Union, Central Baltic Interreg program, funding code CB743, and Vinnova Produktion 2030 testbed project Digilog. The responsibility for the content of this publication lies with the authors.

References

1. Abdirad M, Krishnan K (2020) Industry 4.0 in logistics and supply chain management: a systematic literature review. Eng Manag J (2020). https://doi.org/10.1080/10429247.2020.1783935
2. Hauge JB, Zafarzadeh M, Jeong Y, Li Yi, Khilji WA, Wiktorsson M (2020) Digital and physical testbed for production logistics operations. In: Lalic B, Majstorovic V, Marjanovic U, von Cieminski G, Romero D (eds) Advances in production management systems. The path to digital transformation and innovation of production management systems, vol 591. IFIP Advances in Information and Communication Technology. Springer, Cham, pp 625–633. https://doi.org/10.1007/978-3-030-57993-7_71
3. Baalsrud Hauge J, Zafarzadeh M, Jeong Y, Li Y, Ali Khilji W, Wiktorsson M (2020) Employing digital twins within production logistics. In: Engineering, technology and innovation (ICE/ITMC) 2020 IEEE international (2020)
4. Woschank M, Pacher C (2020) Teaching and learning methods in the context of industrial logistics engineering education. Procedia Manuf 51:1709–1716. ISSN 2351-9789. https://doi.org/10.1016/j.promfg.2020.10.238
5. Romero D, et al (2020) The operator 4.0: towards socially sustainable factories of the future. Comput Ind Eng 139:106128. ISSN 0360-8352. https://doi.org/10.1016/j.cie.2019.106128
6. Romero Gázquez JL, Bueno Delgado MV, Ortega Gras JJ, Garrido Lova J, Gómez Gómez, MV, Zbiec M (2021) Lack of skills, knowledge and competences in higher education about industry 4.0 in the manufacturing sector. RIED Revista Iberoamericana de Educación a Distancia 24(1):285–313. https://doi.org/10.5944/ried.24.1.27548

7. Zafarzadeh M, Wiktorsson M, Baalsrud Hauge J, Jeong Y (2019) Data-driven production logistics - an industrial case study on potential and challenges. Smart Sustain Manuf Syst 3 (1):53–78
8. Sharma R, Jabbour CJC, Lopes de Sousa Jabbour, AB (2020) Sustainable manufacturing and industry 4.0: what we know and what we don't. J Enterprise Inf Manag. https://doi.org/10.1108/JEIM-01-2020-0024
9. Hauge JMB, Stefan IA, Stefan A (2017) Exploring pervasive entertainment games to construct learning paths. In: Munekata N, Kunita I, Hoshino J (eds) Entertainment Computing – ICEC 2017, vol 10507. Lecture Notes in Computer Science. Springer, Cham, pp 196–201. https://doi.org/10.1007/978-3-319-66715-7_21
10. Oliveri M et al (2019) Designing an IoT-focused, multiplayer serious game for industry 4.0 innovation. In: 2019 IEEE international conference on engineering, technology and innovation (ICE/ITMC). IEEE (2019). https://doi.org/10.1109/ICE.2019.8792680
11. Lim T, Louchart S, Suttie N, Hauge JB, Stanescu IA, Ortiz IM, Moreno-Ger P, Bellotti F, Carvalho MB, Earp J, Ott M, Arnab S, Berta R (2014) Narrative Serious Game Mechanics (NSGM) – insights into the narrative-pedagogical mechanism. In: Göbel S, Wiemeyer J (eds) Games for Training, Education, Health and Sports, vol 8395. Lecture Notes in Computer Science. Springer, Cham, pp 23–34. https://doi.org/10.1007/978-3-319-05972-3_4
12. Baalsrud Hauge JM et al (2016) Analysis on educating mechanical engineers through serious games using pervasive technologies. In: Proceedings of the ASME 2016 international design engineering technical conferences & computers and information in engineering conference, [IDETC2016–59826]. American Society of Mechanical Engineers, pp 1–8
13. Söbke H et al (2017) Prime example ingress reframing the pervasive game design framework (PGDF). Int J Ser Games **4**(2) (2017). https://doi.org/10.17083/ijsg.v4i2.182
14. Callaghan M (2017) Mapping learning and game mechanics for serious games analysis in engineering education. IEEE Trans Emerging Top Comput 5(1):77–83. https://doi.org/10.1109/TETC.2015.2504241
15. Juraschek M et al (2018) Mixed reality in learning factories. Procedia Manuf 23:153–158. ISSN 2351–9789
16. Quandt M, Stern H, Grudpan S, Beinke T, Freitag M, Malaka R (2020) Evaluation of human-computer-interaction design in production and logistics by using experimental investigations. In: Freitag M, Haasis H-D, Kotzab H, Pannek J (eds) Dynamics in Logistics. Lecture Notes in Logistics. Springer, Cham, pp 554–566. https://doi.org/10.1007/978-3-030-44783-0_51
17. Carvalho MB et al (2015) An activity theory-based model for serious games analysis and conceptual design. Comput Educ 87:166–181. ISSN 0360-1315.https://doi.org/10.1016/j.compedu.2015.03.023
18. Baalsrud Hauge J, Söbke H, Stefan IA, Stefan A (2020) Applying and facilitating serious location-based games. In: Nunes NJ, Ma L, Wang M, Correia N, Pan Z (eds) Entertainment Computing – ICEC 2020, vol 12523. Lecture Notes in Computer Science. Springer, Cham, pp 104–109. https://doi.org/10.1007/978-3-030-65736-9_8

IoT – Smart Materials and Systems

Artificial Intelligence System for Predicting Cardiovascular Diseases Using IoT Devices and Virtual Instrumentation

Horia Alexandru Modran[1(✉)], Doru Ursuțiu[1,2], Cornel Samoilă[1,3], and Tinashe Chamunorwa[1]

[1] Transilvania University of Brașov, Brașov, Romania
{horia.modran,udoru,csam, chamunorwa.tinashe}@unitbv.ro

[2] Romanian Academy of Scientists, Brașov, Romania

[3] Romanian Academy of Technical Sciences, Brașov, Romania

Abstract. According to a study conducted by the World Health Organization, cardiovascular diseases (CVDs) are the first cause of death globally. About 17.9 million people die each year from CVDs, which represents an estimated 31% of all deaths worldwide, and more than 75% of the deaths occur in low and middle-income countries. Therefore, it is very important to develop new methodologies that could be used in the future for diagnosing CVDs. They should be accessible both in terms of price and technologies, so that they can be available for low and middle-income countries as well. Because of the recent technological progress, Artificial Intelligence (AI) algorithms along with big data produce very accurate results even in the healthcare field, and we believe they can help in solving this problem.

The study presented in this paper is based on using a cheap and powerful Internet of Things (IoT) programmable PSoC6 device to acquire blood pressure data from people and the data is then processed using the LabVIEW virtual instrumentation environment using a machine learning model developed in the Python programming language. For choosing the most appropriate machine learning model for our case, we compared the most widely used algorithms and selected the one that had the best performance indicators. The model was trained using a data set of 70.000 records of patient data and predicts 5-year risk of developing CVDs very accurately.

Keywords: Artificial Intelligence · Machine learning · Virtual instrumentation · IoT · PSoC

1 Introduction

Thanks to the rapid technological progress, Internet of Things (IoT) and Artificial Intelligence (AI) have become much more popular. Algorithms, developed by human programmers, are trained on massive data sets, collected by the IoT devices, and produce very accurate results in many complex tasks [1]. People could take advantage of Artificial Intelligence and IoT to help them in their daily lives, as well as to detect

M. E. Auer et al. (eds.), *Online Engineering and Society 4.0*, Lecture Notes in Networks and Systems 298, https://doi.org/10.1007/978-3-030-82529-4_28

changes in their condition, predict heart or cardiovascular diseases or warn them about any health problems they may have.

According to a study conducted by the World Health Organization, around 17.9 million people die each year from cardiovascular diseases (CVDs), being the number 1 cause of death globally[1] [2]. This disease accounts for about 31% of deaths worldwide, while around three-quarters of these deaths occur in low or middle-income countries. Unfortunately, the mortality of the CVDs is increasing each year and, therefore, it is necessary to develop new methodologies for diagnosing the CVDs in the future. These methodologies should be accessible both in term of price and technologies so that can be used in developing countries as well. The field of cardiac tissue engineering is a possible remedy, through development of bioartificial tissue constructs that provide support to the damaged tissue and restore their function [2, 3].

The approach used in this study is based on developing a system for collecting and processing blood pressure data using machine learning algorithms. The blood pressure data is acquired using a powerful and cheap IoT device based on the PSoC6 microcontroller integrated circuit from the Cypress/Infineon company. The acquired data is then processed inside a Virtual Instrument (VI) created in Laboratory Virtual Instrumentation Engineering Workbench (LabVIEW) that is using a machine learning model developed in Python. We compared the most widely used machine learning algorithms that are suitable for a classification problem and chose the one that has the best evaluation metrics. Thus, our system is using a multilayer perceptron neural network to predict if an individual has a 5-year risk of developing cardiovascular diseases. It was trained using a dataset that consists of 70.000 medical records and was imported in LabVIEW using the Python Node module. Compared to similar studies for predicting CVDs [4], our machine learning model obtained better accuracy and performance indicators.

2 Blood Pressure IoT Acquiring System

The system for acquiring medical record data was developed using the PSoC6 BLE microcontroller integrated circuits produced by the Cypress/Infineon company. PSoC6 covers the gap between low-performance microcontrollers and costly, power-consuming application processors. The architecture of the very low power PSoC6 microcontroller provides the processing performance required by IoT devices, getting rid of the trade-offs between performance and power. It has a dual-CPU architecture, containing an ARM Cortex-M0+ for executing low-power tasks and an ARM Cortex-M4 for high-performance tasks [5].

We programmed the PSoC6 BLE microcontroller to acquire blood pressure data in PSoC Creator using the C programming language. For the schematic (Fig. 1) we used a BLE for sending the data using Bluetooth 5.0 and an UART component for printing out the data through the computer serial port. Processors communicate with each other using a mutex to control access to a shared variable. With built-in security offered by PSoC6, the IoT system is thus protected.

[1] World Health Organization website, [Online] at https://www.who.int/health-topics/cardiovascular-diseases/, last accessed 2020/12/12.

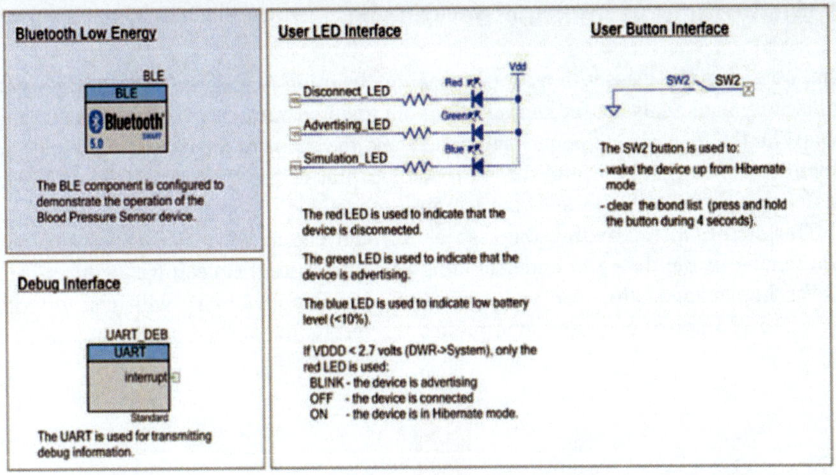

Fig. 1 Blood pressure PSoC schematic

The data acquisition can be controlled via Bluetooth by using an Android or IoS smartphone. The acquisition can be started and stopped by using the Cypress CySmart mobile application, while the systolic and diastolic blood pressure can also be tracked on the smartphone in real time (Fig. 2).

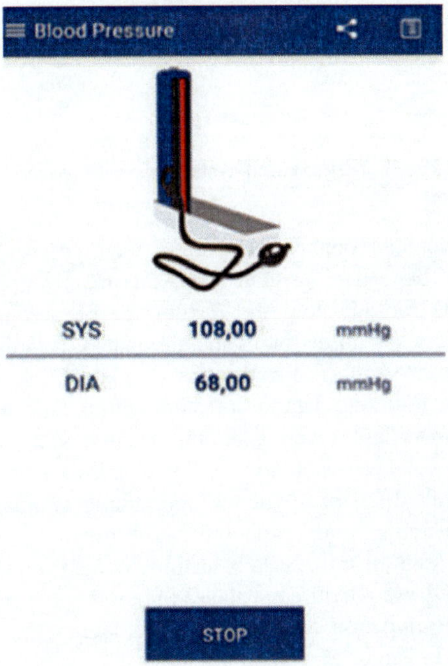

Fig. 2 Blood pressure android application

3 Analyzing and Splitting the Dataset

After obtaining the dataset to train our machine learning model, we should perform an exploratory data analysis, for summarizing the main characteristics of the data through visual charts. This process helps us to understand the shape of the data set and to decide which features might be useful for the model. Exploring the training data set is critical for developing an accurate machine learning model.

This step involves getting the distribution and counts of all variables, analyzing each feature of the data and understanding the correlation between the features [6].

We implemented this step in Python using the Pandas library with just writing a few standard lines of code and then plotted the results in overlaid histograms (Fig. 3).

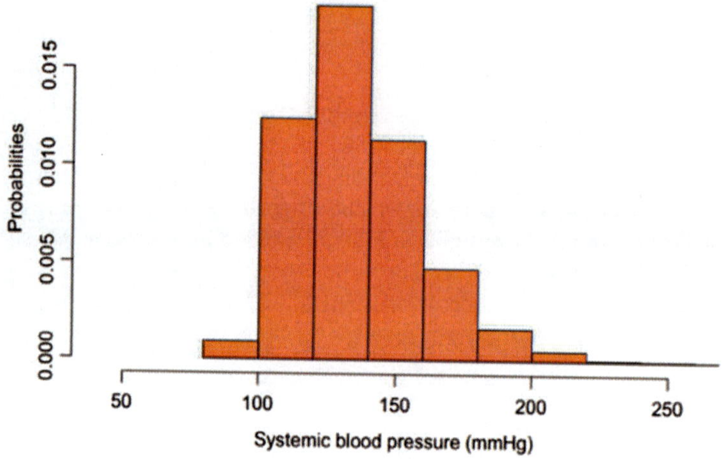

Fig. 3 Histogram of systemic blood pressure

Because the dataset can contain inaccurate, incorrect or irrelevant data, it is important to perform also a data cleaning step. The main purpose of this step is to remove irrelevant data for adjusting the features to be acceptable for the model. Missing or duplicate data should be found and eliminated from the dataset. We should only select the significant features to reduce the amount of data we will use to fit the model. We also replaced missing data for a feature with the mean global value for that feature, while repetitive variables were taken out of the model. Regardless of the type of the data, the outliers and errors should be also cleaned up from the data set [7].

This step was handled in the Python programming language with the help of Pandas and NumPy, which is a very straightforward process.

The data we use to feed the model can sometimes come from multiple distributions. There are three main sets used in different stages of building a machine learning model: the training dataset, the validation dataset and the test dataset, therefore we should split our dataset into those 3 categories.

The size of each of these three datasets can be chosen according to the typology of the problem [8], but as a common approach it is advised that the training set should contain around 60% of the initial dataset, while the validation and training subsets should each comprise around 20% each.

If the data splitting step is not performed correctly, the model can underfit or overfit the data, a situation in which the model will not be able to generalize accurately [9].

The data splitting step was implemented in Python, and then the Python code was imported and run in LabVIEW using the Python Node module. As the *train_test_split()* function from the scikit-learn library accepts only two parameters, the split into three categories was done in two steps. For this process, only a few lines of Python code were needed, as illustrated below:

```
X_train_set, X_test_set, y_train_set, y_test_set =
train_test_split(features, labels, test_size=0.4,
random_state=42)

X_test_xet, X_validation_set, y_test_set, y_validation_set
= train_test_split(X_test_set, y_test_    _set, test_size=0.5,
random_state=42)
```

4 Acquiring Data

The data acquisition process can be controlled through the mobile application connected via Bluetooth to the PSoC microcontroller. The PSoC6 kit was programmed to acquire blood pressure data and print it out, using the Universal Asynchronous Receiver-Transmitter (UART) debug component in the PSoC Creator Schema, at regular time intervals via the serial port. For enabling the debug port, we set the macro *DEBUG_UART_ENABLED* to ENABLED.

The UART component can be used to debug and visualize the acquired data, and can thus be used to integrate to acquired data via PSoC in LabVIEW. The Rx and Tx pins of the UART block were associated with the KitProg on the PSoC 6 kit on ports P12(6) and P12(7) (Fig. 4). The UART data will use the USB interface and there will be a port emulated as a peripheral on the PC side, which can then be used to acquire data in LabVIEW, Hyperterminal or PuTTY.

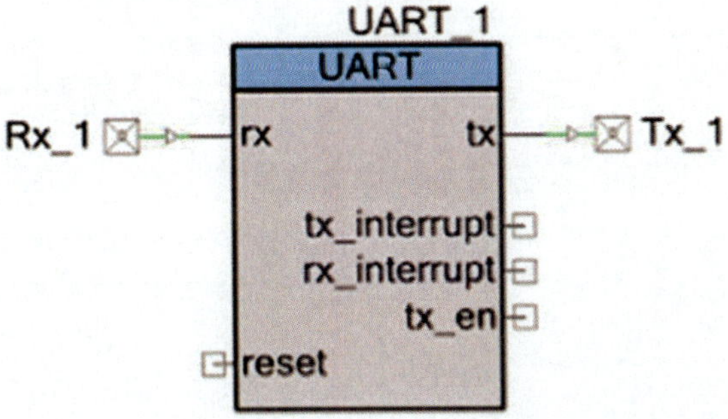

Fig. 4 UART debug block

The see the output we can connect to the serial port by using Putty and it will print out the acquired values:

```
CY_BLE_EVT_GAP_DEVICE_CONNECTED: connIntv = 36 ms
CY_BLE_EVT_GAP_CONNECTION_UPDATE_COMPLETE: connIntv = 6 ms
BLS event: 10041, Blood Pressure Measurement is Enabled
Blood Pressure Ind sys: 100 mmHg, dia: 60 mmHg
BLS event: 10041, Blood Pressure Ind sys:100 mmHg, dia: 60 mmHg
BLS event: 10041, Blood Pressure Ind sys: 100 mmHg, dia: 60 mmHg
BLS event: 10041, Blood Pressure Ind sys: 100 mmHg, dia: 60 mmHg
BLS event: 10041, Blood Pressure Ind sys: 102 mmHg, dia: 60 mmHg
BLS event: 10041, Blood Pressure Ind sys: 102 mmHg, dia: 62 mmHg
BLS event: 10041, Blood Pressure Ind sys: 102 mmHg, dia: 62 mmHg
BLS event: 10041, Blood Pressure Measurement Indication is
Disabled
```

To make this acquired data to be available in the LabVIEW environment as well, we developed a Virtual Instrument (VI) that is using the VISA function to connect to the serial port of the computer. In this VI we have to set some parameters, like the VISA resource name (the serial port on which the PSoC6 is connected), the delay before each read and the serial setting we used for programming the PSoC kit: a baud rate of 115.200, no parity, 8 data bits, one stop-bits and no flow control. We stored all those variables in a cluster data structure (named Serial Settings), which was then unbundled using the "Unbundle by Name" function. Then all those individual variables are sent as parameters to the VISA Serial subVI, which will open a connection, send a data read request, and receive the data from the UART debug block as a response [10].

After reading the values from the serial port, the VISA resource should be closed to release the serial port, to avoid a memory leak. The block diagram of this Virtual Instrument is illustrated in Fig. 5.

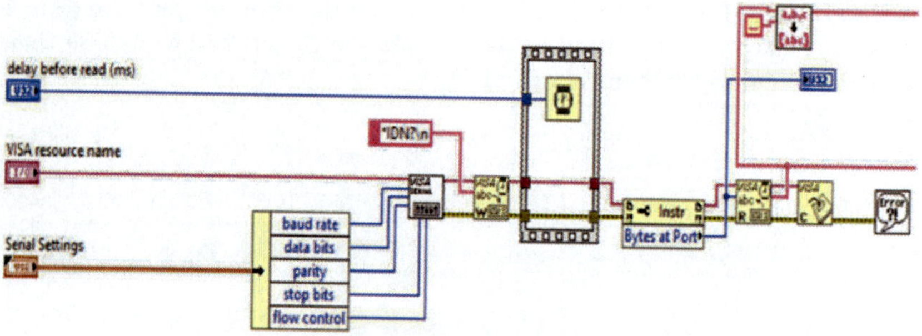

Fig. 5 Virtual Instrument for connecting to the Serial Port

5 Building the Machine Learning Model

For developing the machine learning model, we used the Python Programming language together with the NumPy and Pandas libraries. Because we are training the model with a dataset that contains labeled data, we should use a supervised learning algorithm.

For choosing the appropriate supervised learning algorithm for this model, we compared the most widely used algorithms. Because our prediction problem is of the classification type, we considered only the algorithms that are suitable for classification problems [11]: logistic regression, support vector machines (SVM), k-nearest neighbor (KNN), random forest and Multilayer perceptron neural networks (MLP NN). For this comparison we've taken into consideration the following indicators: the training speed, the prediction speed and the average and the evaluation metrics indicators (accuracy, F-1 score and area under the ROC curve - AUC).

Table 1 Comparison of supervised learning algorithms

Algorithm	Training speed	Predictive speed	Accuracy	F-1 score	AUC
Logistic regression	Fast	Fast	68%	73%	0.65
Support vector machines	Slow	Moderate	88%	94%	0.86
KNN	Fast	Depends on the parameter n	85%	86%	0.84
Random forest	Slow	Moderate	77%	77%	0.74
MLP NN	Slow	Fast	92%	95%	0.89

Accuracy is the percentage (ratio) of the predictions that our model got right:

$$Accuracy = \frac{Number\ of\ correct\ predictions}{Number\ of\ total\ predictions} \tag{1}$$

The F1-score is used to measure the test's accuracy and is calculated from the precision and recall of the test:

$$F1\ score = \frac{TP}{TP + \frac{1}{2} * (FP + FN)} \tag{2}$$

where TP – true positives, FP – false positives, FN – false negatives.

The Area under the ROC Curve (AUC) is an aggregated indicator of performance across all possible classification thresholds [12]. It indicates the probability that the model ranks a random positive example higher than any other negative example.

As we can see in Table 1, the multilayer perceptron and the support vector machine have both very good evaluation metrics. As the neural networks have a better predictive speed than the SVMs, we selected the multilayer perceptron (MLP) neural network for our model [13] (Fig. 6). The model we are using has several features, equal to the number of the features of our train data - of which the most important are age, systolic and diastolic blood pressure, and 3 hidden layers. The main advantage of this type of neural network is that it doesn't have any vanishing or exploding gradient.

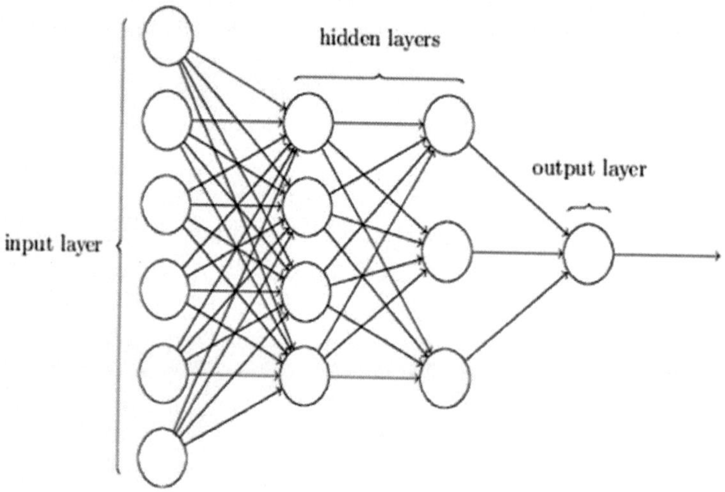

Fig. 6 The architecture of the Multilayer perceptron NN

To import the machine learning model in LabVIEW, we created a subVI that is using Python Node to get a Python session. The Python function takes three input parameters – the systolic and diastolic blood pressure and the age of the patient, and is returning a Boolean flag to indicate if the person is prone to cardiovascular diseases over the next five years. In addition to these, we also have to provide the path of the module and the name of the function as parameters to the Python Node. The errors that can occur are also handled by using an error cluster (*error out*). While the default threshold for interpreting probabilities for the labels in classification problems is usually 0.5 [14], in medical diagnosis is very important that the predictions are as accurate as possible. Therefore, we've set the threshold to be equal to 0.7 (Fig. 7).

We used then the subVI in the main Virtual Instrument that has the purpose to acquire the data from the IoT device, send it to the Python module and display the prediction made by the machine learning model. The front panel of this VI (Fig. 8) is displaying also the response received from the PSoC6 chip and the bytes read from them for debugging purposes. Furthermore, on the right side we are displaying the systolic and diastolic blood pressure as well, together with the predicted output.

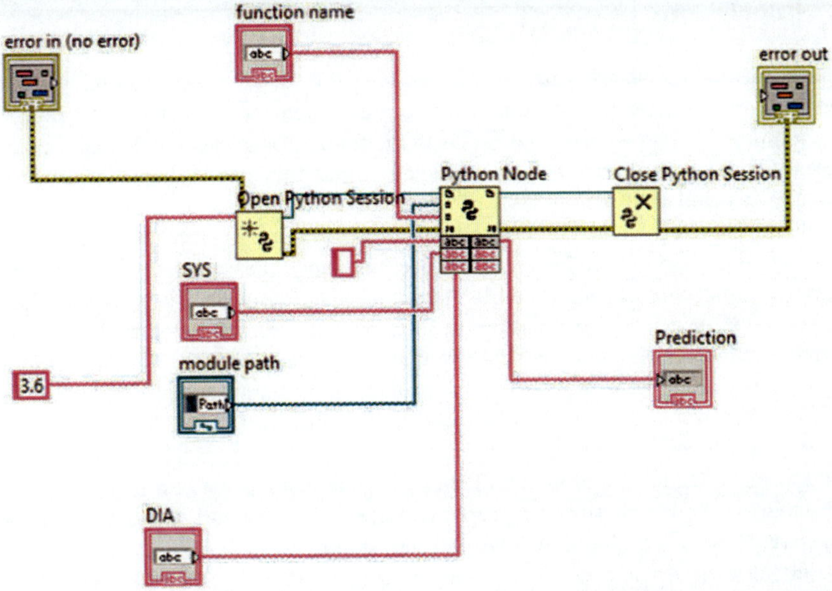

Fig. 7 The architecture of the LabVIEW Neural Network VI

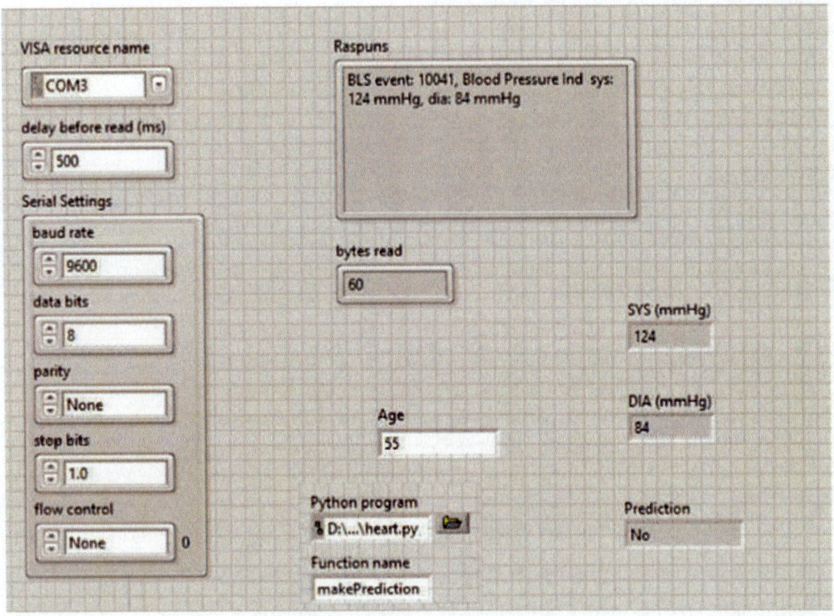

Fig. 8 Front panel of the prediction VI

6 Conclusion

Because cardiovascular diseases (CVDs) are the first cause of death globally and the mortality of this type of disease is increasing each year, it is crucial to develop new methodologies for diagnosing the CVDs in the future. These methodologies should be accessible in terms of price and technologies, so that they can available in developing countries as well.

The approach presented in this study is trying to solve this problem and is based on developing a system for collecting and processing blood pressure data using a machine learning model. We acquired the blood pressure data by using a very accessible and powerful IoT device developed using the PSoC6 microcontroller integrated circuit from Cypress/Infineon. The acquired data is then processed inside a Virtual Instrument (VI) created in the Laboratory Virtual Instrumentation Engineering Workbench (LabVIEW) that uses a machine learning model developed in Python. We compared the training and predictive speed, together with some evaluation metrics for each widely used machine learning algorithm and chose the one that has the best indicators. Thus, our model is using a multilayer perceptron (MLP) neural network to predict accurately the 5-year risk of developing CVDs and was trained using a dataset that consists of 70.000 medical records, which was imported in LabVIEW using the Python Node module. In comparison with similar studies for predicting CVDs, our machine learning model got better performance indicators.

Acknowledgements. We would like to express our great appreciation to National Instruments for providing us free LabVIEW license, thus facilitating this study and making it possible. Special thanks goes to the Cypress/Infineon company as well, for giving us several PSoC kits free of charge. Their generosity and collaboration were greatly appreciated.

References

1. Lucci S, Kopec D (2016) Artificial intelligence in the 21st century: a living introduction, second edition. In: Mercury learning and information. ISBN: 978-1-942270-00-3
2. Theus AS, et al (2019) Biomaterial approaches for cardiovascular tissue engineering. Emer Mater
3. Sharma D, Ferguson M, Kamp TJ, Zhao F (2019) Constructing biomimetic cardiac tissues: a review of scaffold materials for engineering cardiac patches. Emerg Mater
4. Krittanawong C, Virk H, Bangalore S et al (2020) Machine learning prediction in cardiovascular diseases: a meta-analysis. Sci Rep 10:16057
5. PSoC 6 CY8CKIT-062 Pioneer Kit, Cypress/Infineon – DigiKey. https://www.digikey.com/en/product-highlight/c/cypress/psoc-6-cy8ckit-062-pioneer-kit. Accessed 17 Dec 2020
6. Alpaydin E (2020) Introduction to machine learning, 4th edn. MIT Press Ltd., ISBN 9780262043793
7. Tae K, Roh Y, Hun Oh Y, Kim H, Whang S (2019) Data cleaning for accurate, fair, and robust models: a big data - ai integration approach. arXiv:1904.10761v1
8. Xu Y, Goodacre R (2018) On splitting training and validation set: a comparative study of cross-validation, bootstrap systematic sampling for estimating the generalization performance of supervised learning. J Anal Test 2:249–262

9. Langley P (2011) The changing science of machine learning. Mach Learn J 82: 275–279
10. NI VISA User Manual. https://www.ni.com/pdf/manuals/370423a.pdf. Accessed 17 Dec 2020
11. Hassan AU, Khan MS, Shah MA (2018) Comparison of machine learning algorithms in data classification. In: 24th international conference on automation and computing (ICAC), newcastle upon Tyne, United Kingdom, pp 1–6
12. Powers D (2010) Evaluation: from precision, recall and F-measure to ROC, informedness, markedness & correlation. J Mach Learn Technol 2:37–63
13. Naraei P, Abhari A, Sadeghian A (2016) Application of multilayer perceptron neural networks and support vector machines in classification of healthcare data. In: 2016 future technologies conference (FTC), San Francisco, CA, pp 848–852
14. Omary Z, Mtenzi F (2010) Machine learning approach to identifying the dataset threshold for the performance estimators in supervised learning. Int J Inf (IJI) 3(3)

Remote Experiment and Balance Between Monodisciplinarity and Pluridisciplinarity

C. Samoila[1,2(✉)] and D. Ursutiu[1,3]

[1] Transylvania University of Brasov, Brasov, Romania
csam@unitbv.ro
[2] Academy of Technical Sciences of Romania, Bucharest, Romania
[3] Academy of Scientsts of Romania, Bucharest, Romania

Abstract. The paper aims to analyze the balance between mono-disciplinarity, trans-disciplinarity and multi-disciplinarity when is using the remote experiment for education. Most of the axioms of trans-disciplinarity will be addressed, namely: ontological axiom, logical axiom, epistemological axiom. It also deals with the type of competencies that remote experiment educates as a result of mono, pluri and transdisciplinary approaches, namely: methodological, meta-cognitive competencies, pragmatic skills and the generated positive attitude.

The paper highlights the fact that the statements of Jaques Delors, made in the Report of the International Commission on sec. XXI Education, prepared for UNESCO, can be put into operation more easily when virtual environment are used and especially when remote experiment is involved. Here we refer at: learning to know, learning to do, learning to live with others and learning to be.

The main conclusion is that today there is great confusion in the use of the terms multi-disciplinarity, pluri-disciplinarity and trans-disciplinarity and that the use of remote experiment, as a pedagogical method of introducing the virtual environment in education, helps to be done methodological clarifications together with knowledge transfer.

Keywords: Mono-disciplinarity · Multi-disciplinarity · Trans-disciplinarity · Remote experiment

1 Introduction

Remote experiment (R.E.) initially appeared as a solution for the transfer of laboratories from the real environment to the virtual environment so that e-learning can be applied to disciplines that cannot exist without the support of experimental applications. If, at the time of the emergence of this technology, pedagogical opinions were divided between enthusiasm and vehement denial over time, especially due to the predominantly positive opinions of students, technology has gained more and more ground. The maximum pedagogical limit was reached when prestigious universities (Stanford, Princeton, MIT, Harward, etc.) adopted the teaching principle designated by the acronym MOOC (Massive Open Online Courses) which showed how, a technology of virtual learning environment, can to break the numerical limits of enrollment in

M. E. Auer et al. (eds.), *Online Engineering and Society 4.0*, Lecture Notes in Networks and Systems 298, https://doi.org/10.1007/978-3-030-82529-4_29

higher education, until then concentrated in campuses, and to globalize learning without hitting by the limitations in time and geography (in space).

In summary R.E. it reflects the balance that the teacher must maintain between theory and practical applications, using the virtual environment. The theory offers the accumulation of knowledge and the experiment refines this theoretical knowledge, bringing it closer to real life, production, research. Strong of R.E. is that it takes place in the virtual environment, the experiment no longer depends on the level of endowment of the laboratory and allowing contact with sites that offer remote experiments, which so strongly diversifies experiential learning (Kolb) that the limits of desire to know as quickly as possible are reached.as much as in a field.

From the point of view of the theoretical pedagogical perspective, R.E. it is best situated in constructivism (Piaget, Vygotsky) the one who asks students to filter information and, through experiments, to incorporate it into pre-existing mental structures (tacit knowledge). Therefore, student-centered learning is promoted. In improving learning of the type described above, the virtual environment brought into system by the R.E. has a major importance because it allows the combination of individual learning with social learning, because the virtual environment does not isolate the student but puts him in contact with an extremely large mass of students open to exchange information, opinions, experiments.

2 Disciplines Potentially Components of R. E.

In principle an R.E. is designed to provide online experiments to a discipline from curriculum. Let's call it "*pivot discipline.*" In order to arrive at the experiments required by the "*pivot discipline*" it is necessary to use several collateral domains for the composition of the experiment such as:

- The field of programming software, with the finalization in a software brought in the executable stage;
- The field of experimental data processing with finalization in tables, graphs, statistical processing, error analysis, depending on the requirements of the "*pivot discipline*"
- The field of hardware with finalization in an experimental, real or simulated assembly (virtual instrumentation);

In order to illustrate the vastness of the fields mentioned above (and to emphasize that RE cannot be the creation of a single person), a list of some programming disciplines is provided (Table1) and a list of possible programming languages that could be involved in creating a remote experiment (Table 2). It is mentioned that such a list could continue, but it goes beyond the scope of this paper. It is important to note that an R.E. aggregates many disciplines and hence the problem of its approach, Mono, Multi, Pluri, Cross, Inter or Transdisciplinary.

Table 1 Computer programming disciplines

Fundamental disciplines	Specialized disciplines
Computer programming	Computer networks
Data structures	Database
Object-oriented programming	Computer architecture
Programming techniques	Operating systems
Software engineering	Artificial intelligence

Table 2 Programming languages

BASIC (TEO)		C#	Java script	Scriptol	FoxPro	Adobe flash
ALEN (basic mofificat)		Clips	Perl	Tcl/Tk	Visual FoxPro	HTML
BASIC	Quick	Cobol	PHP	MATLAB	SQL	Simulink
	GW	D	Prolog	SmallTalk	Action Script	LabVIEW
	Visual	Fortran	Python	Borland	ASP	
C		Haskel	Ruby	Aigol	ASP NET	
C++		Java	Seed	Simula	MIVA Script	

NOTE. Are marked the most used

3 A Holistic Approach of R.E.

R.E. was created following intense applied research, First as a support for laboratory experiments in some disciplines and later, for the creation of intelligent things and the realization of "Internet of things" (higher level).

Although research in general is a collaborative process, R.E. emphasized this aspect. Only if one reads the disciplines required to create a remote experiment, it is concluded that a single person cannot cover all the knowledge required to realize it. Combining several disciplines into one R.E. it can be done in several ways (Jenseniu - 2012):

- Intradisciplinary (monodisciplinary) - when the knowledge of a single discipline is used: de. ex. Electrical circuits, and R.E. is a succession of circuit cases in a logical sequence;
- Multidisciplinary - when several disciplines contribute to the creation of R.E. each discipline keeping its characteristics and R.E. in the final form it will serve the pivotal discipline that generated the collaboration action;
- Crossdisciplinary - in this case by R.E. knowledge is transferred from two or more collaborating disciplines to the pivotal discipline in which R.E will function, for example by learning of the principles of measurement, or an expert software, which are not taught in the pivot discipline (Fig. 1);

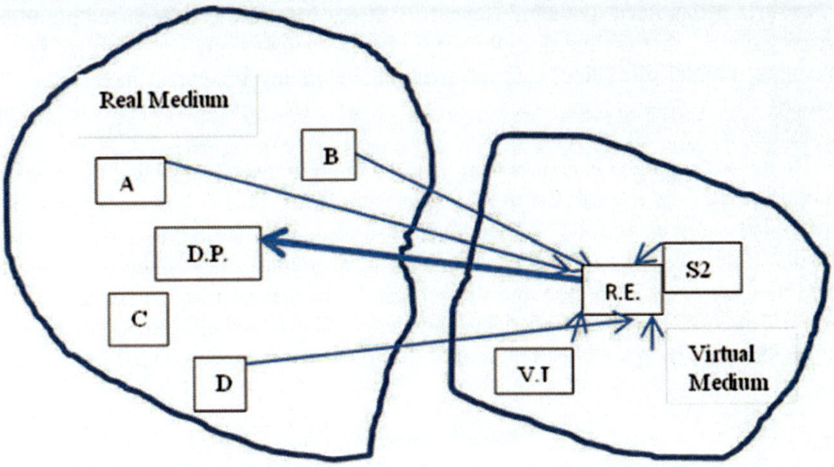

Fig. 1 Pluridisciplinarity. D.P. – pivot discipline, A, B, C, D, – disciplines, V.I. – virtual instrumentation, S1, S2 – softwares

- Interdisciplinary - this character appears when the integration in R.E, of several knowledge and methods produces the creation of new disciplines. It is the case of IoT that incorporates in each object an R.E. but which has become an independent field, with its own laws, its own procedures and its own principles;
- Transdisciplinary - R.E. does not reach this stage. He has not yet determined a philosophical vision on the reality and none of the axioms of transdisciplinarity (ontological, logical and epistemological) apply to him. The finalization of trans-disciplinarity is the understanding of the present world through the unity of human knowledge (Fig. 2).

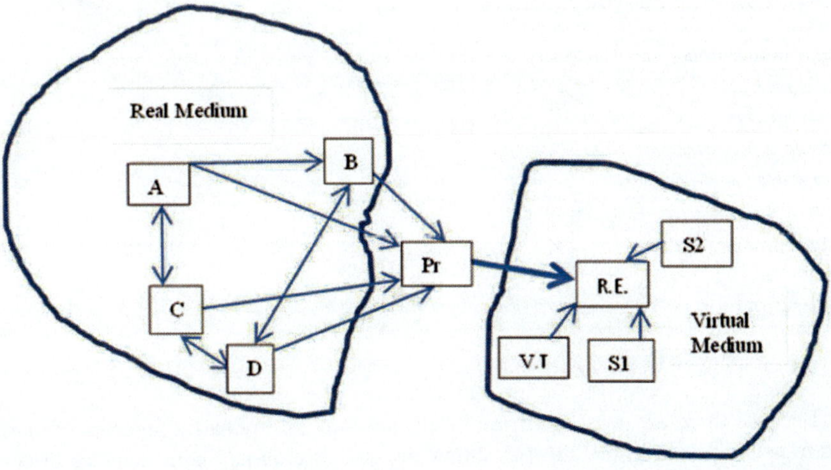

Fig. 2 Interdisciplinarity. A, B, C, D, – disciplines, Pr – project, V.I. – virtual instrum., S1, S2 – softwares

From a pedagogical point of view, R.E. it fits best in the concept of pluridisciplinarity and in that of interdisciplinarity, because it obviously serves to study of a discipline (pivotal discipline) through methods belonging to several disciplines. The nuances that determine the above affiliation of R.E, will be accentuated in the following.

The basic feature of R.E. it is also that, although he uses knowledge from several disciplines, he does not ask the user to master them all. This is how as the notion of "embedded knowledge", i.e. "embedded" knowledge in R.E. but which are only support for the pivotal experiment. The student manipulates experimentally using the notions of the pivot discipline and draws useful conclusions to fix these notions. He may also be interested in "embedded knowledge" but is not obliged to correlate the knowledge he acquires within the pivotal discipline with this embedded knowledge.

Table 3 Competences

Transdisciplinarity	Monodisciplinarity	Pluridisciplinarity/Interdisciplinarity (R.E.)
Methodological (general)		
Observation	X	X
Experimentation	X	X
Graphic representation	X	X
Data interpretation	X	X
Metacognitive		
Strategic planning	–	–
Behavioral monitoring	–	–
Personal learning techniques	–	–
Evaluation of results	X	X
Attitude		
Realism	–	–
Interest in learning	–	–
Tolerance of contradictions	–	X
Personal performance	X	X
Pragmatic skills		
Personal initiative	–	X
Concentration	X	X
Task solving	–	X
Work skills	–	X

Because there are great confusions between multidisciplinarity, interdisciplinarity and transdisciplinarity, the Table 3 shows the transdisciplinary competencies in comparison with the plury and transdisciplinary ones (those of which R.E. is part).

RE. started from serving monodisciplines, bringing everything that is positive in the educational process supported by the virtual environment. However, with the use of R. E., it has become clear that the approach to learning in a monodisciplinary style leads to a linear, gradual way of accumulating knowledge. Precisely because R.E. determine the interaction of two or more monodisciplines, the transversal/integrated competences that they develop have emerged. It thus allowed the transition from the gradual acquisition of knowledge to learning through multidisciplinary interaction and correlation.

R.E. allows easier centering of the instructional process on the student and allows the development of active pedagogies, which takes place in two environments: real and virtual.

The fact that the virtual environment allows access to R.E. from various universities, makes the processes of interaction and multidisciplinary correlation to be extremely diverse, easy exceeding the limits given by the vision of a single professor. This makes the transversality of competencies much more accentuated and much more beneficial for students.

4 Remote Experiment and Creativity

Interdisciplinarity, which has become the basis for the modernization of teaching processes, involves working on the intersections of two or more disciplines. The most convenient way to achieve these intersections is to create a large teams, consisting of specialists in the disciplines involved. The most successful R.E. were the result of the work of such teams.

In turn, R.E., offered to students for handling, leads to teamwork, a facility offered by the virtual environment. What advantages derive from the interdisciplinary character of R.E.?

- Many research topics are located at the intersections between monodisciplines;
- Creativity is educated through interdisciplinary interactions and correlations;
- Enriching the content of monodisciplines using the new knowledge created by R.E. is another advantage;
- R.E. as a result of the work of mixed teams it can be a good solution to combat excessive specialization.

There are other arguments to support the fact that R.E. it is a means of educating creativity. So:

- Diversity of purposes:
    - o The purpose of creating an R.E. is the serving, with a practical application developed in the virtual environment and accessible on the Internet, of a monodiscipline. Because the experiments necessary to understand the theoretical part are diverse and the purposes for which the R.E. are created will be diverse. As a result, the disciplines that will collaborate in creating an R.E. they are also of great diversity, which is favorable to the correlations that students will make;

- Variability of criteria:
  - Each R.E. it is based on a set of composition and evaluation criteria. For the same reasons stated above, these criteria are variable and allow students to develop the elasticity of thinking in approaching an experiment and in drawing the conclusions that the experiment brings;
- Interaction with the economic environment:
  - In the disciplines of the higher years of study, there is a strong connection between the experiments of the specialized courses and the realities of the economic life. The complex approach of an R.E. it resembles the complex approach of an industrial application, where a team of specialists is formed when knowledge from several fields intersects and where solutions with immediate applicability are required. This resemblance leads to the education in students of the method of correct approach to a new problem;
- Transparency of solutions:
  - Even if, as stated before, the student does not have to learn the "embedded knowledge" of an R.E. however, the curiosity to see what was used to create it to achieve the end result works. This is the first degree of transparency. There is also a second degree of transparency when, in order to better understand the experiment, the student accesses similar experiments from the Internet. They reflected the degree of endowment and the level of the school that created them so that the student can see and understand other ways of thinking and approaching. For the development of creativity, this level of transparency is decisive, because it serves to develop divergent thinking, the one responsible for creating the new.

5 Final Comments

Many times R.E. it is considered to belong to multidisciplinarity because it aggregates several disciplines in its creation. The consideration is wrong because in multidisciplinarity, the disciplines are only placed next to each other, without cooperation or coordination, each having its own path.

Closer to reality would be the framing of R.E. in multidisciplinarity, where are formed clusters of disciplines that collaborate with each other but do not have a common coordination, the disciplines continuing to strive towards achieving separate goals.

Compared to crossdisciplinarity, R.E. it is well defined only in terms of defining the field of work. It is located in an intersection that generates a valid issue within each discipline participating in the intersection, but the participating disciplines do not change their final purpose.

In interdisciplinarity R.E. it is best found because the clusters of disciplines they form for its creation are coordinated towards a common goal, which can lead either to changes in the superior understanding of the foundations of one or several disciplines, or to finding solutions that go beyond the goal of a single discipline. By R.E. the methods and approaches specific to each contributing discipline can be integrated to a different extent.

There are several degrees of interdisciplinarity, namely:

- The applicative one: it starts with it because R.E. is directly related to application, by experiments. All the experimental procedures of the disciplines cluster will contribute to find the way needed to perform the experiment in the virtual environment;
- The epistemological one: R.E. will appeal to these knowledge of the disciplines of cluster to support the experiment in the virtual environment;
- The generating one: the transfer of the qualities and possibilities of some systems programming generated the idea of transforming objects in intelligent objects and so the IoT device.

6 Conclusions

R.E. is in a strong connection with all the concepts addressed in the paper i.e. multi, pluri, cross and interdisciplinary. This is possible (the connection because these concepts are not antagonistic but complementary. They have a common objective, namely the understanding, at the higher level, of reality. And reality to be well understood needs experiment.

The theological development brought about by the Industrial Revolution 4.0 and globalization, contribute to the creation of a complex reality that is increasingly difficult to be described monodisciplinary. Multidisciplinarity and pluridisciplinarity emphasize the completeness of reality because it puts people in front of a volume of knowledge that is difficult to grasp.

As a result, integrated approaches have emerged that detect from the multitude of knowledge those knowledge that lead to the desired solution. R.E. is such an integrated approach.

Acknowledgements. The paper was sustained by the CVTC – Center of Valorization and Competence Transfer of the Transylvania University of Brasov-Romania, and the Creativity Laboratory.

References

1. Jantsch E (1972) Towards interdisciplinarity and transdisciplinarity in education and innovation. In: Apostel L et al (ed) Interdisciplinarity: problems of teaching and research in universities paris, france: organization for economic cooperation and development, pp 97–121
2. Dede C (2004) Enabling distributed learning communities via emerging technologies - Part two. T H E J 33(3):16–26
3. Boud AC, Baber C, Steiner SJ (2000) Virtual reality: a toolfor assembly? Presence 2(5):486–496

4. Al-Ashaab A, Flores M, Doultsinou A, Magyar A (2011) A balanced scorecard for measuring the impact of industry–university collaboration. Prod Plan Control 22:554–570
5. Newell WH (2011) Best practices in interdisciplinary studies. Unpublished seminar handout, international network for interdisciplinarity and transdisciplinarity (INIT) symposium, Utrecht, Netherlands
6. Newell WH (2011) The road from interdisciplinary studies to complexity. World Futures. J General Evol **67**: 330–342 http://www.tandfonline.com/toc/gwof20/67/4-5
7. Repko A (2012) Interdisciplinary research: Process and theory, 2nd edn. Sage Publications, Los Angeles
8. Repko A, Newell WH, Szostak R (2012) Case studies in interdisciplinary research. Sage Publications, Los Angeles
9. Szostak R (2002) How to do interdisciplinarity: Integrating the debate. Issues Integrative Stud 20:03–122
10. Szostak R (2002) Intuition and interdisciplinarity: reply to mackey. Issues Integrative Stud 20:131–138

Electronic Educational Laboratory Platform for Students

Tinashe Chamunorwa[1(✉)], Doru Ursuţiu[1,2], Cornel Samoilă[1,3], Horia Hedesiu[4], and Horia Alexandru Modran[1]

[1] Transilvania University of Brasov, Braşov, Romania
{chamunorwa.tinashe,udoru,csam, horia.modran}@unitbv.ro
[2] Academy of Romanian Scientists, Bucharest, Romania
[3] Technical Sciences Academy of Romania, Bucharest, Romania
[4] Technical University of Cluj-Napoca, Cluj-Napoca, Romania
horia.hedesiu@emd.utcluj.ro

Abstract. This paper outlines the implementation of an Internet of Things (IoT) based laboratory system for electronic engineering students. The first stage was the design of a prototype circuit on a Diligent breadboard adapter for NI myRIO. The circuit has a PmodALS ambient light sensor and a PmodTMP3 temperature sensor based on SPI and I2C serial bus connections. The system allows students to access sensor data over the internet and display it on the SystemLink Cloud web application and LabVIEW front panel. Different sensors can be connected on the same Pmod connector making the system flexible to perform various laboratories. The second stage was the establishment of bidirectional communication between LabVIEW software and the SystemLink Cloud platform. SystemLink is a data management system from NI which was used to display LabVIEW data and control remote hardware. Students can acquire hardware and software knowledge as they work through sensor interface protocols, myRIO hardware, LabVIEW and SystemLink cloud.

Keywords: LabVIEW · NI myRIO · SystemLink · Web service

1 Introduction

Internet of Things (IoT) is among the trending technologies projected to play a crucial role in advancing the understanding of practical engineering concepts. IoT refers to many physical devices globally connected to the internet, all sharing and collecting data [1]. The IoT ecosystem encompasses all the components that allow governments, businesses, and consumers to connect with IoT devices. The IoT ecosystem includes remote hardware, cloud dashboards, networks, gateways, data storage, analytics, and security [2]. Thus, the learning of IoT ecosystems in engineering courses brings a diverse spectrum of knowledge to students and has become unavoidable. One of the methods recommended for learning electronic laboratory courses is a case study or [3] Project-Based Learning (PBL). PBL is a teaching method where students learn by actively participating in real-world and personally significant projects. This method engages learners to have deep, long-lasting and inspired learning and personal

M. E. Auer et al. (eds.), *Online Engineering and Society 4.0*, Lecture Notes in Networks and Systems 298, https://doi.org/10.1007/978-3-030-82529-4_30

connection with their academic experience [4]. This paper is an outline of a project-based laboratory. Central to this study is the design of a daughterboard attached to a myRIO development board. Several engineering colleges have enthusiastically adopted educational boards for their lab courses. Some colleges have also designed their daughter boards to go with the development boards [5]. IoT's nature requires an understanding of programming languages, electronic circuits, and sensors to interface with the physical world and networking platforms [1].

This study addresses the following concepts:

1. SystemLink cloud application design for displaying LabVIEW measurements on the web and control LabVIEW processes using a web interface.
2. Student can learn G-programming in LabVIEW, which accelerates design time as compared to text-based coding since its graphically based
3. Understanding of MyRIO (FPGA and ARM) hardware architecture so that they can match with their implementations and develop their systems
4. Understanding of a variety of sensors and interface protocols like Universal Asynchronous Receiver-Transmitter (UART) hardware device, (Inter-Integrated Circuit) I2C protocol, and SPI (Serial Interface Protocol).
5. MyRIO internet configuration and connection through a wireless router
6. Understanding of Simulation-based design process.

2 Related Work

2.1 StudentEDEA and myDAQ in Education and Research Laboratory

This study is the design of a student educational device for electronic applications (StudentEDEA). The design is from the Centre for Valorization and Transfer of Competence (CVTC) at the Transylvania University of Brasov. The device is low-cost and flexible, with many laboratory options and facilities. It consists of PSoC1, myDAQ connector, and a prototyping area offering extra resources for more complex experiments and applications. The developed device is compatible with LabVIEW and PSOC designer software. Some of the functionalities include Measuring environment data (many adaptable sensors for IoT), PWM in LED lighting applications, and Generation, measurement, and analysis of analog signals. The system developed in this study differs in that the processing part resides in the breakout daughterboard, while in the myRIO resides the ARM and FPGA chip, which are the processors [6].

2.2 Remote Laboratories Based On LabVIEW Web Services

This paper described the implementation of LabVIEW web services based remote laboratory. The laboratory covered in the paper is for physical fields and analog electronics. The client web interface's basis is web programming techniques HTML, CSS, JavaScript, and PHP. From the hardware perspective, the concept of "multi-tasking hardware" is presented, allowing more clients to connect to the same platform. NI myRIO board is the hardware for performing the experiments. The real-time processing capability of MyRIO allowed for pipelined access control to the resources. At any given time, one student gets access to the resources, and then the other. This access cycle is repeated among the present students and will not notice the interruptions [7]. This study makes use of a NI SystemLink cloud service, which can be accessed remotely via the internet.

3 Design Process

The design process divides into the following processes:

- My RIO extension circuit prototype design
- LabVIEW Programming
- SystemLink Cloud web hosting
- Mobile Application Design, as shown in Fig. 1.

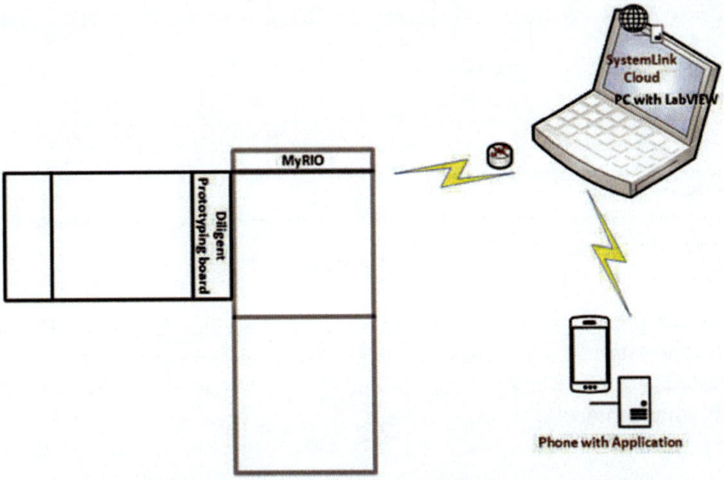

Fig. 1 Laboratory system block diagram

3.1 MyRIO Extension Circuit Prototype Design

Multisim is the circuit design software, and the circuit prototype resides on a Diligent breadboard. The myRIO embedded device served as the central controller of the laboratory platform. It provides reconfigurable I/O that allows for the teaching and implementation of multiple design concepts with one device. It also provides internet-based solutions delivering captivating ways to learn and design creative capstone projects. MyRIO is a reconfigurable I/O (RIO) device [8] used for diverse design concept implementation. MyRIO also connects to multiple sensors using diverse interfacing protocols. The digital and analog I/O connections include UART, SPI, PWM, UART, GPIO.I2C as well as audio. It has a dual-core ARM Cortex A9 processor, which ensures crucial embedded skills like C/C++ and Dataflow Programming in LabVIEW graphical language. Shown in Fig. 2 is the circuit designed in Multisim and prototyped on a Diligent breadboard. In Multisim, there is the use of myRIO Single MXP daughter board template, which provides the 34-pin connection to myRIO. The components in the circuit included sensors for temperature and ambient light. The Pmod connectors shown in the circuit adapt to other sensors with similar interface protocols.

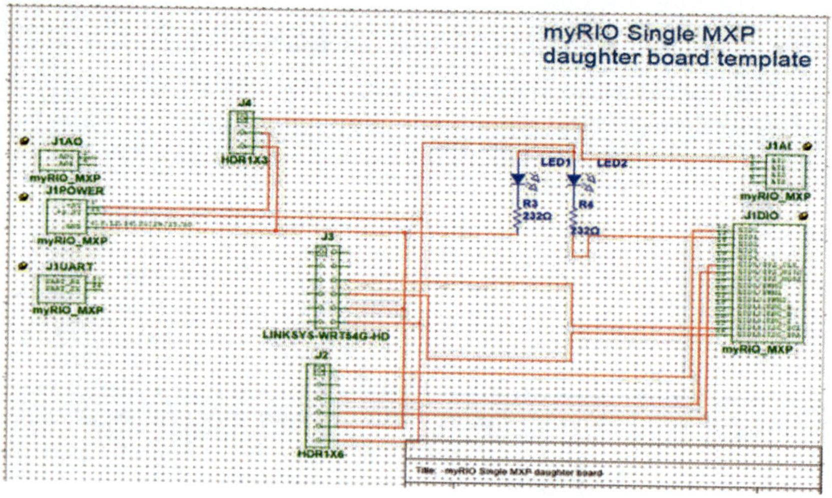

Fig. 2 MyRIO daughter circuit board

The prototype platform is the Diligent breadboard adaptor for NI myRIO, shown in Fig. 3. This prototype implements two applications, which are temperature and ambient light measurement. Sensors used can are interchangeable with other sensors of similar communication protocols.

Fig. 3 The prototype for testing the circuit

The ambient light sensing shown in Fig. 4 is has a PmodALS, which integrates a phototransistor sensor with an 8-bit analog-to-digital converter (ADC) that communicates via the serial peripheral interface (SPI) serial bus.

Fig. 4 Integrated ambient light sensor [7]

The temperature sensor shown in Fig. 5 is part of the design. It has a Microchip TCN75A integrated temperature sensor with I2C bus serial communication. The principal objective was to read and interpret temperature and proving an alert before it exceeds a limit.

Fig. 5 PmodTMP3 temperature sensor [8]

3.2 LabVIEW Programming

The LabVIEW myRIO toolkit is a software add-on for LabVIEW, providing the necessary tools and drivers to rapidly develop and deploy applications to the myRIO Student Embedded Device. The toolkit includes express and advanced I/O VIs, which can be configured for each I/O type on the myRIO device. The toolkit helps students to prototype ideas and creates stand-alone applications.

The following is the outline for LabVIEW myRIO programming.

- The project was opened in LabVIEW 2019 myRIO toolkit using the LabVIEW Project explorer

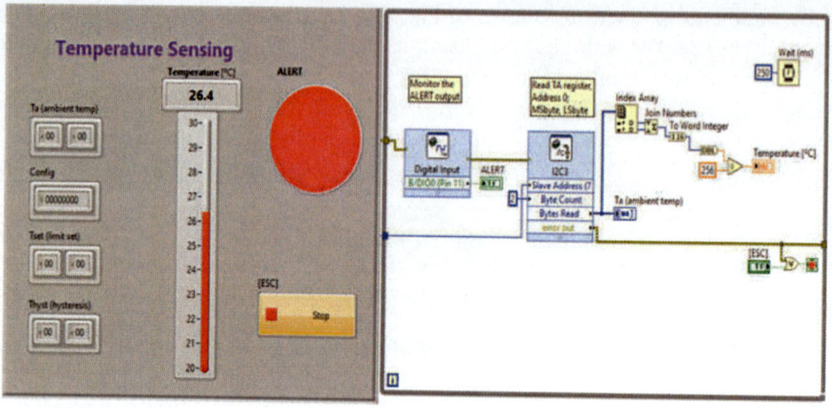

Fig. 6 The block diagram and front panel for temperature measurements

- The block diagram programming uses the myRIO LabVIEW palette, and it consists of PWM, I2C, UART, SPI, and analog–digital I/O configurations.
- The myRIO SPI and I2C interface blocks were used to configure the communication between sensors and myRIO.
- The configuration was done to the following parameters: mode, speed, frequency, frame length, clock phase, clock polarity, and data direction.
- The rest of the block diagram was connected, including respective front panel display components, and Fig. 6 shows part of the block diagram and results in the front panel. The same procedure is applicable for ambient light measurement.

Fig. 7 Cloud and key configurations

3.3 SystemLink Cloud web hosting

The SystemLink interface was configured on the website, and instruments were assigned tags corresponding to a LabVIEW path, establishing communication. The graph and thermometer are outputs receiving data from LabVIEW, and the switch and horizontal slide controller are inputs controlling the gauge and LED in the LabVIEW front panel.

- In The block diagram, add an Open Configuration VI and Open Tag VI in the setup code before the data-producing loop in an existing LabVIEW.

- Go to Data Communications >> SystemLink >> Configuration >> Open Configuration/Close Configuration.
- The Open/Write/read and Close Tag VI are found on: Data Communications >> SystemLink >> Tags.
- The connection and the configurations of the added VIs are seen in Fig. 7 and Fig. 8.

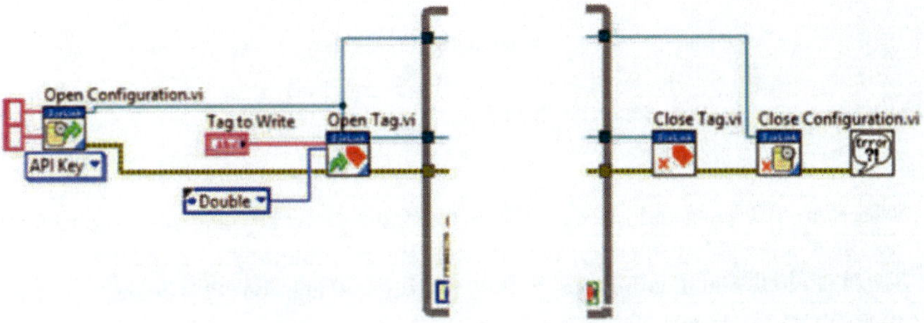

Fig. 8 Open and closing configurations

- Add a Write Tag VI inside the loop and wire the current data output into the VI.

In this example, the double-precision data type corresponds to the simulated signal data type from a simple random number generator simulating temperature data. The Get-Date/Time In Seconds VI can be used to write a timestamp with the data. Using this timestamp, one can know when the measurement occurred, even if there is a bit of delay in receiving the web data. After the block diagram design, connect the display and controls in the front panel to the respective terminals.

Creating the SystemLink Cloud interface is as follows:

- Go to SystemLink Cloud interface homepage (see Fig. 9) and Log on with NI account

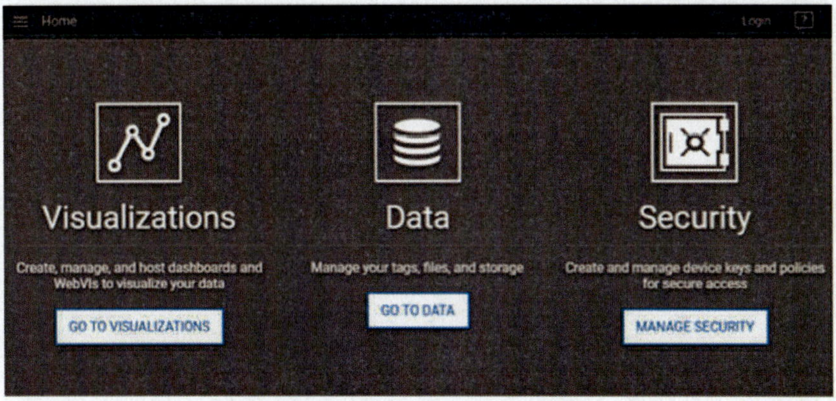

Fig. 9 SystemLink cloud home interface

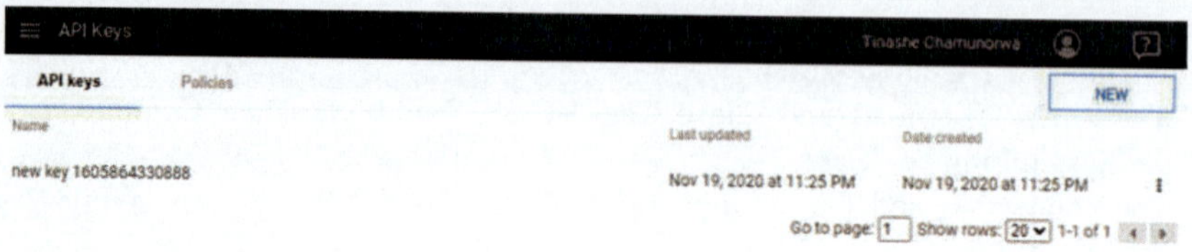

Fig. 10 Creating an API key

- Click on >> Manage Security as in Fig. 10

- Select on **API keys** and click on **NEW** to create a new key and copy the generated key. It is advisable to paste it somewhere. Rename the key and update.
- Select **Policies** and configure **Policy settings**. Click on **Privileges** and assign appropriate privileges as desired.
- Assign the created policy to the API key.
- On the Home Page, click on **Data** and then **Tags** as shown in Fig. 11.

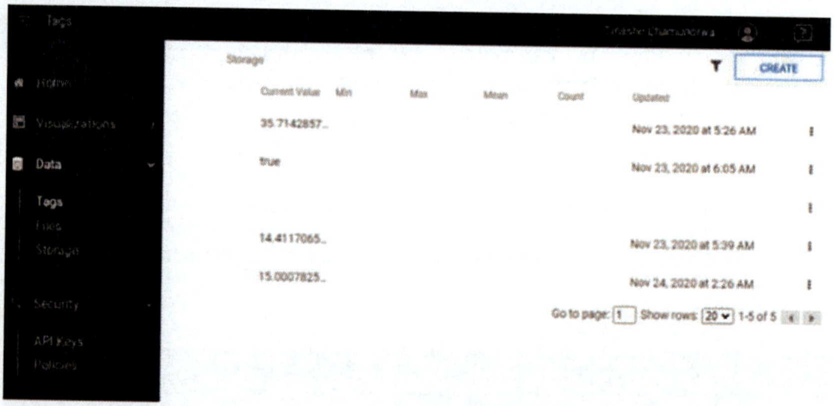

Fig. 11 Creating a tag

- Create a new Tag and assign an appropriate data type.
- Write the name of the tag on the **path control** in the front-panel.
- Go to **Visualizations** and click on **Dashboards** and create a new dashboard. Choose a name and **free-form** type.
- Add components to the dashboards according to the functionality of the LabVIEW program.

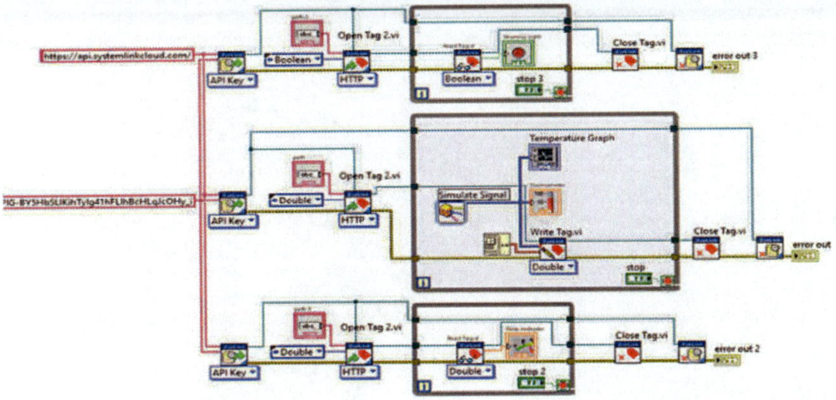

Fig. 12 Block diagram enabling the simulation, display, and control of LabVIEW measurements and processes

- In the simulated signal example Display like gauges and graphs of data type double will be appropriate.
- Most important, drag the appropriate tag to a particular display in this case of data type double.
- Run the LabVIEW and the dashboard to see results.

The signal simulated on the LabVIEW block diagram in Fig. 12 is displayed on the LabVIEW front panel and depicted on the thermometer scale and graph on the SystemLink Cloud interface. An On–off switch pressed on the SystemLink website controls the LED on LabVIEW, and a slider's motion on the website is reflected on the LabVIEW front panel gauge as shown in Fig. 13 and Fig. 14.

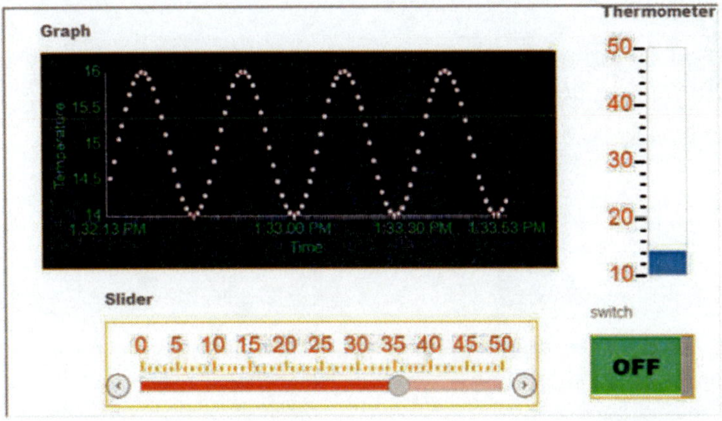

Fig. 13 SystemLink cloud interface for controlling and display of LabVIEW process

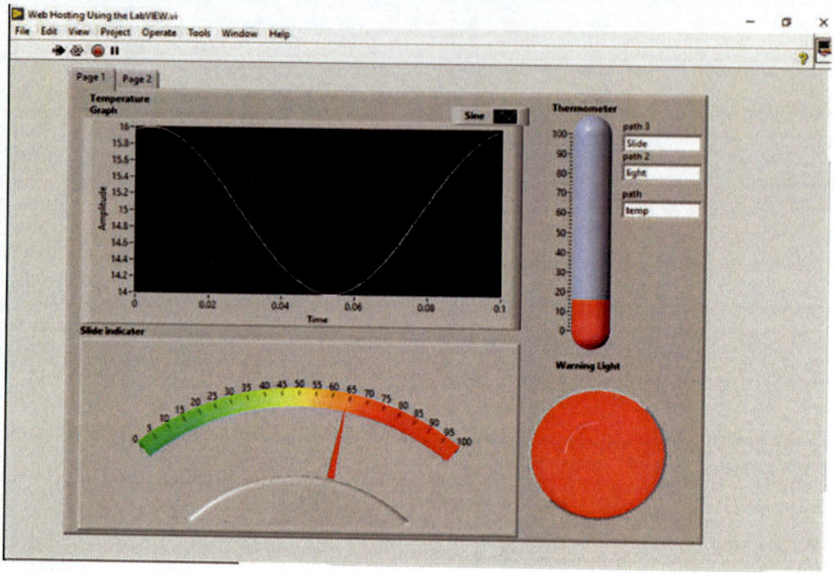

Fig. 14 LabVIEW display sending and receiving information from SystemLink cloud

3.4 Mobile Application Design

LabVIEW sent simulated data to the NI Data Dashboard android application. A switch on a mobile application was able to stop the LabVIEW running program. The connection was enabled by using shared variables over a wireless connection. Figure 15 shows the block diagram, the front panel, and the mobile display.

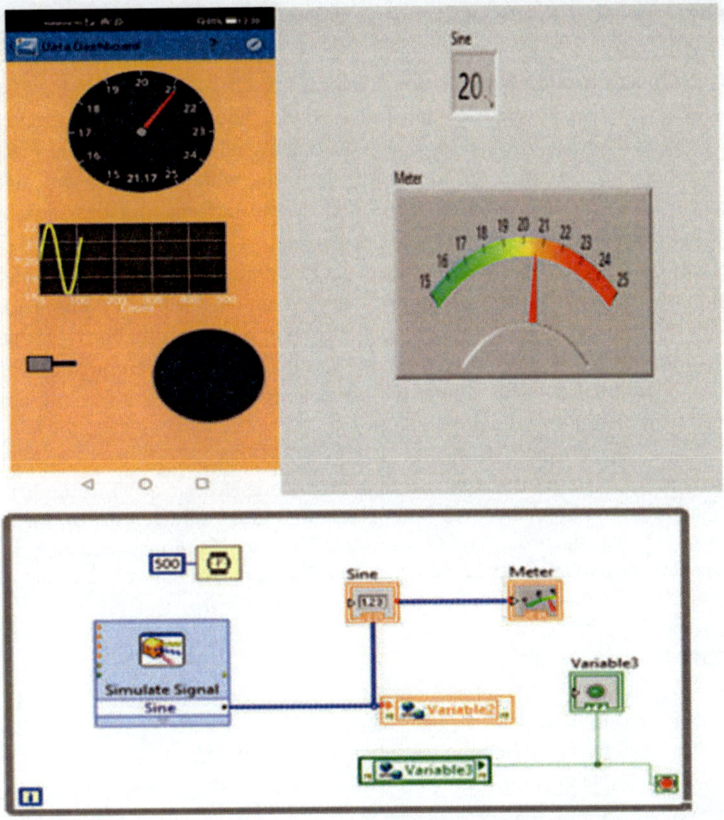

Fig. 15 The block diagram and the displays for the mobile application and front panel

4 Conclusion

The designed prototype circuit was combined with the NI myRIO to provide a useful experimental platform for undergraduate electronics classes. It is a versatile and flexible design because different sensors can be interchanged through the Pmod connectors. The integrated sensors provided by Diligent offer easy connections during prototyping. The myRIO expansion port (MXP) connector has internal circuit protection; students only concentrate on top-level circuit design. The presence of the ARM and FPGA processor in myRIO makes it possible to implement various functionalities with varying complexity.

LabVIEW is excellent for learners as it provides an intuitive graphical programming environment. The visual approach to programming augurs well with engineering students used to model solutions using process diagrams. It also offers a relatively easy way to FPGA designing by abstracting much detail and allowing students to focus on their algorithms. SystemLink platforms offer a suitable platform for conducting laboratories over the internet and utilizing SystemLink cloud service and Dashboard mobile application.

The combination of LabVIEW and SystemLink Cloud in this project is an ideal ecosystem for remote laboratories and IoT designs. The ecosystem is useful for students with limited access to expensive lab equipment and it is an excellent platform for remote laboratories. NI real-time hardware platforms with more than 512 MB RAM can connect with LabVIEW and SystemLink cloud platform to setup remote educational laboratories and industrial process control.

The approach described it this paper could be regarded as a proof-of-concept, opening the appetite for similar projects serving a wide range of engineering and technology students all over the globe. The easy-to-program, both hardware and software, combined with the easy-to-connect solutions, are prerequisites for a post SARS-COV2 education world, where the need for information transfer and absorption is facing the challenge of social distancing.

Acknowledgements. Heartfelt gratitude goes to Prof. Horia Hedesiu of NI Romania to support and provide MyRIO 1900 module and Diligent modules to use in this research. I also wish to extend gratitude to my supervisor, Prof.Dr.Doru URSUTIU for his efforts in securing the kits and unwavering support throughout the development of this report writing of this report.

References

1. ZDNet Homepage. https://www.zdnet.com/article/anu-develops-vegetation-monitoring-system-to-help-firefighters-on-the-frontline/. Accessed 10 Oct 2020
2. https://www.7wdata.be/internet-of-things/what-is-the-internet-of-things-iot-meaning-definition/
3. Paquet VL, Bisantz A (2001) Case study based laboratories for an undergraduate human factors engineering curriculum. Paper presented at 2001 annual conference, Albuquerque, New Mexico. https://doi.org/10.18260/1-2-8984
4. Pblworks.org. https://www.pblworks.org/what-is-pbl. Accessed 10 Oct 2020
5. IIT homepage. https://www.ee.iitb.ac.in/~wel_iitb/resources_development_boards.html. Accessed 27 Oct 2020
6. Ursutiu D et al (2013) Student educational device for electronic applications (StudentEDEA). In: 2013 international conference on interactive collaborative learning (ICL), pp 552–555
7. Cotfas PA, Cotfas DT, Machidon OM (2016) Remote laboratories based on LabVIEW web services. In; Edulearn 16 proceedings, pp 3683–3689
8. NI.com. https://zone.ni.com/reference/en-XX/help/373925C-01/myriohelp/myrio_hardware_bd/
9. Diligent.com. https://reference.digilentinc.com/reference/pmod/pmodals/start?redirect=1
10. Diligent Homepage. https://store.digilentinc.com/pmod-tmp3-digital-temperature-sensor/

Localized Vineyard Monitoring for Specific Diseases

Petru Epure[1(✉)], Andra Perju[1], Doru Ursutiu[2,4], Cornel Samoila[2,3], and Petru P. Epure[1]

[1] EPI-SISTEM SRL, str. Livezii nr. 17A, Sacele, Romania
petru.epure@epi.ro
[2] Universitatea Transilvania din Brasov, Blv. Eroilor nr. 29, Brasov, Romania
[3] Academia de Stiinte Tehnice din Romania, Brasov, Romania
[4] Academia Oamenilor de Stiinta, Brasov, Romania

Abstract. This paper presents an embedded system that measures the botrytis cinerea infection in vineyard. The first stage of development of the system integrates a potentiostat for electrochemical measurements performed on screen printed electrodes upgraded with GPS coordinates to localize the danger. The infection risk is favored according to several climatic factors. Monitoring the vineyard every week become important according to humidity, temperature, sun exposure and many other parameters. In this way we can predict the possible appearances of the disease in the incipient phase.

Keywords: Winemaking · Disease · Electrochemistry · Biosensors · Measurement & control · Embedded C · IoT

1 Introduction

The biosensors market was valued at $17,500.0 million in 2018 and is expected to reach $38,600.2 million by 2026, registering a CAGR (Compound Annual Growth Rate) of 10.4% from 2019 to 2026 [1]. Biosensors are analytical devices that can convert a biological response into an electrical signal. Biosensors majorly comprise three parts, including a component that identifies an analyte and produces signal, a signal converter combined with a digitalization device. Biosensors find their applications in food & beverages, environmental monitoring, home care diagnostics. and various nano mechanical biosensors. In addition, biosensor applications are also widely dominant in the field of plant biology to study metabolic processes [2].

The wearable market was in 2019 about 32.63 billion USD and will evolve with a CAGR of 15.9% between 2020 and 2027. Just for the smartwatch devices the sales in 2019 were at 148 million units having a grow of about 30–35%. The evolution will continue for the next 5–10 years in the same pace. The IoT (Internet of Things) market will reach 1.1 trillion USD between 2018 and 2023 giving a significant impact on the global internet network. It is estimated that by 2025 over the world will exist more than 21 billion connections dedicated to IoT [3]. The evolution of this huge infrastructure involves big efforts to be made by companies from various sectors providing hardware and software parts or modules.

M. E. Auer et al. (eds.), *Online Engineering and Society 4.0*, Lecture Notes in Networks and Systems 298, https://doi.org/10.1007/978-3-030-82529-4_31

Applications of agriculture automation consist in monitoring and controlling the crop evolution process using new wearable devices and IoT monitoring systems. The precision agriculture has the particularity of a low degree type of automation due to the fact that large areas of crops are spread throughout agricultural areas within a region. The goal was to develop and manufacture a cost-effective, more flexible monitoring system for real-time evaluation of specific infections in vineyard [4]. On the first step it was developed and build a specific sensor having localized coordinates allowing to take preventive measures just in the area of infection. The device is using biosensors functionalized with different methods for botrytis cinerea and laccase activity [5, 6].

2 System Description

The measurement point has several characteristics determined by the scope of measurement. It's important to determine the measurement position, the temperature and humidity recorded at the measurement time and the electrochemical response of the biosensors [7]. According to these specifications it was established a target architecture for this device.

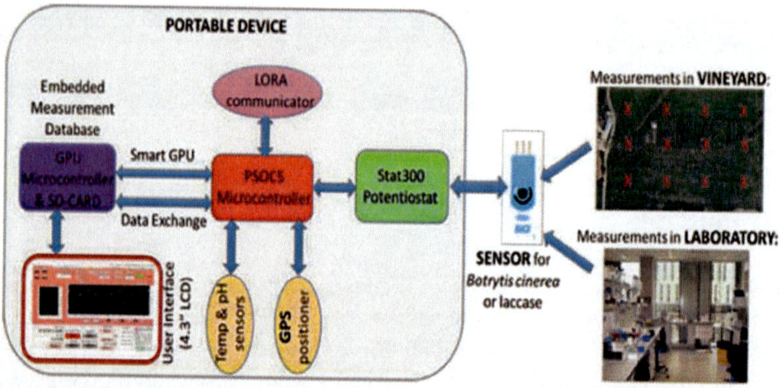

Fig. 1 Target architecture of the portable device

In Fig. 1 is presented a potentiostat µstat300 connected to a PSOC 5 (Programmable System-on-a-Chip generation 5) microcontroller having a 4.3″ LCD display and several sensors around.

Having in mind the miniaturization goal, it was introduced **AduCM335** (analog microcontroller based on ARM M3) from Analog Devices instead of µstat300 and the architecture gained also the possibility to perform also impedance measurement. The temperature and humidity sensor it's a DHT11 (combined temperature and humidity sensor) device with digital response similar to one wire. For GPS coordinates was integrated NEO-6 M (GPS sensor) version and the communication is also digital.

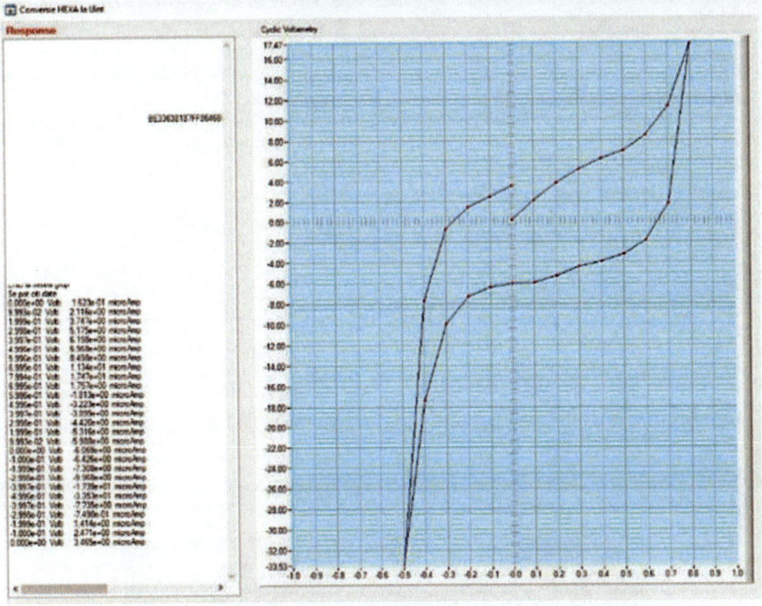

Fig. 2 Cyclic voltammetry performed via INTEL NUC

The PSOC 5 microcontroller is a CORTEX-M3 (ARM microcontroller version) on 32 bits manufactured by Cypress Semiconductor, actual part of Infineon. It includes a bundle of digital and analogic modules and provide communication facilities and interfaces. The AduCM335 microcontroller was upgraded by Analog Devices with a firmware made in collaboration with Palmsens from Netherlands [8]. This firmware version includes voltammetric and amperometric electrochemical methods and also an impedance method designed mainly for electrochemical interfaces. It was made a control version via INTEL NUC and the results are presented in Fig. 2.

The potentiostat provides a resolution down to 5.5 pA on lowest current range and an accuracy of 0.5% on applied potential and measured current. In terms of concentration a current of 10nA corespond to 1 U/ml (units per milliliter). The linear domain of the sensor developed for this measurements is between 0.8 until 15.6 U/ml.

The presented architecture can be the frame for any measurement application based on biosensors or electrochemical sensors. In this case the configuration use screen printed electrodes manufactured on alumina substrates and containing WE (working electrode), RE (reference electrode) and CE (counter electrode). An example of such electrode is presented in Fig. 3.

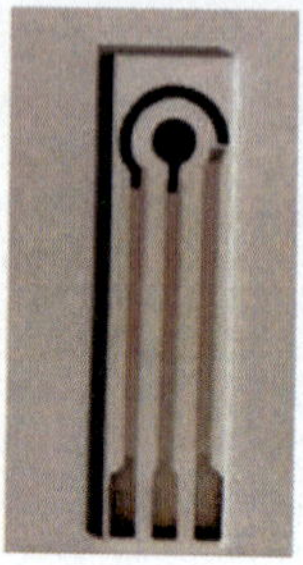

Fig. 3 SPE typical configuration

For portable devices was developed and implemented an other control version around PSOC 5 and the results are displayed on a smart GPU (graphical processing unit) like in the figure below [9]. Measured points are marked with circles and then the evolution between points is marked with thin lines.

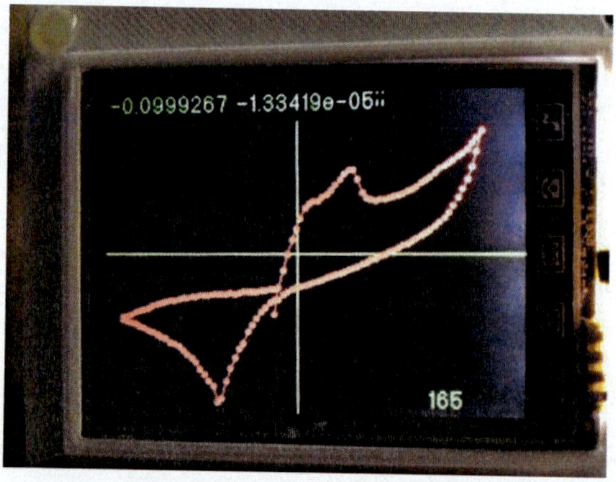

Fig. 4 Cyclic voltammetry performed via PSOC

For this version the special icons on the right of the screen in Fig. 4 are used to perform several screens and procedures described in chapter 3. In this example a cyclic voltammetry method was performed on a carbon SPE using a pH = 6.3 buffer.

3 Software Description

At the beginning the development was started with a 2.4″ version of Smart GPU and based on this configuration we implemented several user interfaces like in Fig. 5.

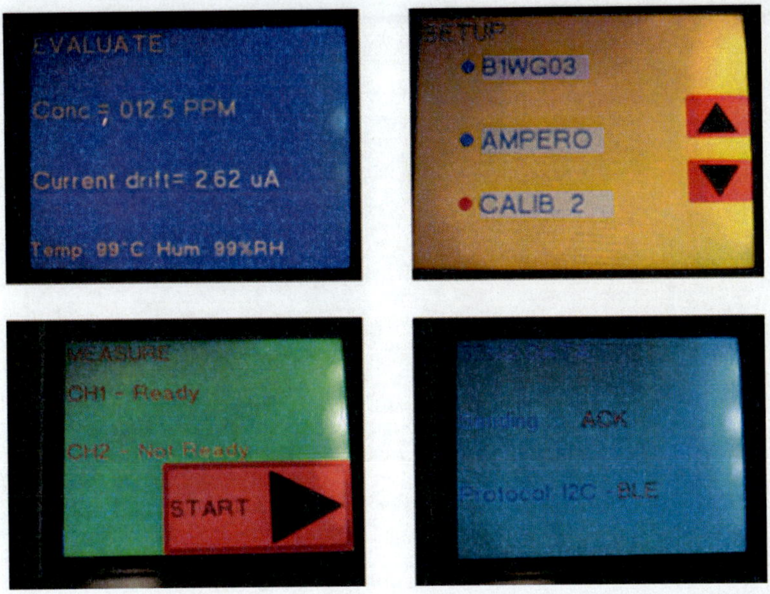

Fig. 5 GUI (Graphic user interface) versions for 2.4″ SmartGPU

All sections of GUI are performing different functionalities: SETUP, MEASURE, EVALUATE and SEND DATA. SETUP section allow end-user to choose a label, a method and a calibration curve. MEASURE section display availability of electrochemical channels and allow end-user to start the electrochemical measurement on channel 1 or channel 2. In some special configuration the measurement can be performed simultaneously on both channels having the same method: amperometry with sampling on minimum 2 s. EVALUATE section allows end-user to read all data before sending it to cloud via an external GPRS (General Packet Radio Service) device.

The electrochemical measurements are compensated by temperature and humidity factors. In this situation the calibration curve is changed mainly because of pH changing due to temperature and humidity. Calibration curve is affected mainly in a slope changing, the increase of acidity is giving a bigger slope of the calibration curve. In this case a current variation is giving a lower variation of concentration when the acidity is higher. All this characteristics are available for the linear range of the sensor. The other factors like soil temperature or seasonal factors have indirect implication and does not affect the performance of the measurement system. Those factors only favor the conditions for infection appearance.

There is a specific frame box made using the 3D-DESIGN technologies and then 3D-PRINT prototyping facilities. The box volume is adjustable and fit the dimensions of the SmartScreen (Fig. 6).

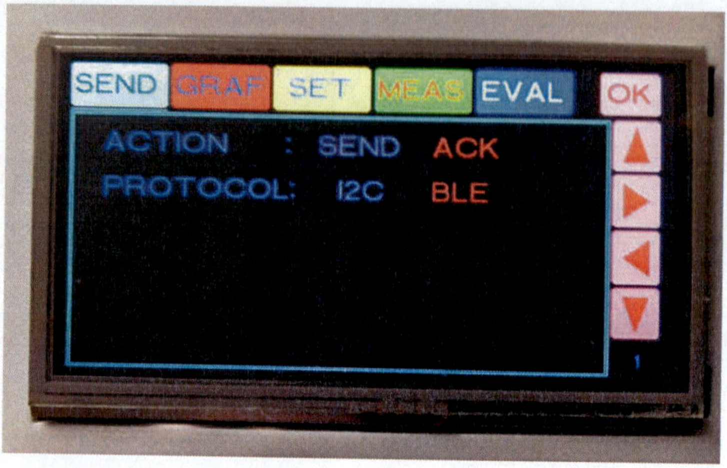

Fig. 6 Box models for 4.3″ GPU

The Vizic SmartGPU includes touch functions and can perform complex actions including audio—video content. Those functions allow future development of the measuring device including alarms or video tutorials regarding operating mode.

Software development is made in PSOC Creator based on Keil C programming language. At the beginning the user needs to configure the hardware modules, then needs to generate analogic and digital path and then implement the software links to sensors, potentiostat, touch screen or any other peripheral [10].

All data are uploaded in via internet in cloud using GPRS communication. The existing database on the cloud is updated with new data every year and creates a solid fundament for a decision software. Based on latest information gathered from vineyard the decision software calculates the imminence of infection appearance and determine the hot zone where active measures should be taken. In this manner every risk encountered will generate an alarm, will be managed by competent engineers according to the recommendations issued by the software. The measuring device have in this architecture the role of IoT and provide important information about the degree of infection and where is localized. The usage of this system generates important economies in the case of an infection on large vineyards and saves important time to recover the situation.

There is also a version that uses communication with LORA (Long range Wi-Fi 868 MHz communication) and is made using SPI (Serial Peripheral Interface) and then the IQRF (Intelligent protocol based on Radio Frequency) mesh consolidate the information towards the access point AP (Fig. 7).

Fig. 7 Measured data taken into field

Data collected from measuring points are stored in a local SQL database. From there a decision software will generate a map where active measures are proposed to be taken to prevent infection spread. This type of architecture is useful also for other types of applications like agriculture, forestry, fire prevention or river delta.

4 IoT Mapping

Starting from an area of 10 ha (100000 m^2) located in Valea Calugareasca—Prahova County—Romania, it was scheduled to make determinations on an area presented inside the red polygon drawn in Fig. 8 where the project is focused in 2020 and 2021. The measurements will be continued also in Spain in Victoria county, Basque Country.

According to Spanish partners measurements will be made there on a much larger area of around 100 ha (1 million m^2) and therefore localization of measurements will be more important in that case.

Based on this data a decision software will recommend some concrete actions to be taken to prevent the disease spread in vineyard [7, 8]. Using this decision tool the stakeholders will avoid huge loses of grapes or wine generated by botrytis cinerea and laccase.

Fig. 8 Measurement area on Riesling and Sauvignon Blanc

5 Conclusion

Measurements made on large areas can be configured more easily this days and allow monitoring in forest, agriculture crops, smart cities and many other. Based on integration of several microcontrollers like PSOC 5, AducM355, SmartGPU2, NEO-6 M in the same 3D-Design and 3D-Printed frame we can adjust the bio electrochemical measurements according with the temperature and humidity values. User can choose a specific calibration curve and can find out the risk evaluation in a precise position on vineyard. By consolidation of several measured points using biosensors on screen printed electrodes the user will have a risk map for botrytis cinerea and laccase. Based on the risk map a decision software will recommend actions to be taken just in the risk areas. Using this methods the end-user will record important reduction of loses due to the preventive localized actions in vineyard.

Combining classical sensors with other biosensors or electrochemical sensors measurement devices allows the user a better evaluation of the environmental status or health risks on crowded urban areas or in unbuilt territories. Consolidated data can be accessed on web providing a real time evaluation. There is also the portability advantage by comparison with automated lab methods [11]. The portable localized biochemical sensor based devices will have more and more impact on the next period to come.

Acknowledgements. This work was supported by a grant of the Romanian National Authority for Scientific Research and Innovation, **CCCDI – UEFISCDI**, project name ERANET – MANUNET-III-P3-**WINBIOTOOL** within **PNCDI III**, contract no.150/2020.

The Romanian project team members will continue to collaborate on providing each other valuable support in order to develop and optimize the monitoring system.

References

1. Grandview webpage. https://www.grandviewresearch.com/press-release/global-biosensors-market. Accessed 28 Nov 2020
2. Bilkiss M, Shiddiky MJA, Ford R (2019) Advanced Diagnostic Approaches for Necrotrophic Fungal Pathogens 458 of Temperate Legumes With a Focus on Botrytis SPP. Front Microbiol 10
3. Statista webpage. https://www.statista.com/statistics/1101442/iot-number-of-connected-devices-worldwide/. Accessed 28 Nov 2020
4. ERA-LEARN webpage. https://www.era-learn.eu/network-information/networks/manunet-iii/manunet-call-2019/portable-biosensor-based-device-for-mapping-the-risk-of-grape-infection-by-botrytis-cinerea-in-the-vineyard. Accessed 28 Nov 2020
5. Titoiu AM, Porumb R, Fanjul-Bolado P, Epure P, Zamfir M, Vasilescu A (2019) Detection of allergenic lysozyme during winemaking with an electrochemical aptasensor. Electroanalysis 31:2262–2273
6. Vasilescu A, Fanjul-Bolado P, Titoiu AM, Porumb R, Epure P (2019) Progress in Electrochemical (Bio)Sensors for Monitoring Wine Production. Chemosensors 7:66
7. Grassin C, Dubourdieu D (1989) Quantitative determination of Botrytis laccase in musts and wines by the 462 syringaldazine test. J Sci Food Agric 48:369–376
8. Analog Devices webpage. https://www.analog.com/en/products/aducm350.html#product-overview. Accessed 28 Nov 2020
9. Vizic Technologies webpage. https://www.vizictechnologies.com/downloads. Accessed 28 Nov 2020
10. Cypress Semiconductor webpage. https://www.cypress.com/documentation/development-kitsboards/cy8ckit-059-psoc-5lp-prototyping-kit-onboard-programmer-and. Accessed 28 Nov 2020
11. Urbano Cuadrado M, Pérez-Juan PM, Luque de Castro MD, Gómez-Nieto MA (2005) A fully automated 467 method for in real time determination of laccase activity in wines. Analytica Chimica Acta 553: 99–104

Real-Time Monitoring of Indoor Healthcare Tracking Using the Internet of Things Based IBeacon

Mohamed Zied Chaari(✉), Rashid Al-Rahimi, and Abdulaziz Aljaberi

Qatar Scientific Club, Doha, Qatar
chaari_zied@ieee.org

Abstract. This research will present the utility of using the Real-Time Monitoring of indoor healthcare tracking utilizing the Internet of Things-Based I-Beacon. According to the fast and widespread COVID-19 contagion globally, the demand for ventilators and PPE materials is very h h to provide safety equipment to healthcare professionals and doctors. The existing Powered Air-Purifying Respirator (PAPR) it's not designed to COVID-19 and social distance. With the limitations of the current navigation systems such as Global Positioning System (GPS) in indoor settings (Hospital, Quarantine, etc.), alternative PAPR tracking methods need to developing for such environments. This project aims to design a system that uses Bluetooth for PAPR tracking. Bluetooth Low Energy modules (BLE) serve as beacons, and the hospital will have information about their position. They will continuously advertise this information about the passing doctors. When the PAPR, which is attached to a Bluetooth module, passes by a beacon on the hospital corridors, it will receive the position data and send it directly to the ESP32 circuit. In turn, the ESP32 device will send the data to be the control room. So will add IoT technology in the respirator to guarantee physical distance, healthcare sta tracking, and recording.

Keywords: Powered Air-Purifying Respirator (PAPR) · iBeacon · BLE · Embedded · IoT · COVID-19 pandemic

1 Introduction

Satellite navigation systems such as the GPS provide accurate and reliable positioning for tracking outdoor locations. Many research types have proven the significance of GPS is revolutionizing current practices of monitoring and managing traffic in relatively open areas. However, GPS is highly dense in urban and indoor settings due to the deflection and distortion of satellite signals by neighboring buildings, degrading GPS performance. The impact of Coronavirus has influenced every corner of life. COVID-19 outbreak has caused hundreds of millions of lives around the world. To control the COVID-19 from propagating, many states, including the USA, European Union, U.K., and others, are

M. E. Auer et al. (eds.), *Online Engineering and Society 4.0*, Lecture Notes in Networks and Systems 298, https://doi.org/10.1007/978-3-030-82529-4_32

currently in some level of lockdown, with the closure of companies, gyms, airports, universities, and schools [1–3]. Physical distancing means keeping space between doctors and other people inside the hospital and Quarantine. It is a vital step in helping to slow the spread of this virus. Unfortunately, the current PAPR device does not fabricate for the Coronavirus, and it does not save use 100% of the virus threat. Since the Coronavirus outbreak, scientific groups are upgrading a lot of existing PPE response to the COVID-19 pandemic [4]. This research will present the utility of adding a tracking system into PAPR device for healthcare staff tracking recording, and physical distancing [5,6]. We are not interested in making a new device; we are just thinking of upgrading the exit device to be compatible with the pandemic [7–10]. With Real-Time monitoring of indoor healthcare tracking using the IoT-Based I-Beacon, we make a high-efficiency PAPR device in COVID-19. This prototype proposes and evaluates the possibility of using a new approach to allocate doctors. This research is toward creating IoT smart devices, as shown in Fig. 1.

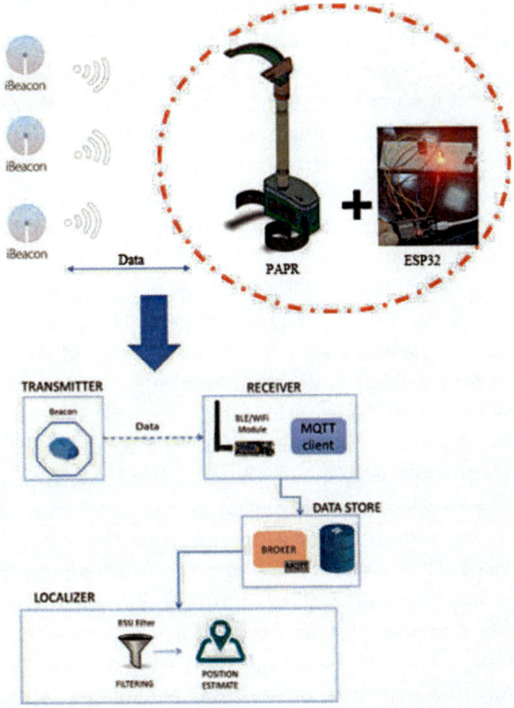

Fig. 1. The upgrade PAPR respirator with locality BLE Beacon

2 Real-Time Monitoring of Indoor Healthcare Tracking Using the Internet of Things-Based I-Beacon

Diffusion of the COVID-19 virus can occur by direct contact with an infected person or indirect contact with surfaces in the immediate atmosphere or with items used on the infected people. A new device implanted inside the air-purifying respirator could register the details of other PAPR that come into its small area (diameter of 50 cm). The specification of the PAPR real-time localization system (RTLS) is below:

- Wireless indoor pallet tracking system,
- Accurate location of moving & stationary healthcare staff,
- Accuracy of 50 cm × 50 cm meters for locating indoor objects,
- Asset monitoring system,
- No electrical or wiring or networking cables need to get laid,
- Physical distancing,
- Strengthening of security surrounding the clinical area.

This section aims to develop new indoor doctors tracking to improve positioning accuracy [11,12]. However, the rapid technology improvement opens for us new ways, which might be more effective. Unfortunately, there are no methods for indoor tracking that is approved. Nevertheless, this project improved and proves that an indoor doctor's tracking system is essential in the Covid pandemic. Moreover, offering a new solution to assist doctors. The purpose of this research is to design a system that uses Bluetooth for indoor track. The device will provide the doctors with the most accurate latitude and longitude inside the hospital, as shown in Fig. 2. The importance of tracking the doctors in the hospital is to increase the matter of doctor's safety. The low strength Bluetooth uses low radio frequency, which has a lower health hazard to the surrounding individuals.

The prototype is to design a doctor's tracking system for indoor settings (hospitals, laboratories, and Quarantine) based on Bluetooth technology. The device will identify the current position of a PAPR with a Bluetooth module attached to it. Multiple Bluetooth low energy modules will broadcast the position data that will serve as beacons placed across the hospital. Whenever the PAPR passes by a beacon, the ESP32 module on the ventilator receives the place data and gives it to the data store to evaluate. Kalman filter will use the received data for more precise results [13,14]. BLE provided 40 channels. Use Bluetooth modules as beacons to advertise the location of the nearby PAPR. Bluetooth acts as an aided localization system where it uses the electronics to allocate the doctors. The BLE communication technology will broadcast a radio frequency in a low range. The passing doctors will receive these signals. The Low range radio frequency of 40 channel signals will result in maximum location precision. The iBeacons will propagation its location, and it will do their job to evaluate the place by the distance gap. The Kalman filter will use in this phase to ensure the estimation value is well calculated. Use a relatively cheap technology to develop a tracking PAPR. It compares alternative technology based on their precision, energy consumption, and compatibility with the local area. The alternative technique will be compared based on previous features. Bluetooth found PAPR tracking could be the best smart low cost. It is cheap to implement and maintain during a crisis. It operates along with the same band as 802.11 b/g/n WiFi, the 2.4 GHz ISM; although, there is no significant R.F. interference because of the low broadcast energy. There are three main channels for Bluetooth to propagate (39, 38, and 37). These channels have selected not to interfere with the WiFi channels (11, 6, and 1). This technique deployed numerous sensor nodes at fixed spots inside corridors. This target transmits a unique code every 20 s via a beacon sensor. A location manager for correction views the nodes of the sensor. This centralized manager technique calculates the active badge's real position and provides location information to the doctors [15,16].

The system's primary function is to determine the current position using multiple BLE-enabled modules located along with the corridors and act as beacons that broadcast the position information to be received and interpreted by the microcontroller. The HM-10 module can use the advertisement principle, which allows the module, as a BLE device, to broadcast data to adjoining devices periodically. The HM-10 module will be programmed using AT commands that will send through using UART pins on the Arduino Nano (TX and R.X.), as shown in Fig. 6.

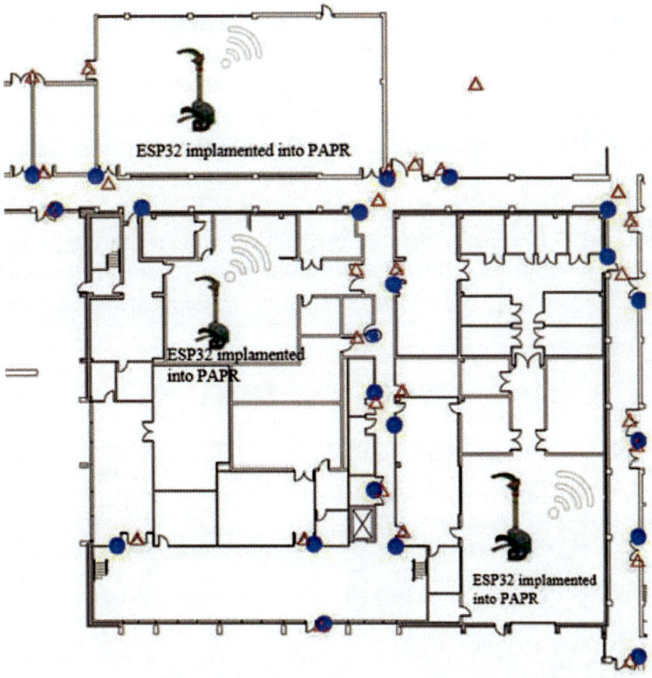

Fig. 2. Example of the quarantine floor plan with indoor navigation and tracking system

The ESP32 micro-controller on the tracking system must receive the beacon's advertisement communication to find the beacon location data. The location data include the beacon's name and the latitude and longitude values embedded in the advertisement, as shown in Fig. 3. TX power value advertised by the module is a hexadecimal value that represents the RSSI (Received Signal Strength Indication). This value is required to determine the proximity of the beacon to the tracking device. Therefore, the system will identify the nearest beacon whenever more than one beacon is available, as shown in Fig. 4. AT commands are simple commands used to communicate with specific devices. For this prototype, the controls will communicate with the Bluetooth modules to configure them as beacons and assign the UUID to each HM-10. Such configuration is done by connecting R.X. and TX pins of the beacon to pins 9 and 8 of the Arduino Nano to create UART communication. When the tracking device discovers multiple beacons available nearby, it has to decide which beacon to receive the position data. This technology is successful by comparing the RSSI value of each detected HM-10 beacon sensor. The RSSI value is negative and indicates the strength level of the received signal. The additional the beacon is from the tracking system, the higher the RSSI absolute value. So, when the tracking system discovers multiple beacons nearby, it will accept location data from the beacon with the lowest RSSI value. The device received the advertised data from the beacon and converted the hexadecimal data into decimal values that can interpret as position data. Bluetooth Low Energy (BLE) components serve as beacons, placed along the corridors will have information about their location. They will continuously advertise this information to the passing healthcare staff wear the air ventilator system. When the healthcare staff, which is attached to a Bluetooth module (ESP32), passes by a beacon on the hospital corridors or quarantine area, it will receive the position data and send it directly to the micro-controller. The gateway is placed around the PAPR device. They connect via WiFi. The gateways detect module implemented already in the upgraded PAPR device and send this information (PAPR Name, PAPR Serial number) to the computer for recording and diagnostic. The RTLS circuit received the advertised data from the beacon that interpret as location data. Moreover, the system can receive the location

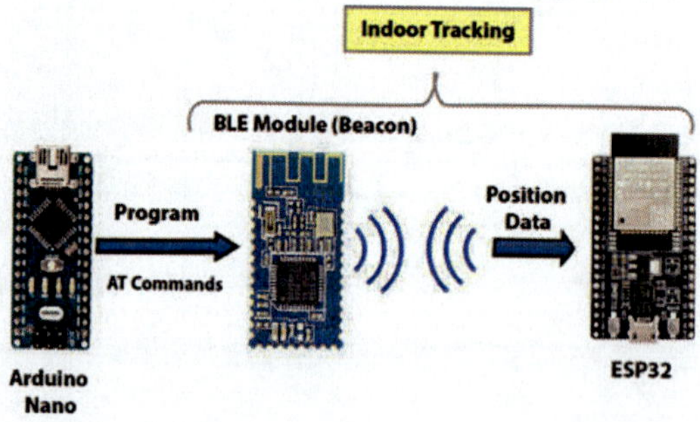

Fig. 3. Tracking system directly with one beacon

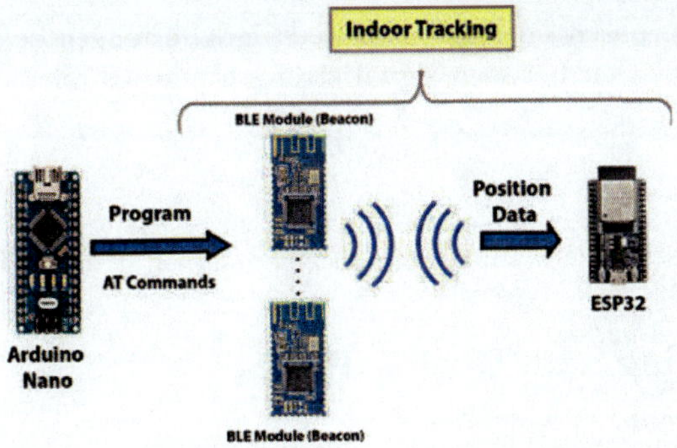

Fig. 4. Tracking system with multi-beacon

Fig. 5. The architecture of the Bluetooth PAPR tracking and recording system

data from the absolute nearest beacon tested to ensure accuracy. The ESP32 will scan for nearby beacons and receive the closest beacon location data, as shown in Fig. 5. The Bluetooth ability of ESP32 will allow the device to scan the area for nearby beacons [17–21]. Ranging is the process of determining the cap distance between two beacons and PAPR without any angle. It is also the base used for alteration. Several methods to accomplish this exist. The two most common radio ranging are based on either received signal strength or flight [22,23]. In free space, the relation between signal strength and distance can be expressed as:

$$d = 20^{(RSSI+a)} \tag{1}$$

Where:
a: is an offset based on the maximum received signal strength for the Beacon

3 Prototyping and Testing Phase

After understanding the BLE and the possibility of integration with the PAPR, we start the fabrication and prototyping process. ESP32 is a powerful IoT device having inbuilt support for Bluetooth and WiFi. The BLE option can use for optimized battery applications like Bluetooth proximity advertisements or beacons. In this prototype, we will build a BLE iBeacon using HM-10, where ESP32 will act as a server, and the iBeacon fixed inside the building will serve as a client. Every beacon consists of a BLE module and battery holder with the 3 V battery to power the module, as shown in Fig. 6.

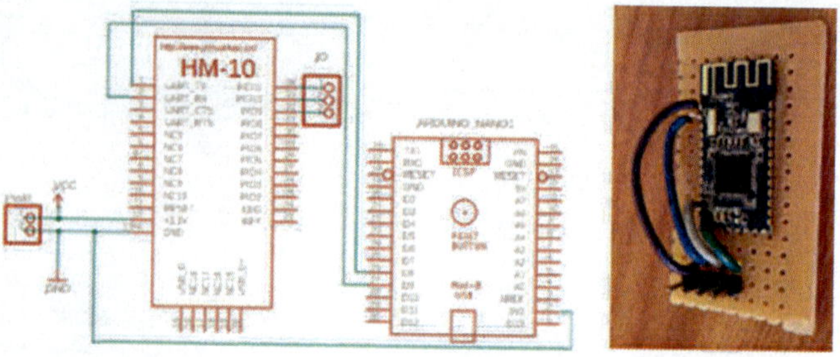

Fig. 6. iBecon consists of a BLE module and battery 3 V

In this prototype, we can easily track and keep the social distance between the PAPR use. After examining the iBeacon, we will see the RSSI, UUID, Major, and Minor of iBeacon. The RSSI will modify if we take iBeacon away from each other. Here on this screen, the RSSI is (−37 dB). Other than this, there is some information such as device type and the local name of the ESP32 module. Here it is, ESP32 as micro-controller and HM-10 as iBeacon. You can change the local representation in the system, as shown in Fig. 7.

Fig. 7. ESP32 will fix it in the PAPR for real-time location system

We add another option in the system when the ESP32 detect any beacon near it; it will automatically activate the speaker's warning alarm for keeping the social distance, as shown in Fig. 8.

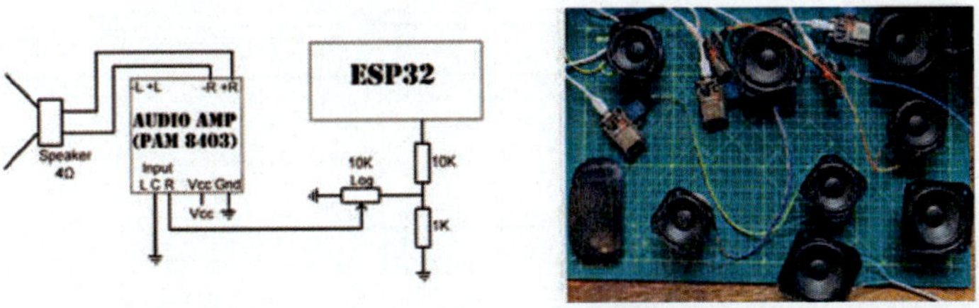

Fig. 8. ESP32 warning alarm with a speaker

Significantly, this project offers a better solution, including lower prices and high efficiency. Bluetooth low energy (BLE) is inexpensive to install and programming. After fabrication, the tracking PCB will fix it inside the PAPR package for testing and validation, as shown in Fig. 9.

As is shown in Table 1, we did many experiments. It's because the result seems very good (compare to the work measured by my android phone). We are doing two experiments for two beacons target, and each experiment is performed every 20 s, the experiment distance from 50 cm to 7 m. Besides, we try to put the beacon and the EPS32 together (0 space), but it looks like the RSSI change significantly.

Fig. 9. PAPR with Bluetooth iBeacon of the real-time location system and physical distance

Table 1. Effect of obstructing objects on RSSI and accuracy

Distance (m)	iBeacon1 RSSI (dB)	iBeacon2 RSSI (dB)
1	−56	−56
2	−59	−58
3	−62	−60
4	−66	−65
5	−69	−69
6	−62	−62
7	−63	−58

4 Conclusion

We are adding IoT technology in the respirator for physical distancing and healthcare staff tracking in this device. Therefore, this project is intended to provide an alternative tracking method when the GPS signal is weak or lost based on Bluetooth technology. In this research, we approved that we are repurposing a new solution to ensure the social distance between the doctors and healthcare staff while fighting the Coronavirus local with indoor healthcare tracking. In this research, we made and tested the additional PCB circuit, the essential tool of the first defense line, to improve the social distance. Furthermore, the extra part into PAPR capable of receiving the exact location from the absolute nearest beacon. We tested, and we ensure the system's position accuracy. We proved that the solution could reduce contact with the doctors inside the operation area by reducing Corona cases. In future work during the system testing, we have observed that the position estimation was occasionally inaccurate given that the doctor's current position and the inquiry duration.

References

1. Miles LF et al (2020) The design and manufacture of 3D printed adjuncts for powered airpurifying respirators. Anaesth Rep 8(2):84–86. https://doi.org/10.1002/anr3.12055
2. Holland M et al (2020) COVID-19 Personal Protective Equipment (PPE) for the emergency physician. Vis J Emerg Med 19:100740. https://doi.org/10.1016/j.visj.2020.100740
3. Ranney ML et al (2020) Critical supply shortages - the need for ventilators and personal protective equipment during the Covid-19 pandemic. N Engl J Med 382(18):e41. https://doi.org/10.1056/NEJMp2006141
4. Chaari MZ, et al (2020) Use of reverse engineering method for respirator devices in COVID-19 crisis. In: 2020 2nd international conference on electrical, control and instrumentation engineering (ICECIE), pp 1–4. https://doi.org/10.1109/ICECIE50279.2020.9309686
5. Choi B et al (2020) Persistence and evolution of SARS-CoV-2 in an immunocompromised host. N Engl J Med 383(23):2291–2293. https://doi.org/10.1056/NEJMc2031364
6. Emanuel EJ et al (2020) Fair allocation of scarce medical resources in the time of Covid-19. N Engl J Med 382(21):2049–2055. https://doi.org/10.1056/NEJMsb2005114
7. Sohail A, Nutini A (2020) Forecasting the timeframe of 2019-nCoV and human cells interaction with reverse engineering. Prog Biophys Mol Biol 155:29–35. https://doi.org/10.1016/j.pbiomolbio.2020.04.002
8. Ullah S et al (2020) Reusability comparison of melt-blown vs nanofiber face mask filters for use in the coronavirus pandemic. ACS Appl Nano Mater 3(7):7231–7241. https://doi.org/10.1021/acsanm.0c01562
9. Wang CJ et al (2020) Response to COVID-19 in Taiwan: big data analytics, new technology, and proactive testing. JAMA 323(14):1341. https://doi.org/10.1001/jama.2020.3151
10. Yu L et al (2020) Catching and killing of airborne SARS-CoV-2 to control spread of COVID-19 by a heated air disinfection system. Mater Today Phys 15:100249. https://doi.org/10.1016/j.mtphys.2020.100249
11. Singh H, Pallagani V, Khandelwal V, Venkanna U (2018) IoT based smart home automation system using sensor node. In: 2018 4th international conference on recent advances in information technology (RAIT). IEEE. https://doi.org/10.1109/rait.2018.8389037
12. Kodali RK, Jain V, Bose S, Boppana L (2016) IoT based smart security and home automation system. In: 2016 international conference on computing, communication and automation (ICCCA). IEEE (2016). https://doi.org/10.1109/ccaa.2016.7813916
13. Faragher R (2012) Understanding the basis of the Kalman filter via a simple and intuitive derivation [lecture notes]. IEEE Signal Process Mag 29(5):128–132. https://doi.org/10.1109/msp.2012.2203621
14. Pop MD, Protean O, David TM, Protean G (2020) Hybrid solution combining Kalman filtering with Takagi-Sugeno fuzzy inference system for online car-following model calibration. Sensors 20(19):5539
15. Lu M, Chen W, Shen X, Lam HC, Liu J (2007) Positioning and tracking construction vehicles in highly dense urban areas and building construction sites. Autom Constr 16(5):647–656. https://doi.org/10.1016/j.autcon.2006.11.001

16. Bahl P, Padmanabhan V, (2000) RADAR: an in-building RF-based user location and tracking system. In: Proceedings IEEE INFOCOM 2000. Conference on computer communications. nineteenth annual joint conference of the IEEE computer and communications societies (Cat. No.00CH37064). IEEE. https://doi.org/10.1109/infcom.2000.832252
17. Khalid A, Memon I Bluetooth-based traffic tracking system using ESP32 microcontroller. In: Advances in machine learning and computational intelligence, pp 737–746. Springer
18. Fujihara A, Yanagisawa T (2015) Proposing an extended iBeacon system for indoor route guidance. In: 2015 international conference on intelligent networking and collaborative systems pp 31–37. IEEE, Taipei (2015). https://doi.org/10.1109/INCoS.2015.72
19. Giuliano R et al (2020) Indoor localization system based on bluetooth low energy for museum applications. Electronics 9(6):1055. https://doi.org/10.3390/electronics9061055
20. Goh BS et al (2020) IoT based indoor locating system (ILS) using bluetooth low energy (BLE). In: 2020 IEEE international conference on consumer electronics (ICCE), pp 1–4. IEEE, Las Vegas. https://doi.org/10.1109/ICCE46568.2020.9043108
21. Misal SR et al (2020) Indoor positioning system (IPS) Using ESP32, MQTT and bluetooth. In: 2020 fourth international conference on computing methodologies and communication (ICCMC), pp 79–82. IEEE, Erode. https://doi.org/10.1109/ICCMC48092.2020.ICCMC-00015
22. Takahashi C, Kondo K (2015) Indoor positioning method for augmented audio reality navigation systems using iBeacons. In: 2015 IEEE 4th global conference on consumer electronics (GCCE), pp 451–452. IEEE, Osaka. https://doi.org/10.1109/GCCE.2015.7398636
23. Lin X-Y et al (2015) A mobile indoor positioning system based on iBeacon technology. In: 2015 37th annual international conference of the IEEE engineering in medicine and biology society (EMBC), pp 4970–4973. IEEE, Milan. https://doi.org/10.1109/EMBC.2015.7319507

Cloud Platform for Software Development: Review

Anzhelika Parkhomenko(✉), Yaroslav Zalyubovskiy, and Andriy Parkhomenko

National University Zaporizhzhia Polytechnic, Zhukovskogo Str. 64, Zaporizhzhia 69063, Ukraine
parhom@zntu.edu.ua

Abstract. The peculiarities of the existing models of cloud services provision and the possibilities of their usage for software development are investigated in this work. It is shown that SAP Cloud Platform is an enterprise cloud platform that allows companies to develop their applications, extensions, or add-ons for existing SAP products or create a hybrid IT landscape by integrating cloud solutions. The analysis of the limitations and services of SAP Cloud Platform, including DevOps methodology and CALMS framework is performed. The method of business application development in the Cloud Foundry environment of SAP Cloud Platform is presented.

Keywords: Intellectual enterprise · IT landscape · Cloud service · DevOps methodology · Business application

1 Introduction

Extensive data management, the Internet of Things (IoT), analytics, machine learning, robotics of routine operations are modern technologies that can be used by enterprises and companies to implement innovative scenarios in business. Ready-made cloud platforms, assembled by large suppliers of software and services provided by subscription, help in building modern digital enterprise-scale solution [1].

Infrastructure-as-a-Service (IaaS), Platform-as-a-Service (PaaS), and Software-as-a-Service (SaaS) are existing cloud service delivery models. Their ratios are often represented as a pyramid with different levels of information control. The top is the end-user who works with personal data presented in a program or service with user-friendly interface. The second level is the program or service deploys on a specific technological platform. The base of the pyramid is infrastructure: virtual servers, computing power, data storage and communication channels [2, 3].

Studies have shown that SaaS is the most understandable and convenient model of providing cloud service from the end user's point of view. It is often more comfortable and more efficient to use the ready-made SaaS service that is already configured for specific requirements. Nevertheless, ready-made solutions do not always exist, and in this case, PaaS and IaaS models are recommended. There is no universal approach for introducing cloud technologies to improve the enterprise`s IT landscape. The selection

M. E. Auer et al. (eds.), *Online Engineering and Society 4.0*, Lecture Notes in Networks and Systems 298, https://doi.org/10.1007/978-3-030-82529-4_33

of a model and the process of migrating to the cloud are non-trivial tasks. The implementation of new technologies requires considerable effort, and they are poorly integrated with existing solutions at the enterprise. Usually, information about them is scattered in many Internet sources, and sometimes it is not relevant. Therefore, the topic of the review is of current interest.

2 Analysis of Cloud Solutions for Development and Operation of Software

2.1 Overview of SAP Cloud Technologies

German company SAP is one of the world's market leaders in enterprise applications. The company helps organizations of all sizes and specializations to increase the efficiency of their business management. A big number of customers use SAP solutions and services. The company's advanced technologies guarantee high profitability, promote continuous adaptation and sustainable growth. SAP company is also one of the well-known suppliers of integrated solutions for banks, enterprises and companies in Ukraine [4].

The company offers the whole line of solutions for the IoT, which are included into the digital innovation system SAP Leonardo. Significantly, SAP Leonardo platform combines various software capabilities: machine learning, the IoT, Big Data, analytics, and blockchain on SAP Cloud Platform (SCP) together with the design thinking methodology.

SCP provides the ability to connect people, objects, and businesses, as well as technologies for implementing the Internet of Things and Industry 4.0 in digital logistics, manufacturing, and asset management.

SAP IoT Bridge is a digital command center that allows operations managers to ensure business processes' transparency and enables them to act in real-time.

SAP Global Track and Trace is a cloud application for unified tracking, monitoring, and reporting objects and business processes in network supply channels.

SAP Digital Manufacturing Insights is a cloud solution for centralized production management and workflow optimization.

SAP Asset Manager is a cloud-based mobile application that allows to monitor employees' health, the state of equipment, manage maintenance and security.

SAP Leonardo IoT Edge is a cloud software for computing, data storing and business semantics, that allows devices to continue participate effectively in business processes almost in real-time without access to a company's data center [5].

Thus, SAP cloud technologies open up endless opportunities to develop the company's resources, document management tools, and mobile cloud services. They also give the green light to the usage of machine learning, big data, the IoT, and other innovative technologies.

2.2 SAP Cloud Platform Architecture and Functionality Investigation

Studies have shown that PaaS is a cloud platform that provides runtime environments for developing, testing, and managing applications. PaaS vendors offer a complete infrastructure for application development, while developers are responsible for the code.

According to G2 Scores [6] SCP is one of the best Cloud Platform as a Service Software. It gives a lot of possibilities for fast and efficient solutions of various tasks [7]. SCP provides a full range of service directories, including database, data storage, and backup, report service, and transaction-level for cross-platform software development. SCP clients can use this cloud environment to control software development or they can also use a hybrid model based on the cloud environment and in their own development environment [8].

The SCP can be integrated with SAP applications, third-party applications, or internal company solutions for data receiving and software development. The platform provides creation, integration, and expansion of cloud applications without spending time and financial costs for IT infrastructure maintenance. Some companies deploy SAP digital solutions in their physical infrastructure because they fear that their cloud data is not secure enough. However, remote SAP repositories are better protected than local infrastructure as well as their usage is cheaper. SAP ensures complete cloud data protection, using best practices in IT security, strict compliance with the country's laws in which it operates, and the best infrastructure. High-reliability data centers Tier-3 and Tier-4 do not stop working even in case of failures and errors. Also, double backup and duplication of the system are realized [9].

SAP cloud platform is based on open-source standards and gives the developers flexibility and control over clouds, applications and environments deploying. SCP uses different development environments, including Neo and Cloud Foundry, and provides several programming languages [7].

SAP cloud platform's main limitation is that it is designed to support cloud-to-cloud integration, but not for integration between local systems. It makes the tools less useful for integrating SAP applications and third-party applications that reside on local servers [7].

Thus, the two main tasks that the SCP platform solves are development and operation of IT solutions (hosting, monitoring, scaling, etc.) [10].

There are several categories of SAP Cloud Platform services:

- Services for data processing and storage: in-memory, SQL, NoSQL, Big Data Services.
- Services for application development: Java, XS/JS (internal SAP language for programming under SAP HANA), as well as support for BYOL model (Bring Your Own Language) using any programming language, such as PHP, Python.
- A set of ready-made technological services for creating business applications: blockchain services, integration bus, the IoT, etc.
- A set of ready-made business services: shopping cart, loyalty program, etc. [10].

A generalized map of SCP services is presented in Fig. 1. The research has shown that SCP is the cloud platform that provides runtime environment for applications developing, testing and managing. One of SCP services is DevOps set of practices for software development and operational efficiency improvement, which supports the principles of Agile and Continuous delivery of software [11].

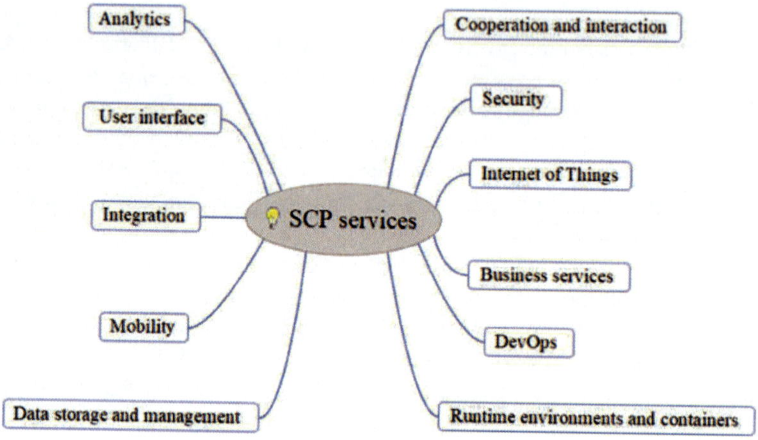

Fig. 1 SCP cloud services map

CALMS framework allows assessment of transition to DevOps readiness and can be used as a model to identify possible improvements in the software development, implementation and maintenance processes [12]. Also, an important feature of SCP is the support of BYOL (Bring Your Own Language) model, which provides the usage of any programming language, such as PHP, Python, and others.

3 Method of Business-Application Development Based on Cloud Technologies Usage

Analysis of works [1, 13–15], in which examples of development in SCP environment for different tasks are provided, made it possible to form a generalized method for business application developing. It is shown in the form of the flow diagram in Fig. 2.

It is based on the usage of global standard for cloud application development Cloud Foundry (CF), which allows developers to concentrate more on the tasks and facilitates the process of application deployment and management.

The core technologies and services used for software development are the following: PostgreSQL for cloud storage; Application Logging for creating, viewing, and analyzing application logs; Application Autoscaler for automatic application scaling.

Step 1. Preparation of Integrated development environment (Fig. 3). After authorization and logging in, you need to select Cloud Foundry Trial on the Home page (Fig. 4). An account, a subaccount, an organization and space in the selected region are provided after selecting a region. Then you need to go to the created space by clicking

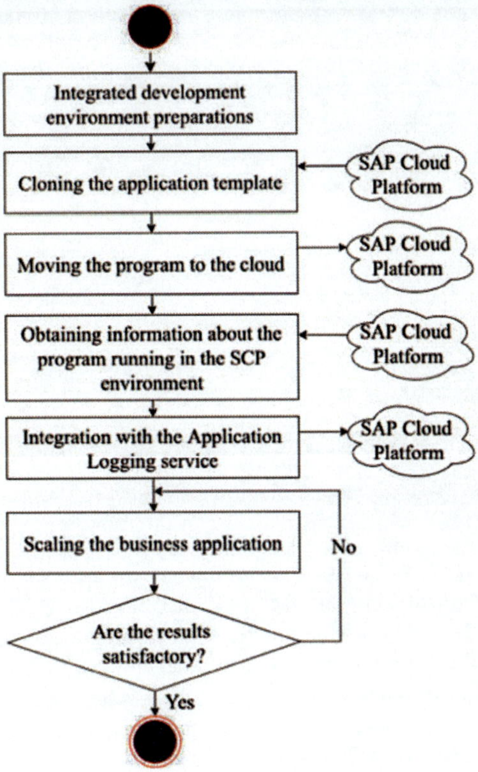

Fig. 2 Flow diagram of business application development method

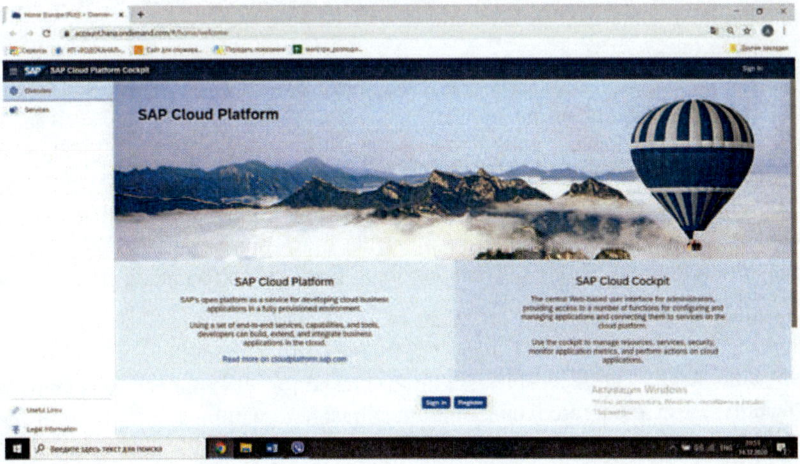

Fig. 3 Start page of SCP

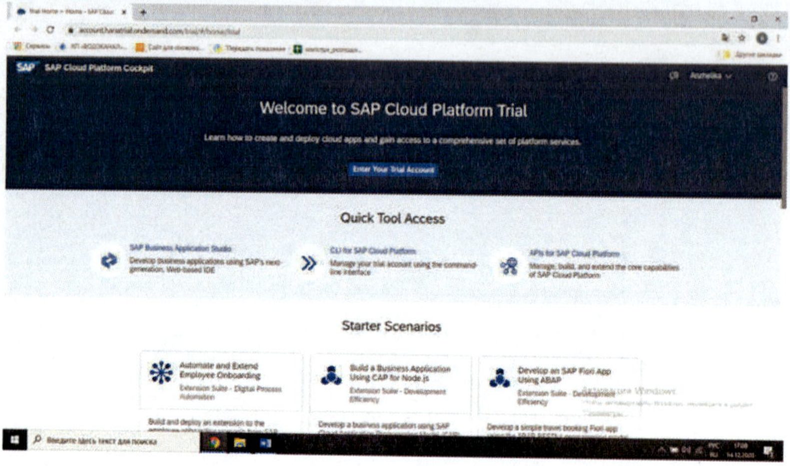

Fig. 4 Page of SCP trial

Go to Space. To configure the local development environment, you need to install Cloud Foundry Command Line Interface (CF CLI) on your PC.

Step 2. Cloning the application template. After cloning the application template and its importing into the working environment, developing and modifying based on BYOL model, and further assembling are performed. After successful compilation, the application should be downloaded to the account in the SCP platform's CF environment. This can be done using CF CLI.

Step 3. Moving the program to the cloud. Before loading the application into the CF environment, you need to set the program parameters into the command line for the push command, which allows to run the application in the cloud or use the application descriptor file. Setting the parameters for moving in this file prevents entering them again each time using the push command. The application is now running in the CF environment of the SCP platform.

Step 4. Obtaining information about the application running in the SCP environment. The information about the applications and services running in your CF account, as well as basic metrics and logs (CPU, memory and disk usage), is available both in CF CLI and the SCP control panel.

The checking of program performance is a monitoring process that constantly checks the status of an application running in CF. In CF, the default timeout is 60 s and the maximum timeout is 180 s. There are three types of performance checking available in CF:

- http validation is performed using GET request to the configured http endpoint; when the health check receives HTTP 200 response, the application is considered "healthy"; the configured endpoint must respond within 1 s to be considered "healthy."

- port is the default check; this type of authentication makes TCP connection to the port or ports configured for the application (default 8080); TCP connection must be established in 1 s for the application to be considered "healthy."
- process Diego ensures that any process declared for application will work. If the process completes, Diego will stop and delete the application instance.

The type of performance checking can be specified when moving the application to the cloud with the push command.

Step 5. Integration with the Application Logging service. For this, the library of logging for CF should be supported in the application source code. To generate logs, it is enough to call the application via the URL in the SCP control panel.

Step 6. Scaling the business application. System load, number, and nature of tasks are the factors, that can change disk space and memory, that are used by the application. For most applications, the increasing of available space and memory, or running of more instances, can improve overall performance. Similarly, the running of additional program instances may allow the application to adjust load increasing and the number of simultaneous requests. This adjustment is called application scaling. It is possible to scale the application for its adaptation to traffic changes manually via CF CLI or the SCP control panel. There is also a possibility to configure the automatic use of the SCP platform's Application Autoscaler service. In particular, the vertical scaling of the application changes the disk space limit or memory limit that CF applies to all program instances.

Horizontal scaling creates or deletes instances of the application. Incoming application requests are automatically adjusted by the load in all instances of the application, and each instance processes the task in parallel with other instances. Adding instances allows the application to handle more traffic and requests. There are two ways to scale the program vertically: using CF CLI or via SCP control panel.

The disadvantage of manually scaling the application is that these actions are performed only if the developer notices that the performance of the application has deteriorated. There is some risk that the application will fail if scaling is not performed in a timely manner. Therefore, sometimes it is better to be able to scale the application automatically using the Application Autoscaler service of the SCP platform.

Step 7. Verification of results. After checking the application's compliance with the requirements, in case of satisfactory results, the development process is completed, otherwise, it is necessary to return to the scaling stage to make adjustments.

Therefore, SCP platform is an excellent solution for developing a cloud business application that allows to add monitoring, scaling and update capabilities [16].

Since ABAP is a key language for developing next-generation SAP applications [17–19], its availability as an add-on environment on SAP Cloud Platform enhances software development capabilities for a variety of business applications [20].

4 Conclusion

As a result of the performed research, the urgency of cloud technologies implementation for improvement of IT landscape of the intellectual enterprise is proved. It is shown, that SCP refers to PaaS-class cloud platforms which provide runtime environment for software development, testing, and management. The important feature of SCP is the support of BYOL model, which provides the use of any programming language that expands software developers' capabilities. To increase the efficiency of software development and operation processes, it is recommended to use DevOps practice set and CALMS framework, which allows to assess the readiness of the transition to DevOps and it can be used as a model to identify possible improvements.

References

1. How to create an extension for the digital core using SAP Cloud Platform. https://habr.com/en/company/sap/blog/502572/. Accessed 14 Dec 2020
2. Cloud pyramid: IAAS, PAAS and SAAS. https://gigacloud.ua/ru/blog/navchannja/hmarna-piramida-iaas-paas-i-saas. Accessed 14 Dec 2020
3. IBM Cloud service models: IaaS, PaaS and SaaS. https://www.ibm.com/ru-ru/cloud/learn/iaas-paas-saas. Accessed 14 Dec 2020
4. Brig Retail. http://brigretail.com/. Accessed 14 Dec 2020
5. SAP presented new IoT services at the SAP Leonardo Live conference. https://clck.ru/SRbGn. Accessed 14 Dec 2020
6. Best Cloud Platform as a Service (PaaS) Software. https://www.g2.com/categories/cloud-platform-as-a-service-paas. Accessed 25 Dec 2020
7. SAP Cloud Platform. What is SAP Cloud Platform?. https://dataseed.ru/produkty/sap-cloud-platform-chto-takoe-oblachnaya-platforma-sap/#SAP_Cloud_Platform_SDK_iOS. Accessed 14 Dec 2020
8. SAP Cloud Platform Introduction. https://coderlessons.com/tutorials/sap/izuchite-sap-cloud-platform/oblachnaia-platforma-sap-kratkoe-rukovodstvo. Accessed 14 Dec 2020
9. SAP Cloud Platform - cloud CRM system. https://atlantconsult.ru/technology/sap-cloud-platform/. Accessed 14 Dec 2020
10. Launching blockchain applications on SAP Cloud Platform. https://habr.com/ru/company/sap/blog/347556/. Accessed 14 Dec 2020
11. DevOps: Breaking the Development-Operations barrier. https://www.atlassian.com/ru/devops/what-is-devops. Accessed 14 Dec 2020
12. CALMS (ex-CAMS) - the basis of DevOps. http://agilemindset.ru/calms/. Accessed 14 Dec 2020
13. Development of multi-tenant applications on the SAP Cloud Platform in the Neo environment, part 1. https://habr.com/ru/company/sap/blog/415351/. Accessed 14 Dec 2020
14. Guide to developing cloud applications using SAP Cloud Platform and Cloud Foundry. https://habr.com/ru/company/sap/blog/350690/. Accessed 14 Dec 2020
15. Lauer R: The modern way to develop apps on SAP. https://www.progress.com/blogs/the-modern-way-to-develop-apps-on-sap. Accessed 14 Dec 2020
16. SAP Cloud Platform Reviews & Product Details. https://www.g2.com/products/sap-cloud-platform/reviews. Accessed 14 Dec 2020
17. Moxon P (2014) Begginers guide to SAP ABAP, 300 p. Kindle Edition

18. Togurage G (2018) Programming with ABAP/4 - The world of SAP: Coding & Programming for beginner & professional, vol 1, 344 p. ShortWord Publication
19. Kobyliatskiy AG, Parhomenko AV, Galushko IA, Novosela PA, Rybakova OV (2009) Adaptation system SAP R/3 for solve business tasks. Radio Electron Comput Sci Control 1:124–128
20. How to prepare the arrival of ABAP on SAP Cloud Platform?. https://www.soapeople.com/blog/abap-on-sap-cloud-platform. Accessed 14 Dec 2020

Monitoring Air Decontamination by Photocatalysis

Petru Epure(✉), Andra Perju, and Petru P. Epure

EPI-SISTEM SRL, Str. Livezii nr 17A, Sacele, Romania
petru.epure@epi.ro

Abstract. This paper presents a monitoring system capable to display main parameters using remote technologies. The first stage development of the system integrates several sensors, a 3D-PRINT photocatalytic reactor, a gas sample chamber and a gas pumping system. The goal is to evaluate remotely the degree of CO2 contamination or any other VOC. Monitoring the air quality changing allow a better understanding of decontamination process and perform efficiency evaluation on new material testing. In this way we can follow the air quality change during long processing.

Keywords: Air · CO2 · Photocatalytic · Measurement & control · Embedded C · IoT

1 Introduction

The environmental quality standards establishes the quantitative limits to the priority pollutants. Also it was defined as the type of pollutants and the acceptable concentration level in the field of water policy. There is a list of priority substances and updated the environmental quality standards to be met by 2021 [1]. From all of these contaminants, from the perspective of environmental protection, the system proposed is focusing on emergency pollutant detection (pesticides, phenolic compounds, pharmaceutical, heavy metals). The categories of contaminants are monitored due to their adverse effects lead to various cancers in the population exposed to content exceeding the allowed limit and/or persistence in the ecosystem. One of the most challenging tasks confronted by the wastewater treatment technologies from agriculture and pharmaceutical industries is the incomplete removal of recalcitrant pollutants in a trace amount, which makes severe environmental pollution [2].

Advanced oxidation processes (particularly photocatalysis) to remove emergency pollutants has become an emerging area of research and development. The proposed system will become an efficient, cost-effective and environmentally friendly method for degradation of specific target pollutants. The wastewater and air treatment for reuse is an alternative source of clean environment and in most cases it is demonstrate that require lower investment costs and energy [3].

M. E. Auer et al. (eds.), *Online Engineering and Society 4.0*, Lecture Notes in Networks and Systems 298, https://doi.org/10.1007/978-3-030-82529-4_34

New degradation photocatalytic methods are actively contributing to reducing greenhouse gas emissions and is considered a major priority. The system proposed is based on a photocatalytic reactor built using new 3D carbonaceous materials. To obtain better performances the carbonaceous materials need to have an enhanced specific surface capable to provide more efficient photocatalytic activity [4]. Starting from low active carbon materials made from hard wood like acacia or oak the carbonaceous materials are enhanced with TiO2, phosphor, reduced graphene oxide and other materials. The resulting high-surface area activated carbonaceous material will be part of the new modular photocatalytic reactor. Based on this elements the system includes a measurement system capable to evaluate the efficiency of each material involved in reactor functionalization.

2 System Description

The 3D-Photocat measuring system is dedicated in this stage to air decontamination and it is built according to most important parameters to manage: gas flow, measurement chamber volume, CO2 concentration, VOC concentration [5]. It is important also to control the temperature, humidity, electrochemical response. It's important to monitor the evolution in time of all these parameters according to the materials used. The modules used for system functionalization are:

Photocatalytic reactor—it's a small volume chamber designed for air decontamination and capable to accommodate the active material. Gas inlet is linked to gas precision pump and for evacuation a peristaltic pump is connected to outlet. The prototype is build using 3D-DESIGN and 3D PRINT technologies (Fig. 1).

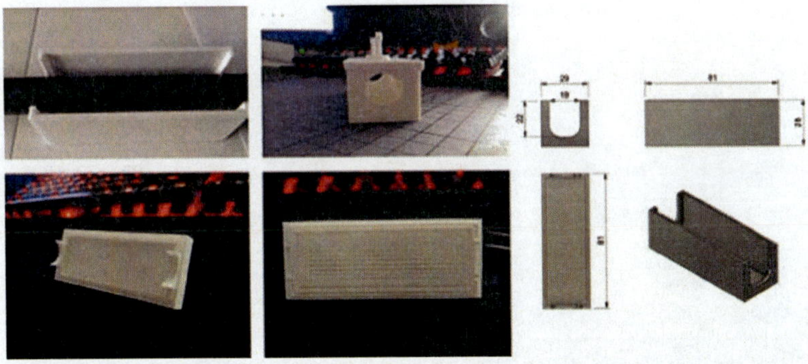

Fig. 1 Reactor prototype using 3D-PRINT

The 3D-PRINT technologies will be also used for reactor functionalization with carbonaceous materials. It will be o combined material printing or a DUAL material printing.

Fig. 2 High-surface area activated carbonaceous material

All materials presented in Fig. 2 were obtain in laboratory and manifest different efficiency in terms of decontamination capacity of the air.

The peristaltic pump will be used for evacuate the gas inside the reactor and the flow will be controlled according to internal pressure of the reactor. The gas pump it's a precision gas pump and it is used to create the desired catalytic pressure inside the reactor by injecting contaminated air. Gas pump it's P1500 model from Xavitech controlled by a specific controller like AFC (analog flow controller) or by a digital channel from Labjack U3 DAQ [7]. The air comes from a measurement chamber having a volume of about 5 L and capable to accommodate all the sensors that system needs: temperature, humidity, CO2, VOC (volatile organic compound) [6] (Fig. 3).

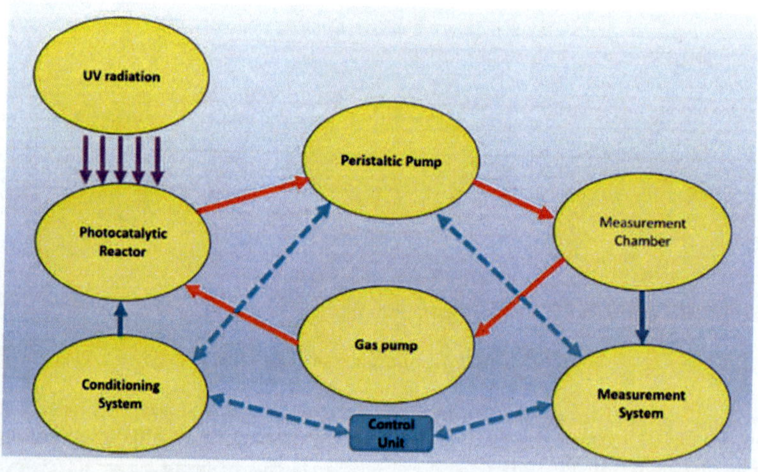

Fig. 3 Target architecture of the 3D-Photocat system

For temperature and humidity the system includes a digital sensor based on DHT 11 version. For CO2 sensor the system includes the Gravity IR sensor manufactured by DFRobot and for VOC gases the system incorporate MQ-2, MQ-8 and MQ135.

Measurement system is based on a Labjack U3 DAQ (data acquisition card) capable to acquire analog and digital data but also capable to control using a PS12DC a combination of peristaltic pumps or valves if needed. The system includes also a PSOC 6 (Programmable system on a chip generation 6) from Cypress Semiconductor especially for sensors measurements.

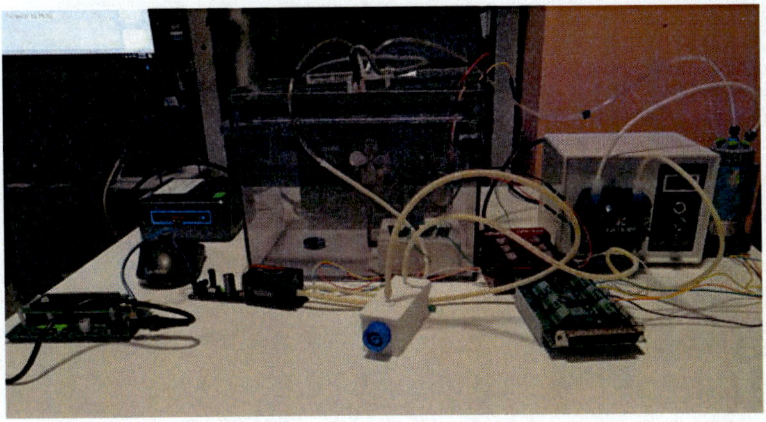

Fig. 4 Measurement chamber and setup

All the 3D-Photocat system is controlled via a small factor PC using Windows 10 and a software developed in LabWindows/CVI—the C version of National Instruments. In Fig. 4 the pumps are connected to reactor and to measurement chamber and the switching valve can adjust the content inside the chamber. The flow speed can be controlled for each pump and sensors are connected directly to Labjack or via conditioning boards. Additional wi-fi sensors are mounted inside the measurement chamber. A BLE temperature and humidity sensor, a 2.4 GHz WI-FI sensor and a PM2 sensor are located inside for additional info for the measurement system.

3 Software Description

At the beginning the development was started with a GUI (graphic user interface) to allow end-user to monitor data evolution and based on existing configuration the user can start pumps and valves according the needs.

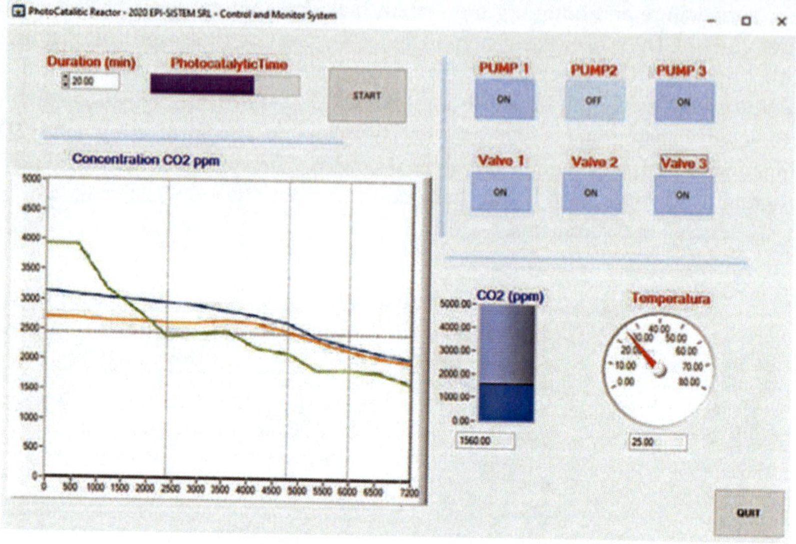

Fig. 5 GUI (Graphic user interface) version 2020

The interface presented in Fig. 5 it's available in the control unit screen and it can be used also remotely. For situation when the data need to be monitored on web we implemented a version of NI Systemlink graphic page (Fig. 6).

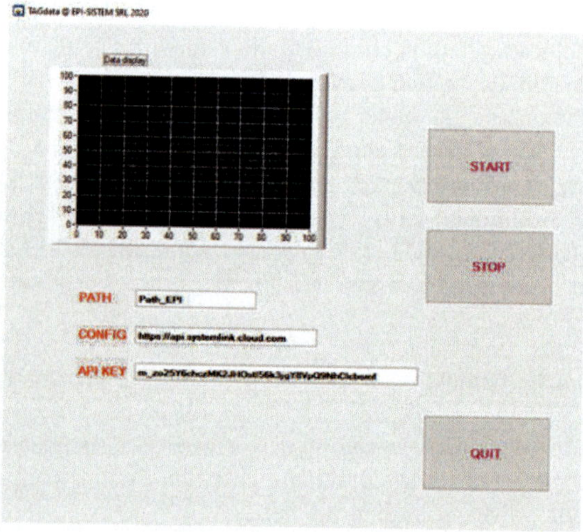

Fig. 6 GUI for SystemLink Client

Every time a communication session with NI SystemLink is started there are some security items to settle in order to obtain access to remote webpage. There are some flows of data called PATH and for every path you can choose a security key and an associated policy. All these items are mandatory for data communication remotely and are part of National Instruments technologies.

Data collected from sensors are stored and some of them are displayed graphically. When the decontamination process is a slow one and the duration is too long then the remote monitoring is highly appreciated. In that case the duration can be up to 2900 min meaning 2 days. Because the process cannot be supervised all the time by a human operator then the system can be remotely monitored via NI SystemLink and controlled using techniques similar with a virtual node client.

4 Long Period Evaluation

Because of decontamination process which can take over 2 days it is mandatory to have a very good and reliable software developed under professional platforms. The software provided for 3D-Photocat is written under LabWindows/CVI and includes well known National Instruments versatility and reliability properties. In this respect and based on previous experience regarding long time testing systems the actual solution performs exactly the desired measurement scenario and can implement different protocols according to end-user needs. Sensor measurements are made at different sampling rates and every 10 min the average values are displayed or plotted graphically (Fig. 7).

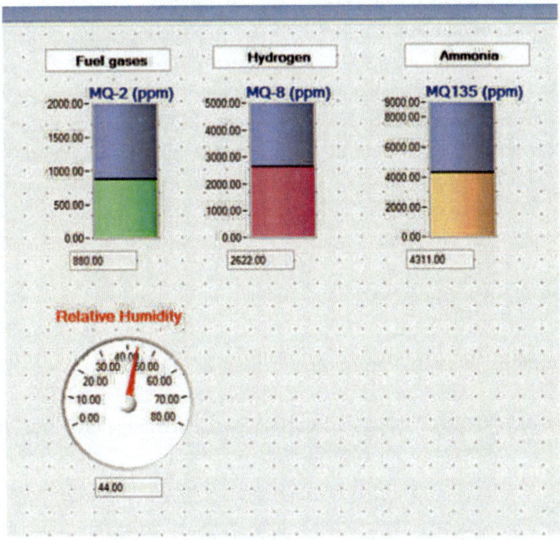

Fig. 7 Sensor representation screen

Long period evaluation are more suitable for VNC (Virtual node client) as the SystemLink can change the security policy after 24 h. Updating the API key every 12 h can be a solution for distance evaluation on decontamination process.

5 Conclusion

Measurements made over large periods presents several key issues: stability and flexibility. The controlled system should be able to change some functional parameters without stopping the measurement process and storing meanwhile all the data acquired and averaged. Using several sensors like temperature, humidity, MQ-2, MQ-8, MQ135 and also additional microcontrollers like PSOC 6 or data acquisition cards like Labjack U3 confers a certain degree of complexity and flexibility [8]. The software application helps the end-user to evaluate the materials efficiency and also to apply fine tuning if the process evolution requires. User can choose a specific behavior of the gas flow and also can find out if there are critical situations to confront. In the next phase the 3D-Photocat system will be able to be shutdown remotely in order to avoid emergency alarms.

Combining analogic modules and digital modules the software is able to integrate additional modules like syringe pumps, power relays, air conditioning, heating or temperature regulation, programmable power supplies, 6 way valves, spectrometric detectors, electrochemical detectors or colorimetric detectors. Consolidated data can be analyzed after the decontamination process and data can be exported in a specific format [9]. The 3D-Photocat is a lab version measurement system but it can evolve to a portable version according to de user needs. In the near future the system will have also another version for liquid photocatalytic decontamination, mainly for water, having the goal to eliminate pesticides, pharmaceutical compounds or heavy metals [10].

Acknowledgements. This work was supported by a grant of the Romanian National Authority for Scientific Research and Innovation, **CCCDI—UEFISCDI**, project name ERANET–M-3D-PHOTOCAT-2 within **PNCDI III**, contract no.170/2020. The project team members will continue to collaborate on providing each other valuable support in order to develop and optimize the monitoring system.

References

1. FAO webpage. http://www.fao.org/faolex/results/details/en/c/LEX-FAOC177462/. Accessed 28 Nov 2020
2. EU webpage. https://eur-lex.europa.eu/LexUriServ/LexUriServ.do?uri=OJ:L:2001:331:0001:0005:EN:PDF. Accessed 28 Nov 2020
3. EEU webpage. https://www.eea.europa.eu/publications/the-eu-emissions-trading-system-1. Accessed 28 Nov 2020
4. Isac L, Andronic L, Visa M, Enesca A (2019) Selective photocatalytic degradation of organic pollutants by CuxS/ZnO/TiO2 heterostructures. Ceram Int. https://doi.org/10.1016/j.ceramint.2019.10.147

5. Winsen webpage. https://www.winsen-sensor.com/sensors/combustible-sensor/mq2.html?gclid=EAIaIQobChMIgo-X4KbO7QIVokiRBR0iEgS1EAAYASAAEgLU8fD_BwE. Accessed 28 Nov 2020
6. Winsen webpage. https://www.winsen-sensor.com/d/files/PDF/Semiconductor%20Gas%20Sensor/MQ-8%20(Ver1.4)%20-%20Manual.pdf. Accessed 28 Nov 2020
7. Labjack webpage. https://labjack.com/support/datasheets/u3. Accessed 28 Nov 2020
8. Cypress Semiconductor webpage. https://www.cypress.com/documentation/development-kitsboards/psoc-6-blc-prototyping-kit-cy8cproto-063-blc. Accessed 28 Nov 2020
9. LabWindows/CVI webpage. https://www.ni.com/pdf/manuals/373552k.pdf. Accessed 28 Nov 2020
10. M-era.net webpage. https://m-era.net/joint-calls/joint-call-2019/list-of-funded-projects_call-2019-1.pdf. Accessed 28 Nov 2020

Automation of IoT Based Services Using Digital Twin

Daniel Anghel and Titus Constantin Balan(✉)

"Transilvania" University of Brasov, Str. Politehnicii, nr.1, Brasov, Romania
titus.balan@unitbv.ro

Abstract. Digital Twin is the digital representation of any physical asset that can be addressed, belonging to real world. This new concept provides a new abstraction layer which serves as a common operating system, offering an enhanced experience named Device-as-a-Service, being based on calling and manipulating the assets by their function, with increased usage scenarios mainly in industry, but also in home scenarios. The purpose of this paper is to demonstrate the main benefits of digital twin implementations: using a common and vendor agnostic IoT operating system, the independence from the IoT protocol to be used and from the transmission media, user profile persistency. The described implementation is based on the open source project called Ditto.

Keywords: Digital Twin · IoT · Common industrial operating system · Device as a Service · Ditto

1 Introduction

Digital Twin (DT), as the name suggests, is the digital representation of any addressable physical asset or process, being based on real-time communication. Sensors and actuators are used in order to collect data that is going to be processed and used to build a virtual avatar of the underlying device, while reflecting the real-time changes that may occur. Thus, it provides an accurate live one-on-one monitoring digital avatar that mirrors any behavior change or predicts further states based on collected information so far.

In addition to this, in order to deliver not only a less complex way to manage our physical assets, but also a more customized experience to the customer, we have to refer to the digital avatar, that is, a digital profile that reflects the customer's preferences. By relying on this, all the predefined parameters complying with user preferences persist even when accessing a new environment, maintaining the stability of the configured user-specific profile, under which the underlying commands may occur.

Among virtualization technologies and exponential growth of IoT usage within public and industrial fields, Digital Twin becomes one of the top ten most innovating trends that may improve and optimize our daily living, as being stated by consulting technology experts from Gartner Inc. [1].

This paper is presenting a Digital Twin implementation based on the open source project called Ditto [2] having the aim of demonstrating the main advantages introduced by the Digital Twin concept, presented in paragraph 2. There are several

M. E. Auer et al. (eds.), *Online Engineering and Society 4.0*, Lecture Notes in Networks and Systems 298, https://doi.org/10.1007/978-3-030-82529-4_35

practical scenarios implemented, using different IoT elements, protocols, communication technologies, but also describing from user perspective the interaction with objects addressable via digital twin methods. These illustrative scenarios from home environment can be further used and enhanced with additional functionality for real life implementations.

2 Digital Twin Concept

During the last few decades, the user-technology relationship became not only more and more ubiquitous and embedded within our lives, but also more interactive and responsive. Leaning towards the integration of more connected devices that interact with their customers, did nothing but amplified the technology experts' thrive to bring it to a more personalized experience through an evolutionary approach. By defining customized preferences that form a unique persistent digital avatar, customers do not need to adapt to foreign environments but command these newly discovered assets as they were used to [3].

Interacting with the virtual representations of real-life smart devices on a per function basis, highlights the trend's versatility and flexibility that were long awaited. The need of having a universal model that orchestrates all the connected assets in an effortlessly manner has become more and more by interest as the growth and complexity of information systems is more visible. Ease of access and use is one of the foundational aspects a system has to preserve along its lifetime, labeling it whether is a reliable solution for the issues it addresses.

One major drawback being addressed by digital twin trend is the interoperability among vendor-specific hardware and software solutions, that is, each company providing their most efficient and convenient approach which, in most of the times, does not necessarily go along with universal interchangeable solutions. This aspect brings up issues that may arise in heterogeneous environments where different proprietary technologies have to integrate their capabilities in order to solve more complex problems.

Large scale applications and business intelligence models require long term viable solutions in order to decrease the complexity of systems within industrial IoT ecosystems. Aggregating bulk of data without performing analysis and making predictions while accounting the considerable proportions upon industrial field is built, it might impact the performance of the system and raise concerns regarding the efficiency. Digital twin has the potential to address these undesirable outcomes by: modeling the real life assets' digital profiles in order to simulate properties and behaviors under certain circumstances, synchronizing the corresponding real world parameters to the virtual representation in order to define a more accurate simulation, mapping the physical objects' measured or simulated properties with aiming of doing a three dimensional representation.

Concluding an in-depth analysis of the available technologies [3, 4] that implement the concept of digital twin, the DT solutions are diverse and dedicated to different situations. There are several industrial [5] proprietary initiatives, one of the most influential being Siemens MindSphere proprietary technology [6], offering support for digitizing the physical assets as described above. For Digital Twin implementations in home office environment we can mention OpenMiHome [7], a plugin provided by Xiaomi in order to deliver all the necessary requirements to build and maintain a favorable digital twin medium. Unfortunately, it does not comply with one of the fundamental characteristics of this new trend, that being the lack of interoperability between multiple vendors. One of the most mature open source options is Ditto, the project used for the implementations presented in this paper.

Since digital twin concept become more popular and widely used in IoT, industrial fields reconsidered their position and started leaning towards integration of digital replicas in different domains. As described in [8], operations and maintenance of vertical transportation systems could be improved by creating virtual twins that could provide information otherwise not available helping to test and estimate the installation parameters, using the measurements of a scaled bench test. Digital twin's blueprint becomes more and more prominent [11, 12] within the development of smart manufacturing, so mentioned in [9]. In [10] it is described a research based on digitalizing a building lifecycle by using more than 25,000 sensors which were used to read, collect and process the acquired data in order to create and test a digital twin of an office building facade. [1] and [2] extend the DT perspectives within healthcare domain.

We have identified a series of elements that we consider defining for the DT concept:

- common vendor-agnostic operating system
- independence of IoT protocol and communication media
- transparent user perspective, seamless integration of underlaying solutions, and profile preservation (configuration, preferences, environment)

3 The Ditto Open Source Project

Ditto is an open source project provided by Eclipse that does not only deliver a containerized system for digital twin deployment, but also ensures the validity of all the characteristic advantages of this new trend. The scope of this project [2] is to simplify the process of representing and digitizing virtual profiles of the real-world assets and providing an interoperable platform in order to motivate experts to further contribute to developing adjacent technologies and solutions with regards to digital twin.

Thus, Ditto system [13] provides all the necessary capabilities digital twin requires: providing a new abstraction layer that ensures the transparency of enrolled devices, proven interoperability within an heterogeneous environment consisting of different vendor-specific items, modern architecture based on microservices, connection flexibility and versatility to the system. As visible in Fig. 1, the system defines thinks, policies, methods for things-search and an increased variety of IoT protocols, including IoT protocols (MQTT, AMQP, Kafka). Also gateways to other systems are defines via web sockets or HTTPS.

Ditto defines a set of HTTP APIs in order to implement different operations and is offering the possibility to develop user interfaces (mainly graphical) that could discover and program things, based on their function.

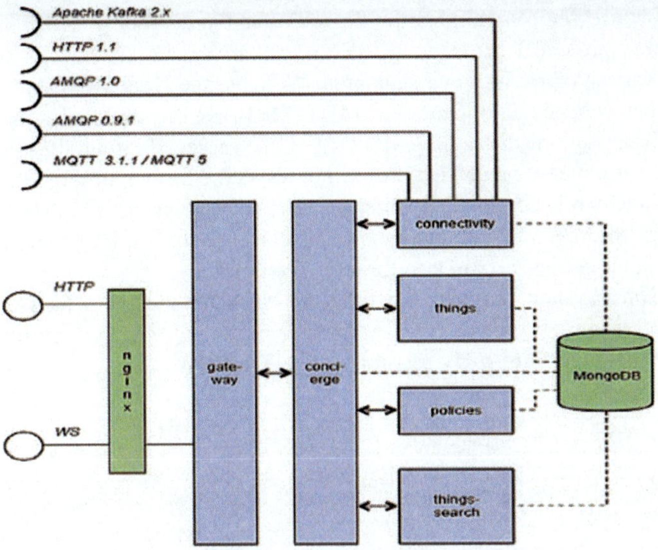

Fig. 1 Architecture of Ditto system [2]

4 Implemented Scenario

The scope of the experimental setup, presented in Fig. 2 is to emphasize the advantages of digital twin concept in practical implementations; elements were organized in the figure as "real world" elements, controlled via their "digital world" representation in Ditto.

The scenario was based on programming and integrating a unified web GUI, meant to provide end users a common interface where they can manipulate system-enrolled digital entities, which in turn, due to Ditto system's capacity to route requests and queries to the corresponding physical assets, are reflected into the real world. Thus, customers are communicating with their devices via digital twins that are registered into Ditto, providing a new abstraction layer that goes along with system's capability to operate within heterogeneous environments and highlighting the Interchangeable-Virtual-Instrument (IVI) alike approach [14], successful concept from instrumentation that defines generic drivers, with subsequent custom mappings from each vendor to the generic driver.

The end user accesses a foreign environment, be that any Ditto-enabled medium that has at least one enrolled asset and may request all the discoverable things to introduces themselves and their capabilities. The user might also opt to customize its searching based on specific filters, for instance listing all the addressable devices that are available on the same floor as the user resides. From now on, the customer might freely choose any of the discovered assets, which in turn the selected twin reveals its capabilities. Thus, the communication with the physical device is transparent, leaving the vendor-specific concerns behind, and providing a centralized model to control the IoT devices.

Our in-house setup highlights the flexibility and versatility of the communicating between peer protocols provided by Ditto system, that is establishing connections through a variety of well-known standards. We have the NodeMCU Arduino family-based device connected to Ditto via MQTT IoT-specific protocol, assisted by the specialized message-mediator HiveMQ [15] online server. By doing this, the customer might send certain commands, with respect to a specific format, through the configured live channel which is continuously interrogated by Ditto system. On the other hand, in order to test the reliability of supported protocols, we have a Raspberry Pi 4 single-board-computer connected to Ditto through WebSocket.

Thus, Ditto system provides not only the characteristics of a likely digital twin solution but encourages the use of multiple protocols and ways to enroll the physical assets, even through third-party servers such as HiveMQ.

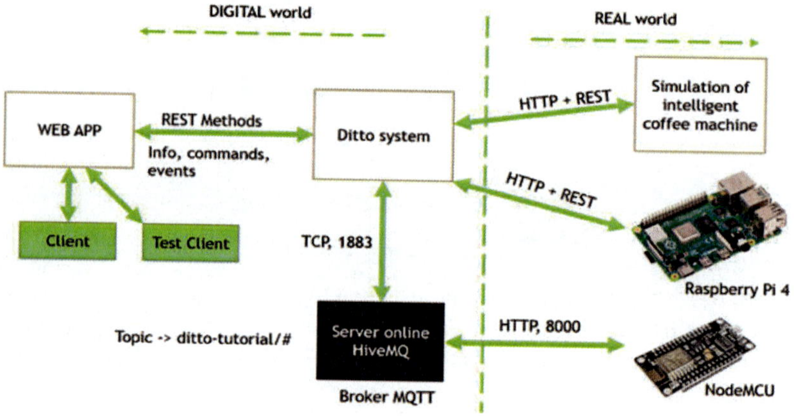

Fig. 2 Ditto experimental implementation using different hardware elements, different IoT protocols (MQTT, HTTP) and clients with profile persistency for the implemented GUI management interface

5 Experimental Outcomes

First of all, testing the functionality and availability of Ditto system seemed to be the foundational step of a structured approach. The explained examples on the official repository [2] were meant to assure the validity of concept while providing a starting point for any user that decides to customize and digitize its environment. Among some scenarios that have been tested, one that includes a virtual instance of a smart coffee machine (based on the Ditto offered example) demonstrates the profile persistence aspect.

Regarding the interoperability within heterogeneous environments, we decided to use the two physical assets mentioned above in order to demonstrate the ease of use on customer's behalf, where he only has to discover the virtual entity and simply interact with its capabilities in a seamlessly manner.

Naturally, since the didactic examples smoothly work, we decided to implement a web universal GUI using JavaScript to integrate multiple use cases. By using this interface, customers could not only discover Ditto-enrolled things meant to be available for each user, based on a basic authentication process, but also select any of the digital twins in order to see what they are capable of and interact with them through the displayed buttons that suggest intuitive actions.

The interoperability aspect is addressed in a scenario where the customer could command each of the 2 physical assets, both performing the same actions but underlining the different vendor provenance. By commanding the NodeMCU to turn on its built-in LED, the specific payloads were routed through the use of a third-party mediator HiveMQ, whereas, on the other end, Ditto system continuously polls the queue with the same topic name where messages might arise. On the other hand, the Raspberry Pi is directly connected to Ditto through WS protocol, highlighting the flexibility Eclipse's solution has to offer. Thus, the common general operating system Ditto relies on, provides a vendor-agnostic feature while offering transparency and making the proprietary specifications irrelevant on the customer's behalf (see Fig. 3 and Fig. 4).

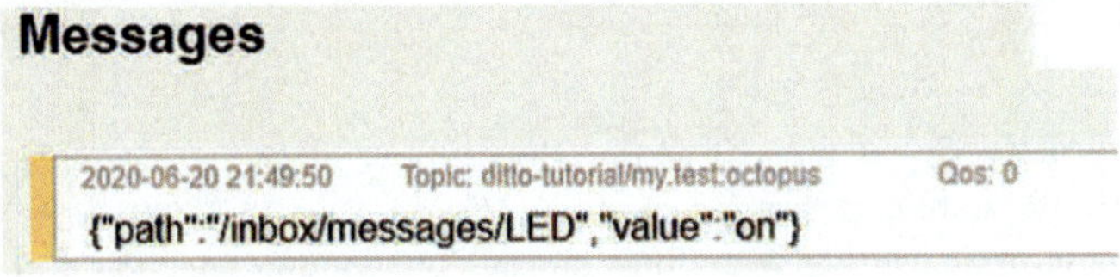

Fig. 3 The inbound message sent from web GUI and received on HiveMQ's graphical interface

Taking advantage of the already implemented devops-based features within Ditto was one of the most efficient approaches in order to manage the system while it is up and running. We have used a POST method and sent it to Ditto's IP address followed by */devops/piggyback/connectivity? = 10*, hence the latter URL part announces the system that the correspondent content provided by this operation has to be interpreted as an administrative feature implementation.

The POST method is presented below (in JSON format) and it is triggering to the connectivity API:

```
{"targetActorSelection": "/system/sharding/connection",
"headers":  {"aggregate": false},
"piggybackCommand": {
"type": "connectivity.commands:createConnection",
"connection": {"id": "mqtt-example-connection-123",
"connectionType": "mqtt",
"connectionStatus": "open",
"failoverEnabled": true,
"uri": "tcp://broker.hivemq.com:1883",
"sources": [{ "addresses": ["ditto-tutorial/#"],
"authorizationContext": ["nginx:ditto"],
"qos": 0,
"filters": []}],
"targets": [{"address": "ditto-tutorial/{{ thing:id }}",
"topics": ["_/_/things/twin/events",
"_/_/things/live/messages"],
"authorizationContext": ["nginx:ditto"],

– "qos": 0  }] } } }
```


Thus, by delivering a request as described above, Ditto will have established a connection to the public messaging platform HiveMQ, an MQTT broker, and will be listening to any topic starting with ditto-tutorial/ and routing back the proper response on the same topic while embedding the physical asset's proper ID, thereby all the commands seamlessly reach to their intended destination.

In order to command the two physical elements (Raspberry Pi4 and NodeMCU) we needed to have some mappings between the generic Ditto routines and the specific commands that should be implemented on each device performing the same function. This proves that the success of DT technology and the implementation to be used is dependent on the number of vendors that are willing to adhere a common DT program and develop plug-ins towards the common generic interface.

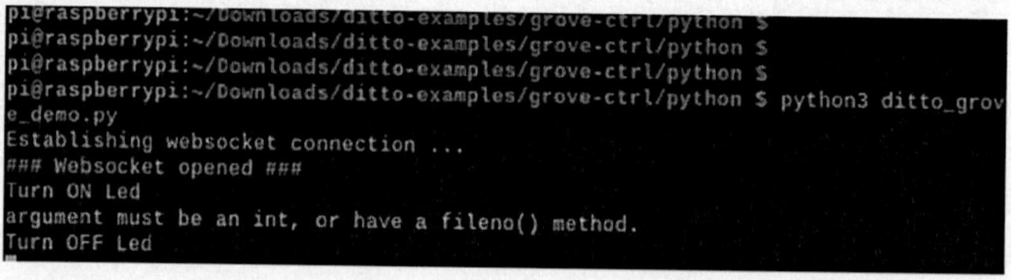

Fig. 4 The inbound message sent from web GUI and received on Raspberry Pi's behalf

Aiming to prove the concept of a persisting profile based on custom preferences when accessing a new environment, we defined two users with different coffee choices.

These options will be encapsulated inside the payload, sent to Ditto system which will forward the messages to the proper asset, in our case the virtualized smart coffee machine, interpret the values and perform certain actions within the received parameters on device's behalf. See Fig. 3. in order to spot the differences between payloads and implicitly the preferences, belonging to different users, that are maintained even when accessing a foreign medium (Fig. 5).

```
▸{cups: 1, strength: 0.4, amount: 230, captcha: "ditto"}        index.js:349
▸{cups: 1, strength: 2.3, amount: 268, captcha: "ditto"}        index.js:277
```

Fig. 5 Customized preferences sent by each user when interacting with the same smart coffee machine. Preference is retain also when connecting to a different element with the same function

In this case, by preference we are referring to the type of coffee the user prefers and, in case the users will enter a foreign environment where a device with the same function is discovered (a coffee machine, as per our example), the preference/settings/profile for that specific user are transmitted to the newly discovered device with the same function (in our case the function is coffee machine and a preference for coffee can be "double expresso with no sugar").

This type of approach can be continued into a "universal remote control", very useful for people that are and not technology proficient and would not like to learn how to use a new device with its new set of features and preferences (e.g. the expresso machine from the hotel). The same generic unified interface can be used, no matter of vendor, provided that the devices are enrolled to the digital twin common operating system.

During our implementation we have encountered a couple of issues related to the web server (Nginx) services being unreachable due to gateway errors. Since Ditto project is under continuous development and has not yet met maturity as a system. However, its versatility and connection flexibility provided through HTTP, WebSocket, multiple IoT protocols, such as MQTT, AMQP, Kafka, alongside its modern microservices-based architecture, recommends this open source alternative as a solution to satisfy DT trend's requirements.

6 Conclusion

Digital Twin is gaining popularity and is credited among technologies of the future (besides AI, deep learning and fog/edge computing). Due to its capabilities and by bridging the gap between the real and virtual worlds, it represents a cost-efficient solution, providing a flexible and reliable alternative to the new era of industrialization.

The paper presented methods for development of relevant DT scenarios, using the Ditto open-source implementation, based on a unified web graphical user interface in order to facilitate ease of use on end-user's behalf while relying on characteristics of digital twin concept.

DT is based on the innovating idea of providing a new abstraction layer which serves as a common operating system meant to establish a higher-level communication between the user and the real-world asset. Thus, rendering likely the possibility of interoperability within heterogeneous environment due to this new layer, the physical smart things are commanded on a function-based manner while being vendor-agnostic.

In the above described scenario, the user might enter a new medium and discover all the present smart devices that are enrolled within Ditto local system. Furthermore, the user may opt to filter the results based on certain criteria such as location, floor, capabilities and so on. Among other discovered assets, two of interest are the NodeMCU and Raspberry Pi that both perform the same function. The primary proven aspect is the vendor-independent feature which renders customers to transparently interact with any physical asset, despite the diversity of their provenance, while operating in a seamlessly manner. The second major feature treated along this article is the persistence of users' customized preferences that are maintained even when accessing a new environment.

Despite the provenance of the physical entity connected to Ditto, it may be commanded on a function-based policy while having its resources and abstracted communication protocol. Thereby, having the idea of a universal interface, the scenarios themself can be extended to broader implementations where any type of user, be it technical or lacking technical knowledge, may take advantage of this concept's ease of use and access.

References

1. Top 10 trends and strategy technologies for 2019. https://www.gartner.com/en/newsroom/press-releases/2018-10-15-gartner-identifies-the-top-10-strategic-technology-trends-for-2019. Accessed 25 Oct 2020
2. Ditto official repository. https://github.com/eclipse/ditto. Accessed 25 Oct 2020
3. Barricelli BR, Casiraghi E, Fogli D (2019) A survey on digital twin: definitions, characteristics, applications, and design implications. IEEE Access 7:167653–167671. https://doi.org/10.1109/ACCESS.2019.2953499
4. Fuller A, Fan Z, Day C, Barlow C (2020) Digital twin: enabling technologies, challenges and open research. IEEE Access 8:108952–108971. https://doi.org/10.1109/ACCESS.2020.2998358
5. Assawaarayakul C, Srisawat W, Ayuthaya SDN, Wattanasirichaigoon S (2019) Integrate digital twin to exist production system for industry 4.0. 2019 4th technology innovation management and engineering science international conference (TIMES-iCON), Bangkok, Thailand, pp 1–5. https://doi.org/10.1109/TIMES-iCON47539.2019.9024430
6. https://siemens.mindsphere.io/en. Accessed 25 Oct 2020
7. OpenMiHome project. https://github.com/openmihome
8. González M, Salgado O, Croes J, Pluymers B, Desmet W (2020) A digital twin for operational evaluation of vertical transportation systems. IEEE Access 8:114389–114400. https://doi.org/10.1109/ACCESS.2020.3001686
9. Lee J, Azamfar M, Singh J, Siahpour S (2020) Integration of digital twin and deep learning in cyber-physical systems: towards smart manufacturing. IET Collab Intell Manuf 2(1):34–36. https://doi.org/10.1049/iet-cim.2020.0009

10. Khajavi SH, Motlagh NH, Jaribion A, Werner LC, Holmström J (2019) Digital twin: vision, benefits, boundaries, and creation for buildings. IEEE Access 7:147406–147419. https://doi.org/10.1109/ACCESS.2019.2946515
11. Barricelli BR, Casiraghi E, Gliozzo J, Petrini A, Valtolina S (2020) Human digital twin for fitness management. IEEE Access 8:26637–26664. https://doi.org/10.1109/ACCESS.2020.2971576
12. Laamarti F, Badawi HF, Ding Y, Arafsha F, Hafidh B, Saddik AE (2020) An ISO/IEEE 11073 standardized digital twin framework for health and well-being in smart cities. IEEE Access 8:105950–105961. https://doi.org/10.1109/ACCESS.2020.2999871
13. Bosch's blog about Digital Twin using Ditto. https://blog.bosch-si.com/developer/eclipse-ditto-1-0-a-framework-for-digital-twin-builders/. Accessed 25 Oct 2020
14. IVI Foundation - Interchangeable Virtual Instruments. https://www.ivifoundation.org/. Accessed 25 Oct 2020
15. MQTT Broker – HiveMQ. https://www.hivemq.com/. Accessed 25 Oct 2020

Purple Team Security Assessment of Firmware Vulnerabilities

Lucian Florin Ilca and Titus Balan(✉)

Transilvania" University of Brasov, Str. Politehnicii, nr.1, Brasov, Romania
titus.balan@unitbv.ro

Abstract. The purple team concept in offensive and defensive security is related to a set of methods, processes or activities that will combine red and blue team to cooperate in order to create a solid security operations team for improving due-care and due-diligence policies in a company or institution. The accelerated evolution of communication technologies that reached also industrial areas, as well as the complexity rising of new security threats have led to the need of security assessment at component and system level. In this paper we present the purple team methodology illustrated by a firmware security analysis of a communication component form the red team perspective, presenting also an analysis report of the threat related information that we were able to collect. We conclude by proposing recommendations for mitigating attacks from the blue team perspective and presenting the integrative approach of the purple team.

Keywords: Purple team · Reverse engineering · Access point firmware · Information security · Firmware analysis · Backdoor · Exploit

1 Introduction

With a view to achieving best security practices, network equipment's rely heavily on data security, the process of transforming best practices into a secure environment, using due care and due diligence processes. Failure to secure data or detecting malicious activities in an equipment (network, IoT), may allow a bad intended person who is taking malicious activities behind the network's infrastructure, to create and use different attacks against legit users connected to the system.

Embedded firmware is present in every product that we use day by day, such as: IoT, smartphones, gaming consoles, industrial robots, network equipment, printers and so on. If the data or the configuration interface is exposed to the Internet, they become "smart" and remote controllable but also susceptible to security attacks. The importance of rapid problems detecting becomes even more crucial for the propagation of sensitive data, so before going into the production environment all individual components and the complete solution should go through an auditing process.

One newer security concept used for vulnerability assessment of a system is the purple team concept. The purple team concept resides at the intersection of red and blue team concepts [1], terms inherited from military lexicon, representing attackers and defenders as part of simulated exercises, terms adopted also in the cyber security

M. E. Auer et al. (eds.), *Online Engineering and Society 4.0*, Lecture Notes in Networks and Systems 298, https://doi.org/10.1007/978-3-030-82529-4_36

industry. The red team is the offensive side, the attackers, there are not focus only on penetration testing, they could have knowledge in exploit development, reverse engineering, get initial access, get information about any infrastructure with any personal or public methodology (NIST, OWASP) used, and, there is the defensive side, the blue team. The blue team is related to security operations, security analysts, verifying and watching the network, endpoints, protecting those systems and it could include in the blue team, engineers that are doing forensics, malware analysis, breaking down a binary or something malicious to really understand what, where and why a software is trying to achieve in a network, what data could be exfiltrated, checking for indicators of compromises, verify suspicious traffic.

Red and blue tram examples are available in illustrative papers for different industry domains: malicious circuit identification [2], hardware elements assessment [3], processor design [4], but also at network level [5–7]. However, the purple team concept is newer and brings some advantages.

The purple team is a combination of a red team member and a blue team member, working together for security effectiveness. The red team members from the purple team cross-check with the blue team possible security gaps and start investigating vulnerabilities. Also they perform verification of the real time detection methods implemented by the blue team. The blue team members start verifying logs, intrusion detection or prevention system, checking for malicious traffic coming from the red team. A blue team should think as a red team and vice-versa, because, creatively and innovation is very important in information security. The purple team in an organization has the advantage of continuous feedback between vulnerability report and remediation process. The purple team fills-in the gap between blue and red team by reducing the duration of analysis and hardening of the system.

In this paper we will illustrate the purple team methodology for analyzing the firmware of a network component (a dual band ceiling access point firmware was chosen for demo purposes). An access point is a device that provides interconnectivity using Wireless LAN (WLAN) usually in an office or large building distribution network and further connects to a router or switch. This device was chosen for concept demonstration, but the same methodology, detailed in the following chapter, can be applied for assessment of other systems/elements in commercial or industrial area. The paper will further focus on the empirical description of firmware analysis techniques using dedicated tools, emphasizing the vulnerability areas.

2 Purple Team Roles and Methodology

Figure 1 presents the methodology describing the red team and blue team co-working in order to improve security metrics for an embedded device (firmware). For improving security design for an embedded device, Red-Team members will start firmware assessment following the next steps:

1. Get information about the embedded device and firmware version
2. Extract firmware from the device or download from the vendor website

3. Using reverse engineering tools such as: IDA Pro [8], Ghidra [9] for inspecting binaries, such as: httpd, busybox and other firmware libraries
4. Detecting errors and getting recommendations on code formatting from a source code static analyser; checking if the source code corresponds to the code formatting standard and security standards. The source code analyser tool will concentrate on detecting potential bugs and performance issues.
5. Security code review for existing files is required to map and find more details about the embedded device. Finding an issue requires additional steps to track down the code that requires remediation. Most of these cases also require documentation for error replication and understanding. These fixes are often rolled into a new or subsequent firmware release.

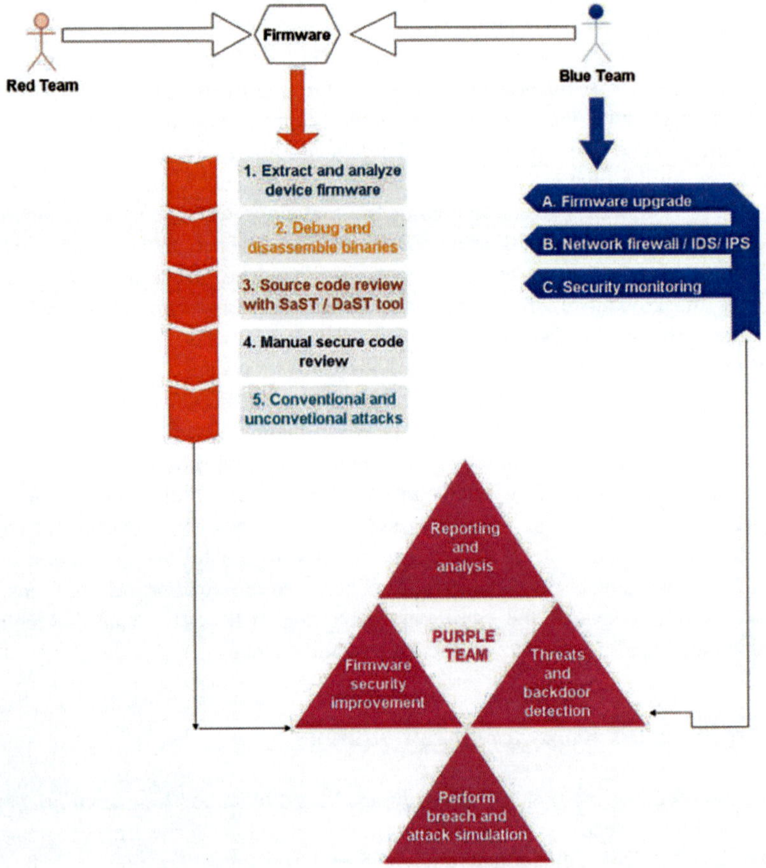

Fig. 1 Purple team methodology for firmware analysis

There are different types of Conventional and Unconventional vulnerabilities, such as:

- Information disclosure - system management mode rootkits, UEFI Firmware Rootkits
- Denial of service - interrupt communication from access point to the controller
- Maliciously input - lead to buffer overflows to perform code injection
- Data tampering - modifying controller variables (Secure Boot, Configuration, etc.)
- Web management controller - Code injection, cross site scripting or other web vulnerabilities could have a huge risk for the access point scenario
- Backdoor - Many firmware contain backdoors which could send data to external command and control servers

The steps followed by the Blue team (part of the purple team) are the following:

1. Get information about version and new releases and perform firmware upgrade when a new software was released
2. Perform network analysis (from the access point in the distribution network to the Internet connection) and check malicious traffic by inspecting open ports and firewall logs
3. Perform security monitoring from the controller side and use an alert system to announce if there exists any potential data exfiltration or new malicious connection between attacker and equipment.

The Purple team should have an overview of the system or equipment which they tested and gather all details for future analysis. In this case, the recommended steps made by the purple team are the following:

1. Improve firmware security - Upgrade firmware with new released software from the vendor website or modify the current firmware if it is possible with extra security controls.
2. Report new threats to the vendor website and request a CVE ID (Common Vulnerabilities and Exposures identifier) for the new discovered vulnerability from Mitre [10].
3. Put in practice cyber-simulation exercises which are recommended as preventive measure for managing cyber security incidents. Simulation exercises provide an opportunity for purple team members to take part in cyber security offensive and defensive exercises, designed to test the security operations team capability to respond to incidents and increasing the level of awareness.
4. The Purple team members are recommended to be apart from the Red and Blue team and should report in the daily or weekly meeting the results and reports related to any new vulnerability or information of interest.

3 Firmware Assessment

3.1 Firmware Analysis Tools

The methodology followed for this project is based on the usage of firmware analysis tool (BinWalk) and binary reverse engineering tool (Ghidra), using an access point firmware for demo purposes and perform reverse engineering analysis on firmware httpd, busy box binaries to gain knowledge about the system.

The following tools will be used in order to perform assessment analysis for a firmware:

A. Ghidra [9] is a software reverse engineering (SRE) framework. This framework includes a suite of full-featured, high-end software analysis tools that enable users to analyse compiled code on a variety of platforms including Windows, macOS, and Linux. Capabilities include disassembly, assembly, decomplication, graphing, and scripting. Ghidra supports a wide variety of processor instruction sets and executable formats and can be run in both user-interactive and automated modes. Ghidra was built to solve scaling and teaming problems on complex SRE efforts, and to provide a customizable and extensible SRE research platform.
B. Binwalk [11] - is a simple Linux tool for analyzing binary files for embedded files and executable code. It is mostly used to extract the content of firmware images. Binwalk is a tool for searching a given binary image for embedded files and executable code. Specifically, it is designed for identifying files and code embedded inside of firmware images. Binwalk uses the libmagic [12] library, so it is compatible with magic signatures created for the Unix file utility. Binwalk also includes a custom magic signature file which contains methods for improved signatures for files that are commonly found in firmware images such as compressed/archived files, firmware headers, Linux kernels, bootloaders, filesystems, etc.

3.2 Firmware Binary Analysis

The main option of binwalk is firmware signature scanning. Binwalk tool will scan a firmware image to search for different embedded file types and file systems. Binwalk use the libmagic library to identify file signatures. Libmagic is the library that commonly supports the file command on Unix. The library handles the loading of database files that describe the magic numbers used to identify various file types, as well as the associated MIME types (Multipurpose Internet Mail Extensions). The library also handles character set detections. With binwalk we can decompress filesystem images from firmware and search for specific password files (passwd, shadow, etc.), and further find password hashes or breaking them. It is possible to analyze binaries to find possible security issues and also binwalk could be used to perform data entropy analysis to search for compressed data or hardcoded crypto keys.

We will start the first process from the red team perspective (1), extraction and analysis device framework using binwalk (Fig. 2).

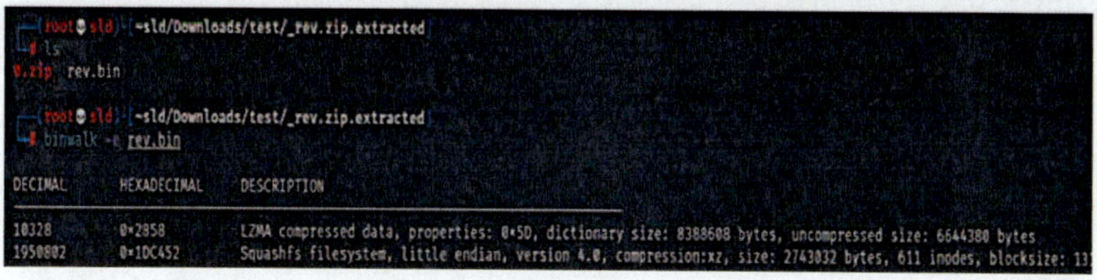

Fig. 2 The *binwalk* command used for decompressing access point firmware: binwalk -e firmware.bin

After *binwalk -e* command we could see information about extracted filesystem which is Squashfs filesystem. Squashfs is a highly compressed read-only filesystem for Linux. Squashfs compresses both files, inodes and directories, and supports block sizes up to 1Mbytes for greater compression it uses Mksquashfs and Unsquashfs utility tools for compressing files or embedding one filesystem into another or extract same from another file. We know that some vendors for LZMA (empel–Ziv–Markov chain algorithm) compressed Squashfs file system use the magic number "shsq". So, we can perform a HEX dump on our firmware and filter using *grep* command for "shsq", to obtain the starting address of the firmware. The binwalk extraction will create a folder called _rev.zip (in this example). Squashfs-root will have the entire filesystem extracted so the next step could be analysis of individual binaries or using *find*, grep, awk commands for searching in the configuration files for identifying new vulnerabilities (Figs. 3 and 4).

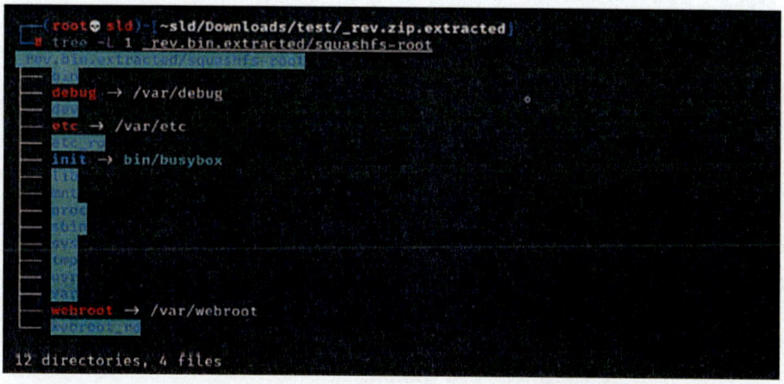

Fig. 3 Using tree -L 1 for directory listing from the squashfs-root directory

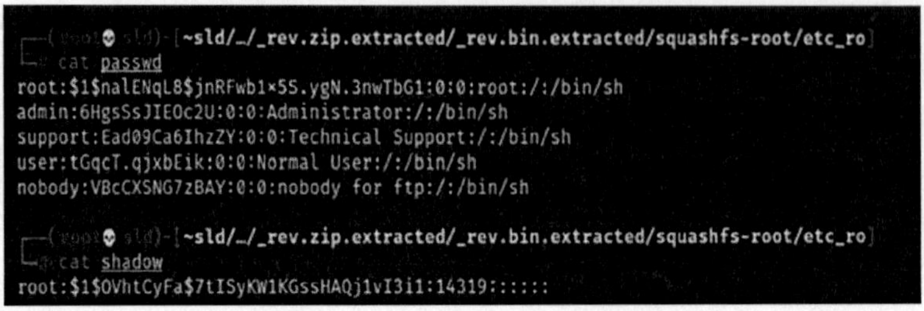

Fig. 4 In the above image we can see that default information from the /etc_ro/shadow, /etc_ro/passwd which contains a list of the system's accounts, giving for each account some useful information like user ID, group ID, home directory, shell

3.3 Binary Disassemble

The next step will be step (2) debug and disassemble a binary. For this article, we will disassemble httpd binary located in: _rev.bin. extracted/squashfs-root/bin. Via the disassembly operation applied on the binaries we will restore the software application code in a readable and understandable format. The assembly code file can be used in reverse engineering processes to establish the logical flows of the computer program or its vulnerabilities in a working environment.

In this example, the binaries use externally controlled format strings in printf-style functions (see Fig. 5), which can lead to buffer overflows vulnerabilities or data representation problems. Format string attacks alter the flow of an application by using string formatting library features to access other memory space. Vulnerabilities occur when user-supplied data are used directly as formatting string input for certain C/C++ functions (e.g. fprintf, printf, sprintf, setproctitle, syslog). If an attacker passes a format string consisting of printf conversion characters (e.g. "%f", "%d", "%n", etc.) as a parameter value to the web application, they may: execute arbitrary code on the server, read values off the stack, cause segmentation faults/software crashes.

In this case the: logLevel = iVar2 is possible to contain uncontrolled format string vulnerability which can lead to a potential integer overflow.

When using sprintf, it's up to the developer to make sure the size of the buffer to be written to is large enough to avoid buffer overflows. Integer overflows can cause the program to crash at a minimum. At worst, a carefully crafted overflow can cause malicious code to be executed. sprintf () or printf () will write the past array as it writes in the string, and therefore invokes no known behaviour. The function will write over the first bytes on the stack or cause a runtime error.

Noncompliant: sprintf (str, "%s", message); vs Compliant: snprintf (str, sizeof(str), "%s", message);

An integer overflow occurs when you attempt to store inside an integer variable a value that is larger than the maximum value the variable can hold. The C standard defines this situation as undefined behaviour (meaning that anything might happen). In practice, this usually translates to a wrap of the value if an unsigned integer was used and a change of the sign and value if a signed integer was used (Fig. 6).

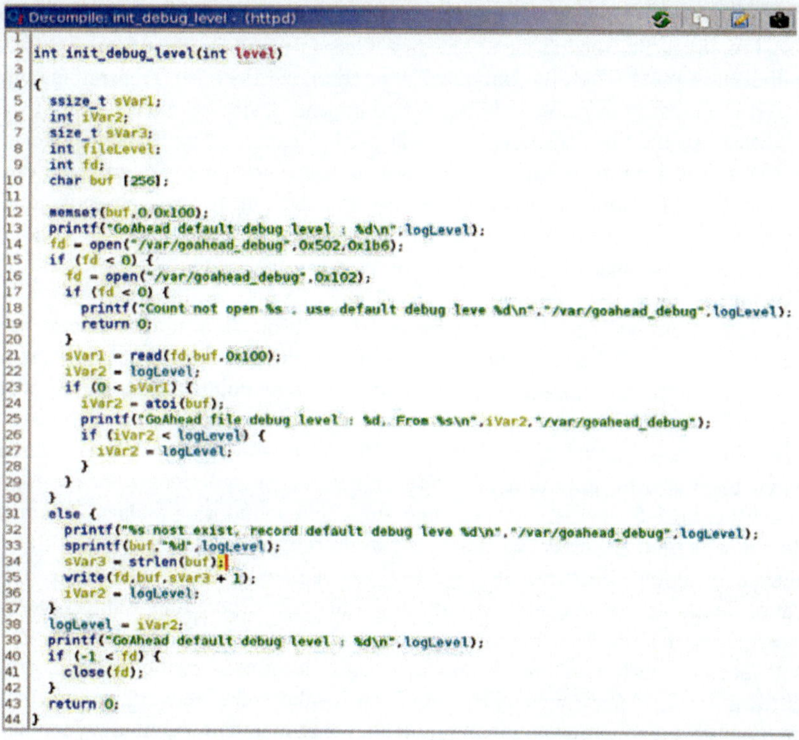

Fig. 5 In the above image we can see that the application is using potentially dangerous function on line 13, 18, 25, 32, 33, 39 which could lead to integer overflow. Integer overflows are the consequence of "wild" increments/multiplications, generally due to a lack of validation of the variables involved

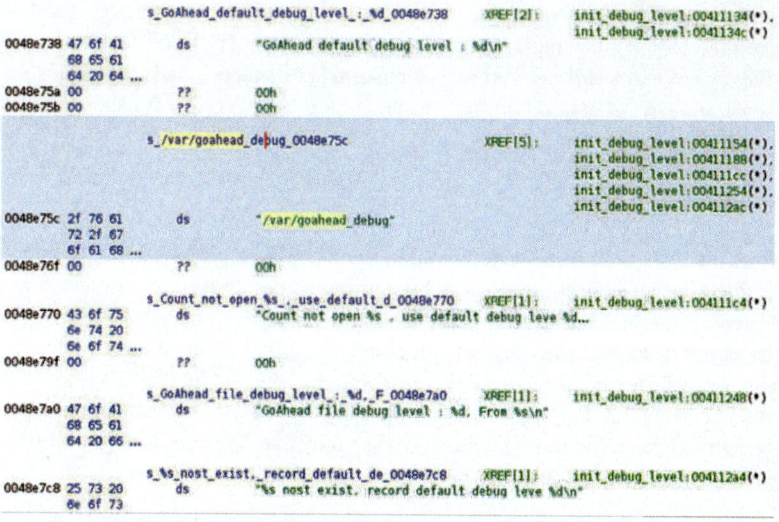

Fig. 6 The assembly view of the http binary can be used to evaluate different system calls or vulnerable code problems

The analysis part of squashfs filesystem with reverse engineering tools like Ghidra or IDA-Pro [8] is the second step. The first recommended is identification of serial communication ports (UART - Universal Asynchronous Receiver Transmitter) which are present in most of the embedded access points and devices and using them to gain shell access to the device. UART stands for Universal Asynchronous Receiver Transmitter. It is used for serial communications over a computer or peripheral device serial port. UART communication makes use of baud rate to maintain synchronism between two devices, where, the baud rate is the rate at which information is transferred in a communication channel. Another possibility is to unsolder the SPI flash and extract the firmware.

The blue team will need to provide real time protection against attacks using predefined actions and rules for firewalls, intrusion prevention systems. The efficient method is to unplug the vulnerable access point or access points and to use alternative wireless distribution until a new firmware upgrade will be available from the vendor.

The blue team part from purple team will cover vulnerability management and firmware upgrade (A), detection of inside or new suspicious connections using log alerts and I.D.S/I.P.S/Firewall rules (B) and will check monitoring tools (C) in order to identify new potential threats (see Fig. 1). The unifying theme is getting the Red and Blue team to agree on their shared goal of organizational improvement and not to introduce yet another entity into the mix. SOC (Security Operations Center) members responsible for data flow or members from the blue team should use automation and SOAR (Security Orchestration, Automation and Response) methods for triggering alerts toward an incident management platform for the entire network.

The purple team will write a full report that will include a detailed threat map of each technique's execution status and analysis from both red and blue team, as well as a detailed guide for the implementation of any recommended related risk that were not fully addressed during the exercise [13]. Using offensive techniques, the blue team alongside with the red team are simultaneously monitoring the logs and systems. If a technique is successful, the purple team current capabilities will be to create and use new prevention/detection mechanism for each successfully techniques used in the exercise [14, 15]. By leveraging every category of the MITRE ATT&CK framework [16], the purple team will need to map a custom set of tactics and techniques that are risk-based, industry standards in order to create a clear overview for the organization.

4 Conclusions

Firmware analysis is an important operation that should be performed in production by the vendors, but also during operation, by administrators or security engineers in order to discover different security vulnerabilities.

The paper described the purple team concept applied on an illustrative case using firmware analysis techniques. Besides describing the role and methodology used for the security assessment in the purple team format, the focus of the paper was on describing some empirical steps for firmware analyses (in our case for an access point firmware), using dedicated tools for decompression and disassemble.

Some common implementation errors were emphasized in the demonstrating scenario, the paper also describing the recommended steps that should be taken by the members of the purple team in case of a vulnerability exposure.

References

1. Veerasamy N (2009) High-level methodology for carrying out combined red and blue teams. In: 2009 second international conference on computer and electrical engineering, Dubai, pp 416–420. https://doi.org/10.1109/ICCEE.2009.177
2. Waksman A, Rajendran J, Suozzo M, Sethumadhavan S (2017) A red team/blue team assessment of functional analysis methods for malicious circuit identification. In: 2014 51st ACM/EDAC/IEEE design automation conference (DAC), San Francisco, CA, pp 1–4
3. Rajendran J, Jyothi V, Karri R (2011) Blue team red team approach to hardware trust assessment. In: 2011 IEEE 29th international conference on computer design (ICCD), Amherst, MA, pp 285–288. https://doi.org/10.1109/ICCD.2011.6081410
4. Bresch C, Michelet A, Amato L, Meyer T, Hely D (2017) A red team blue team approach towards a secure processor design with hardware shadow stack. In: 2017 IEEE 2nd international verification and security workshop (IVSW), Thessaloniki, pp 57–62.https://doi.org/10.1109/IVSW.2017.8031545
5. Olivero MA, Bertolino A, Dominguez-Mayo FJ, Escalona MJ, Matteucci I (2019) Security assessment of systems of systems. In: 2019 IEEE/ACM 7th international workshop on software engineering for systems-of-systems (SESoS) and 13th workshop on distributed software development, software ecosystems and systems-of-systems (WDES), Montreal, QC, Canada, pp 62–65.https://doi.org/10.1109/SESoS/WDES.2019.00017
6. Rastegari S, Hingston P, Lam C-P, Brand M (2013) Testing a distributed denial of service defence mechanism using red teaming. In: 2013 IEEE symposium on computational intelligence for security and defense applications (CISDA), Singapore, pp 23–29.https://doi.org/10.1109/CISDA.2013.6595423
7. Mirkovic J et al (2008) Testing a collaborative DDoS defense in a red team/blue team exercise. IEEE Trans Comput 57(8):1098–1112. https://doi.org/10.1109/TC.2008.42
8. IDA Pro tool webpage. https://www.hex-rays.com/products/ida/. Accessed 25 Nov 2020
9. Ghidra webpage. https://ghidra-sre.org/. Accessed 25 Nov 2020.
10. Mitre CVE webpage. https://cve.mitre.org. Accessed 25 Nov 2020
11. Binwalk webpage. https://github.com/ReFirmLabs/binwalk. Accessed 25 Nov 2020
12. Libmagic manual webpage. https://man7.org/linux/man-pages/man3/libmagic.3.html. Accessed 25 Nov 2020
13. Everson D, Cheng L (2020) Network attack surface simplification for red and blue teams. In: 2020 IEEE secure development (SecDev), Atlanta, GA, USA, pp 74–80.https://doi.org/10.1109/SecDev45635.2020.00027
14. Andreatos AS (2017) Designing educational scenarios to teach network security. In: 2017 IEEE global engineering education conference (EDUCON), Athens, pp 1606–1610.https://doi.org/10.1109/EDUCON.2017.7943063
15. Kim J, Kim K, Jang M (2017) Cyber-physical battlefield platform for large-scale cybersecurity exercises. In: 2019 11th international conference on cyber conflict (CyCon), Tallinn, Estonia, pp 1–19. https://doi.org/10.23919/CYCON.2019.8756901
16. Mitre Attack webpage. https://attack.mitre.org/. Accessed 25 Nov 2020

Automated Testing of GUI Based Communication Elements

Andreea Nica, Alexandra Balan(✉), Corneliu Zaharia, and Titus Balan

Transilvania University of Brasov, Braşov, Romania
alexandra.balan@unitbv.ro

Abstract. This paper describes the principles and a concrete implementation scenario for telecommunication test automation in an industrial system using an enterprise VoIP PBX (private branch exchange) and real IP phones.

Keywords: Automated testing · SIP · VoIP · Selenium · Continuous integration

1 Introduction

Industrial equipment evolution has produced also a change in the human machine interaction. Increased complexity of the set of commands, dependencies between modules, rich set of features and sub-features made the CLI (command line interface) configuration almost impossible, or usable only in debug phases.

The CLI was replaced with a graphical user interface GUI, most of them including web interfaces, thus becoming easier to choose options that trigger hundreds of CLI commands in the background. This change was visible in most industries; thus, the manipulation of complex industrial equipment can be performed by less engineering skillful personnel. This paper used the telecommunication industry as example for the methods implementation, however the same approach can address other industries or elements.

We can exemplify the CLI to GUI change in a very illustrative example based on Cisco equipment: though IOS commands are still learned by engineers and still in use, for more complex application like the VoIP elements of complex firewalls like e.g. Cisco ASA, Cisco Firepower Threat Defense, different GUI options were introduced (some based on Java Virtual Machines, but most of them evolving into web interfaces).

From automation point of view, there should be established or adapted new procedures, automation procedure being in many cases correlated with remote operation of the equipment. The SSH based CLI automation and scripting should be replaced by GUI automation and new specific methods should be implemented.

Modern GUI, accessible also via Internet are implemented via web services, so the configuration interface is accessible in browsers and not using other technologies.

Selenium is a portable framework for testing web applications, providing an Integrated Development Environment (IDE) for describing functional tests. The industrial digitization is marked by the abundance of RPA (Robot Processing

M. E. Auer et al. (eds.), *Online Engineering and Society 4.0*, Lecture Notes in Networks and Systems 298, https://doi.org/10.1007/978-3-030-82529-4_37

Automation) tools, that are capable to automate processes spanning across interfaces like legacy systems, virtual machines and also across multiple applications and systems. Robot Automation is a sequence of steps that gives result in a meaningful action without interference of human being but can also automate processes having unstructured data like pdf, images or documents.

Because RPA tools are generic enough to automate almost everything, for some specific cases they can be prone to errors or might have lower response time or load data time. Also, the open-source maturity for RPA tools is not comparable to the commercial versions. In our case, it was considered easier and efficient, to use Selenium for telecom automation, also considering that the test was part of a Continuous Integration/Continuous Development (CI/CD), so these were the specific test to be performed, and no other elements, tool, platforms should be automatized to justify the use of an RPA tool. Though RPA is more complex, in our case, Selenium was fit-for-purpose, following the statement form the software's web page: "Selenium automates browsers. That's it!"

This paper describes the principles and a concrete implementation scenario for telecommunication test automation in an industrial system using an enterprise VoIP PBX (private branch exchange) and real IP phones. Besides the implementation of the VoIP tests, the implemented test bench will also detail the failed or passed tests.

2 Test Automation Concept

As software testing is increasing complexity, manual software testing becomes overwhelming. Recently, automation is as a major factor in reducing the testing effort by many researchers [1].

Test automation is more than a mechanism for running a consistent and repetitive set of tests without human intervention. This involves a process of designing test programs. In software testing, test automation should meet one or the other several of the following requirements: 1) establishing some test preconditions and using software tools design for control and configuration; 2) running tests; and 3) comparison of current results with predicted results [2].

Over time many researches have been directed toward automatic testing of telecommunications systems. Designing automated tests for telecommunication systems is not a simple matter because it is not possible to create a unique test-suite for all systems. The complexity lies in the fact that these systems have little in common, e.g., at first glance a radio system seems quite different from a wired network system [3]

A test tool and a test framework that can simulate (SIP) client and SIP server behaviors is presented in [4]. In [5] the authors present an experimental approach for automated testing/validation of SIP scenarios in (IMS) networks. For that purpose, an automation framework is proposed allowing to replicate the configuration of SIP equipment from the production network and submit such equipment to a battery of tests in the testing network.

Open Source Robot Framework Automation solution for SIP services is proposed in [6]. The authors present in [7] a novel logic-based approach to test the protocol performance through real execution traces and formally specified properties.

The SIP Protocol and the background needed for developing SIP applications using Java programming language are described in [8].

Selenium sources related on how to implement the automated tests are provided in [9–11] and [12].

The present paper describes and implement another approach of automated tests for telecommunications systems that can be used also in Continuous Development process. Test automation has a great benefit during development phases, in the so called Continuous Integration/Continuous Development process.

CI/CD is intended to be used in combination with automated unit tests(Fig. 1) written through the practices of test-driven development. This is done by running and passing all unit tests in the developer's local environment before committing to the mainline.

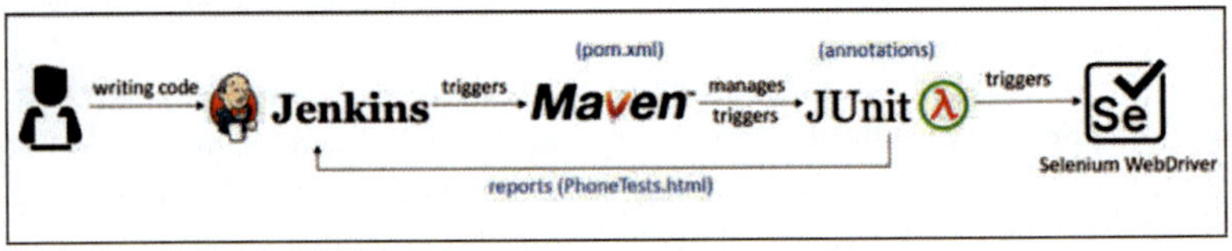

Fig. 1 Integrated Selenium with CI/CD

3 System Architecture

The architecture of the automated testing system designed for communications is presented in Fig. 2.

The automation of the tests is performed with the help of the framework Selenium and ChromeDriver, which are special for performing these functions, for web applications. Google Chrome is the web browser through that the communication is done with the Internet. The scripts are written in the integrated development environment, IntelliJ IDEA [13]. At their base is the Java programming language. The structure of the software stack is presented in Fig. 3. The DUT (devices under test) are represented by an experimental local network where SIP phones and industrial telephone exchange equipment (Unify OpenScape4000 for this experimental case) are connected.

OpenScape Desk CP200 is a phone based on the Session protocol Initiation Protocol (SIP). It supports the entire range OpenScape Voice of feature, has a headset socket and four programmable function keys. The keys are set to perform the functions like ‘conversations’, ‘redial’, ‘forwarding’ and ‘people’. In addition, the phone has fixed buttons for ‘hold’, ‘transfer’ and ‘conference’ to interact quickly with the caller.

OpenScape Desk Phone IP 55G is an innovative and top-quality phone intended especially for those who hold managerial positions to answer their needs.

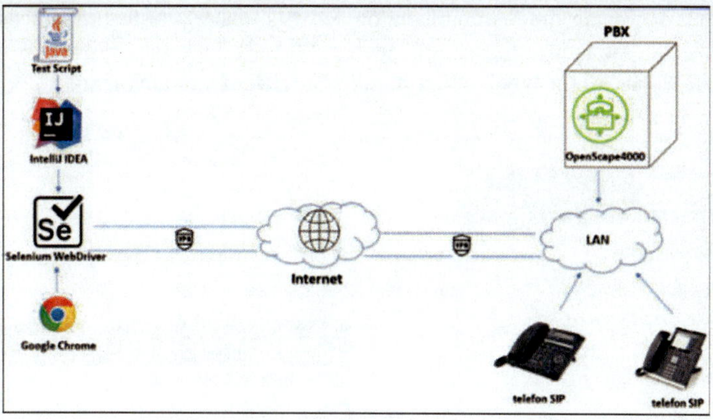

Fig. 2 System architecture

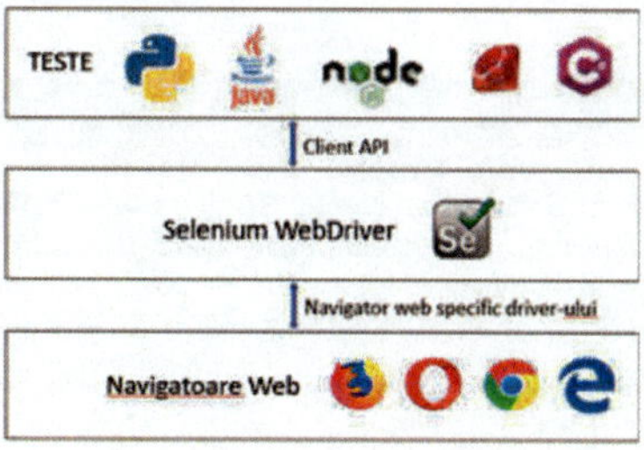

Fig. 3 System architecture

OpenScape Desk Phone IP 35G Eco can be connected to a SIP proxy server through its LAN interface. It performs a lot of functionalities and it is a good choice for business people. The phone has improved energy efficiency and can be used with SIP or HFA software.

The system that connects the phones for testing is called Siemens HighPath 4000. This is a private IP-based telephone exchange (IP-PBX), using time-division multiplexing and the Internet protocol. A circuits or packet switching system contains common components of this telephone exchange and it uses both an internal switching network and an external network. The system implements an architecture distributed over IP, with circuit switching, which can simultaneously support IP, split multiplexing in time and asynchronous transfer mode, ATM.

4 Automated Configuration Scenario of a SIP Phone

A configuration scenario of a SIP type phone that is made through accessing its graphical interface, OpenScape Desk Phone PC200 is described below in Fig. 4 and 5. For phone configuration is required to login with the password in the field "Admin login". After accessing the "Administration settings" page, the features of the phone are available in "General information" such as "MAC address", "Software version", "Last restart", "Backlight type", "Part number", "UBoot version" (Fig. 6).

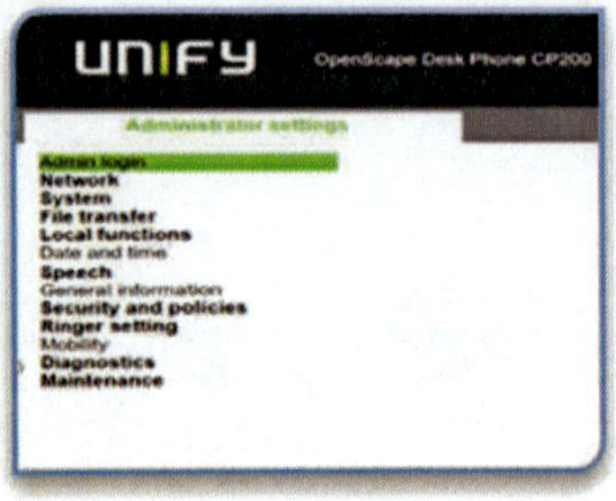

Fig. 4 Administrator settings menu

Fig. 5 Entering the password for admin login

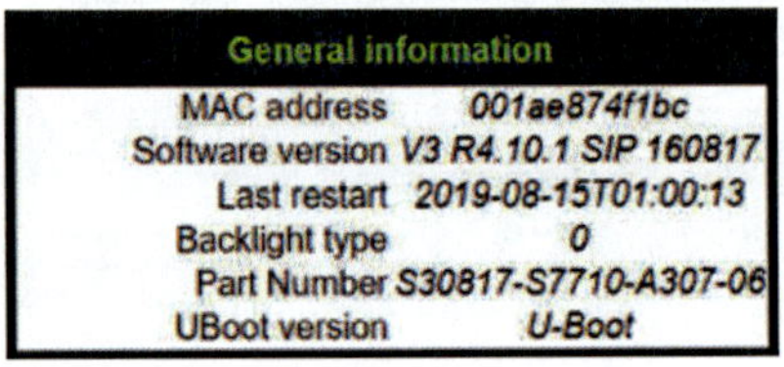

Fig. 6 General information menu

By accessing the "System" field, the "System identity" field is extended and the phone number and the message displayed on the screen could be modified. The following fields were manually completed: "Terminal number", "Terminal name", "Display identity" as shown in the Fig. 7.

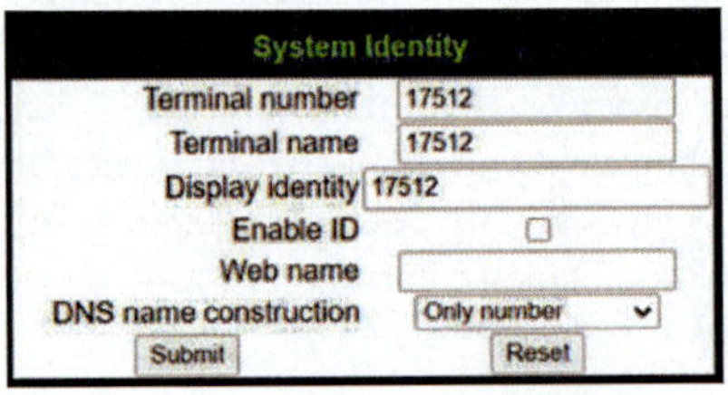

Fig. 7 System identity menu

To complete the information about the physical location, where it is configured the phone number, "Registration" field (Fig. 8) is completed in according to the following configuration requirements: "SIP server address", "SIP registrar address", "SIP gateway address".

This interface in configuration scenario looks like the other two phone interfaces that were tested in the current implementation. The automated tests are dedicated for the configuration of different features for phones and display device information. The functions are independent and these are defined in a class function.

The following automated tests based on the hardware-software configuration presented above were performed:

- authentication tests for user or administrator;

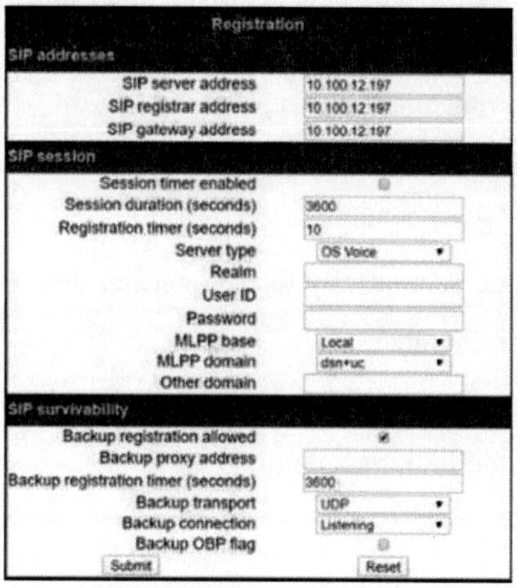

Fig. 8 Registration menu

- tests to change the time automatically depending on the location;
- tests for displaying phone information such as "General information", "Diagnostic Information"
- phone identity configuration tests;
- tests for configuring the phone's recording features;
- tests for changing the password for the user or administrator;
- tests to change the volume for the phone access key • tests to change the recording time of a phone call;

5 The Report of the Tests Performed

After automating the tests mentioned in the previous point, a report for monitoring the final results was created (Fig. 9).

In addition to the final result, if the tests were performed successfully or failed, in this report you can see the parameters of the phones being tested: phone numbers, IP phones and their authentication passwords. The tests are run in ascending order, depending on the index in front of the phone parameters.

In order to be able to see more details about the tests, by extending a test it will show detailed information, such as the date and time of running, all the messages displayed in functions by line code "System.out.println" and driver information (about which port works, the version used and information about its successful or failed start-up). This report is automatically generated by Intellij IDEA through the export option of test results. The information can also be exported in XML format, not only in HTML format. The tests are independent of each other.

The tests are divided into several categories, depending on the results that it is preferred to have. Following are presented the main tests classification:

- Should login with administrator rights.

  These are successful authentication tests for the administrator.

- Should print general information.

  These are the tests that successfully display general device information.

- Should set the Identity number.

  These are the tests that successfully set up your phone's identity.

- Should set the Registration number.

  These are the tests for successfully setting up phone number registration.

- Should modify the registration timer.

  These are successful tests to change the recording time of a call phone.

- Should not modify the registration timer, wrong timer values.

  These are failed tests to change the recording time of a call due to the wrong value entered.

- Should login with user rights.

  These are successful authentication tests for the user.

- Should not login with user rights—wrong password.

  These are the failed authentication tests for the user due to the introduction wrong password.

- Should modify the Auto Time change option.

  These are the tests for activating the option to change the time automatically in function location.

- Should modify the Key click volume.

  These are the volume change tests for the access key.

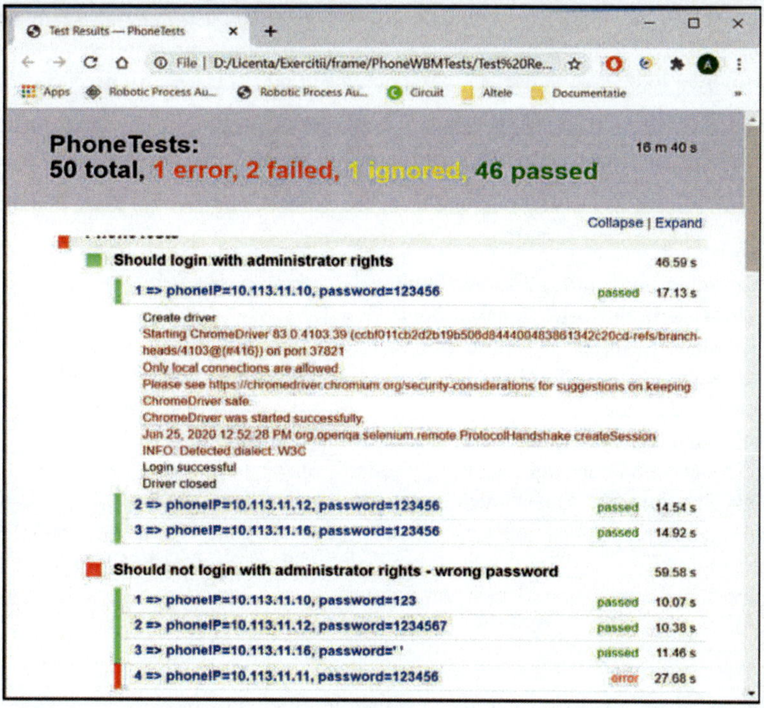

Fig. 9 Test results report

- Should get the diagnostic information.

These are the tests for displaying the information in the “Diagnostic information” window.

- Should change the user password.

These are the password change tests for the user.

6 Source Code Description

The project “Automated testing in communications field” involves writing code in the IntelliJ IDEA utility that contains two main files: DeviceTests and resources. DeviceTests is a file divided into three classes. These are:

- BaseClass;
- Functions;
- PhoneTests.

The "BaseClass" class has two functions, the "setUp" function required for starting and ChromeDriver initialization and the "tearDown" function required to close the driver. In order to be able to use the two functions in other classes, we used inheritance:

- public class PhoneTests extends BaseClass (for PhoneTest);
- public class Functions extends BaseClass (for Functions).

In the "Functions" class it is wrote all the actual code for the tests that will be automated. The defined functions are the following:

- public static boolean logInU(String phoneIP, String password);
- public static boolean getGeneralInfo();
- public static String setSystemIdentity(String phoneNr);
- public static String setRegistration(String server_address, String reg_address, String gateway_address);
- public static boolean modifySIPSession(String reg_timer);
- public static boolean logInA(String phoneIP, String password);
- public String modifyDateTime();
- public static String changePassword(String current_pwd, String new_pwd, String confirm_pwd);
- public static String keyClick(String option);
- public static boolean getDiagnosticInformation();

In the "PhoneTests" class are defined all the tests that use the functions in the "Functions" class. Because each test contains the same annotations, the structure of a test administrator login is presented above:

```
@DisplayName("Should login with administrator rights")
@ParameterizedTest(name = "{index} => phoneIP={0}, pas
word={1}")
@CsvFileSource(resources = "/testLogInAdmin_data.csv")
@Order(1)
public void testLogIn(String phoneIP, String password) throws
Exception{
assertTrue(Functions.logInA(phoneIP, password));
}
```

Code 1 - Source code for administrator authentication in the "PhoneTests" class

The "resources" file contains a multitude of csv files (Fig. 11), which are made up only of the values of the parameters defined in the "Functions" class, which will be used by the tests. In the case of the example above, csv file looks like the Fig. 10 below.

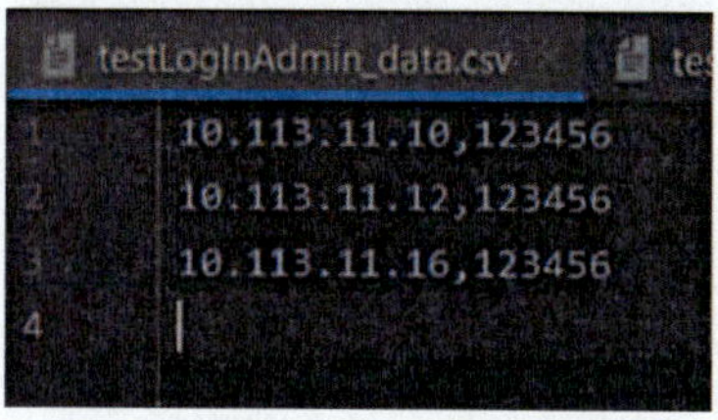

Fig. 10 Defined values for authentication tests

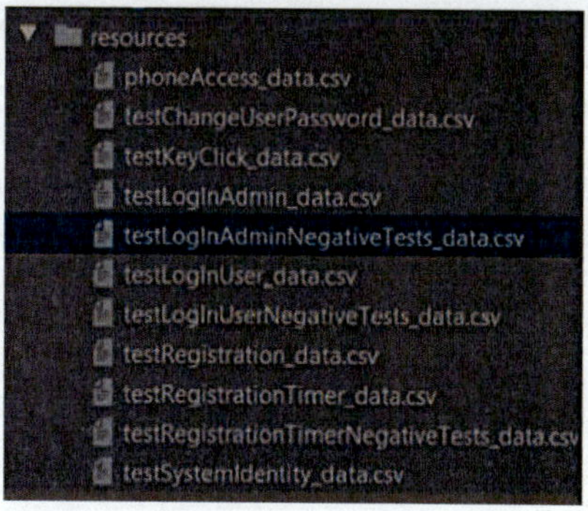

Fig. 11 Csv file

7 Conclusion

In conclusion, the project described above meets the established purpose: to implement, test and most importantly to automate various tests for SIP phones. Therefore, all the programs required for operation have been successfully installed. By Selenium, we managed to automate a lot of tests to configure different phone features and display information about them. The experimental verification of the project operation went through several stages and subsequent adaptations, which led to the achievement of the proposed objectives and identification of possible further developments.

References

1. Hanna M, El-Haggar N, Mostafa M-S (2014) A review of scripting techniques used in automated software testing. Int J Adv Comput Sci Appl 5. https://doi.org/10.14569/IJACSA.2014.050128
2. Hayes L (2004) Automated testing handbook, 2nd edn. Software Testing Institute
3. Lopez AG, Viela IM. An automated model generator for Model – Based Testing, Degree Level in Communication Systems, Second level, KTH Information and Communication Technology, Sweden, Stockholm (2012)
4. Kikagac Y, Simsek S, Yavas D (2016) SIP automated test platform. In: Proceedings of the eleventh international network conference (INC), pp 153 157
5. Gonçalves D, Amaral A, Costa A et al (2016) Towards automated test and validation of SIP solutions. Telecommun Syst 61:579–590. https://doi.org/10.1007/s11235-015-0015-8
6. Appari KK (2017) Open source automation solution – SIP services (RFC 5359). Int J Sci Eng Technol Res 06(07):1464–1466
7. Che X, Maag S (2013) A formal passive performance testing approach for distributed communication systems. ENASE 2013:74–84
8. The SIP Servlet Tutorial (2010) Overview of Session Initiation Protocol (SIP) Application Development. https://docs.oracle.com/cd/E19502-01/821-0203/gflsc/index.html
9. Selenium Webdriver. https://www.selenium.dev/documentation/en/webdriver/

10. Selenium Webdriver for Beginners. https://www.blazemeter.com/blog/selenium-webdriver-for-beginners
11. Webdriver S (2017) Selenium WebDriver. https://www.tutorialspoint.com/selenium/pdf/selenium_webdriver.pdf
12. Advanced Level Test Automation Working Group: Bakker B, Bath G, Born A, Fewster M, Haukinen J, McKay J, Pollner A, Popescu R, Schieferdecker I (2016) Test Automation Engineer. Introduction and Objectives for Test Automation
13. Jemerov D (2008) Implementing refactorings in IntelliJ IDEA. In: Proceedings of the 2nd workshop on refactoring tools, October, pp. 1–2. https://books.google.ro/books?hl=en&lr=&id=LdUGBgAAQBAJ&oi=fnd&pg=PT2&dq=intelij+idea&ots=9hfVHG2mQc&sig=pzlrYtSDzmLSSHoUFzICyU3aR8w&redir_esc=y#v=onepage&q=intellij%20idea&f=false

Biomedical Engineering and Computing

Large Scale Covid-19 Detection with Blood Platelets Using Artificial Neural Network

Ahmed Tamer(✉) and Andreas Pester

The British University in Egypt, Sherouk Governate, Cairo 4920211, Egypt
Contact@ahmed-elmahdy.com, andreas.pester@bue.edu.eg

Abstract. Detection of COVID-19 by using patient's blood platelets using an artificial neural network to perform large-scale automated testing. The deep learning model uses the CBC test results to detect the virus with an accuracy of 89.26%, which achieve higher accuracy than the widely used RT-PCR test. The model has achieved accuracy with only a thousand rows, which assures that the model can achieve better results with more data.

The objective of the paper is to provide an automated and accurate method to detect COVID-19 using machine learning to apply it in large-scale testing, to help in containing the virus and prevent the spread of the virus. The scope of the paper is on the detection methodology using the blood platelets values extracted from the complete blood cycle (CBC) testing and discussing the future possibilities of improving the automated prevention method.

The paper uses a dataset provided by Hospital Israelita Albert Einstein in Brazil, which has over 50 features of different tests as Influenza A, Influenza B, Uria, etc…., many of these features had missing values or wasn't relevant to the existence of the virus in the body; therefore, the feature selection led to filtering the 17 blood features. The K-Nearest Neighbour (KNN) Imputation imputed the missing values in the 17 filtered values.

The analysis & visualization of the blood features concluded that the complete blood cycle test values highly correlated with each other, and there is a change occurs due to the presence of the virus is clear, especially to Eosinophils and Leukocytes, at which they represent the white blood cells forming the immune system that reacts against any virus or bacteria attracting the human body.

The paper has applied five different machine learning algorithms: logistic regression, support vector machine (SVM), random forest, K neighbours and decision tree. Meanwhile, in the second approach, used an artificial neural network with intensive use of network optimization and hyperparameter tuning to achieve the highest accuracy possible.

Keywords: Deep learning · COVID-19 · Blood platelets · Deep neural network

1 Introduction

The detection of Covid-19 in Large-scale is one of the most significant challenges in facing the virus, as the process needs to be automated to be able to perform thousands of tests daily to discover all the cases to be insulated and receive the suitable treatment

M. E. Auer et al. (eds.), *Online Engineering and Society 4.0*, Lecture Notes in Networks and Systems 298, https://doi.org/10.1007/978-3-030-82529-4_38

in early stages. The most widely used testing methodology is Reverse transcriptase-polymerase chain reaction (RT-PCR), which has an accuracy of 59% [1], it causes a high false-negative result. As a result, some countries require each person to have two RT-PCR tests in their medical protocol, and in case of contrasting results a third test would be performed, which is expensive to implement; however, it is currently the best solution to do large-scale testing. Chest CT scans are highly accurate; however, it needs to be scanned by the medical team; machine learning solved this issue and made the detection of Covid-19 with the CT images automated. The scanning of the patients using Chest CT is time-consuming and will not be suitable in mass-testing, as it increases the chances of healthy persons to be infected. The most accurate method is using blood test results, as it is semi-automated as a specialized device processes the results; however, the results need to be examined by a doctor. The proposed solution is to automate scanning the results process with machine learning, which can achieve higher accuracy than RT-PCR.

2 Data Exploration and Filtration

The complete blood count test is performed to help in the detection of infections, anaemia or diseases that affect the immune system [2]. The immune system reacts to COVID-19 at the first moment the virus goes into the body so that the blood test will give accurate results in the early stages of the COVID-19.

The complete blood count test is taken with a needle and collected into a test tube, the process does have less pain than the RT-PCR, and it is also faster than RT-PCR as the taking a blood sample takes less than a minute [2].

COVID-19 affect the white blood cells that are responsible for the immune system, which defends the body against COVID-19; the Leukocytosis count was found to be developed as a reaction against the virus [3].

The complete blood cycle test covers 17 blood cells which are classified to Red blood cells & white blood cells. The red blood cells help in the indication of the oxygen levels, which is heavily affected by COVID-19 in case of severe conditions. The white blood cells do represent the immune system, which acts against any bacteria or virus that infects the human body, and they do increase in number when the COVID-19 is present inside of the body to fight the virus. Hematocrit, which is a white blood cell, is found to be higher in number in the blood of infected patients than susceptible ones.

2.1 Data Analysis and Cleaning

The data cleaning and preparation phase was crucial as dataset suffered from a significant imbalance between the negative and positive COVID-19 cases. The negative and positive cases had a ratio of 5:1, as well as multiple rows suffered from missing values.

First, the process started with the analysis of the 50 features that present in the dataset, the columns that had more than 50% of dataset missing were dropped.

Second, the modification of the data to be in a suitable format for feeding the machine learning model, the modification is presented as normalization and standardization of the data.

Finally, Imputing the missing values in the rows using the KNN imputation algorithm, which is a KNN model that is trained using the existing values to predict and substitute the missing values.

2.2 Feature Selection

Feature selection is the process of deciding the input variables that will feed the predictive model during training, the reduction of the input variables is to reduce the computational cost as the model will be less complicated and to improve the performance of the model by selecting the most critical and useful features.

There were nearly 34 features/columns after the cleaning process, which were complete blood cycle CBC test results and other tests like Urea, sodium and more.

The features were selected based on statistical selection, at which the observation of the relationship between the features takes place by using plots. The correlation and dependence between the input variables are essential to choosing the most relevant features.

Figures 1 represents the plotted correlation between Eosinophils and Leukocyte. The red represents the COVID-19 infected values, and the blue is free off COVID-19. They are values retrieved from the CBC, which led to the decision of choosing the 17 CBC features as input variables to the model.

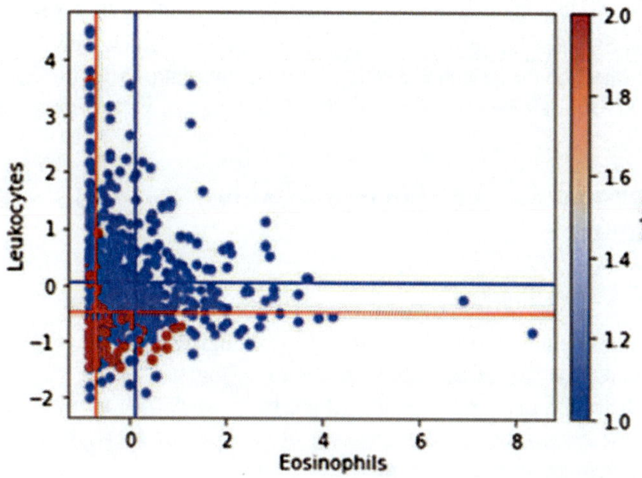

Fig. 1 Correlation between Eosinophils and Leukocyte. The red dots represent the positive COVID-19 patients and the blue represent the negative ones

2.3 Feature Scaling

The dataset does follow the Gaussian distribution, as the values of the blood platelets features do not have a fixed range. The possibility of having outlines in blood platelets is significantly high as a disease might cause an extraordinary shift in the count of blood cells. The suitable feature scaling in his condition is the standard normalization as it also does not have a bound range, which prevents any effects that occur from outliers to the training results.

2.4 Data Balancing

Following the data cleaning & feature selection, a vast data imbalance has been noticed between the positive COVID-19 value and negative ones. Many trials were done with the model with downsampling and oversampling techniques.

The NearMiss algorithm was used for down-sampling in the trial; it aims to balance the dataset by random elimination for the majority class data. The downside of the NearMiss algorithm is that there is a possibility of significant information loss and in the proposed solution case, the dataset was too small as it led to clear underfitting.

The NearMiss algorithm at first finds the distances between the given instances of the majority class and the minority class ones. Moreover, if the expectation is K instances in the minority class, then there will be K·N instances of the majority class because of the nearest distance method's instances. The NearMiss algorithm will pick the N instances in the majority class that have a small distance with those in the minority class.

The most used algorithm for oversampling is Synthetic Minority Oversampling Technique (SMOTE) [4] which was the best fit for the models as the data helped in achieving higher results. The K-Nearest Neighbours (KNN) plays a role in SMOTE, as the KNN of each minority instance is obtained and randomly selected by SMOTE, as the calculation of linear interpolation takes place to produce a new minority instance in the neighbourhood of the selected KNN. The synthetic data based on the similarities between the existing minority instance on the feature space is then generated.

3 Training and Testing Different Machine Learning Algorithms

3.1 Classical Algorithms

The cleaned and imputed data had been trained with logistic regression, SVM, Random Forrest, K Nearest Neighbor and decision tree algorithms.

Machine learning algorithms results which have about 700 training row and 200 testing rows, the results include the accuracy of the Positive class, the Negative class, the weighted accuracy and the total accuracy (Table 1).

Table 1 Machine learning accuracy table

Method	Total accuracy	Positive accuracy	Negative class accuracy	Weighted accuracy
Logistic regression	0.847571	0.753086	0.862403	0.807745
SVM	0.814070	0.740741	0.825581	0.783161
Random forest	0.897822	0.506173	0.959302	0.732738
K Neighbors	0.775544	0.629630	0.798450	0.714040
Decision tree c	0.807370	0.555556	0.846899	0.701227

3.2 Deep Learning

The proposed artificial neural network is a multi-layer fully connected neural network. Figure 2 shows the deep learning network for the detection of COVID-19. The input layer, labelled in grey, has seventeen neurons connected with the first hidden layer, and the seventeen neurons represent the seventeen selected features from the dataset which are values retrieved from the Complete Blood Cycle (CBC) test.

There are five hidden layers in the model, labelled in red, each layer's neurons connected with every neuron in the following layer. The initialization of the hidden layers' neurons is random, and each iteration refines them by the backpropagation process. The model uses the ReLu (Rectified Linear Unit) activation function after each hidden layer.

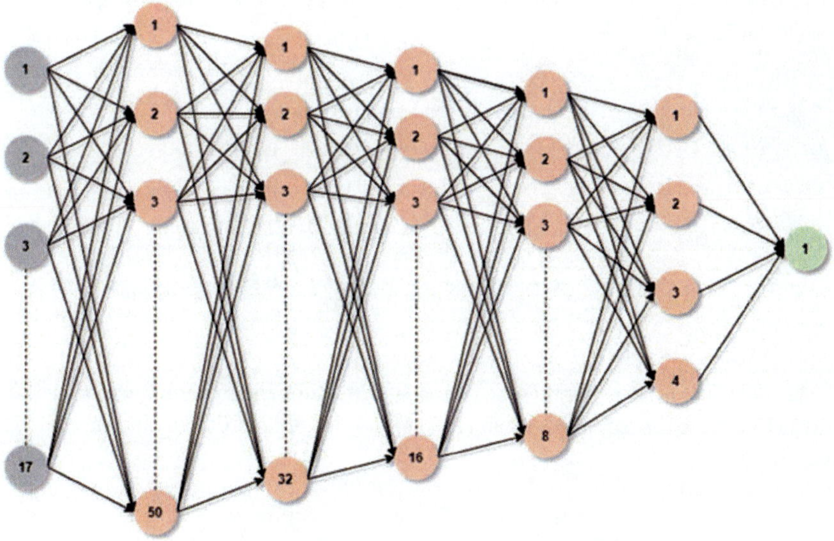

Fig. 2 Artificial neural network architecture

Table 2 Deep learning model's Loss and accuracy results

	Loss	Accuracy
Training	0.0106	98.95%
Validation	0.0111	98.80%
Testing	0.1055	89.25%
After optimization training	0.0060	99.40%
After optimization testing	0.1034	89.26%

The model has a single output layer since the output is binary (0 or 1), which represents infected with COVID-19 or not. The model uses the sigmoid activation function to transform the input signal into a binary signal.

The deep learning model achieved results better than expected since the amount of the data fed to it is considerably small. Table 2 explains the accuracy results, which is the most intuitive performance measure.

The deep learning model has 70% of the data fed to the training & validation phase and 30% to the testing phase. The training and validation are split to 80/20. The testing accuracy results are the real indication of the model's performance and accuracy, and they are always expected to have lower results that training and validation. The optimization is done by parameter-tuning which led to an improvement of $\sim$ 1.6% in the model's accuracy.

As metrics also precision, recall and F1 measure had been used because the data are imbalanced and sensitive to the True Negative. The results are represented in Table 3.

Precision, recall and F1 score are highest for the predicted class 0 (Negative), but not for the predicted class 1 (Positive). The second highest values for precision, recall and F1 measure have the weighted average. The predicted class 1 has the lowest precision, recall and F1 measure.

Table 3 Deep learning report

	Precision	Recall	F1-Score
0	0.92	0.95	0.94
1	0.67	0.56	0.61
Macro Average	0.80	0.75	0.77
Weighted Average	0.89	0.89	0.89

The deep learning network was tweaked with different hyperparameters and the optimal constellation of hyperparameters found for this problem are presented in Table 4.

Table 4 Deep learning model's parameter trails

Parameters	Trials	Optimal parameter
Batch size	10,25,30,50	25
Epochs	50,100,200,500	500
Learning rate	0.0001, 0.001, 0.01. 0.1	0.0001
The hidden layer activation function	ReLU, leaky ReLU, softmax	ReLU
Loss function	Mean Squared Error, Binary cross-entropy	Mean Squared error
Optimizer	ADAM, SGD, RMSPROP	Adam

4 Conclusions

The traditional Random Forest Classifier machine learning algorithm has outperformed the other algorithms and the deep learning model, however, it was expected to achieve this result as the traditional machine learning outperform the deep learning architectures only when the data available for training is limited to a lower number like in our case. With more data, this situation can change.

The deep learning model has gone through along stage of optimization and hyperparameter tuning to be able to achieve the best results possible.

Acknowledgements. We thank the faculty of Informatics and Computer Science of The British University in Egypt for their support during the work on that project. Our special thanks go to Walid Hussein, who gave valuable pieces of advice concerning the use of different algorithms and feature engineering.

References

1. Xie X, Zhong Z, Zhao W, Zheng C, Wang F, Liu J (2020) Chest CT for typical 2019-nCoV pneumonia: relationship to negative RT-PCR testing. Radiology 296(2):E41–E45
2. Beverly G, Parker K (2003) Understanding the complete blood count with differential. J Perianesth Nurs 18(2):96–117
3. Berber I, Cagasar O, Sarıcı A, Kirici Berber N, Aydogdu I, Yildirim A, Gocukara Bag HG, Acun Delen L, Ulutas O (2021) Peripheral blood smear and COVID-19 severity. Mediterr J Hematol Infect Dis 13(1):e2021009. https://doi.org/10.4084/mjhid.2021.009
4. Chawla NV, Bowyer K, Hall L, Kegelmeyer WP (2002) SMOTE: synthetic minority oversampling technique. J Artif Intell Res 16:321–357

Cloud-Based Data Storage System for eHealth Smart Devices

Paulo Abreu(✉) and Maria Teresa Restivo

LAETA-INEG, Faculty of Engineering, University of Porto, Porto, Portugal
{pabreu, trestivo}@fe.up.pt

Abstract. The availability of cloud storage and data processing providers opens the opportunity to use them with instrumented smart devices in the field of eHealth and eRehabilitation. This work presents the current development of a system comprising a software and a prototype of an instrumented device for assessing the force–time behaviour during a handgrip strength test. The hardware dynamometer connects wirelessly with a software application running on an Android device. The developed App provides the user interface to operate with the system, allowing the graphical view of the test results and connecting to a cloud-based data storage. The Firebase platform, from Google, is used. The App provides the definition of the handgrip test time and the graphical view of the force-time curve. In the end of the test, other force-time parameters, computed from the acquired data, are displayed. This system provides a start concept for an application that can be expanded to support other instrumented devices. It is used both as an interface to operate a device and for data cloud storage, easing the procedure to store, process and share data.

Keywords: eHealth · Cloud-based data storage · Handgrip strength-time curve · Muscle function

1 Introduction

In the healthcare areas it is common the use of multiple instrumented devices, both for evaluation and training. The current developments and incorporation of electronics and microcontroller based systems, coupled with communications capabilities, allows the development of a prototype concept of a smart instrumented device prepared for eHealth. Such device concept can make use of smartphones and tablets as the main user interface, simplifying the implementation and reducing the cost of the instrumented device. This approach allows the storage of the acquired data and exchange of information in an automatic way, simplifying the process of collecting and processing data. In health fields, the availability of historic data is regarded as very valuable asset to assess and to identify trends. In the rehabilitation field, the availability of data regarding the progress of a patient on conducting a given set of exercises or evaluation exams using smart instrumented devices, allows the therapist to evaluate the progress and adjust the therapeutic program. The use of computer based systems, where data can be stored, processed and easily accessed, both by the therapist and the patient, contributes to the widespread use of this type of eHealth systems.

M. E. Auer et al. (eds.), *Online Engineering and Society 4.0*, Lecture Notes in Networks and Systems 298, https://doi.org/10.1007/978-3-030-82529-4_39

This work presents a system architecture for such a system and describes the implementation of the used hand dynamometer. The system has the particularity of assessing the force-time behavior during a handgrip isometric strength test. The user interface, to process and present the data from the test, runs on a handheld Android device (tablet) that connects to a cloud-based data storage. The Firebase platform, from Google [1], provides the functionalities to use the dynamometer with different users that are allocated with distinct clinics. The functionalities of the developed application are presented.

Hand dynamometers are widely used to assess the handgrip strength (HGS) that has been recognized as an indicator of body muscle strength [2–4] and a marker of nutritional status [5], frailty [6, 7], and many other health issues. For example, several studies have also shown that the assessment of HGS, together with other screening measures for undernutrition, can be a predictor of hospital length of stay [8–10].

Work has been done on evaluating and comparing results obtained with distinct dynamometers [11–13], some custom-made others commercially available and recognized by regulatory agencies. Typically, most of the devices are only able to measure the maximum value of the handgrip force, while very few are able to continuously register the force along the test. The Jamar hand dynamometer, that is considered to be the golden standard, has evolved from a design using a hydraulic manometer to visualize just the maximum grip force, to a new design using an electronic load cell and continuous monitoring of force (Jamar® Smart Hand Dynamometer). These devices are made to operate on their own, or have particular proprietary applications, being their integration with other computer systems difficult or even not possible to implement.

The possibility to evaluate the force during a HGS test opens new opportunities to study muscle function behavior than just the maximum HGS. Several studies have been conducted, proposing different parameters to characterize the explosive isometric muscle contraction during a HGS test [14–16], and studies carried out to identify particular behavior of specific group of users such as patients or sport athletes [17–20].

These lead to the development of a cloud-based data storage application to deal with smart instrumented devices. As a pilot study, the smart device uses a hand dynamometer configured to provide the capability to communicate wirelessly with a handheld Android device. The capability to measure the force-time behavior during the handgrip strength test opens the opportunity to define distinct parameters related to the test. The rate of force development up to the maximum force, that some authors identify as the explosive force [16], or the time to reach maximum force, or the rate of force decay after reaching the maximum force, can be computed. In this way, the developed system can be used to explore these parameters and study their interest on their own or related them with specific pathologies the patients may have [16].

2 System Development

2.1 System Architecture

The framework to implement the use of a smart instrumented device is centered on the use of a handheld Android device (smartphone/tablet) that provides the user interface and the communications capabilities both for the instrumented device and for the cloud-based data storage (Fig. 1). The adopted framework uses BLE to connect with the Arduino based hand dynamometer and Wi-Fi to communicate with the cloud database.

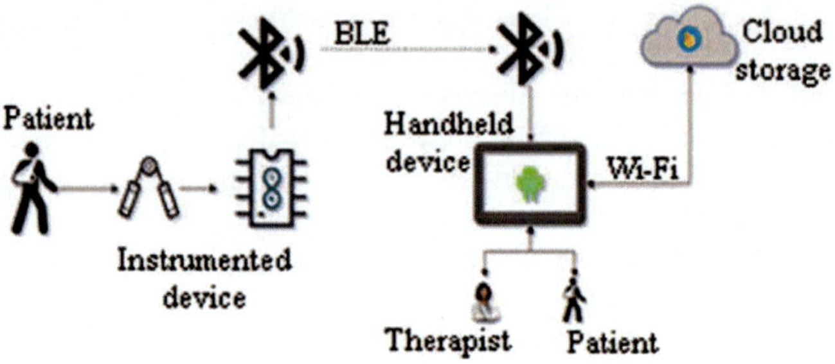

Fig. 1 System framework

2.2 Hand Dynamometer

The pilot smart device comprises a hand dynamometer with an electronic interface (Fig. 2). The hand dynamometer uses a standard beam load cell (50 kg, 2 mV/V), where two machined handles, covered with a rubber band, are attached to provide the support for the hand. The electronic interface is implemented using a standard Arduino Uno microcontroller, with a HC-05 Bluetooth module for communications and a load cell amplifier module HX711. The microcontroller is programmed to acquire and send data at a rate of 10 Hz.

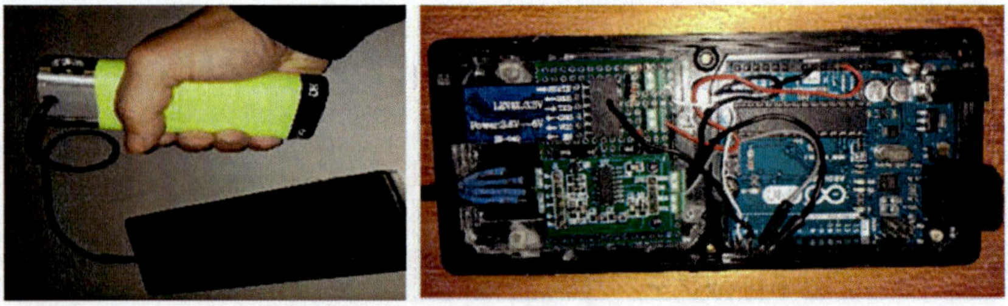

Fig. 2 Hand dynamometer (left) with electronic interface (right)

2.3 Software Development

The developed App provides the user interface to operate the hand dynamometer and saves the data on a cloud-based data storage. The service is provided by Firebase, from Google. The services used from Firebase include Authentication, for the login process, Cloud Firestore Database, to store recorded data, and Hosting, to allocate the webpages that make possible to get and save data, from and to the Database. Other functionalities available on Firebase such Machine Learning Kit and Analytics are planned to be used in future developments.

The App was developed to be used by different therapists and patients, allocated to distinct clinics and to cope with distinct instrumented devices. It supports the use of more than one hand dynamometer and the current implementation allows the use of a second instrumented smart device prototype, the Wrist evaluation system (West). Other devices could be added with minor changes to the structure of the software. The current App structure is depicted in Fig. 3, which only shows the part related to the hand dynamometer (handgrip).

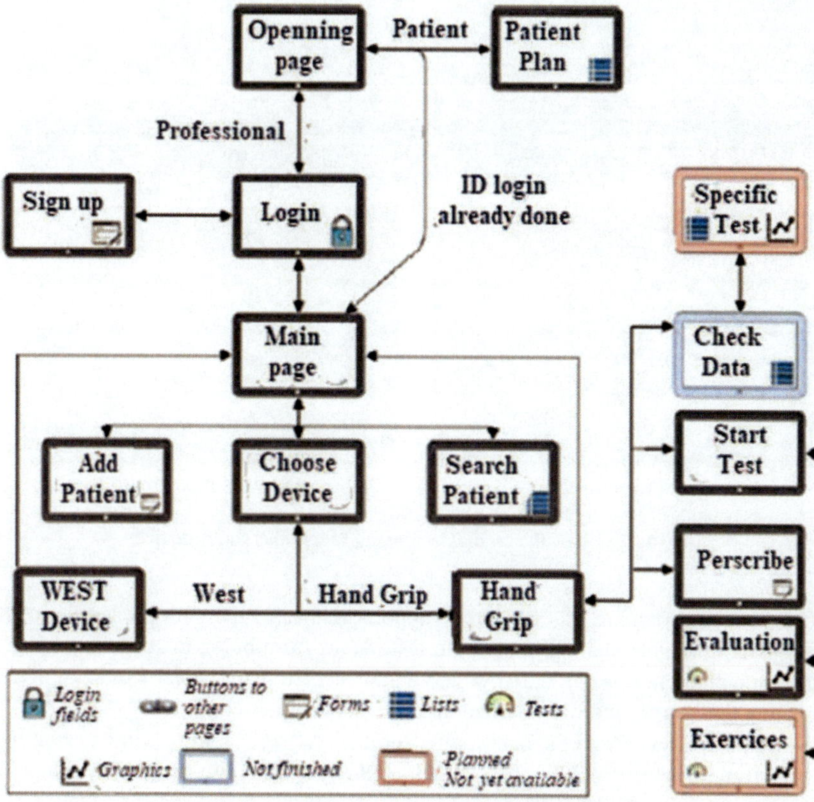

Fig. 3 Structure of current developed App to deal with the hand dynamometer

Figure 4 presents the steps to initiate the App. with the correct device, to select test, to introduce the user ID, to start the test and the respective data storing process in the database during the test time interval. Finally, to present the results in a graphical form.

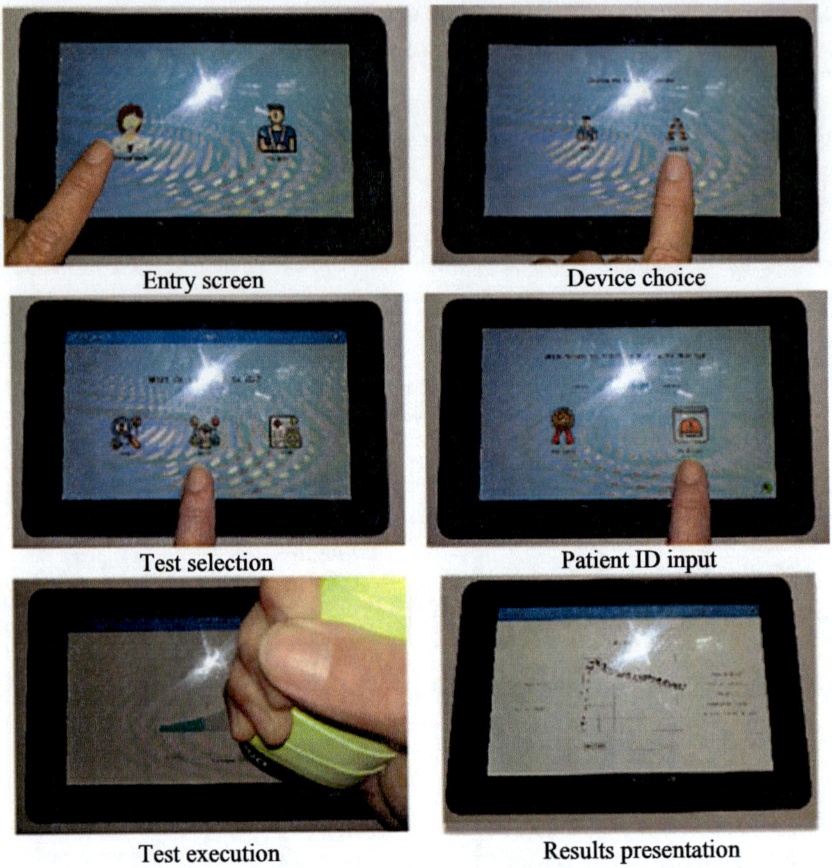

Fig. 4 Some of the available screens to use the smart device

One particular feature that this App supports relates to the data processing of the sensing data during the handgrip strength test. Apart from being possible to configure the test time interval, it is computed, stored and presented in graphical window different parameters that can be used to characterize the handgrip strength test.

Those parameters are: i) the maximum exerted force (F_{max}); ii) the time instant F_{max} occurs; iii) the time to reach 63.2% (1−1/e) of F_{max}; iv) the rate of force development up to reaching F_{max}; v) the sustainability rate (determined between the instant of F_{max} and that for the test ending – defined by the test interval), the average force and the value of the area under the force-time curve [16].

For each test, these parameters provide a base to study, characterize and assess individuals or groups of users such as athletes, being the data available in a database with easy access, not only with the App, but also through a web browser. Data stored in the database for a handgrip strength test is shown in Fig. 5.

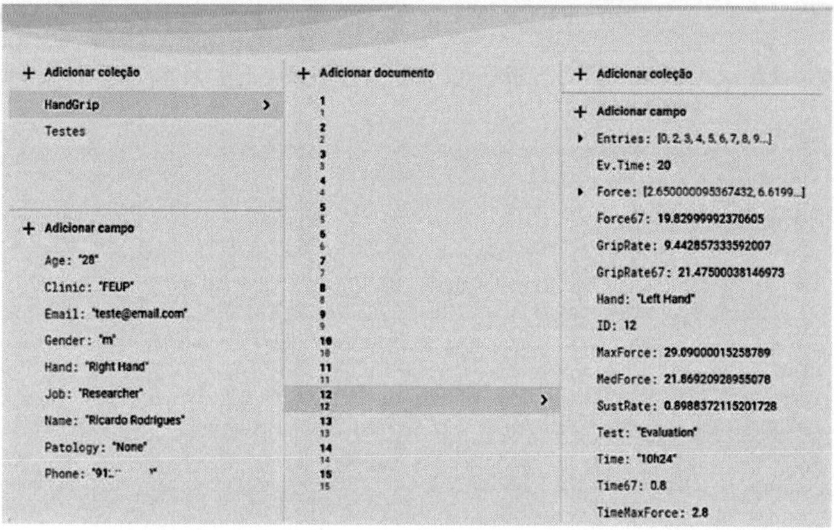

Fig. 5 Data stored in the database from handgrip strength test

3 Conclusions

This work presented the current developments of a conceived system prototype comprising a smart device in the form of a hand dynamometer with a cloud-based data storage functionality. The adopted system architecture takes advantage on the use of a handheld device as the user interface to operate a hand dynamometer and communicate to the cloud storage database. The use of smart instrumented devices, build around common available microcontrollers, provide the required functions of data acquisition and wireless communication. This approach allows to implement a simpler and portable instrumented device by shifting the user interface to a handheld device, while still providing communications and processing power that require more energy to operate.

The structure of the App developed to deal with the hand dynamometer provides the user interface and the communications with the instrumented device and the access to cloud-based data storage. The Firebase from Google was chosen, providing the service functionalities to implement a database, allowing the management of the users, devices and historic data that were implemented. The stored data along the time will be a valuable source for developing knowledge using intelligent algorithms, learning from the experience and inferring new insights. These functions are of particular importance, easing the process of conducting tests or exercises with the instrumented devices. The automatic registration of results, that are made available on the cloud, constitute an advantage both for the patient and therapist that can access the data.

It is now intended to fine tune the developed application and start using the system as a way to collect data from multiple users with a view to characterize specific group of users and explore the force-time parameters that the App can compute.

Funding. This research was funded by of Project NORTE-01-0145-FEDER-000022 - SciTech - Science and Technology for Competitive and Sustainable Industries, co-financed by Programa Operacional Regional do Norte (NORTE2020), through Fundo Europeu de Desenvolvimento Regional (FEDER) and by the Portuguese Foundation for Science and Technology, under Project LAETA - UIDB/50022/2020.

References

1. Google Firebase Homepage. https://firebase.google.com/. Accessed 07 Jan 2021
2. Lauretani F, Russo CR, Bandinelli S, Bartali B, Cavazzini C, Di Iorio A, Ferrucci L (2003) Age-associated changes in skeletal muscles and their effect on mobility: an operational diagnosis of sarcopenia. J Appl Physiol 95(5):1851–1860
3. Rantanen T, Volpato S, Luigi Ferrucci MD, Eino Heikkinen MD, Fried LP, Guralnik JM (2003) Handgrip strength and cause-specific and total mortality in older disabled women: exploring the mechanism. J Am Geriatr Soc 51(5):636–641
4. Richards L, Palmiter-Thomas P (1996) Grip strength measurement: a critical review of tools methods and clinical utility. Crit Rev Phys Rehab Med 8:1–2
5. Norman K, Stobäus N, Gonzalez MC, Schulzke JD, Pirlich M (2011) Hand grip strength: outcome predictor and marker of nutritional status. Clin Nutr 30(2):135–142
6. Syddall H, Cooper C, Martin F, Briggs R, Aihie Sayer A (2003) Is grip strength a useful single marker of frailty. Age Ageing 32(6): 650–656 (2003)
7. Reeve TE, Ur R, Craven TE, Kaan JH, Goldman MP, Edwards MS, Hurie JB, Velazquez-Ramirez G, Corriere MA (2003) Grip strength measurement for frailty assessment in patients with vascular disease and associations with comorbidity, cardiac risk, and sarcopenia. J Vasc Surg 67:1512–1520
8. Guerra RS, Fonseca I, Pichel F, Restivo MT, Amaral TF (2015) Usefulness of six diagnostic and screening measures for undernutrition in predicting length of hospital stay: a comparative analysis. J Acad Nutr Diet 115(6):927–938
9. Guerra RS, Fonseca I, Pichel F, Restivo MT, Amaral TF (2014) Handgrip strength cutoff values for undernutrition screening at hospital admission. Eur J Clin Nutr 68(12):1315–1321
10. Flood A, Chung A, Parker H, Kearns V, O'Sullivan TA (2014) The use of hand grip strength as a predictor of nutrition status in hospital patients. Clin Nutr 33(1):106–114
11. Kim M, Shinkai S (2017) Prevalence of muscle weakness based on different diagnostic criteria in community-dwelling older adults: a comparison of grip strength dynamometers. Geriatr Gerontol Int 17:2089–2095
12. Guerra RS, Amaral TF, Sousa AS, Fonseca I, Pichel F, Restivo MT (2017) Comparison of jamar and bodygrip dynamometers for handgrip strength measurement. J Strength Cond Res 31:1931–1940
13. Turusheva AV, Frolova EV, Degryse J (2018) Comparison of measurement results are obtained with dynamometers DK-50 and JAMAR® Plus. Russ Fam Doct 22(1):12–17
14. Demura S, Nagasawa Y, Yamaji S, Ikemoto Y, Shimada S (2001) Force developmental phase and reliability in explosive and voluntary grip exertions. Percept Mot Skills 92 (3_suppl):1009–1021 (2001)
15. Ikemoto Y, Demura S, Yamaji S, Minami M, Nakada M, Uchiyama M (2007) Force-time parameters during explosive isometric grip correlate with muscle power. Sport Sci Health 2 (2):64

16. Urbano D, Restivo MT, Amaral TF, Abreu P, Chousal MF (2020) An attempt to identify meaningful descriptors of handgrip strength using a novel prototype: preliminary study. Information 11(12):546
17. Schettino L, Luz CPN, de Oliveira LEG, de Assunção PL, da Silva Coqueiro R, Fernandes MH, Pereira R (2014) Comparison of explosive force between young and elderly women: evidence of an earlier decline from explosive force. Age 36(2):893–898
18. Borges LS, Fernandes MH, Schettino L, da Coqueiro RS, Pereira R (2015) Handgrip explosive force is correlated with mobility in the elderly women. Acta Bioeng Biomech 17 (3):145–149
19. De Dobbeleer L, Beyer I, Njemini R, Pleck S, Zonnekein N, Mets T, Bautmans I (2017) Force-time characteristics during sustained maximal handgrip effort according to age and clinical condition. Exp Gerontol 98:192–198
20. Cronin J, Lawton T, Harris N, Kilding A, McMaster DT (2017) A brief review of handgrip strength and sport performance. J Strength Cond Res. 31(11):3187–3217

Impact of Biofeedback in the Motor Rehabilitation of Patients with Acquired Brain Injury

Paulo Abreu[1(✉)], Maria Teresa Restivo[1], and Helena Sousa[2]

[1] LAETA-INEG, Faculty of Engineering, University of Porto, Porto, Portugal
{pabreu, trestivo}@fe.up.pt
[2] CIR, Center for Rehabilitation Research, School of Health, Polytechnic of Porto, Porto, Portugal
hes@ess.ipp.pt

Abstract. This work describes the system SHaRe, an instrumented device that communicates wireless with several computer gaming applications, developed in cooperation of hand rehabilitation experts to focus its use on handgrip strength control for daily use manipulation tasks, in patients with acquired brain injury. In the current implementation, the SHaRe takes the shape of a glass, instrumented with force and inertial sensors. The sensing elements allow the user to monitoring his/her applied handgrip strength to the glass and its orientation, by visual augmented feedback. The grip strength drives the visual deformation of the glass model and auditory effects, added by the glass orientation, providing biofeedback. The task-oriented approach, together with the use of games-type applications, engages the patients on performing the exercises. The system can assume many different shapes based on the designed sensorization, reproducing many different daily use objects to re-educate handgrip strength. SHaRe has been used in pilot studies conducted by hand rehabilitation experts that showed to have impact on the success of rehabilitation activities.

Keywords: Smart device · Augmented feedback · Hand rehabilitation · Biofeedback

1 Introduction

The concepts of biofeedback, augmented feedback and gaming in rehabilitation are jointly included in SHaRe smart device prototype development focused on re-education of fine control of handgrip strength in patients with acquired brain injury.

Biofeedback techniques have matured considerably incorporating new technologies since 1960 and "Biofeedback in Rehabilitation" is precisely the title of chapter 5 of ref. [1]. The biofeedback uses multiple technologies including electromyography (EMG), electroencephalography (EEG) and other sensing devices coupled with visual and auditory resources to provide the user with the relevant information. The Biofeedback effect has been used in many studies but all of them are always restricted to small samples directly related with the need of homogeneity of the studied groups and the special care required when performing studies in situations where drawbacks are not

M. E. Auer et al. (eds.), *Online Engineering and Society 4.0*, Lecture Notes in Networks and Systems 298, https://doi.org/10.1007/978-3-030-82529-4_40

expected. However, the conclusion of reference [2], "larger studies are needed to determine whether EMG BF assisted exercises provide additional benefits", is also common to many of the literature in the rehabilitation area [3].

Augmented feedback is also based on techniques used for a long time, which have been enriched with the development of new technologies. Since the 1990s, published work has been focused in physiological movement and neurophysiology, and shows that patients under treatment, using augmented feedback, can improve motor learning and increase their overall performance [4]. This last reference defends theoretically the benefits of incorporating augmented feedback, but it also consider that results are difficult to reach due to the shortness of the convenient samples, the short duration of tests and the need of a long-term follow-up.

Reference [5] presents a review on "augmented feedback in the rehabilitation of lower extremity musculoskeletal dysfunctions". It investigates the effect of verbal augmented feedback in this type of rehabilitation. Additionally, it also studies the effect of the verbal augmented feedback on motor learning, on stages of acquisition, retention and transfer in the patient process [5]. In conclusion, the authors consider, once again, the need of future studies conveniently designed to confirm the obtained results.

In another study "Augmented visual feedback-aided interventions for motor rehabilitation in Parkinson's disease: a systematic review" [6], the authors clearly state that "feedback appears to be useful" in motor rehabilitation but they conclude again that "high-quality, rigorous studies remain limited".

Recent studies show that the use of computer games custom-designed for one particular training tool and pathology/group of patients will contribute to engage them in physical exercise, transforming many routine exercises into pleasant training and allowing data record to evaluate and follow patients' progress [7–10]. In occupational therapy, the use of those games also contribute to preserve cognitive skills [11].

Therefore, the use of "smart devices" in eHealth has been receiving interest worldwide in many different areas, namely in rehabilitation and occupational therapy.

The System for Hand Rehabilitation in Dexterous Manipulation of Daily Objects (SHaRe) is a device shaped as an object of daily use (in this case, a drinking glass). It is instrumented and capable of interacting wirelessly with computer applications in order to enhance the cognitive process of re-learning control on the capability to hold and manipulate daily objects. The handgrip strength exerted when holding the glass can be measured and applications can provide visual and auditory feedback during the object manipulation promoting sensory re-education. The data from the sensors and the conducted activities can be stored for later evaluation. Most of the applications are developed in a gaming approach.

The concept of this tool for Hand Rehabilitation in Dexterous Manipulation of Daily Objects, was oriented to make it a very low price equipment at the same level of cost of common home blood pressure monitors, electric mucus suction pumps, nebulizers, etc. Therefore, being a kind of affordable equipment, patients could have them for home training therapy and its inherent features allow experts to follow up the patients' evolution and so, to reshape their therapy plan if required.

SHaRe has been used in different contexts with positive results and arousing great user interest and motivation. It has been used by patients with stroke sequels in Institute CUF, Porto, in rehabilitation of patients with multiple sclerosis at ANEM

(National Association for Multiple Sclerosis) and in School of Health of Polytechnic of Porto. In this work, the system will be briefly described and two different conducted studies with preliminary results will be summarized.

2 Study Description

2.1 The System

The SHaRe System assumes a glass shape object according to hand rehabilitation experts' suggestion due to the fact that is a very familiar daily object. Typically, a glass can be found to be made from different types of materials such as flexible plastic, hard plastic and glass, presenting a distinct behavior while being manipulated, so that the handgrip strength has to be properly adjusted. The need to re-educate the magnitude of the handgrip strength on the manipulation of objects made of different materials in patients with acquired brain injury is a common requirement that this system addresses by making use of augmented feedback to mimic the object deformation.

In fact, considering that the developed glass is made from hard plastic, the way to simulate different materials requiring distinct hand strengths makes use of augmented feedback. A model of the glass is presented on the computer screen and the shape of the model is deformed in reaction to the exerted force on the physical glass. For this purpose, the physical glass is instrumented with a force sensor and an inertial sensor measurement unit (IMU) integrating accelerometers, gyroscopes and magnetometers. Therefore, the orientation of the glass and its augmented deformation can be presented on the visual model within the computer applications. In this way it is possible to enhance the cognitive process of re-learning control of his/her capacity to apply strength and manipulate an object, by providing visual feedback during task performance. SHaRe communicates wirelessly with the computer software. Most of the applications are developed using a gaming approach. Auditory feedback is also added to this visual feedback through sounds modulated in amplitude by one of the measured quantities. The applications also provide recording the relevant data from the sensors and tasks performed for later evaluation The data can be stored in a database on the computer itself or in the cloud. The therapist may access this data and check the training history along the series of pre-defined activities, which will make it possible to assess the impacts from the used therapies and, eventually, introduce new approaches to the patient recovery program.

The prototype is produced with a 3D printer and its instrumentation allows the production of the hardware system at a very low cost. Figure 1 presents the prototype.

Due to the developed electronics and sensing architecture, the system can assume many different shapes for easily reproducing many different daily use objects.

For this system, different games were developed based in virtual environments with visual and auditory feedback [8, 12].

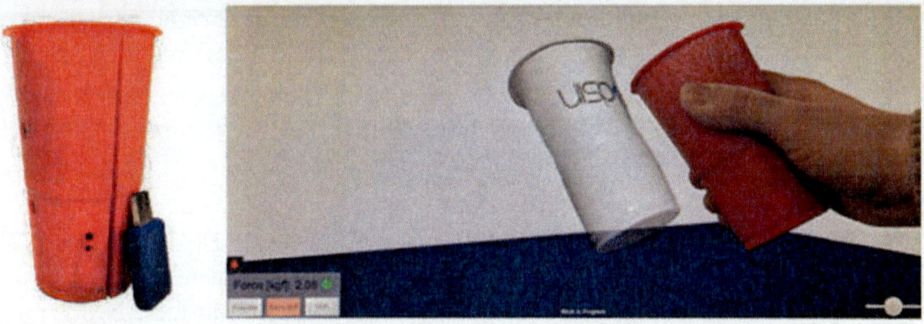

Fig. 1 SHaRe device (left) and an illustration of visual augmented feedback, allowing to select material on the lower left corner of the user interface (right)

2.2 The Study

The present study aims to evaluate the efficacy of an intervention based the use of the SHaRe system, in the control and maintenance of distal segment strength in adults with acquired brain injury, namely in a group of patients with stroke sequelae and in a patients suffering from multiple sclerosis (MS).

Stroke is one cause of acquired neurological injury that affects a large number of young adults. It is the second major cause of death and disability worldwide with over 13 million new cases every year. Stroke is defined as a clinical syndrome with rapidly developing clinical signs of focal disturbance of cerebral function (or global in case of coma), lasting more than 24 h or leading to death with no apparent cause other than a vascular origin [13].

Despite a high mortality rate, many patients survive but with disabilities. These sequelae may be temporary or permanent, compromising the performance and, consequently, affecting Activities of Daily Living (ADL) such as eating, dressing, personal hygiene. The use of daily equipment is also affected and some activities such as preparing meals, money management, laundry, among others [14].

Multiple Sclerosis (MS) is another cause of acquired brain injury. It is a chronic autoimmune inflammatory disease characterized by the presence of multifocal lesions in the Central Nervous System where demyelization and axonal loss predominate. Besides limitations in mobility, individuals with MS often have difficulty in reaching, grabbing and holding objects, which further affects their professional and social life [15].

Sample

The sample of this study follows the non-probabilistic and convenience method, consisting of individuals over 18 years with sequelae of stroke recruited from an elderly day center in the Porto region. A total of 25 individuals of both sexes, 85.7% male participated in the study. The mean age was 55.71 years (±11.19 standard deviation), the oldest participant was 72 years old and the youngest 37.

The sample of the study with patients with MS involved 7 participants, recruited from the National Association of Multiple Sclerosis of Porto (ANEM). The inclusion criteria consisted in adults older than 18 years diagnosed with sclerosis multiple. Exclusion criteria were the presence of a pathology other than MS that compromises upper limb movements, upper limb paralysis, use of prostheses or orthosis in the upper limbs, severe visual deficits and severe cognitive deficits. The sample consisted of 5 females (71.4%) and 2 males (28.6%). The mean age was 56.0 years (±7.21), with the oldest participant being 67 years old and the youngest 43. The MS was diagnosed on these patients for more than 18 years.

Procedures

The study was approved the Ethics Council of the School of Health of Porto. The selected participants signed the Informed Consent Form statement. The study took place at the two locations mentioned, an elderly day centre and at the ANEM.

An initial evaluation was performed in which a sociodemographic questionnaire was applied, with data related to gender, age, marital status, pathologies, use of support products, visual deficit, weight, height and time of diagnosis. Five evaluation instruments were also applied, at an initial and final stage, which were: FIM (Functional Independence Measure), DASH (Disabilities of Arm, Shoulder and Hand), MMSE (Mini Mental State Examination) Handgrip strength evaluation (Jamar Dynamometer) and final satisfaction questionnaire. With the SHaRe system, it was evaluated the capability to support the glass within the hand during the tasks.

The study took four weeks, involving three sessions per week, with a total of 12 sessions. Each session was guided by an Occupational Therapist. Apart from the initial and last session allocated to evaluation, in each session the patient carried out, the ShaRe system was used with distinct games during approximately 15 min.

For both studies, the SPSS Statistics 26 software package was used. For the tests, the significance level of 0.05 was considered. In the descriptive analysis, the results of the qualitative variables were demonstrated through absolute (n) and relative (%) frequencies, and the quantitative variables results were analyzed through the measures of central tendency (mean) and of dispersion (standard deviation).

Then, inferential analysis was used in order to compare means for the two distinct evaluation moments (pre and post intervention). Shapiro–Wilk test was used for testing variables normality (sample $n < 30$).

For the variables presenting normality the parametric test t-Student was used for paired samples. For the others the non-parametric test for paired samples Wilcoxon was used.

Results

From comparing the data in the two evaluation moments with both type of patients, it was possible to notice that integrating ShaRe in the intervention of individuals, improves the control of movement in the upper limb and hand strength.

From the results of the FIM assessments, it was verified an increase in the medium values obtained with this pilot study in both groups: in the stroke group, it was observed an increase in the order of 3%, and of 0.4% for the MS group. For both cases these results show an increase on the hand function, that are encouraging results, considering that in the MS group, the patients have been suffering from MS for more than 18 years and improvements are not usually expected at this evolution stage of the disease.

It should also be pointed out that a significant result was found on the patient capability to sustain the glass during the performance of multiple tasks, with an average increase of 67% (from 52 to 87 s).

Through statistical inference of results, it was not possible to draw conclusions about efficacy of this intervention, taking into consideration the small size of the involved samples. However, through the measures of central tendency, it was observed an improvement in the functionality, maximum manual strength and support time of an object in the hand.

3 Conclusions

These pilot studies show that the SHaRe system [12] can be a promising tool for hand strength and manipulation rehabilitation. Its hardware architecture allows being included in a variety of daily use objects shapes, and it communicates wirelessly with a computer/mobile device. The developed software applications take advantages of graphical interfaces enhanced by visual and auditory feedback, allowing the users to gain a clear perception of the manipulation activities they perform with the object. The set of the developed applications incorporating games illustrates the potential of using this instrumented device in e-rehabilitation.

The system has been used in a pilot study with two different groups of patients.

Within the group of 25 stroke patients, the results showed improvements in the functionality of the hand in what concerns the increase of the magnitude of the maximum force exerted and in the glass holding time. Patients' motivation on its use has also been very positive.

One of the biggest difficulty in people with stroke sequelae is the hand strength control. In this study, it was observed an improvement in this variable.

Through the satisfaction questionnaire it was possible to assess that 85% of the participants of both pilot studies were pleased with this type of intervention, and this can be explained, possibly, by the interactive dynamics that these games provide. This makes the intervention a fun and motivating activity, leading to a bigger adherence to the rehabilitation program.

In fact, the results are very promising for the area of physical rehabilitation. Despite the results obtained, it would be interesting to understand whether there are deficits in tactile and proprioceptive sensitivity and how they may influence the control and maintenance of strength when using the ShaRe system. Future studies should be conducted in order to understand the depth impact of this intervention on acquired brain injury.

This research was funded by of Project NORTE-01-0145-FEDER-000022 - SciTech - Science and Technology for Competitive and Sustainable Industries, co-financed by Programa Operacional Regional do Norte (NORTE2020), through Fundo Europeu de Desenvolvimento Regional (FEDER) and by the Portuguese Foundation for Science and Technology, under Project LAETA - UIDB/50022/2020.

References

1. Thompson SBN, Morgan M (1990) Occupational therapy for stroke rehabilitation https://doi.org/10.1007/978-1-4899-3083-5
2. Arpa S, Ozcakir S (2019) Does Electromyographic biofeedback improve exercise effects in hemiplegic patients? A pilot randomized controlled trial. J Rehabil Med 51(2):109–112
3. Giggins OM, Persson UM, Caulfield B (2013) Biofeedback in rehabilitation. J Neuroeng Rehabil 10:60. https://doi.org/10.1186/1743-0003-10-60.PMID:23777436;PMCID:PMC3687555
4. Hartvel A, Hegarty JR (1996) Augmented feedback and physiotherapy. Practice 82(8):480–490
5. Storberget M, Grødahl L, Snodgrass S, van Vliet P, Heneghan N (2017) Verbal augmented feedback in the rehabilitation of lower extremity musculoskeletal dysfunctions: a systematic review. BMJ Open Sport Exerc Med 3(1):e000256. https://doi.org/10.1136/bmjsem-2017-000256
6. Kearney E, Shellikeri S, Martino R, Yunusova Y (2019) Augmented visual feedback-aided interventions for motor rehabilitation in Parkinson's disease: a systematic review. Disabil Rehabil 41(9):995–1011. https://doi.org/10.1080/09638288.2017.1419292
7. Hung YX, Huang PC, Chen KT, Chu WC (2016) What do stroke patients look for in game-based rehabilitation: a survey study. Medicine 95(11):e3032. https://doi.org/10.1097/MD.0000000000003032
8. Carneiro F, Tavares R, Rodrigues J, Abreu P, Restivo MT (2018) A gamified approach for hand rehabilitation device. Int J Online Biomed Eng (iJOE) 14(01):179–186
9. Bonnechère B (2018) Serious games in rehabilitation. In: Serious games in physical rehabilitation. Springer, Cham. https://doi.org/10.1007/978-3-319-66122-3_4
10. David L, Bouyer G, Otmane S (2017) Towards a low-cost interactive system for motor self-rehabilitation after stroke. Int J Virtual Reality 17(02):40–45. ffhal-01634944f
11. Perry JC, Andureu J, Cavallaro FI, Veneman J, Carmien S, Keller T (2011) Effective game use in neurorehabilitation: user-centered perspectives. In: Handbook of research on improving learning and motivation through educational games: multidisciplinary approaches. IGI Global, pp 683–725
12. https://youtu.be/mt0CMLe7cz0?t=10. Accessed 07 Dec 2020
13. Chugh C (2019) Acute ischemic stroke: management approach. Indian J Crit Care Med 23 (Suppl 2):S140–S146
14. Schepers V, Ketelaar M, Port I, Visser-Meily JMA, Lindeman E (2007) Comparing contents of functional outcome measures in stroke rehabilitation using the International Classification of Functioning Disability and Health. Disabil Rehabil 29:221–230
15. Abdullah EJ, Badr HE, Manee F (2018) MS people's performance and satisfaction with daily occupations: implications for occupational therapy. OTJR Occup Part Health 38(1):28–37. https://doi.org/10.1177/1539449217719867

A Different View on Artificial Intelligence Applications for Cardiac Arrhythmia Detection and Classification

Dragoş-Vasile Bratu(✉), Maria-Alexandra Zolya, and Sorin-Aurel Moraru

Department of Automatics and Information Technology, Transilvania University of Braşov, Braşov, România
{dragos.bratu, smoraru}@unitbv.ro, maria-alexandra.zolya@student.unitbv.ro

Abstract. Ischemic heart disease and stroke are the world's highest killers. These diseases have continued the foremost cause of death globally in the last 15 years. An electrocardiogram (ECG or EKG) is a record of the electrical activity of the heart over a period of time, represented by one-dimensional data. In this paper, different methods such as Wavelet Transform will be applied to the ECG signal to increase accuracy on arrhythmia detection. The main arrhythmia disorders studied in this work were the following: atrial premature and supraventricular beat, the fusion of ventricular and normal beat, isolated QRS-like artifact, ventricular escape beat, and premature ventricular contraction. The annotation on each sample was made by certified cardiologists and each database was uploaded on Physionet (Research Resource for Complex Physiologic Signals). The input data used was 2D images instead of classical time-series data to demonstrate that the presented system can be used to identify with success different arrhythmias directly on images. In other words, a similar algorithm can be used in mobile devices such as smartphones.

The test accuracy obtained demonstrates the efficiency of the system that could be applied to an electrocardiogram to easily detect any specific arrhythmia disorders.

Keywords: ECG · Arrhythmia · Convolutional Neural Network · Physionet · Wavelet Transform

1 Introduction

Nowadays, machine-interpreted ECGs (produced by different devices) have diagnostic accuracy constraints [1] and require special human examination and supplementary interpretation. The availability of extremely precise and completely automated electrocardiogram evaluation utilizing machine learning algorithms has the potential to substantially strengthen the precision as well the time associated with the analysis of medical ECG. Taking into consideration the accessibility of on-demand electrocardiogram diagnostics (such as AliveCor, Apple, or some other portable ECG [14]),

M. E. Auer et al. (eds.), *Online Engineering and Society 4.0*, Lecture Notes in Networks and Systems 298, https://doi.org/10.1007/978-3-030-82529-4_41

accurate and automatically generated ECG analysis methods are now becoming more and more necessary [1].

The large volume of ECG data collected every day, in-home and hospital may prevent data analysis by human operators. Machine learning-based methods for ECG monitoring and interpretation range from traditional machine learning to deep learning, and their combination [2, 5]. In this paper, is reviewed how to improve [2, 11] learning for detecting arrhythmias.

Arrhythmia may lead to psychological distress and mortality. For example, as one of the most common arrhythmia, atrial fibrillation (AF) is associated with an increased risk of a heart attack. Simplicity and reduced cost make an ECG the most common method for identifying cardiac arrhythmias [2].

Deep learning has generated a fundamental shift in industry and academia in past years, with healthcare being one of the most significantly influenced domains.

Nowadays, technological developments in mathematical (statistics) regarding learning techniques and high computation power are capable of recognizing correlations in large amounts of data. In accordance with the already mention fields, the availability of a vast number of medical data offers an opportunity to evaluate automated medical diagnostics.

This paper proposes an advanced deep learning strategy for the identification of ECG arrhythmias using different methods: raw one-dimension ECG images signals and Wavelet Transform.

2 Methods and Apparatus

2.1 Data Collection and Preprocessing

The data that was used in this paper was annotated by certified cardiologists and each database was uploaded on Physionet (Research Resource for Complex Physiologic Signals) [15]. The origins of the ECGs included in the MIT-BIH Arrhythmia Database is a series of more than 4000 records of an ambulatory electrocardiography device (Holter) that have been acquired between 1975 and 1979 from the Beth Israel Hospital Arrhythmia Laboratory. Around 60% of these records have been made from hospital patients. The database contains twenty-three recordings randomly selected which are labelled from 100 to 124 and the other thirty-five records (labelled from 200 to 234 inclusive) are selected from the same set to include a variety of significant, important, and rare anomalies not well reflected by a small random selection of Holter recordings.

The MIT-BIH Arrhythmia Database presents 48 half-hour samples (each of the 48 records is slightly wider than 30 min) of two-channel ECG reports from 47 subjects aged between 23–89. The patients that were subjected were twenty-five men aged 32–89 years old, and twenty-two women aged 23–89 years old with the remark that 201 and 202 records were from the same male subject.

The databases used from Physionet are:

- MIT BIH Arrhythmia Database
- MIT BIH Supraventricular Arrhythmia Database

- St. Petersburg Institute of Cardiological Technic 12 lead Arrhythmia Database
- Sudden Cardiac Death Holter Database

In this study, for the first part, a Deep Neural Network was trained to detect arrhythmias using, seven diagnostics selected from all diagnostics: Normal beat (N), Atrial premature beat (A), Fusion of ventricular and normal beat (F), Isolated QRS-like artifact (|), Premature ventricular contraction (V), Ventricular escape beat (E) and Supraventricular premature or ectopic beat (atrial or nodal) (S). This selection was made on the one hand because of the high amount of data that had to be inputted into the neural network and on the other hand, it was easier to filter the data manually, process, and train only with seven classes. The raw signals are shown in Fig. 1. As can be seen in the already mention figure the ECG signals are time-series signals represented in a one-dimensional manner (1D) [8, 13]. The shape of the signal differs in terms of duration, amplitude, or frequency. Having this into consideration, using a Deep Neural Network to detect these different patterns is a good idea.

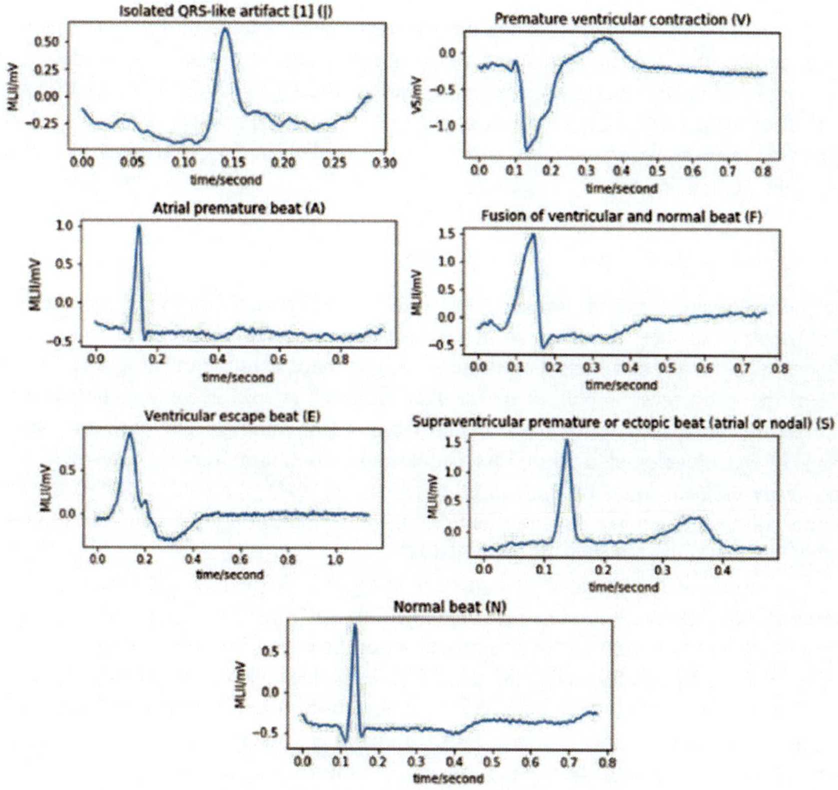

Fig. 1 Raw ECG Signals plotted, 7 classes

The first step before training the network was to create a Python script that will create all the samples as a .PNG file. This type of approach was chosen instead of just working with arrays of numbers because it was easier for the author of this paper to select "good signals" from bad ones. A bad signal means that the output contains noise, or it is far more different from the majority, or it was a mistake from the annotator point of view as can be seen in Fig. 2.

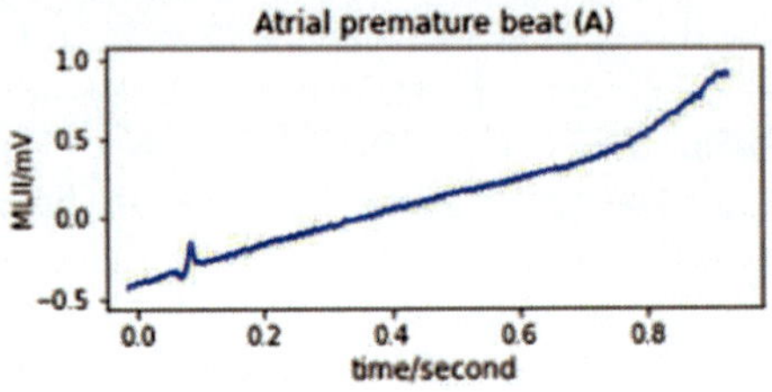

Fig. 2 Noisy Atrial premature beat

The window frame is most of the time approximately one second long (depending on the subject that was analyzed and where the R-Peak annotation was made). As a reference in order to detect the interesting part of the signal, R-Peak annotation was used. That means that signals are containing only one R-Peak per window and the signal was made by cropping the signal from 150 ms before current R-Peak to 150 ms before the next R-Peak.

2.2 Convolutional Neural Network Approach

The following subsection describes the architecture of the network, which was used in all of the experiments. The framework used for this study was Keras and it was chosen because of the areas in which the platform weighs more than the existing alternatives. One of the main reasons behind that is that in Keras simple or very complex architectures of a neural network can be made within a few minutes. The pythonic way in which it was developed, makes this framework more user-friendly alongside with modularity consequently, different prototypes can be made in a short period of time. Besides all the advantages already mentioned, Keras comes on top on another platform named TensorFlow [16] with all its benefits.

The design of the model that had been used is a Sequential one because and it allows the user to create models layer-by-layer, so each layer is connected to the next one. For the compilation part of the model an optimizer, a loss function and a list of metrics needed to be chosen by the user. The optimizer used was Adam [12] with a learning rate set at 0.0008. This optimization algorithm (used instead of the traditional Stochastic Gradient Descent) was used to update network weights iterative based in training data. Moreover, because it is computationally efficient, well adapted for challenges that are large in terms of data and parameters and hyper-parameters require little tuning [12]. The architecture has 12 layers and it looks as in Fig. 3 and the different types of available wavelets filters are shown in Fig. 4.

```
Layer (type)                 Output Shape              Param #
=================================================================
conv2d_9 (Conv2D)            (None, 275, 105, 32)      320
_________________________________________________________________
max_pooling2d_9 (MaxPooling2 (None, 137, 52, 32)       0
_________________________________________________________________
conv2d_10 (Conv2D)           (None, 135, 50, 64)       18496
_________________________________________________________________
max_pooling2d_10 (MaxPooling (None, 67, 25, 64)        0
_________________________________________________________________
conv2d_11 (Conv2D)           (None, 65, 23, 128)       73856
_________________________________________________________________
max_pooling2d_11 (MaxPooling (None, 32, 11, 128)       0
_________________________________________________________________
conv2d_12 (Conv2D)           (None, 30, 9, 128)        147584
_________________________________________________________________
max_pooling2d_12 (MaxPooling (None, 15, 4, 128)        0
_________________________________________________________________
flatten_3 (Flatten)          (None, 7680)              0
_________________________________________________________________
dropout_3 (Dropout)          (None, 7680)              0
_________________________________________________________________
dense_5 (Dense)              (None, 512)               3932672
_________________________________________________________________
dense_6 (Dense)              (None, 7)                 3591
=================================================================
Total params: 4,176,519
Trainable params: 4,176,519
Non-trainable params: 0
```

Fig. 3 Convolutional Neural Network—Network architecture

The Adam's learning rate parameter (also referred as an alpha parameter) was chosen by just tuning it until a satisfactory accuracy was met. The other parameters of the optimizer were left untouched as in recommended by the paper (beta_1 = 0.9, beta_2 = 0.999, epsilon = 1e-08, decay = 0.0) [12]. The loss function was selected to be Mean-Squared-Error (L2 Loss or Quadratic loss) as it is sensible to outliers and yet gives a more stable and closed-form solution. The list of metrics was set to "accuracy" as this is a classification problem.

The model requires to know how the input shape will look like to know what to expect. The first layer in the Sequential model (and only the first one) needs to receive information about its input shape since the following layers can do automatic shape inference so it was set to (277, 107, 3) which represents the size of the image in pixels and the channel (3 is for colour mode RGB). The next layer is a Conv2D layer, which is a convolution layer. As the entire network is a **C**onvolutional **N**eural **N**etwork (CNN) specialized for working in this case with two-dimensional image data (can be used for one-dimensional or three-dimensional data too). The Convolution Layer from this network is applying a linear operation. This involves specific multiplication of a set of weights with the input, similar to a traditional **N**eural **N**etwork (**NN**). In other words, this layer creates a convolution kernel making a bond with input layers, which helps produce a tensor of outputs. The kernel is smaller in size than the input data and the type of multiplication utilized within a filter-sized patch of the input and the filter is a dot product. As the activation function, non-linear function **ReLU** (**R**ectified **L**inear **U**nit) was preferred. The main advantage of it is that because of the 0-value on the negative axis the network will run faster. In other words, the computation load is

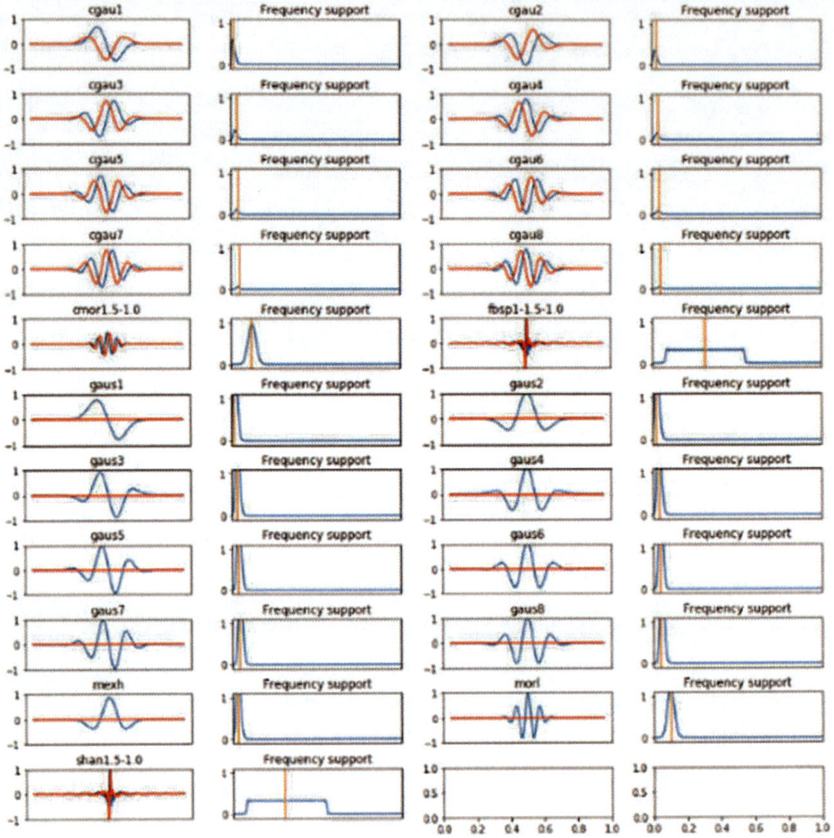

Fig. 4 Different types of wavelets from scaleogram python library

smaller than other computation of activation functions (such as Sigmoid or Hyperbolic Tangent functions). The next layers contain Maximum Pooling, or Max Pooling, which is a downsampling strategy (sample-based discretization process) consisting in a pooling operation that calculates the maximum, or largest value in each patch of each feature map. That result in decreasing its dimensionality and is allowing assumptions to be made about features included in the binned (grouped) sub-regions.

In order to avoid overfitting and high computational cost besides the Max Pooling layer, the Dropout layer was utilized which withdraws 50% of nodes during training.

In the last layer, which is a Dense one, where all the inputs are connected to the output by a weight, the activation function was chosen Sigmoid. The main reason why the Sigmoid function was preferred is that it exists between (0–1). Consequently, it is especially applied for models where the probability needs to be predicted as an output. Considering this, the probability of anything exists only between the range of 0 and 1 is the main reason why Sigmoid was the right choice.

For the generator, as a callback function besides "ModelCheckpoint" callback, "EarlyStopping" callback was applied which as the name suggests will halt the training of the neural network at the right time avoiding overfitting or underfitting if the training does not increase by "val_acc" metric for 20 consecutive epochs.

Wavelet Transformation Approach
Wavelet Transform [9, 18] is a mathematical method for performing signal analysis when signal frequency ranges over a period of time. Wavelet analysis provides increased precise information about the signal data than other signal analysis techniques such as **F**ourier **T**ransform (**FT**). This type of transform can be used in speech, image and audio [17] processing and nevertheless in the biomedical domain [2, 3, 8]. The Wavelet Transform was preferred over other transforms because it is using a series of functions named wavelets, each with a distinct scale. The Fourier Transform applies a range of sine-waves with various frequencies to analyze a signal which means that the signal is reproduced by a linear combination of sine-waves.

The foremost difference implies that a wavelet is localized in time and the sine-wave (FT) is not limited in time which means that it stretches out from −infinity to +infinity. This provides an advantage of the Wavelet Transform to concern time-information and frequency information. Similar to a kernel in image processing the wavelet is applied, as in a convolution process, on the signal from the start towards the end of it.

As it can be seen in Fig. 5, the Wavelet Transform of the one-dimensional ECG data signal will have two dimensions. The second dimension outputted after applying the Wavelet Transform is the time-scale representation of the signal in the form of a scaleogram. There are two types of transforms the Continuous and the Discrete Wavelet Transform. For this study, the first one was chosen.

Continuous **W**avelet **T**ransform (**CWT**) is an implementation of the wavelet transform using random wavelets and scales. The wavelets used can be treated as statistically independent and the data produced by this transform being deeply correlated. The CWT operates by applying directly the definition of the wavelet transform, i.e. we are computing a convolution of the signal with the scaled wavelet. For each scale, we obtain an array of the same length as the data signal has.

The decision of the wavelet that is used for time–frequency decomposition is important as it can affect the time and frequency resolution of the result. For the Morlet wavelet [18] for example, (real part is the damped cosine function) high-frequency resolution can be exacted as this wavelet is very well localized in frequencies. In contrast, if **D**erivative **o**f **G**aussian (**DOG**) wavelet is used that will end on sufficient time localization, but smaller in frequencies. Fourier Transform just has one kind of transformation but Wavelet Transform can have many kinds of transformation (the possibilities of the kind of transformation are infinite).

Wavelet **T**ransform (**WT**) is broadly applied [3, 4, 7, 8] to represent the ECG morphology characteristics because it is sturdy to turbulences such as noises and can efficiently describe information about time and frequency in diverse resolutions.

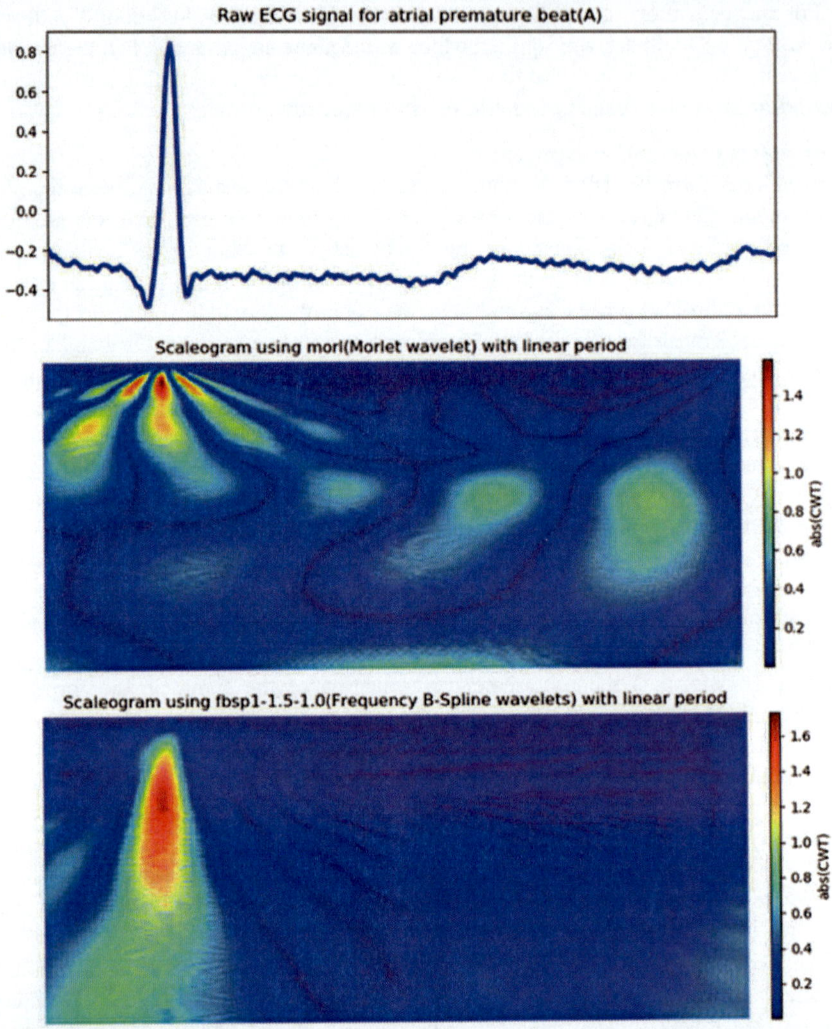

Fig. 5 Plot signal of an atrial premature beat and transformation to wavelet transform (morl and fbsp)

By applying WT onto the ECG, differences between arrhythmias signals will be easier to be detected by the Neural Network. The proposed system utilises Morlet wavelet transform and Frequency B-Spline wavelet transform to identify better the arrhythmias from ECG signal. Taking this into consideration, both of the transforms are more perceptive to the variance of non-stationary signal and can be considered a better compensation.

The ECG data were transformed into a scaleogram which is a visual representation of the spectrum of frequencies of the ECG signal while it modifies with time. For that "scaleogram" Python library was used to generate the images alongside with "Matplotlib" library. The scaleogram image looks like in Fig. 5.

3 Results and Discussions

In this section, the findings from the experiments are being presented. As it was already mentioned, in this study four notable different experiments were driven (Table 1).

Table 1 Training results - seven classes raw signal Gray images filtered (1000 balanced samples)

Epoch	Network metrics				
	Loss	Acc	val_loss	val_acc	Time (s)
1	0.1228	0.2200	0.0401	0.8247	9
2	0.0435	0.8122	0.0166	0.9253	8
3	0.0303	0.8663	0.0194	0.9155	8
4	0.0241	0.8888	0.0122	0.9513	8
10	0.0094	0.9594	0.0080	0.9696	8
11	0.0097	0.9566	0.0135	0.9416	8
30	0.0034	0.9869	0.0068	**0.9805**	**8**

The input data samples were images (size 277 × 107 pixels) having different representation and a distinctive number of samples.

3.1 First Experiment 1000 One-Dimensional ECG Filtered Samples – 7 Balanced Classes

For the first experiment, a total number of 1000 samples were used for training the network. From this 1000 samples, 20% were used for testing and the remaining 80% were used for testing. The signals were arranged in seven distinct classes (as mention in second chapter) and the images contained only the signal in one-dimension format (Fig. 1). As it can be seen in Table 1, in 30 epochs the Convolutional Neural Network reached 98.05% detection rate. In this test, the input data, similar to Fig. 1 was chosen manually removing any noisy and non-similar samples. After epoch numbered 30 the model did not improved its "val_acc" metric so the training was halted.

3.2 Second Experiment 1000 Scaleograms ECG Filtered Samples Using Morlet Wavelet – 7 Balanced Classes

In the second experiment the same input data was used, the exception was that signal instead of being a one-dimensional one it was transformed into a scaleogram with the Morl Wavelet Transform (Fig. 5). What is interesting to see here is from the first epoch we have a higher value to the "acc" and "val_acc" metric, which shows the benefit of Continous Wavelet Transform. Another interesting part is that is getting in the 6th epoch the best value for "val_acc" and the model stops training as it gets no better (Table 2).

Table 2 Training results - seven classes wavelet images filtered (morl) (1000 balanced samples)

Epoch	Network metrics				
	Loss	Acc	val_loss	val_acc	Time (s)
1	0.0720	0.8549	0.0221	0.9594	16
2	0.0326	0.9399	0.0403	0.9115	15
3	0.0234	0.9579	0.0263	0.9469	15
4	0.0211	0.9612	0.0323	0.9311	15
6	0.0166	0.9702	0.0105	**0.9869**	**15**

It is impressive to see that this is more than two times better than the first experiment but having in mind that the time spent on an epoch was almost double in the second experiment.

This is because in the first experiment the network was tuned as the input images colour mode is grayscale and in this experiment was set to RGB mode to apply a convolution kernel that takes care over different colour needed for WT. For the first two experiments, data were obtained from MIT BIH Arrhythmia Database [10], Supraventricular Arrhythmia Database, St. Petersburg Institute of Cardiological Technics 12-lead Arrhythmia Database v1.0.0 and Sudden Cardiac Death Holter Database v1.0.

3.3 Third Experiment – 3000 Unfiltered ECG Samples 1D – 3 Classes

For the third and fourth experiment (Table 3) the samples used were only from MIT BIH Arrhythmia Database and were trained only for three classes: Normal beat (N), Atrial premature beat (A), Fusion of ventricular and normal beat (F). In this particular situation, the number of samples was increased to 3000 samples with the same 20–80% ratio for test and train. Besides that, the main differences were this time that the input data was not filtered by hand anymore so noisy samples (Fig. 2) were inside the train and test folder. Remarkably enough, from the first epoch, the "val_acc" metric hit 93.13% which on the one hand proves that by having more samples the accuracy will increase, on the other hand, proves that noisy data does not have so much importance when the number of samples increases.

Table 3 Training results - three classes raw signal color images unfiltered (3000 balanced samples)

Epoch	Network metrics				
	Loss	Acc	val_loss	val_acc	Time (s)
1	0.1054	0.7789	0.0436	0.9313	23
2	0.0437	0.9194	0.0292	0.9406	19
3	0.0301	0.9493	0.0271	0.9556	21
4	0.0249	0.9568	0.0145	0.9688	20
5	0.0216	0.9619	0.0144	0.9778	21
7	0.0185	0.9686	0.0119	0.9810	20
10	0.0149	0.9760	0.0091	**0.9844**	**20**

In this experiment, similarly as in the first experiment, the input images contained only the ECG signal in one dimension. It is interesting to see that the model stops training at the 10th epoch with an accuracy higher than the first experiment but not much more than the second experiment where the Wavelet Transform was used. Interesting enough, the time spent on an epoch is larger than both first experiments (Table 4).

Table 4 Training results three classes wavelet color images unfiltered (fbsp1-1.5-1.0) (3000 samples)

Epoch	Network metrics				
	Loss	Acc	val_loss	val_acc	Time (s)
1	0.1290	0.7269	0.0981	0.7844	25
2	0.0772	0.8559	0.0637	0.8594	23
3	0.0485	0.9063	0.0481	0.9139	24
4	0.0339	0.9406	0.0335	0.9406	23
5	0.0314	0.9413	0.0297	0.9470	24
7	0.0261	0.9550	0.0197	0.9719	24
43	0.0055	0.9916	0.0033	**0.9938**	**23**

This is because the number of samples was increased and because also the Convolutional Neural Network colour mode parameter was configured as RGB (similar to the second experiment). Another result that needs to be highlighted here is that the third experiment had in the forth epoch the best value for "val_acc" hitting more than 96% (Table 3) compared to the other two experiments.

3.4 Forth Experiment 3000 Unfiltered ECG Two-Dimensional Images Using Frequency B-Spline Wavelet – 3 Classes

Things become more interesting in the forth experiment as shown in Table 4 where the Wavelet Transform was changed to Frequency B-Spline Wavelet (fbsp1-1.5-1) (Fig. 5) using the same 3000 number of samples as in the third experiment. Here the same as in the previous experiment the data was unfiltered. In this case, the training went slower and the number of epochs until it reached the maximum value for "val_acc" metric was much higher than all the experiments. Despite this fact, the current experiment hit the best accuracy score of all the experiments. That proves once again that Wavelet Transformation with a proper wavelet function can have pretty strong accuracy (99.38% in epoch number 43). The increased time per epoch was expected as the number of samples was increased and the complexity of the images was increased by applying Frequency B-Spline Wavelet.

4 Conclusions

In conclusion, the author of this paper trained and design a deep learning model (using a Convolutional Neural Network) for the detection of cardiac arrhythmias from an ECG and managed to obtain superior performances up to 99.38%. In this study is proposed a 12 layer Convolutional Neural Network that depending on the input data will have better results if the data is expressed in a two-dimensional format by using Wavelet Transform such as Morlet Wavelet or Frequency B-Spline Wavelet. Another impressive finding is that by using Wavelet Transform with some low numbers of samples for training the model can have better detection rates than if the number of samples was higher but the training samples were in a one-dimension format (as in time-series).

False arrhythmia detection still poses a challenge [6] as the data collected can be noisy or if the input data samples still are not sufficient to detect most of the arrhythmias. Another challenge is the detection of some uncommon arrhythmias for which the number of samples provided from the database are insufficient for doing proper training and the signal of the shape has some uncommon features. Augmenting those rare conditions can be one of the solutions for lacking of train data but samples from the field should be more effective and chosen with priority.

Automated detection can help cardiologists to give a diagnostic faster as in the case of long-term ECG to detect anomalies will take more time. Furthermore, nowadays, smartwatches, portable ECGs and holters are becoming more popular and affordable for the public [14]. That means that the patient by itself can detect at home by using this type of device when a type of arrhythmia occurred (during high effort, sleep, stress, etc.). Supplementing that it is monitored 24h a day so in case of a problem can discuss with a cardiologist and offer valuable information.

References

1. Rajput KS, Wibowo S, Hao C, Majmudar M (2019) On arrhythmia detection by deep learning and multidimensional representation arXiv:1904.00138v4
2. Parvaneh S, Rubin J, Babaeizadeh S, Xu-Wilson M (2019) Cardiac arrhythmia detection using deep learning: a review. J Electrocardiol
3. Schiff SJ, Aldroubi A, Unser M, Sato S (1994) Fast wavelet transformation of EEG. Electroencephalogr Clin Neurophysiol 91(1994):442–455

4. Addison PS (2005) Wavelet transforms and the ECG: a review. Physiol Meas 26:R155–R199
5. Gupta A, Huerta EA, Zhao Z, Moussa I (2019) Deep Learning for Cardiologist-level Myocardial Infarction Detection in Electrocardiograms. arXiv:1912.07618v1
6. Mousavi S, Fotoohinasab A, Afghah F (2019) Single-modal and multi-modal false arrhythmia alarm reduction using attention-based convolutional and recurrent neural networks. arXiv:1909.11791v1
7. Barmase S, Das S, Mukhopadhyay S (2013) Wavelet Transform-Based Analysis of QRS complex in ECG Signals. arXiv:1311.6460
8. Kiranyaz S, Ince T, Gabbouj M (2016) Real- time patient-specific ECG classification by 1-d convolutional neural networks. IEEE Trans Biomed Eng 63(3):664–675
9. Yu, S-N, Chen Y-H (2007) Electrocardiogram beat classification based on wavelet transformation and probabilistic neural network. Pattern Recogn Lett 28(10):1142–1150
10. Moody GB, Mark RG (2001) The impact of the MIT-BIH arrhythmia database. IEEE Eng Med Biol
11. Li Y, Pang Y, Wang K, Li X (2020) Toward Improving ECG biometric identification using cascaded convolutional neural networks. Neurocomput J
12. Kingma DP, Ba JL (2017) Adam: a method for stochastic optimization. arXiv:1412.6980v9
13. Jero SE, Ramu P, Ramakrishnan S (2015) ECG steganography using curvelet transform. Biomed Signal Process Control 22:161–169
14. Bansal A, Joshil R (2017) Portable out-of-hospital electrocardiography: a review of current technologies. J Arrythmiahttps://doi.org/10.1002/joa3.12035
15. Moody GB, Mark RG (2001) The impact of the MIT-BIH arrhythmia database. IEEE Eng Med Biol 20(3):45–50. PMID: 11446209
16. Abadi M, Agarwal A, Barham P, Brevdo E, Chen Z, Citro C, Corrado GS, Davis A, Dean J, Devin M, Ghemawat S, Goodfellow I, Harp A, Irving G, Isard M, Jozefowicz R, Jia Y, Kaiser L, Kudlur M, Levenberg J, Mané D, Schuster M, Monga R, Moore S, Murray D, Olah C, Shlens J, Steiner B, Sutskever I, Talwar K, Tucker P, Vanhoucke V, Vasudevan V, Viégas F, Vinyals O, Warden P, Wattenberg M, Wicke M, Yu Y, Zheng X (2015) TensorFlow: Large-scale machine learning on heterogeneous systems. tensorflow.org
17. Purwins H, Li B, Virtanen T, Schlüter J, Chang S, Sainath T (2019) Deep learning for audio signal processing. J Sel. Top. Signal Process. 13(2):206–219
18. Unser M, Aldroubi A (1996) A review of wavelets in biomedical applications. Proc IEEE 84 (4)

Assessing mHealth Applications for Medical Prescription: A Case Study

Mario A. Bochicchio[1(✉)], Lucia Vaira[1], Andrea Mortara[2], and Renata De Maria[3]

[1] Department of Innovation Engineering, University of Salento, Lecce, Italy
{mario.bochicchio, lucia.vaira}@unisalento.it
[2] Department of Cardiology, Monza Polyclinic, Monza, Italy
[3] National Research Council, Inst. of Clin. Phys., ASST G.M. Hospital Niguarda, Milan, Italy

Abstract. Smartphones and mobile health applications could be very helpful to support patients with chronic non communicable disease in following the complex protocols needed to manage their diseases. A number of mobile health applications have been developed to ease self-monitoring of symptoms, vital signs and physical activity; regular medication intake; and education of patients and caregivers about the relevant aspects of the disease. However, several barriers make it difficult to fully exploit their technological potential and produce clear clinical evidence of their effectiveness. App providers do not help users distinguish between useless/dangerous apps and valid solutions. The latter are few and often characterized by rapid obsolescence, lack of interactivity and lack of authoritative information. Then, systematic assessments can help physicians and researchers to rank Apps, support their continuous improvement, find the 'best candidate for a specific health problem' in a repeatable manner and pave the way for well-grounded and fruitful discussion on their effectiveness. To this purpose, the authors discuss a flexible method and a supporting tool to assess mobile health application for chronic diseases, based on the Internatinal Measurement System (IMS) scoring system, on the Mobile Application Rating Scale (MARS) and other minor customizations. The method has been co-designed by IT experts, health professional and patients to guarantee the due rigor while including, from the beginning, privacy and confidentiality aspects as well as ease of use, sustainability and significance of results. To exemplify the application of the proposed method, the assessment of a real mobile health application is discussed in the last part of the paper.

Keywords: Digital health · mHealth · Usability assessment · Clinical effectiveness · Heart failure · Chronic care management · Self-care

1 Introduction and Background

Mobile Health (mHealth in the following) is defined by the World Health Organization (WHO) as "the use of mobile wireless technologies for public health" [1] and, in more detail, as the "medical and public health practice supported by mobile devices, such as mobile phones, patient monitoring devices, personal digital assistants, and other wireless

M. E. Auer et al. (eds.), *Online Engineering and Society 4.0*, Lecture Notes in Networks and Systems 298, https://doi.org/10.1007/978-3-030-82529-4_42

devices" [2]. As of October 2020, there were 5.2 billion of unique mobile phone users worldwide, demonstrating a 67% penetration and a 2.0% growth rate compared to 2019 [3], with a huge potential to bring mHealth products and services all over the world. Software applications running on mHealth devices, commonly defined as "mHealth Apps", are targeted to different areas such as wellness management, behavior change, health data collection, disease management, self-diagnosis, and rehabilitation, and medication reminder. Due to the large number of mHealth Apps available from app providers' stores, many assessment techniques have been developed to support continuous improvement of these apps [4, 5], but two mHealth App categories, among others, are subject to special attentions and to specific laws and rules. These are:

- **Digital Therapeutics** mHealth Apps (**DTx** in the following) [6]. These can be used for therapeutic (software-based) interventions to improve the outcome of a patient's disease for which they are indicated by a physician. They are developed and studied through a randomized controlled clinical trial (using the same methodology for studying traditional drugs), follow a path to approval by regulatory agencies, can be prescribed only by a physician and, in specific cases, reimbursed. The concept behind digital therapy is the active ingredient: no longer chemistry and biology, but algorithms and software.
- **Digital Patient Support Programs** (**PSP** in the following). These are interventions of various kinds that do not have therapeutic properties but, rather, the purpose of optimizing the therapy (pharmacological or otherwise) that the patient is taking [7]. PSP Apps are intended to:
  - record patient information, such as clinical diaries or questionnaires,
  - facilitate patient communication with their physician, other health care providers or other patients,
  - offer tools for screening, diagnostic framing or monitoring of disease progression or therapy;
  - encourage patient adherence to therapy, through alerts and alerts related to taking therapy or due dates for scheduled visits.

Indeed, given the therapeutic value of DTx and PSPs Apps, according to the European Medical Devices Regulation [8] they require very different design/assessment pathways than all other Apps. The reason is that they:

1. are intended for medical treatments of patients, whose intellectual and/or physical capacities are potentially impaired by the disease or who (if caregivers) lack in-depth medical knowledge about the disease and treatment techniques. On the other hand, these users are very likely to stop using the prescribed mHealth Apps in case of poor design, buggy implementation, lack of feedback about their effectiveness or other usability issues [9]. Hence, a suitable user experience and a good level of usability are essential to the success of the prescribed medical treatment.
2. must pass a phase of clinical trials that, potentially, imposes functional and usability constraints, quite different from those normally used for the design of the User Experience of normal Apps for mobile devices. Clinical trials are experiments,

performed on volunteer patients, that follow the principles of evidence-based medicine and are governed by very strict criteria to protect the health of the volunteers participating in the experiment and of future patients subjected to that treatment. Their primary purpose is to accurately and repeatably measure the clinical effectiveness of the given treatment and to exclude side effects worse than the treated disease.

Even if new assessment approaches are proposed [10, 11] to deal with mobile Apps in general, and mHealth Apps in particular, nothing exists, to the best of our knowledge, to satisfy the complex requirements connected to the assessment of DTx and PSP Apps. This is probably due to the fact that, on Dec. 2020, only 11 DTx mHealth Apps exists (their names, web sites, producers, dates of approval and indicated diseases are reported in [12]). Moreover, only 3 papers are reported by Google Scholar when searching for "Digital Patient Support Programs" (no scientific papers are reported by IEEE Xplore, ACM Digital Library and MS Academic Search). On the other hand, assessing these mHealth Apps is very important because of the relevance of the two App categories for many patients, physicians and caregivers and for several diseases. The relevance is also confirmed by the list of primary pharmaceutical companies that have signed agreements with the manufacturers of the aforementioned DTx Apps for the production and distribution of new DTx Apps, which includes Roche, Sanofi, Novartis, AstraZeneca, Abvie, Pfizer, Bayer, Novo Norisk and others [12].

For these reasons we decided to start from existing assessment methodologies and extend them to consider in an organic way the requirements synthetically expressed in the above-mentioned points 1 and 2. The description of these aspects is reported in Sect. 2. To verify the sustainability and effectiveness of the proposed method, we applied it to a case study: the PSP App called CuorMio, developed by ANMCO (Associazione Nazionale Medici Cardiologi Ospedalieri). The description of the choices underlying the analysis and the results obtained are reported in Sect. 3. Conclusions and future developments are reported in Sect. 4.

2 The Proposed Approach

In the EU, all medical devices require CE marking, which is the manufacturer's claim that a product complies with all EU medical device directives, is fit for purpose, and is safe [8]. When medical devices are based on digital portable hardware, the related software is generically referred to as "mHealth App". This type of software is currently classified according to its purpose: if it has a medical purpose (e.g. prevention, diagnosis, monitoring, and treatment of any disease), it is considered a medical device. However, if it provides patients with disease-related education, monitors their fitness, health or wellness, stores and transmits their medical data, but is not intended for medical treatment, then it is not considered a medical device. However, it is important to note that the presence of the CE mark does not imply that the application meets best practices or has been tested for accuracy or clinical benefits, since the latter requirements can only be verified by purposely designed clinical trials.

mHealth Apps classified as DTx or PSP can only be approved by the EMA (European Medicines Agency) and by the corresponding national authorities, after a period of actual usage by the intended volunteer patients, under strict medical control after assessment of the clinical evidence of efficacy i.e. that the treatment is beneficial for the patient in the framework of a clinical trial. However patients and caregivers are very likely to stop using the prescribed mHealth Apps in case of poor design, buggy implementation, lack of feedback about their effectiveness or other usability issues [9].

Therefore, a sustainable evaluation methodology should be based on a co-design approach, involving all stakeholders, to consider:

- the needs, expressed by physicians participating in the trial, to minimize the operational complexity and time required to perform clinical evaluations on patients and to analyze the results without affecting the quality of the clinical research;
- the needs, expressed by patients and their organizations, in terms of actual and perceived usefulness, ergonomics, usability, operational simplicity, direct engagement in all decisions about patients 'health, respect for privacy and confidentiality;
- the need, expressed by medical societies, for a correct diagnostic and therapeutic approach, as well as correct relationships among physicians, patients and App providers involved in the medical treatment.

The purpose of this analysis is not to create a new assessment method from scratch, but to leverage existing assessment methods by "assembling" their most useful parts and creating a "balanced mix," with possible additions for requirements not already covered by existing methods. To achieve this goal, after analyzing several assessment approaches [4, 5, 10, 11, 13, 14] and their application to known mHealth Apps, we decided to integrate the IMS scoring system [15], and the MARS assessment [16] approach, to support them with a purposely-developed assessment tool based on Open Source Software and to complete them with complimentary actions and questions for both patients and physicians involved in the assessment. Referring to IMS and MARS, in more details:

1. in 2013 the IMS Institute for Healthcare Informatics proposed a scoring system which has been used to analyze more than 40,000 health Apps available for download from various providers. Then it has been adopted by several researchers for assessment and comparison purposes. As shown in Table 1, the scoring system is based on 11-items. The higher the score, the more complete and potentially helpful the mHealth App is. Unlike other assessment, the IMS scoring system is based on an objective evaluation of app functionality and does not reflect subjective patient or physician judgments [16].
2. In 2015, MARS (Mobile App Rating Scale) was introduced by Stoyanov et al. as a reliable, multidimensional measure for trialing, classifying, and rating the quality of mobile health apps [16]. It includes 5 subscales (engagement, functionality, aesthetics, information, subjective quality), plus an App-specific perceived impact section. Items are scored on a 5-point scale (1 inadequate to 5-excellent).

To perform an effective assessment session on a given DTx or PSP mHealth App, IMS and MARS tests cannot be administered "in isolation". To receive a realistic feedback, in fact, we need to recreate, or at least to simulate, the same scenario experienced by patients, physicians and caregivers during the clinical test. To achieve this goal, given the disease for which the App has been created:

1. we carefully instruct all stakeholders about the steps of the (simulated) trial, which include, for example, a baseline visit, a follow-up period, in which the patient uses the App at home, and a final visit;
2. We create a suitable set of questions (IMS + MARS + personal profile questions) for physicians and patients, to be deployed online on a digital survey platform;
3. We create a suitable set of digital contents (texts, images, videos, animations, tutorials, …), to be deployed online on a content management system platform, to instruct and support physicians and patients in all operative steps;
4. We start the (simulated) trial. During this period, we observe how users behave, by analyzing (if available) the system logs and report generated by the App under test. Then, we collect the user-generated data and we support physicians and patients in case of problems.
5. At the end of the (simulated) trial, we support physicians to complete their data analyses, then we collect and analyze the results of IMS and MARS questionnaires. These data contain useful insights to rank or correct or improve the App under test.

Table 1 IMS Institute for Healthcare Informatics App Functionality Scoring System

Functionality Scoring Criteria	Description
1. Inform	Provides information in a variety of formats (text, photo, video)
2. Instruct	Provides instructions to the user
3. Record	Capture user-entered data
a. Collect data	Able to enter and store health data on individual phone
b. Share data	Able to transmit health data
c. Evaluate data	Able to evaluate the health data entered by patient and provider, provider and administrator, or patient and caregiver
d. Intervene	Able to send alerts based on the data collected or propose behavioral intervention or changes
4. Display	Graphically display user-entered data/output user-entered data
5. Guide	Provide guidance based on user-entered information, and may further offer a diagnosis, or recommend a consultation with a physician/a course of treatment
6. Remind or alert	Provide reminders to the user
7. Communicate	Provide communication between healthcare professionals and patients and/or provide links to social networks

3 Case Study

To verify the sustainability and the effectiveness of the proposed method, we applied it to the PSP mHealth App named "CuorMio", developed by the ANMCO (Associazione Nazionale Medici Cardiologi Ospedalieri) in Italy. Subsect. 3.1 describes the criteria that led us to identify the App for assessment, Subsect. 3.2 describes how the assessment process was carried out, and Subsect. 3.3 describes the results of the assessment process and proposes possible improvements to the CuorMio App that cmcrgcd from the analysis of the results (the App is available for iPad and iPhone at https://apps.apple.com/us/app/cuormio/id1472008868#?platform=iphone; it is also available for the Android Italian market at https://play.google.com/store/apps/details?id=it.cncgroup.scompenso&hl=it).

3.1 CuorMio: A Digital App for HF Patients

Clinical outcomes in chronic diseases are often influenced by patients' cultural level, self-management skills and ability to early identify how symptoms are evolving. Polypharmacy (i.e. the use of five or more medications at a time), multimorbidity and disabilities are common n patients with chronic diseases. This can make adherence to treatment plans complex and require constant monitoring of physiological parameters, as well as a strong pressure to observe strict and stressful prescriptions for diet, physical activity and daily actions. Innovative Patient Support Programs (PSP) can be very helpful to improve the self-management of such complex, costly and relapsing conditions. Evidence of its clinical benefits is still scarce, but the ubiquitous presence of smartphones and wearable devices and the increasing aptitude of patient of all ages and conditions to use them for wellness, health and care reasons makes PSP Apps very interesting for self-management support.

Among other chronic diseases, Heart Failure (HF in the following) patients are excellent candidates to verify how beneficial PSP Apps could be, because these Apps may help reduce fragmentation of care and poor adherence to polypharmacy, which are crucial factors for HF outcomes. On the other hand there are problems and barriers because HF is a disease of the elderly: 80% of prevalent cases in Europe and in the US are aged 65 or more, with a median of 76 years [17]. This implies the presence of physical barriers, including sensory, cognitive and motor changes related to ageing, and psychological barriers such as the poor acceptance of technological devices by the older population [18]. The proportion of older people who own and use a smartphone decreases with increasing age, from 59% (for age 65–69) to 31% (for age 75–79) to just 17% of people aged >80 years [19]. Health problems and specific disabilities are the most common reasons for not using digital technologies. Many older people lack confidence in their ability to learn about and properly use electronic devices: over three-quarters of US subjects aged 65+ reported that they need help by someone to set up and use a new electronic device [19]. This may undermine patients' motivation to engage with PSP Apps, lead patients to poorly rate App effectiveness and limit their perceived value.

These barriers may be impossible to overcome when paticnt have to use poorly designed PSP Apps. Designers should consider very seriously the difficulties that patients are likely to experience and consider the user-centric design features as "first-class citizen" in their Apps [18]. Issues that have been highlighted as particularly important in the literature include easy navigation, restricting the number of items in the menu, streamlining the data entry processes, making recovery from errors clear and simplifying visualizations of data patterns.

Because of all these reasons, the HF working group of ANMCO (Associazione Nazionale Medici Cardiologi Ospedalieri) designed a PSP App to support HF patients in self-care. The App, officially available for download on the Google Play Store, represent an excellent case study for our approach, because it has been created by a well-known and reputed scientific society of cardiologist to respect carefully the clinical protocols related to HF. Moreover, the App (four screenshots are shown in Fig. 1) has to be used by real patients to improve their self-care skills (i.e. to prevent re-hospitalization and major adverse cardiac events after hospital discharge, to improve their clinical stability, autonomy and recovery of function and to allow the continuation of daily activities, including work and caregiving) and to support their adherence to medical treatment.

The objective of the assessment, discussed in the next subsections, is the evaluation of the strengths and weaknesses of the App, as a starting point to design an improved version of the App itself. For comparative purposes, the CuorMio App, with other similar Apps, has been assessed also by [20].

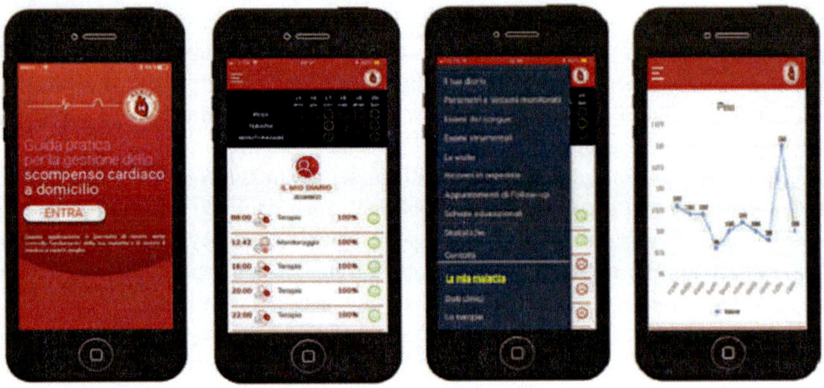

Fig. 1 CuorMio: user interface

3.2 Assessing CuorMio

According to the procedure described in Sect. 2:

Step 1: on September 2019 we designed an experimental assessment based on the voluntary participation of about 20 cardiologists and 40 HF patients. This was not a problem in our case because Since the goal of CuorMio, and of PSP Apps in general, is to support the patient self-management, the experimental design did not foresee collection, storage, processing nor analysis of sensitive data, unlike a clinical trial. Each patient can collect his/her (possibly encrypted) data on his/her smartphone and can share (by manually showing them) data with the attending cardiologist but cannot use the PSP App to transmit data to other participants.

Step 2: we prepared an IMS questionnaire, a MARS questionnaire and a third questionnaire to profile patients according to their technical skills, age and cultural level. Then we published all questionnaires, as shown in Fig. 1, on a purposely configured instance of LimeSurvey available (in Italian) on http://cuormio.unisalento.it/.

Step 3: we created a set of digital contents (texts, images and video tutorials) to train cardiologists and patients in all steps of the experiment and we published it on a purposely configured instance of WordPress. Platform and contents are available at the above-mentioned URL.

Step 4: on November 2019 we started the assessment with 22 cardiologists, expert in HF care (13 M/9 F, age 51 ± 9 yrs.); each of them had to enroll 2 HF patients equipped with smartphones, who volunteered to collaborate. Patients were invited to download the App from online stores. Cardiologists helped patients to set up the App on their devices by entering clinical data, individual HF therapy, and predefined custom ranges of parameters to be monitored.

Step 5: we concluded all experimental activities on Jan. 2020 by collecting and analyzing the answers of the proposed questionnaires and the feedback from the participating cardiologists and patients.

It is worth noting that the procedure is easy to customize to support the assessment of other DTx or PSP Apps. The platform (a page of which is shown in Fig. 2) can also be easily reused and adapted for similar evaluation projects.

Fig. 2 CuorMio assessment platform - questionnaire page

3.3 Results and Discussion

At the study kick-off, on November 2019, 43 patients (M, 90%, age 64 ± 12 years) accepted to volunteer for the assessment project. Their educational level was medium–high (62% high-school or degree). Half of them were able to fully exploit their smartphones using mail, chats, video and digital services. Two patients dropped out during the trial and one was hospitalized on week 2, while 40 patients (93%) successfully completed the study.

During the study, patients were asked to record three parameters (heart rate, blood pressure, weight) on a daily basis. This task was completed, on average, by 62% of patients. 42% of parameters were recorded for more than 80% of the study period. Patients' MARS scores averaged 3.44 ± 0.76 (32% scored $\geq$ 4), with the highest ranking for the perceived impact section (3.74 ± 1.07, 50% scored $\geq$ 4). No relevant correlations were found among MARS score, percent of usage time, socio-demographics level and digital literacy characteristics. Cardiologists' MARS score averaged 3.64 ± 0.44 (20% scored $\geq$ 4), with the highest ranking for the perceived impact Sect. (4.03 ± 0.68, 50% scored $\geq$ 4). Physicians' sex or age had no impact on rating. Figure 3 depicts overall and subscale MARS scores. Patients' and cardiologists' score were directly correlated (rho 0.399, $p = 0.013$). IMS score achieved Inform = 1, Instruct = 1, Record = 1, Collect = 1, Shre = 0, Evluate = 1, Intervene = 1, Display = 1, Guide = 1, Alert = 1, Communicate = 0, for total of 9.

A detailed discussion of the achieved quantitative results and of possible improvements for the CuorMio PSP App are out of the scope of this article, whose twofold focus is to present a novel assessment method for two very important categories of mHealth Apps and to discuss a case study about the proposed method.

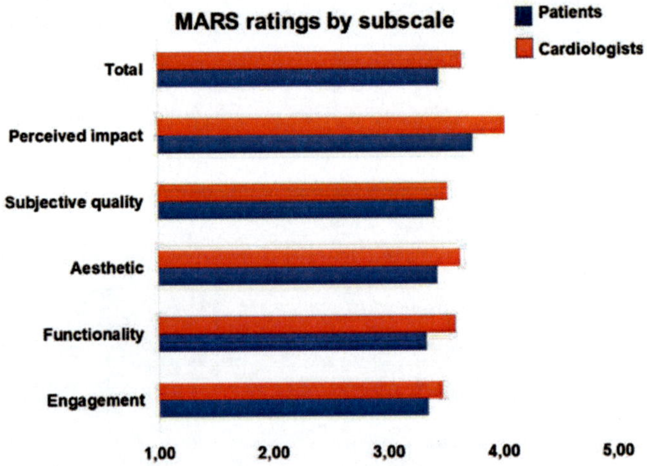

Fig. 3 CuorMio – MARS rating

4 Conclusions and Future Steps

There is great potential for mHealth apps to develop innovative care pathways, to foster patient engagement in self-care and to improve the interaction among patients, caregivers and health professionals in a cost-effective way. To achieve these goals, new tools and new methods must be developed, supporting all phases of mHealth App lifecycle with specific attention to the clinical trial phases and to the assessment procedures.

To contribute to this effort in the article we propose a collaborative method, involving patients and physicians, to assess Digital Therapeutics (DTx) Apps and Digital Patient Support Program (PSP) Apps. Then, we discuss a case study in which CuorMio, i.e. a real PSP App developed in EU by a scientific society of cardiologists for self-management of Heart Failure, is assessed with our approach. The assessment took about 2.5 months and involved 20 cardiologists, 43 patients and 2 ICT researchers.

As well as for the development of new drugs in the pharma sector, the clinical trial approach, complex, long and resource intensive is currently the only accepted way at the regulatory level for approval of DTx Apps according to the new EU rules on medical devices, with the direct involvement of patients and physicians. A sound standardized method to ensure that only DTx Apps with a high and tested usability profile and positive user experience enter the clinical trial process is important to maximize external validity of trials and minimize waste of economic and professional resources.

References

1. Report by the Secretariat, "mHealth: use of mobile wireless technologies for public health." World Health Organization, May 27, 2016. https://apps.who.int/gb/ebwha/pdf_files/EB139/B139_8-en.pdf. Accessed 06 Sep 2020
2. WHO Global Observatory for eHealth and World Health Organization, MHealth: new horizons for health through mobile technologies. World Health Organization, Geneva (2011)
3. "Digital 2020: Social Media Users Pass the 4 Billion Mark as Globl Adoption Soars," We Are Social. https://wearesocial.com/blog/2020/10/social-media-users-pass-the-4-billion-mark-as-global-adoption-soars. Accessed 20 Dec 2020
4. "Developing A Set Of Usability Guidelines For Mobile Health Applications." https://www.meddeviceonline.com/doc/developing-a-set-of-usability-guidelines-for-mobile-health-applications-0001. Accessed 07 Sep 2020
5. Zapata BC, Fernández-Alemán JL, Idri A, Toval A (2015) Empirical studies on usability of mhealth apps: a systematic literature review. J Med Syst 39(2):1. https://doi.org/10.1007/s10916-014-0182-2
6. Kvedar JC, Fogel AL, Elenko E, Zohar D (2006) Digital medicine's march on chronic disease. Nat Biotechnol 34(3). https://doi.org/10.1038/nbt.3495. Art no 3
7. Ganguli A, Clewell J, Shillington AC (2016) The impact of patient support programs on adherence, clinical, humanistic, and economic patient outcomes: a targeted systematic review. Patient Prefer. Adherence 10:711
8. "Regulation (EU) 2017/745 of the European Parliament and of the Council of 5 April 2017 on medical devices." https://eur-lex.europa.eu/eli/reg/2017/745/oj. Accessed 04 Jan 2021
9. "European Commission - GREEN PAPER on mobile Health - 2014.pdf." https://ec.europa.eu/transparency/regdoc/rep/1/2014/EN/1-2014-219-EN-F1-1.Pdf. Accessed 03 Jan 2021
10. Paglialonga A, Lugo A, Santoro E (2018) An overview on the emerging area of identification, characterization, and assessment of health apps. J Biomed Inform 83:97–102. https://doi.org/10.1016/j.jbi.2018.05.017

11. "Xcertia mHelth App Guidellines - 2019 final.pdf." https://www.himss.org/sites/hde/files/media/file/2020/04/17/xcertia-guidelines-2019-final.pdf. Accessed 07 Sep 2020
12. daVinciDTx, "Terapie Digitali e Farmaco, sostituzione o combinazione?," *TerapieDigitali*, July 11, 2020. https://terapiedigitali.davincidtx.com/terapie-digitali-e-farmaco/. Accessed 04 Jan 2021
13. Krebs P, Duncan DT (2015) Health app use among us mobile phone owners: a national survey. JMIR MHealth UHealth 3(4):e101. https://doi.org/10.2196/mhealth.4924
14. Zhou L, Bao J, Setiawan IMA, Saptono A, Parmanto B (2019) The mHealth App Usability Questionnaire (MAUQ): development and validation study. JMIR MHealth UHealth 7(4): e11500. https://doi.org/10.2196/11500
15. Zhang C, Zhang X, Halstead-Nussloch R (2014) Assessment metrics, challenges, and strategies for mobile health apps. Issues Inf Syst 15
16. Stoyanov SR, Hides L, Kavanagh DJ, Zelenko O, Tjondronegoro D, Mani M (2015) Mobile app rating scale: a new tool for assessing the quality of health mobile apps. JMIR MHealth UHealth 3(1):e27. https://doi.org/10.2196/mhealth.3422
17. Bui AL, Horwich TB, Fonarow GC (2011) Epidemiology and risk profile of heart failure. Nat. Rev. Cardiol. 8(1):30–41. https://doi.org/10.1038/nrcardio.2010.165
18. Joe J, Demiris G (2013) Older adults and mobile phones for health: a review. J Biomed Inform 46(5):947–954. https://doi.org/10.1016/j.jbi.2013.06.008
19. Hern, A., et al.: "Infographic: Are seniors as crazy for the internet as the younger generation?," *Techaeris*, March 22, 2019. https://techaeris.com/2019/03/22/infographic-are-seniors-as-crazy-for-the-internet-as-the-younger-generation/. Accessed 05 Jan 2021
20. Mortara A, et al (2020) Would you prescribe mobile health apps for heart failure self-care? an integrated review of commercially available mobile technology for heart failure patients. Card Fail Rev 6. https://doi.org/10.15420/cfr.2019.11

GiroJampa: A Serious Game Prototype for Wheelchairs Rehabilitation

Marieliza Araújo Braga(✉), Liliane dos Santos Machado, Leonardo Wanderley Lopes, Clauirton de Albuquerque Siebra, Elaine Cappellazzo Souto, and Rebeca Raab Bias Ramos

Universidade Federal da Paraíba, João Pessoa, PB, Brasil
liliane@di.ufpb.br, cas@cin.ufpe.br

Abstract. Spinal cord injury is an impairment in the spinal canal, that leads to sensorimotor deficiency, causing dependency by motor impairment, affecting not only the physical aspect, but also the psychoemotional. Serious games are games whose purpose lies beyond entertainment, and they can be applied in a multitude of activities within the health field, including rehabilitation. This work's goal is show the results from the development of a serious game prototype. The game was designed to aid in both the motor rehabilitation and the psychoemotional aspects of patients with thoracolumbosacral spinal cord injury, wheelchair users, through a decision-making model that balance difficulty from user performance. The development process was structured in three stages: the first was the elaboration of project step diagram; the second consisted of the game's design; and the third consisted of defining the prototype tests. The prototype contains the basic elements of a game, with definition of the scenario and aesthetics, technology, narrative, and definition of the decision-making model, in addition to having a multidisciplinary team to define the priority points for approaching the game, attending to the target audience needs.

Keywords: Serious games · Rehabilitation · Wheelchair users

1 Introduction

Spinal cord injury is any impairment in the spinal canal caused by compression or rupture, resulting in sensorimotor deficiency and alteration in the autonomic nervous system (ANS), according to the compromised vertebral level [1, 2]. The injury can have a traumatic and non-traumatic origin: where the first is caused by direct trauma, often secondary to accidents and falls; and the second by different pathologies [3, 4]. In addition to sensory-motor and ANS disorders, the psychoemotionalfactor is also affected in injured individuals [5].

The comorbidities caused by spinal cord injury affects the individual in multiple aspects: motor, interfering in his locomotion and independence, and consequently psychosocial [6]. Due to such limitations, they tend to become insecure, dependent, and isolated, showing fear, difficulty in socializing and locomotion. The dependency factor generates emotional overload, and the accessibility issues can generate social isolation and difficulties of reinsertion in the society [6, 7].

M. E. Auer et al. (eds.), *Online Engineering and Society 4.0*, Lecture Notes in Networks and Systems 298, https://doi.org/10.1007/978-3-030-82529-4_43

Physiotherapy is an important ally in increasing the life of these individuals, contributing to their independence, autonomy, and psychoemotionalcondition. Through physical therapy prescriptions, neuroplasticity and motor learning are favored by the repetition of exercises, generating improvement in the kinetic-functional condition and functional independence measure [8, 9].

Serious games (SG) are a physiotherapeutic prescription option, characterized as digital games with a purpose that goes beyond entertainment. They are used in several areas of health, from the training of professionals, to the use in rehabilitation, with recreation as a primary factor in the good acceptance of the user [10, 11].

Martín-Ruiz, Máximo-Bocanegra and Luna-Oliva [12] used SG for facial motricity training in children and observed that they hear a positive response regarding facial muscle tone. Martins, Carvalho and Soares [13] used SG in the motor rehabilitation process of patients affected by stroke and reported improvement in the individual's participation in the rehabilitation process. Considering SG applied in the rehabilitation of patients with spinal cord injury, Oliveira et al. [14] presented a prototype that uses the Kinect sensor to capture articular movements of the user's wrist. This movement, captured by the sensor, generates the movement of the avatar in the game. These authors point out that there are two benefits of using SG: the availability of its use anywhere (with a Kinect sensor present) and social inclusion.

Several elements are necessary to compose the SG: the narrative, responsible for the story presented by the game; aesthetics, which captures the user's attention with sound and visual resources; the technology, focused on the user's interaction with the game, responsible for capturing the input data generated by the user's action; and mechanics, where the intelligence of the game is found [15, 16]. A decision-making model, implemented in the game's intelligence and based on input data related to the user's performance when performing the actions, can provide feedback to the player, for example, providing new phases and balancing the difficulty [16, 17].

Therefore, this work intends to present the results obtained by the prototype of the serious game GiroJampa, developed to aid in motor rehabilitation and improvement of psychoemotional aspects of individuals with thoracolumbosacral spinal cord injury, using a decision-making model to assess user performance for game difficulty balancing.

2 Methods

The game development process was composed by three stages: project step diagram; game design; and prototype tests.

- 1st Stage: Project Step Diagram
  A project step diagram was elaborated based on the existing theme's literature, in the recognition of the target population needs, and in the necessary steps for the development of an SG. A requirements document was carried out based on the following topics: purpose of using the SG; main subject; target audience; scenario; difficulty gradation. From the requirements document, the necessary elements for the development of the SG were defined [17]. Figure 1 shows the stages of the design and development process for the serious game GiroJampa (see Fig. 1).

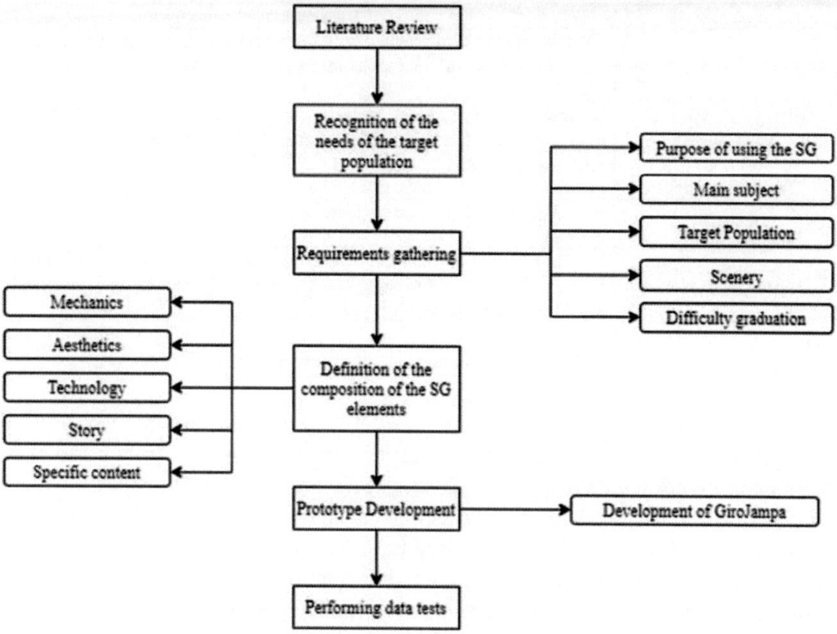

Fig. 1 Project step diagram

The game's name "Girojampa" is a portmanteau word, derived from the Portuguese word "Giro", which stands for the circular movement of the wheels of a wheelchair; and the word "Jampa", a nickname of the city in which the SG scenario was inspired (João Pessoa, capital of the state of Paraíba, in Brazil). From this junction, the term "GiroJampa" emerged.

- 2nd Stage: Game design

  The game design was developed, following the expanded Schell's Elemental Tetrad, considering the elements: narrative, mechanics, aesthetics, and technology, always related to the specific content [17]. The needs and lessons of potential players were mapped from a visit to the Physical Conditioning Laboratory for individuals with spinal cord injury, linked to the Department of Physical Education of the Federal University of Paraíba in Brazil. After that, two requirements of the game were realized based on the existing literature, in the observations of two users of the referred laboratory (wheelchair users), and the elements of the Expanded Schell's Elemental Tetrad [17] necessary for the composition of an SG (see Fig. 2).

After surveying the requirements, the basic elements that compose a serious game were defined [17]. The prototype architecture was developed for mobile devices that use Android as operating system. For visualization and interaction was considered a first person viewing by the use of a smartphone docked to a head up display.

Requirements	Description
Purpose of using the SG	To stimulate the displacement with the wheelchair, in a playful and safe way, of patients with thoracolumbosacral spinal cord injury, wheelchair users.
Main subject	Difficulty of the patient affected by thoracolumbosacral medullary injury, wheelchair users, when moving around with the wheelchair in open environments.
Target Population	Patients with thoracolumbosacral spinal cord injury, wheelchair users.
Scenery	Beachborder of João Pessoa, from the bust of Tamandaré to the Hotel Tambaú.
Difficulty graduation	It will be carried out from the adaptation of a decision model, which will balance the level of difficulty, based on the evaluation of the variables; in addition,a nd overall user rating.

Fig. 2 – GiroJampa requirements

The Borg Scale of Perceived Exertion was developed to assess perceived effort by the user after physical activity [20] and consists of 15 points, graded from 6 (no effort) to 20 (maximum effort). Its use in the SG was necessary for the intelligence module to identify how much each task physically requires from the user. For data collection, line sensing, developed by Souto, Siebra e Silva [18], was used, where a sensor connected to an ergometer without wheels [19] captures information in each task and transmits it to a mobile device. From these data, the decision-making method present in the game's intelligence performs the gradation of the difficulty of the subsequent tasks. This gradation considers the effort perceived from the Borg Scale [20].

Figure 3 shows the design of the GiroJampa with its platform and workflow. The prototype used previously recorded interaction data obtained by the sensor.

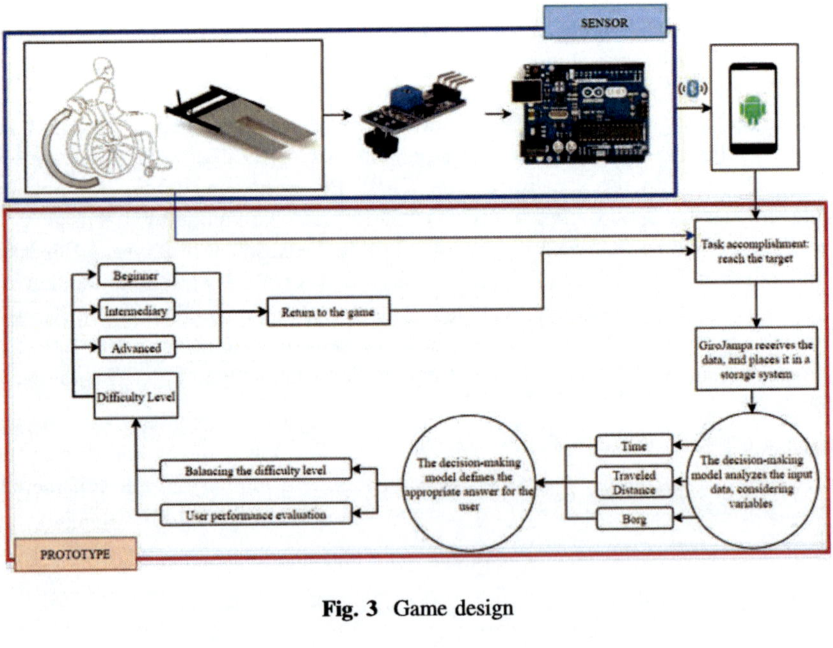

Fig. 3 Game design

- 3rd Stage: prototype test
  This third stage is focused on testing, where several aspects of the prototype are evaluated, such as the efficiency of all programming scripts (avatar movement and camera movement from interaction data); and whether the decision-making model was successful or not in correctly adjusting the difficulty for the game's tasks from interaction data.

3 Results

The narrative was based on a character in a wheelchair who wants to meet his fiends. During the game play, the user must travel by the seafront sidewalk of the main beach of João Pessoa city. The scenario was developed in a cartoonized way and with bright colors. The objective of the character is to reach specific points and meet his friends.

Scripts were developed in C# language to move user's avatar; changing scenes; camera movement and access to external data. The visualization style is in first person because the player visualizes the scene through the eyes of the avatar (see Fig. 4). The user movements with wheelchair are captured by sensor and feed the avatar movements in the game.

Fig. 4 GiroJampa prototype

The ergometer consists of a platform fixed to the floor, and of two cylinders, where the wheelchair is positioned on top and secured. When the user moves the wheelchair's wheels, simultaneously, the ergometer cylinders rotate without moving the wheelchair forward. The rotation of the ergometer cylinders, fixed to the floor, results from the movement of the chair wheels and generates the data captured by a line sensor, sent to an Arduino, which performs the calculations, and sends it to the mobile device where the serious game is installed by bluetooth [18]. From the use of the sensed wheelchair, the object that represents the avatar moves forward. In the serious game there are no opponent players, and the user must overcome their own difficulties.

As the users progresses in a task, they receive new goals to be achieved. The action of trying to reach them, by the movements produced using the sensed wheelchair, will generate input data which will be analyzed by the decision-making model. This model provides a response which considers the user's performance and adjusts the next task difficulty. For this, three variables are used: distance covered, time and perception of effort, according to Borg's perception of effort [20].

There are three levels of difficulty, where the individual can be allocated according to their performance assessment by the decision-making model: beginner, intermediate and advanced. Each level has specific goals and activities to be carried out, as well as transition milestones between the levels of difficulty (see Fig. 5).

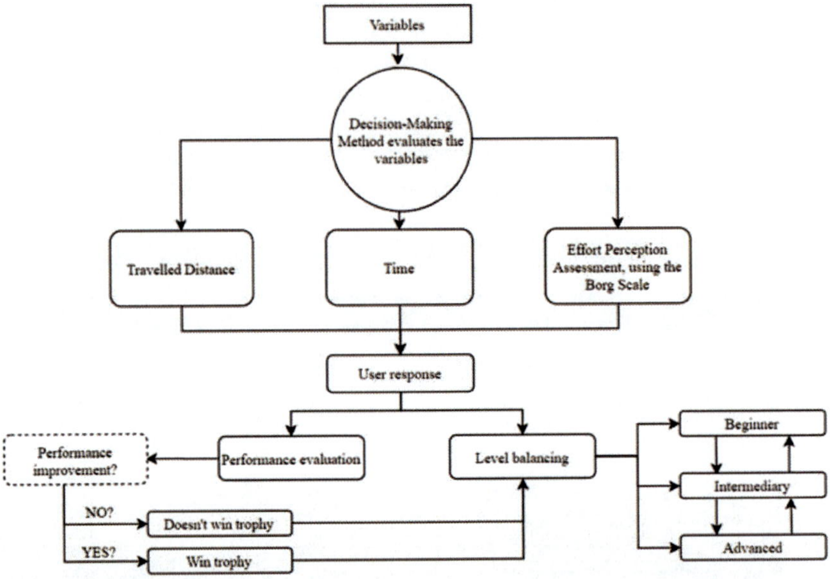

Fig. 5 – Variables and responses generated by the game intelligence decision-making method

Let's consider an example of an activity performed by the individual: the player at the beginner level will have the goal of reaching 300 m, in *X* minutes, with perceived effort, considering the Borg Scale, between 6 (no effort) and 12 (Light). In the game environment, the task will be to move around the seafront setting. In the real environment, this player will be attached to the ergometer and the data will be captured and sent to the mobile device, as described by Souto, Siebra e Silva [18]. When the activity is finished, the player will indicate how much effort was perceived by him, considering the Borg Scale. The variables will be read and analyzed by the decision-making model, which will generate a response in relation to the evaluation of the user's performance and will carry out the balance of difficulty.

If the distance traveled determined as a goal has not been reached, the player remains at the beginner level; if the character reaches the goal, the player will receive the trophy of distance traveled and will be assessed on how long it took to be covered the determined distance. If the time is more than *X* minutes, the player will remain at the beginner level; if the time is equal to or less than *X* minutes, it will be assessed how much this player declared of perceived effort at the end of the activity, considering the Borg Scale. If the declaration of effort is higher than 13 (hard), the beginner level will be maintained; if the declaration of effort is equal to or lower than 12 (easy), the player will advance to the intermediate level (see Fig. 6).

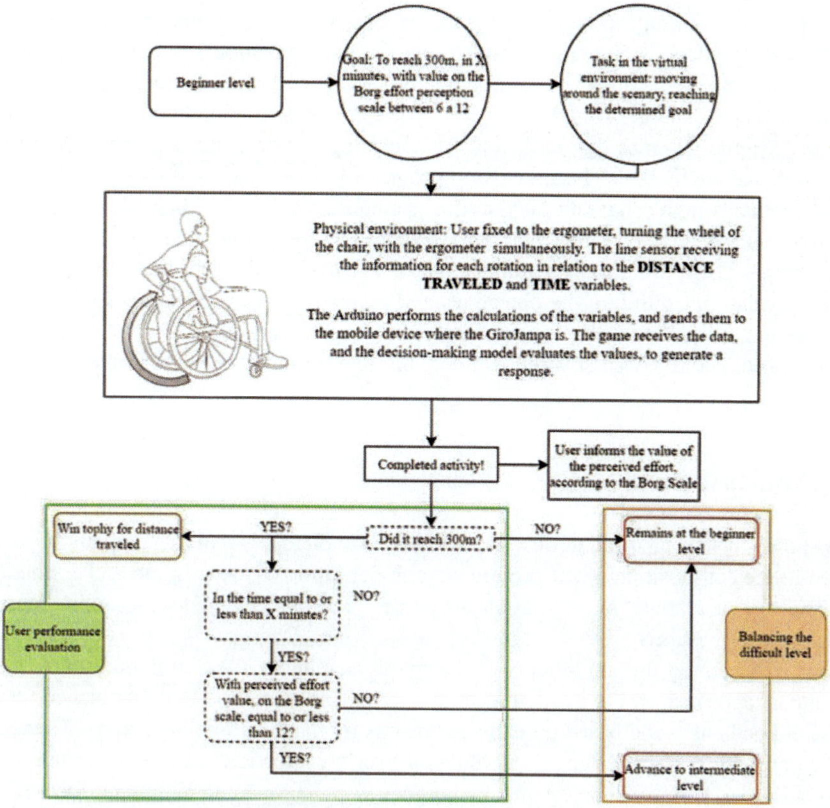

Fig. 6 Game response, made by the decision-making model, to user action

To test the decision-making model in the game, secondary data were used, collected by the system of capture and visualization of physical parameters, developed by Souto, Siebra and Silva [18], including simulated data on perception assessment of effort, considering the Borg Scale. These data were captured from individuals who used the sensed system, still not connected to the game, and integrated for testing the SG prototype. Thus, this data was considered as input from a real user and used to test the prototype and game intelligence. With this, it was possible to verify the probable suitability of SG GiroJampa for motor rehabilitation and improvement of psychoemotional aspects of individuals with thoracolumbosacral spinal cord injury.

In regard to user experience, the game has been developed in accordance to the Game Flow model – a theory that identifies and describes strategies to cause a feeling of enjoyment when playing a game [21]. Sweetser and Wyeth [21] divides the Game Flow model into eight categories, with the objective of assessing enjoyment in games in the use of digital games, they are: concentration, challenge, player skill, control, clear objectives, Feedback, Immersion and Social Interaction. According to such criteria, the game must require concentration of the player, with challenges that hold the user's attention, requiring skills and supporting the development of new ones, providing a sense of control, giving adequate feedbacks, providing immersion experience, and creating social interaction. Considering them, GiroJampa has been designed to provide an environment of concentration for the user through the scenario, the proposed activities and the feedback offered by the SG, encouraging him to overcome the challenges and advance the SG levels, by balancing the difficulty according with user performance evaluation. The immersion provided by the head up display was thought as an elements than can contributed to the user concentration. The balance of difficulty, performance board and target points are examples of elements that can contribute to enjoyment. Then, users will be guided on how to carry out the activities, taking control of the avatar and fulfilling the objectives and goals that will be clear, and will respect the user's ability. As feedback, the user will have a screen presentation of data such as time, speed, and distance covered, while playing, monitoring their performance during use.

4 Conclusions

As shown, it was possible to observe that the basic elements that make up an SG were used in the composition of the prototype, with definition of the scenario and aesthetics, technology, narrative, and definition of the decision-making model that will generate feedback to the user.

It was noticed the importance of the multidisciplinary team to define the priority points of approach in the game, responding to the needs of the target audience, in an efficient and easy, and based on expected results by the health point of view. Then, the adoption of a scale of effort perceived as a user's evaluation parameter is important, since, despite being a digital game, its purpose goes beyond entertainment, focused on rehabilitation. This requires the need for a controlled virtual environment, respecting and adapting to the user's needs and limitations.

The prototype cannot be tested with wheelchair users due to the limitations imposed by the COVID-19 pandemic. However, it is worth noting that the data used in the tests are from real users and were previously been collected using the sensed platform. The final game GiroJampa will be tested in real wheelchair users, respecting ethical aspects, using the GameFlow questionnaire to assess the user's enjoyment [21].

Acknowledgements. The authors would like to thank CNPQ (process 315298/2018-9).

References

1. Ahuja CS, Wilson JR, Nori S, Kotter MRN, Druschel C, Curt A et al (2017) Traumatic spinal cord injury. Nat Rev 3(17018):1–21. https://doi.org/10.1038/nrdp.2017.18
2. Alizadeh A, Dyck SM, Karimi-Abdolrezaee S (2019) Traumatic spinal cord injury: Na overview of pathophysiology, models and acute injury mechanisms. Front Neurol 10 (282):1–25. https://doi.org/10.3389/fneur.2019.00282
3. Buker DB, Oyarce CC, Plaza RS (2018) Effects of spinal cord injury in heart rate variability after acute and chronic exercise: a systematic review. Top Spinal Cord Injury Rehab 24 (2):167–176. https://doi.org/10.1310/sci17-00028
4. Patek M, Stewart M (2020) Spinal cord injury. Anaesth Intens Care 21(8):411–416. https://doi.org/10.1016/j.mpaic.2020.05.006
5. Kang Y, Ding H, Zhou HX, Wei ZJ, Liu L, Pan DY et al (2017) Epidemiology of worldwide spinal cord injury: a literature review. J Neurorestoratol 2018(6):1–9
6. Tzanos IA, Mavrogenis A, Gioti K, Papagelopoulos P, Panagiotopoulos E (2018) Depressive mood in individuals with spinal cord injury (SCI) living in Greece. Spinal Cord 56:883–889. https://doi.org/10.1038/s41393-018-0093-z
7. Lim SW, Shiue YL, Ho CH, Yu SC, Kao PH, Wang JJ et al (2017) Anxiety and depression in patients with traumatic spinal cord injury: a nationwide population-based cohort study. PLoS ONE 12(1):1–14. https://doi.org/10.1371/journal.pone.0169623
8. Harvey LA (2016) Physiotherapy rehabilitation for people with spinal cord injuries. J Physiother 62(1):4–11. https://doi.org/10.2147/JN.S143236
9. Williams TL, Smith B, Papathomas A (2016) Physical activity promotion for people with spinal cord injury: physiotherapists' beliefs and actions. Disabil Rehabil 40(1):52–61. https://doi.org/10.1080/09638288.2016.1242176
10. Batista TV, Machado LS, Valença AM (2019) FarMyo: a serious game for: hand and wrist rehabilitation using a low-cost electromyography device. Int J Ser Games 6(2):3–19. https://doi.org/10.17083/ijsg.v6i2.290
11. Machado LS, Moraes RM (2020) Innovative technologies to support education and training: researches by LabTEVE. Comunicações em Informática 4:50–53. https://doi.org/10.22478/ufpb.2595-0622.2020v4n2.54653
12. Martín-Ruiz ML, Máximo-Bocanegra N, Luna-Oliva L (2016) A virtual environment to improve the detection of oral-facial malfunction in children with cerebral palsy. Sensors 16 (4):1–18. https://doi.org/10.3390/s16040444
13. Martins T, Carvalho V, Soares F (2015) A Serious games for Rehabilitation of Neurological Disabilities: Premilinary Study. In 4th Portuguese meeting on bioengineering, Porto, Portugal. https://doi.org/10.1109/ENBENG.2015.7088843
14. Oliveira LC, Gomes TCF, Fernandes FG, Oliveira EC (2016) Realidade Virtual aplicada no desenvolvimento de um Serious games para Reabilitação de Cadeirantes utilizando Kinect. In XV SBGames, São Paulo, Brasil
15. Schell J (2014) The arte of games design. CRC Press, Boca Raton
16. Costa TKL, Machado LS, Moraes RM (2014) Inteligência artificial e sua aplicação em serious games para Saúde. Revista Eletrônica de Comunicação, Informação & Inovação em Saúde 8(4):525–529. https://doi.org/10.3395/reciis.v8i4.844.pt
17. Machado LS, Costa TKL, Moraes RM (2018) Multidisciplinaridade e o desenvolvimento de serious games e simuladores para educação em saúde. Revista Observatório 4(4)
18. Souto EC, Siebra CA, Silva LHL (2019) Avaliação de um sistema de captura e visualização de parâmetros físicos em um ergômetro para cadeirantes. In XI Congresso Brasileiro de Atividade Motora Monitorada, Alagoas, Brasil

19. Souto EC (2018) Prática de atividade física para indivíduos com lesão medular atendidos no Sistema Único de Saúde de João Pessoa/PB – Fatores intervenientes e influência na aptidão física e saúde. Londrina, Paraná, Brasil. Tese [Doutorado em Educação Física] - Universidade Estadual de Londrina
20. Cabral LL, Nakamura FY, Stefanello JMF, Pessoa LCV, Smirmaul BPC, Pereira G (2020) Initial validity and reliability of the Portuguese borg rating of perceived exertion 6–20 scale. Meas Phys Educ Exerc Sci 24(2):103–114. https://doi.org/10.1080/1091367X.2019.1710709
21. Sweetser P, Wyeth P (2005) GameFlow: a model for evaluating player enjoyment in games. Comput Entertain (CIE) 3(3):1–24. https://doi.org/10.1145/1077246.1077253

Applications and Experiences

Augmented Reality-Based Product Validation to Support Collaborative Engineering of Complex Technical Systems

Daniel Eckertz(✉), Harald Anacker, and Roman Dumitrescu

Fraunhofer Institute for Mechatronic Systems Design IEM, Paderborn, Germany
daniel.eckertz@iem.fraunhofer.de

Abstract. Within the course of a research project, a digital collaboration platform has been developed to exploit the potentials of Additive Manufacturing (AM). Potential users of AM and technology experts are digitally guided through the whole development process of individual AM components on the platform. This paper presents an Augmented Reality-based solution for interactive product validation across locations that supports the collaboration via the platform. It enables the efficient integration of the customer into the development. The solution consists of an AR application for the Microsoft HoloLens 2 for the customer and a visualization application for the expert. To enable a collaborative AR visualization, the two applications are networked and synchronized. 3D models of the developed components can be interactively visualized and experienced at the future application site. The expert can immerse into the customer's view and explain important aspects via audio communication and through interactive animations and further functionalities. On the other side, the customer can give feedback to the 3D model using gestures and speech control and point out aspects to be revised. The solution is integrated into the collaboration platform and the development processes. Based on use cases of two companies from different industries the solution has been evaluated and optimized. This ensures the practical applicability and a functional scope that meets the needs of the industry. Especially in times of Corona, Augmented Reality-based product validation is one way to let customers and developers work together efficiently across locations.

Keywords: Augmented Reality · Remote engineering · Virtual collaboration · Product validation · Technical systems

1 Introduction

Digitization and continuing technological advancements lead to changes in product development and production. Digital tools and methods are increasingly used to develop products individually for customers. One example is the technology Additive Manufacturing (AM). AM enables individualized products with complex component structures and high functional integration. High cost-effectiveness can be achieved even for small quantities. However, due to a lack of knowhow, many companies require support from technology experts in the targeted development of AM components.

M. E. Auer et al. (eds.), *Online Engineering and Society 4.0*, Lecture Notes in Networks and Systems 298, https://doi.org/10.1007/978-3-030-82529-4_44

Increasing geographical separation and flexible work environments and working hours hamper the necessary cooperation between experts and customers. Therefore, a digital collaboration platform has been developed within the course of the research project DigiKAM to exploit the potentials of AM. Initially, potential users of AM are networked with technology experts via the platform according to their needs. Subsequently, the entire development process of AM components is covered by the platform and supported by digital communication and secure data exchange solutions.

To fully make use of the potentials of AM, the development must be highly efficient and carried out under strict consideration of customer wishes and requirements. The customer must therefore regularly validate the current development status to enable targeted development. Classical validation approaches like physical prototypes are not time- and cost-efficient. A validation based on technical drawings or two-dimensional representations is also not sufficient due to the high product complexity. One goal of the project therefore was the use of the innovative technology Augmented Reality (AR) to support the integration of the customer for product validation. AR enables the interactive visualization of virtual content embedded in real environments. Digital data and models can be superimposed and viewed freely in space or precisely and true to scale in the real environment. The user can interact with the displayed content and manipulate it. In the context of product validation, enormous potential results from AR. Products and components can be validated at the future application site using digital models. Through manipulation and animation possibilities, complex product properties and functionalities can be interactively visualized and explained. The user perceives the product three-dimensionally in real size and can interact with it. Thus, user-related aspects like ergonomics can be actively tested. In addition to pure visualization, AR enables the digital networking of distributed stakeholders and immersion into the view of the communication partner. Consequently, AR enables, compared to classical approaches, more efficient and meaningful validation of complex products even over distances. Accordingly, an Augmented Reality-based solution for collaborative product validation has been developed. The solution is integrated into the collaboration platform and supports an efficient integration of the customer into the development process. The overall idea is already presented in [1]. This paper now describes the technical and functional implementation of the developed AR solution.

The remainder of this paper is structured as follows. In the next section related work is presented. Afterwards the technical solution is described, followed by its evaluation. The paper is completed with a conclusion in Sect. 5.

2 Related Work

There are several publications evaluating the potentials and benefits of immersive technologies for design reviews compared to traditional approaches [2, 3]. Albers et al. describe the preparation of AR-based validation environments with focus on the product data [4]. Technical solutions are mainly described for individual use cases or specific products. Uva et al. developed a simple application for mobile devices to visualize products on an AR marker and validate them in 3D [5]. An automotive supplier uses a single user AR application on the Microsoft HoloLens to visualize digital product data interactively on physical mockups of assembly stations. The application enables the early validation of the assembly process [6]. In a Teaching Factory AR enables engineers to envision and evaluate a product design interactively on a Microsoft HoloLens [7]. Further publications focus on annotations to 3D models in different ways without addressing a complete solution for product validation. Chang et al. for example describe a solution for air drawing on AR glasses [8]. A collaborative mobile application for real-time video annotations with multiple clients is described in [9]. Free drawings can be attached in AR and communicated to other users. Bruno et al. present an AR solution for handheld devices, that workers can use at the workplace to annotate design variations they made during their working hours [10]. This is used as a documentation for other workers. Similar functionality is also available by professional software solutions like Vuforia *Chalk*. Furthermore, there is a variety of software solutions available on the market to visualize design data. Autodesk for example offers visualization solutions for AR and VR[1] to experience CAD models interactively in 3D. However, these solutions only work with Autodesk software and can not be used independently. The software *ARES* (Augmented Reality Engineering Space) by Holo-Light enables the visualization of design data on different AR devices and supports single and multi user mode. The existing solutions and publications only cover individual aspects of the functionality that is required for this project. An AR-based solution, that enables collaborative product development integrated into a collaboration platform and the related development processes is not available at present.

3 Augmented Reality-Based Collaborative Product Validation

The developed AR-based solution for product validation should support the collaborative development of Additive Manufacturing components via the collaboration platform. First, the basic concept of the solution is presented, followed by the description of the technical realization of different aspects of the concept.

3.1 Conceptual Design and System Structure of the AR Solution

To allow for a targeted development of the solution, needs and requirements were specified together with companies as the future users of the solution. This led to a higher-level functional concept. The expert provides the visualization data based on the current design via the collaboration platform. The solution then enables the customer to experience the AM design data interactively. This can be at the later application site to check installation dimensions or optics or to test ergonomic aspects. The expert is connected live and illustrates product characteristics and special features to the customer. The customer in return gives feedback that can then be used by the expert to improve the design. Accordingly, four basic subgoals for the solution have been defined (Fig. 1):

[1] https://www.autodesk.de/solutions/virtual-reality.

1) **Preparation of the Visualization Data:** The solution needs to work with changing design data from different projects. The original design data needs to be prepared for the visualization to meet the requirements of the validation and provide all relevant information. If necessary, different design variants and special views or visualizations need to be integrated in this data set. Furthermore, the data is to be loaded into the applications dynamically during runtime.
2) **Customer AR Application**: This application runs on an AR device and is used by the customer to interactively visualize the digital design data. The application provides interaction possibilities to efficiently give feedback to the visualized 3D models and comment on well-solved aspects or highlight things that need to be changed. To exploit the potentials of AR for efficient product validation, the application supports intuitive and natural interaction and provides easy to use feedback functionalities.
3) **Expert Visualization Application**: The application does not feature AR-based visualization but visualizes the design data on a standard screen. It enables the expert to immerse into the customer's perspective. Thus, the expert sees what the customer sees without being on site. The application enables the visualization of different design variants that the expert wants to show the customer. Furthermore, different interaction possibilities to highlight and explain product features interactively are provided.
4) **Platform Integration, Communication, and Synchronization:** A connection to the collaboration platform is needed to efficiently integrate the solution into the development processes and to exchange necessary data. To enable a collaborative AR-based visualization and validation the two applications are networked and synchronized. Via audio communication, the customer and the expert can talk to each other and discuss the design data. Furthermore, the customer's perspective is transferred to the expert's application and all the interactions and the feedback by the customer are synchronized.

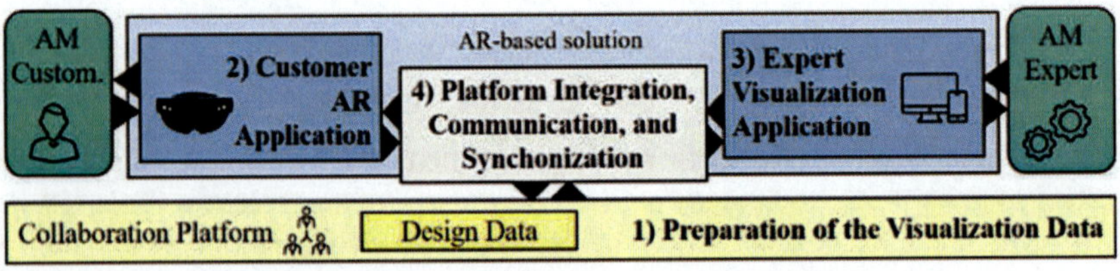

Fig. 1 Higher-level structure of the AR-based solution for product validation

The four individual aspects were further specified according to the respective requirements and needs. Based on research and analysis of available devices, software development kits (SDK) and technical possibilities, concepts for the individual aspects and the overall solution were developed and implemented.

3.2 Preparation of the Visualization Data

The AR-based solution needs to be usable with varying design data sets. These data sets could include different variants of the developed component. Furthermore, special animations or visualizations are necessary to experience specific characteristics or features of the component. Standard animation or manipulation functionalities are provided by the application and can be used on demand. Individual functionalities like the animation of specific features or special views need to be individually prepared in advance. This is either done by the AM expert or by an AR visualization expert who offers data preparation services on the collaboration platform. In both cases the data needs to be prepared in such a way, that it can be loaded dynamically into the applications.

Accordingly, a standardized data structure was developed, that defines how the visualization data is prepared. As data format AssetBundles[2] are used. AssetBundles are archive files created in the game engine Unity[3] that contain 3D models, textures, animations, functionalities and more. Figure 2 shows the defined structure for the AssetBundles and its usage for the dynamic loading of the data into the applications. The AssetBundles are a set of geometries or models hierarchically structured under one *Root* node. To configure functionalities and characteristics, a set of scripts has been developed. These predefined scripts can be attached to the models and provide standardized interfaces to the functionalities. *Animation* scripts provide start and stop functionalities for the related animation. The script *OnlyExpert* means that the respective model is only to be visualized in the expert's application. This is for example the case, when the customer uses a physical part like a grip as an AR marker. The expert does not have this part and it needs to be visualized. The script *SpecialView* indicates that there is a special visualization, like special light conditions, that can be activated using the script.

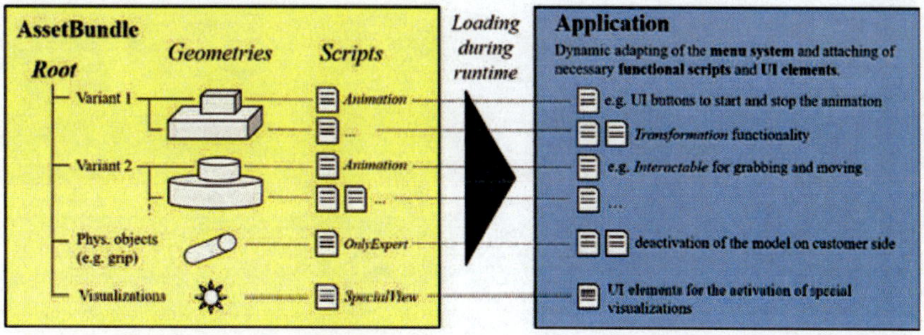

Fig. 2 Standardized AssetBundle structure for the preparation of the visualization data

[2] https://docs.unity3d.com/Manual/AssetBundlesIntro.html.

[3] https://unity.com/de.

The standardized data structure allows the dynamic loading of the data into the applications during runtime and the respective configuration of the user interface and functionalities. The *Root* node acts as the starting point for an iteration through the hierarchy and the analysis of the contained data. Each child of the *Root* node is considered a variant of the design data unless there is a *SpecialView* or *OnlyExpert* script attached. The other child nodes are treated as special visualizations or models not to be visualized in AR respectively. In the application the menu system is dynamically adapted according to the AssetBundle. Additional necessary scripts are attached to the models. User interface (UI) elements are dynamically created to control the animations and visualizations and device-specific scripts are attached to enable functionalities like transformation or grabbing and moving objects. This way the visualization data can be used on different devices without manual changes. The AssetBundles are provided via the platform and loaded directly into the applications at the beginning of a collaborative product validation meeting. Further details follow in Sect. 3.5.

3.3 Customer AR Application

The prepared design data is interactively visualized in the customer AR application. To support 3D visualization and 3D perception of the design data a stereoscopic display system is necessary. Furthermore, in certain validation activities ergonomics are to be tested. This requires the hands to be free to use. Therefore, the new Microsoft *HoloLens 2* has been chosen as the device for the AR application. It is a mobile stand-alone device with a binocular display system that can be used without additional hardware like a computer or controllers. Interaction is provided through gesture and speech control. Thus, it is very well suited for interactive product validation.

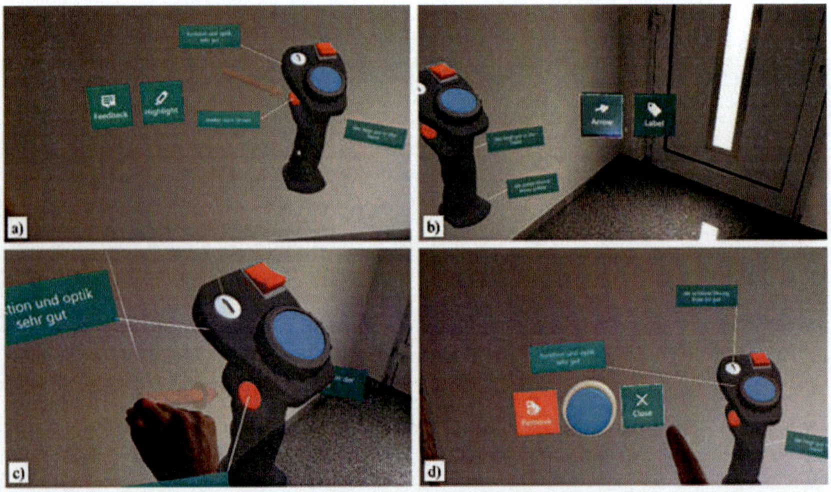

Fig. 3 AR-based visualization from the customer's perspective (captured with HoloLens 2) on the example of a handheld control device. **a** Menu to choose the functionality with the model and already added text labels in the background. **b** Choice of feedback options arrow or label. **c** Gesture-based creation of a 3D arrow. **d** Menu to modify a selected text label

After starting the application on the HoloLens 2 the user enters the ID of the related project using a virtual keyboard with hand gestures. Based on the project ID the visualization data is automatically loaded from the platform and the connection to the expert application is made. Further details follow in Sect. 3.5. The visualization data is positioned relatively to a predefined AR marker. Thus, it can precisely be visualized at the later application site or even in the user's hands if he or she holds the marker. With the HoloLens 2 the customer is free to walk around the virtual content and expect it from any perspective. Using several provided functionalities, the customer can give feedback directly to the 3D models. All menus and user interface items are realized as virtual elements. They are floating three-dimensionally in front of the user and can be interacted with by gestures. By virtually clicking on a part of the 3D model, that part is selected and a virtual menu shows up (Fig. 3 a). Here the user selects to either highlight the selected part or give feedback to it. When highlighting is clicked the part begins to slowly blink in red, both in the customer's and the expert's visualization. To give feedback, two options are offered in a submenu (Fig. 3 b). Option 1 are 3D arrows. The user can add an arrow by virtually clicking at the 3D position the arrow should point at and then dragging the arrow end in the desired direction (Fig. 3c). Option 2 are text labels. After pressing a virtual record button (blue button in Fig. 3 d) the user starts speaking the comment and it is automatically converted into text using *Speech-to-Text* functionality by Microsoft. The conversion stops when the user releases the button. The text is set to a virtual label that is automatically positioned around the selected part of the model (see examples in Fig. 3 a-d). Virtual labels as well as the 3D arrows can be repositioned by drag gestures. By clicking on a feedback item, a menu shows up. This way the items can be deleted or in case of labels, the text can be updated.

3.4 Expert Visualization Application

The expert visualization is not AR-based but a standard visualization of 3D content on a 2D screen. The AR application runs on a HoloLens 2 which is a Universal Windows Platform (UWP) device. To simplify the development, the expert application is a UWP application as well. Thus, the developed application is compatible with any UWP device (e.g. the Surface series). Like the customer, the expert needs to enter the project ID first. Then the visualization data is downloaded and integrated and the connection to the customer's application is made (see Sect. 3.5 for further details). The design data is visualized on a neutral dark background and the camera as well as the orientation and position of the model are automatically synchronized with the customer's application (Fig. 4). The expert's user interface provides several functionalities. At the top of the screen the expert can change between the different product variants that are contained in the visualization data set. Depending on the device the expert can select parts of the model by mouse click or touch. The selected part becomes yellow. Simultaneously an interaction menu appears at the lower right side of the screen. Using this menu the expert can highlight the selected part. This is the same functionality the customer has as well. The highlighted part starts blinking slowly in red. An example is shown in Fig. 5 a). If there is an animation provided for the selected part, the animation button can be used to open the animation. The description of the animation is shown and it can be started and stopped with a button click. This can for example be used to explain the functionality of a key-based control feature (Fig. 5 b). Furthermore, the expert can transform the selected part. The expert can define the step size and move the part along the three coordinate axes using a small menu (Fig. 5 c). This way, some changes requested by the customer can directly be made and checked in the validation session. When the customer is moving around in the room or holding the AR marker in hands, the model and the camera are also moving in the expert's visualization. This hampers the selection of a part by clicking on it. For these situations, the expert can deactivate the synchronization of the perspective. Instead, the expert can independently control the perspective and do the selection with more precision. Additionally the expert application provides a virtual pointer functionality. When this function is activated, the expert can click on any part of the virtual model. A red 3D pointer appears at the clicked position. By dragging, the pointer can be moved on the model dynamically. The virtual pointer directly appears in the customer's application at the same position. It can be used to lead the customer's view to certain points or specific parts of the design data.

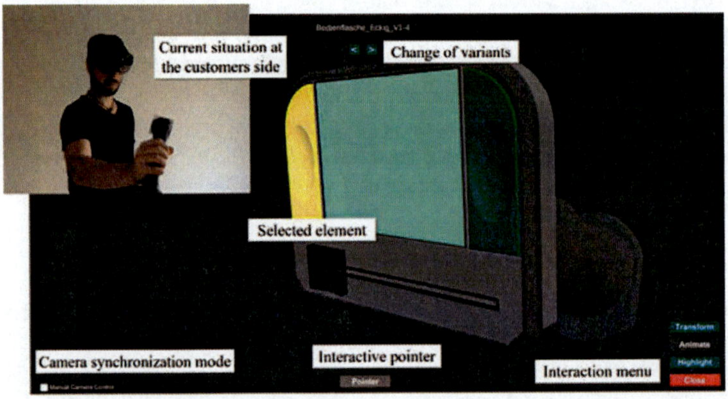

Fig. 4 The expert application with a synchronized perspective and interaction possibilities

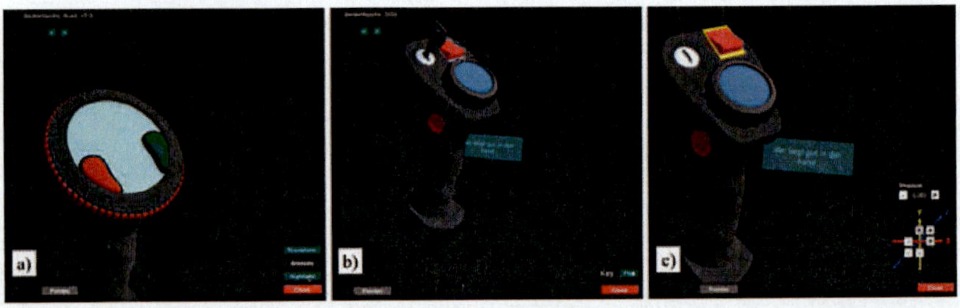

Fig. 5 Interaction possibilities of the expert: **a** Highlighting, **b** Animation, **c** Transformation

3.5 Platform Integration, Communication, and Synchronization

The basic structure of the customer and the expert application is the same. Additionally both applications run on Universal Windows Platform (UWP) devices. Therefore, the two applications are developed in one single UWP project in Unity and can be deployed to different UWP-based devices. Based on *Conditional Compilation Symbols* like *"#if UNITY_WSA_10_0"* the target device is checked and individual instances of the application are created (Fig. 6). This leads to two applications, the customer AR application and the expert visualization application. They share the same basic functionality but differ in their user interfaces (UI) and the type of visualization in the virtual scene.

As described before the customer and the expert both enter the project ID after starting the applications. The design data that is provided on the platform has a *Uniform Resource Identifier (URI)* based on this project ID. Accordingly, the visualization data is accessed using *HTTP*-based *REST* calls and downloaded and integrated into the visualization during runtime (Fig. 6). As described in Sect. 3.2 necessary functional scripts are automatically added and the UI is adapted accordingly.

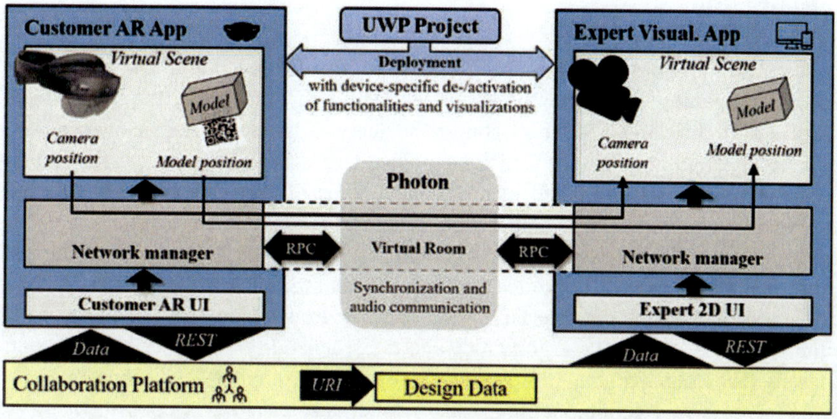

Fig. 6 Two device-specific instances of one UWP project are deployed to different devices as the customer and the expert application. The connection and synchronization is solved using the SDK *Photon* and platform integration is realized based on *REST* calls

Both applications contain a *Network manager.* Based on the SDK *Photon*[4] the network manager handles the communication and the synchronization of all relevant data and interactions. At the beginning of each validation session, a virtual room is created with the project ID as unique identifier. After the project ID is entered at the start of the applications, the connection to the virtual room is made. From then on, the users of the applications can communicate via audio. The camera position and the model position are continuously transferred from the customer's to the expert's

[4] https://www.photonengine.com/pun.

application using the network callback function *OnSerializeNetworkView*. The transformation of the expert's camera and model are set accordingly. So whenever the customer moves his or her head or the AR marker and so the virtual model with it, the perspective is automatically updated on the expert's side. All the changes in the virtual scene that are triggered by the customer or the expert are synchronized using *Remote Procedure Calls* (RPC). RPC is a technique to realize interprocess communication, so function calls in other address spaces or on other machines. When the user changes something in the scene that is to be changed in the other application as well, an RPC is send to the Photon server and from there to the other application. This is used for the control of animations, transformations and all the feedback added by the customer.

After the validation, the feedback given by the customer is saved. This is done in one file that contains the 3D visualization data including changes that have been made and information about all the added feedback items, their position and content. The feedback file is transferred back to the platform via a *REST* call and available for following development iterations. It can be visualized in an individual viewer application.

4 Evaluation

The developed AR-based solution has been tested and evaluated with two different use cases of companies from different industries. The first use case is the redesign of a control device for large industrial storage systems, a handheld device with a set of interaction elements. The former device was large, heavy, and not very handy. Accordingly, the was optimized using the potentials of AM. A grip with individual finger troughs was developed and several different variants of the layout of the interaction and control elements proposed by the expert. In the validation sessions, the customer already got the 3D printed grip which was used as the AR marker. With AR the different variants of the interface were visualized on the grip and validated by the customer (Fig. 3, 4, 5). The ergonomics and the reachability of all the elements could be tested efficiently without printing and delivering protoypes to the customer. The second use case was a component of a dish washer, a so called flow control. Using AM the inner structure of the component and so the material flow could be improved. In this case the AR-based solution was used to motivate the new manufacturing technique to the companies' management. The visualization data contained animations showing the results of fluid simulations. These were overlaid through AR on the original physical component to point out and explain the benefits of the new construction in an interactive and intuitive way (Fig. 7). The use cases have been evaluated with the companies and additionally with visitors at a technical fair. The test persons answered questions about different aspects of the AR-based solution. The evaluation results can be summarized as follows:

- **Usability**: The usability and the user experience of both the expert and the customer application have been rated very good. The user interface on the expert's side is kept simple and straightforward. It is easy to use for all the functionalities. The test persons evaluated the AR interaction by gestures and speech control as very intuitive and especially efficient. There were no problems in the usage of both applications.

- **Functionality**: Overall the functionalities matched the companies' requirements. The expert can efficiently highlight and point out important aspects. It was remarked that complex animations need to be prepared in advance. However, by using the predefined scripts and the standardized data format this is relatively easy to do. The companies stated, that in comparison to the preparation of for former validation activities the necessary effort is low. The test persons rated the feedback functionality of the customer as very appropriate. It provides enough ways to efficiently give all the necessary feedback and attach it in 3D to the visualized design data.
- **Integration**: The companies consider the integration of the AR-based product validation into the development to be very good. The validation activity can be integrated into the processes on the collaboration platform as needed. In addition, the visualization data is loaded directly from the platform and the feedback is transferred back accordingly. Thus, the solution supports a seamless process chain.
- **Benefit**: Overall, the companies see AR-based product validation as a very good tool for collaboration across locations. The companies find this particularly helpful for complex products whose functionalities and properties were previously difficult to comprehend. Visualization at the later application site to check the appearance and fit is also seen as a good support. Still considered critical are the price and form factor of current AR glasses. But due to the ongoing development the companies expect to use AR productively for product validation in the coming years.

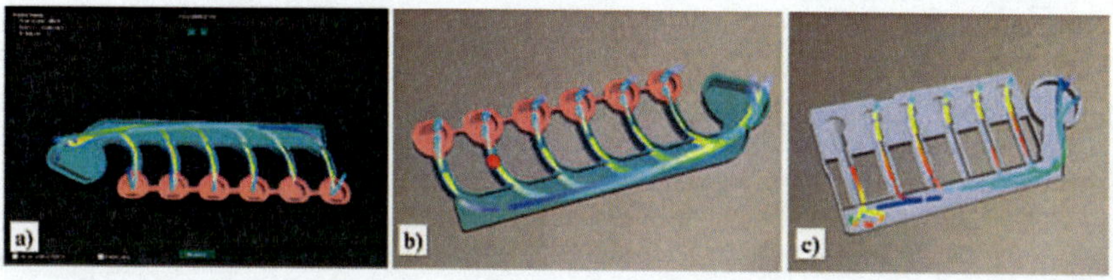

Fig. 7 Visualization of the flow control and the fluid simulation data. **a** View of the expert of the new component. **b** View of the customer of the new component. **C** View of the customer of the former component. (Due to the COVID situation, a visualization on the real physical component at the company's site was not possible during the paper preparation.)

5 Conclusion

In this paper, a technical solution is presented to support the collaborative platform-based development of Additive Manufacturing components by Augmented Reality-based product validation. The practical suitability and the benefits of the solution for companies are demonstrated and evaluated based on two practical use cases in the context of Additive Manufacturing. On the one hand, the evaluation showed that the solution works well and is easy to use. Accordingly, it can also be used by untrained employees without prior knowledge. Besides, the companies consider the solution to be a very good tool to support product development. The companies further acknowledge that the solution can be integrated seamlessly into the development processes on the collaboration platform according to their needs. The solution can be used by customers to experience design data during the development process at the later application site. In this way, the customer gets a better understanding of the developed product or component and can provide feedback that is more meaningful. This leads to an increase in product quality. Besides, physical prototypes are no longer necessary. This leads to lower costs and time savings since digital data can be provided quickly. Eventually, the use of AR for product validation leads to an increase in efficiency.

There is a general shift towards mobile work and flexible working hours in companies all over world. Product development will therefore increasingly take place across locations in the future. In times of Corona, this is intensified by contact bans and travel restrictions. New forms of collaboration are therefore required. Augmented Reality-based product validation is one way to let customers and developers work together efficiently across locations. The developed solution is therefore highly relevant and can be transferred to other industries and use cases. Through further technical advancements, access to the technology, especially to AR devices, will become easier and cheaper in the future, so that more companies can participate in the potentials.

References

1. Eckertz D, Berssenbrügge J, Anacker H, Dumitrescu R (2019) Work-in-progress: enhancing collaboration using augmented reality design reviews for product validation on the example of additive manufacturing. In Cyber-physical Systems and Digital Twins: Proceedings of the 16th international conference on remote engineering and virtual instrumentation. Springer
2. van Goethem S, Verlinden J, Garcia I, Degryse R, Perdieus S, Watts R, Verwulgen S (2020) The use of immersive technologies for design reviews. In Proceedings of the AHFE 2020 virtual conferences on usability and user experience, human factors and assistive technology, USA, Cham, 16–20 July 2020. Springer, Cham, pp 656–662
3. Berg LP, Vance JM (2016) An industry case study: investigating early design decision making in virtual reality. J Comput Inf Sci Eng. https://doi.org/10.1115/1.4034267
4. Albers A, Reinemann J, Fahl J, Hirschter T (2019) Augmented reality for product validation: supporting the configuration of ar-based validation environments. In Chen JYC, Fragomeni G (eds) Virtual, augmented and mixed reality. Applications and case studies, vol 11575. Lecture Notes in Computer Science. Springer International Publishing, Cham, pp 429–448
5. Uva AE, Fiorentino M, Monno G (2011) Augmented reality integration in product development. In Proceedings of the IMProVe 2011 international conference on innovative methods in product design
6. Dyck F, Stöcklein J, Eckertz D, Dumitrescu R (2020) Mixed mock-up – development of an interactive augmented reality system for assembly planning. In Chen JYC, Fragomeni G (eds) Virtual, augmented and mixed reality. Design and interaction, vol 12190. Lecture Notes in Computer Science. Springer, Cham, pp 37–49
7. Mourtzis D, Zogopoulos V, Vlachou E (2018) Augmented reality supported product design towards industry 4.0: a teaching factory paradigm. Procedia Manuf https://doi.org/10.1016/j.promfg.2018.04.018

8. Chang YS, Nuernberger B, Luan B, Höllerer T, O'Donovan J (2017) Gesture-based augmented reality annotation. In: 2017 IEEE Virtual Reality (VR), pp 469–470. https://doi.org/10.1109/VR.2017.7892383
9. Dongxing C, Sangwook K (2020) Augmented reality annotation for real-time collaboration system. J Korea Multimed Soc. https://doi.org/10.9717/KMMS.2020.23.3.483
10. Bruno F, Barbieri L, Marino E, Muzzupappa M, D'Oriano L, Colacino B (2019) An augmented reality tool to detect and annotate design variations in an Industry 4.0 approach. Int J Adv Manuf Technol. https://doi.org/10.1007/s00170-019-04254-4

Collaborative and Contextual Learning Experience, Supported in Virtual Scenarios and Software Defined Radio Tools, for the Development of a Communications Laboratory in Engineering

Mario Andrés Ramos[1(✉)], Paola Andrea Buitrago[1,2], and Raúl Camacho[1]

[1] National Open and Distance University - UNAD, Bogotá, Colombia
{mario.ramos,paola.buitrago,raul.camacho}@unad.edu.co
[2] University of Medellín, Medellín, Colombia

Abstract. The National Open and Distance University—UNAD is an autonomous entity of higher education of a national order, due to its open nature, and at a distance it is supported by virtual learning environments, enabling social inclusion through eight national branches, located in: Amazonía Orinoquia, Centro Boyacá, Caribe, Centro Bogotá Cundinamarca, Centro Oriente, Centro Sur, Occidente and Sur, and has sixty-four educational centers. For our case study, a Software Defined Radio laboratory was developed in the courses: communication systems and antennas and propagation to engineering students nationwide, the total sample was 54 students, divided into two groups, group A from 27 students is the control population that had no interaction with the Software Defined Radio—SDR laboratory, and group B of 27 students is the target population that used the tool and were subsequently evaluated. In addition, group B was applied a laboratory perception instrument, managing to evaluate the pedagogical tool from the contextual experience of each of the students who had interaction with it. Based on the above, the laboratory's impact and relevance validation process was carried out in learning scenarios for engineering courses: communication systems and antennas and propagation.

Keywords: UNAD · SDR · Laboratory · Communication systems · Antennas and propagation

1 Introduction

The National Open and Distance University—UNAD, is the largest public institution in Colombia, with a presence throughout the national territory. UNAD has the ECBTI School of Basic Sciences, Technology and Engineering, which is made up of Training Chains; One of them is the Electronic, Telecommunications and Networks Training Chain (ETR), which has two technological training level programs, two university training level programs and a specialization training level program. The study

M. E. Auer et al. (eds.), *Online Engineering and Society 4.0*, Lecture Notes in Networks and Systems 298, https://doi.org/10.1007/978-3-030-82529-4_45

presented in this article was developed by articulating the Telecommunications Engineering program, which is part of the ETR Training Chain, which is offered through the virtual modality at the national level. The program is based on the Unadista Pedagogical Model supported by e-learning [1, 2], as a guiding and referential framework that allows, in a conscious, intentional, planned and related way, to declare the course and take positions in relation to the way of conceive your educational process.

It is important to bear in mind that each course designed for the Telecommunications Engineering program of the UNAD, is the didactic expression of the curriculum in which the training intentions, the contents, the strategies for academic work and learning, the resources and technological tools and forms of evaluation, which contribute to the development of the purposes of the levels, cycles and training programs [3]. For UNAD, the Practical Component is the opportunity to build the learning process from the perspective of experience, interaction and dialogue in the context of that specific area of knowledge. The UNAD uses three scenarios to develop the practical component, such as: with technological support, in situ and remote; where the student understands the strategies for the academic accompaniment of the training process [4].

This article shows the impact of the experiment carried out with the participation of the students of the Telecommunications Engineering program; analyzing the appropriation of the basic concepts of the Communication Systems and Antennas and Propagation courses, in addition to evaluating the relevance of the laboratory from the students' perception based on practical experience. In Sect. 2 a description of the background related to the research work and the identified needs that support the development is presented, in Sect. 3 the phases of the laboratory implementation model are described along with the application of the instruments in the experimentation of appropriation of basic concepts and of students' perception of the use of the laboratory as a teaching tool. Finally, Sect. 4 presents the conclusions of the research work and possible future work.

2 Background

The UNAD Telecommunications Engineering program is offered in virtual mode and nationwide. One of the challenges that this type of educational attention presents is the availability in their headquarters of significant pedagogical resources for the integral learning of students, taking into account that "doing" is a fundamental part of professional training due to its need to appropriate concepts through contextual experience.

Due to the nature of the concepts used in the courses mentioned, it is complex to put into practice the theoretical concepts addressed, since generally high costs and some complexity are required in the deployment of infrastructure to carry out practices, as well as movement of personnel to places where Communications systems are already deployed. Taking these considerations into account, it is necessary for the proposed laboratory to handle low costs and technologies that facilitate its use from the sites where the students are located, allowing the application of the appropriate concepts and the development of competencies of the Telecommunications engineer in a scenario of actual application. Therefore, the implementation of a laboratory for the courses communication systems and antennas and propagation based on radio defined by software (SDR) is proposed, to evaluate its impact on contextual learning in a higher education environment.

The telecommunications engineering program of the National Open and Distance University UNAD, by having practical component scenarios in situ, provides spaces of great importance to apply the appropriate concepts in a theoretical way, such as, for example, in the Communication Systems courses and the course Antennas and Propagation, which constitute an important part of the curriculum and where it is necessary to appropriate a conceptual base necessary to be applied in subsequent courses in the specific disciplinary field of the program. The laboratories or practices related to the Communications Systems and Antennas and Propagation courses are usually carried out in scenarios with technological support, but it is far from what would be a real implementation.

For some years, Software Defined Radio technology has become popular, as an innovative alternative for the deployment of radio frequency systems, as presented at the Wireless Innovation Forum, SDR is defined as a type of radio communication, in which some or all of the physical layer functions are defined by software [5]. This technology offers significant benefits, including flexibility in configuration and low cost compared to conventional hardware [6]. This possibility opens the doors for the development of practices closer to the real context of application. A strong point of SDR technology is its flexibility, the ability to reconfigure equipment that allows the same devices to be used to deploy different types of systems and configure both analog and digital modulations covering a wide field of application in the telecommunications area.

It is important to consider those new to education, who play an important role in generating new learning spaces, where the student learns by seeing and doing, and where a piece of equipment or device itself is not simply another module of operation, but a learning mechanism supported by a platform that allows you to integrate knowledge. The potential of SDR has been identified to implement flexible laboratories, which allow changing the traditional laboratory models, where the equipment is already configured by a manufacturer and with the limitation that it cannot be modified by a student according to the application required in the themes in development.

Therefore, the use of SDR technology in industry and academia becomes increasingly important, which is why universities worldwide have implemented practical components through laboratories with SDR. Currently, prestigious universities such as Monash University have software-defined telecommunications laboratories; the goal is to be Australia's premier laboratory based on GNU open source radio technology and the OpenFlow Software Defined Network. The laboratory will be open to national and international researchers and industry users who will be able to collaborate with scientific and technical personnel. Research topics will include modulation, encoding, MIMO, and broadcast network services. Among the work carried out, implementations such as bidirectional retransmission networks through digital coding using SDR [7], stand out, it has also allowed to test modern modulation schemes, such as in the implementation of an OTFS modem to experience effects of deterioration of the signal in reception, projecting this implementation to the needs of 5G [8].

On the other hand, the University of Colorado, with its Interdisciplinary Telecommunications Program (ITP), has developed a strategy to complement theoretical radio transmission and digital communications concepts with practical laboratory exercises using SDR. Even SDR technology has been adapted to the research processes of various research groups at the university [9], developing robust implementations for communication systems and telemetry systems through SDR configuration, articulating the work with the Department of Aerospace Engineering Sciences [10].

The University of Notre Dam has established since 2003 the SDR LAB [11], which was implemented with the support of InterDigital and National Instruments. This laboratory has numerous MIMO 2X2 USRP RIO units and LabView is used for the creation of prototypes and flow of FPGA-SDR algorithms, allowing the development and validation of new algorithms. For its part, the University of California from its SDR laboratory mentioned in [12], shows how they have developed several important projects related to the detection of radio frequency signals, location of transmitters, microwave automatic adjustment links, as well as monitoring the electromagnetic spectrum. This work has been a reference for the strengthening of laboratory settings for the Higher Technological School of Engineering of Seville - ETSI, which in turn has been studying SDR technology for a decade and the possibility of articulating it with its academic practices [13].

Also, the Iowa State University since 2009 launched its Software Defined Radio Laboratory [14], dedicated to research processes and academic work. Mainly its researchers address the issues of wireless coding, spectrum sensing and development of cognitive radio networks. In addition, it is emerging to introduce elements of SDR in junior and senior level courses of the University. It is important to mention that it was found in the Wireless Engineering (BWE) curriculum at Auburn University [15] an SDR lab course based on GNU Radio and the Universal Software Radio Peripheral (USRP) platform; which compares existing approaches and presents laboratory examples and their results, for the spring 2013 offering, a small-scale evaluation was conducted with positive student response.

Therefore, a contextual learning laboratory is presented as an alternative to engineering education, placing students in exposure to real-world applications, especially when they are presented in an active and experiential learning environment that increases their interest in learning. as well as pedagogical effectiveness. This idea of contextual learning, is a type of learning designed so that students can perform activities and solve problems in a way that reflects the real-world nature of such tasks, is based on the theory of cognitive learning [16].

This is why the proposed laboratory becomes important, since in engineering general professional knowledge is required to apply in specific contexts and practical knowledge is also needed that will be built from working in many different situations, so interpreting the context and how relates to learning is essential for pedagogical innovation. From the contextual approach, learning occurs only when students process new information or knowledge in such a way that it makes sense to them, in their own frames of reference, that is, finding meaning and usefulness to the information that has been acquired either in the classroom or outside of it; internalizing concepts through the process of discovering, reinforcing and relating in a real context [17].

This is why the proposed laboratory allows students to interact with real communication systems, configure their operation and validate the implementations made. In addition, it encourages the proactive action of students to solve problems and interact collaboratively when addressing challenges posed in the activity guides, achieving evidence of the importance of the theory acquired.

3 Description of the SDR-Based Lab Deployment Model

The design of the laboratory is carried out taking into account the topics addressed from the Communication Systems and Antennas and Propagation courses, in addition to the previous knowledge that must be taken into account for an appropriation of the subject matter developed in the laboratory, that is why designs the laboratory in 4 sequential phases.

UNAD's SDR-based laboratory allowed students to have access to practical challenges raised in guides, which had to be solved by configuring SDR devices. The laboratory distributes the students into different groups to collaboratively approach the development of the practical guides. The operation of the laboratory is currently based on the use of Matlab and its Simulink block tool, since it intuitively promotes the development of laboratory practices for students who are involved with this type of technology for the first time, managing to implement systems complete communications through the use of connection blocks. In this way, the UNAD offers a laboratory that provides students with basic practices around communications systems and gradually deepens practical implementations in the context of the use of radio frequency, giving the possibility to students who are unfamiliar with SDR technology in this field, in addition to strengthening the training process in these telecommunications engineering issues.

In the first measure, there is Phase one (1) or introductory phase for students to the laboratory, it basic topics of communication systems are addressed and it is intended that students carry out two laboratory practices related to analog and digital modulations. For this phase, a mathematical software tool was used, in this case Matlab and its Simulink block tool. Once the basic concepts have been addressed, three additional phases are developed, phase two (2) focuses on topics of digital signal processing and spectral analysis with guide number 3, where the software defined radio tools are already used, mainly radio receivers for the analysis of signals in the frequency domain and the scanning of nearby transmitting sources.

For phase three (3), concepts of antennas and propagation are involved, leading the student to delve a little more into the applicability of radio communications and the use of SDR devices, for this, laboratory guides 4 and 5 are presented. Guide 4 aimed at mounting an FM radio receiver and guide 5 focuses on the implementation of a basic radio link, at this point It is necessary to implement a transmitter and a transmitter device for the deployment of the radio link, for this purpose, use is made of the Simulink functional blocks that represent digital transmitters and receivers. In phase four (4), the student is led to practice with the implementation of software-defined radio applications to explore its potential, for this purpose, the implementation of a basic aircraft radar is proposed in guide 6 and transmitting images to Through a point-to-point radio link, for this, the Matlab Communications Support Toolbox was used and the support provided by Mathworks was used [18]. Next, Fig. 1 shows the general structure of the laboratory with its respective sequence.

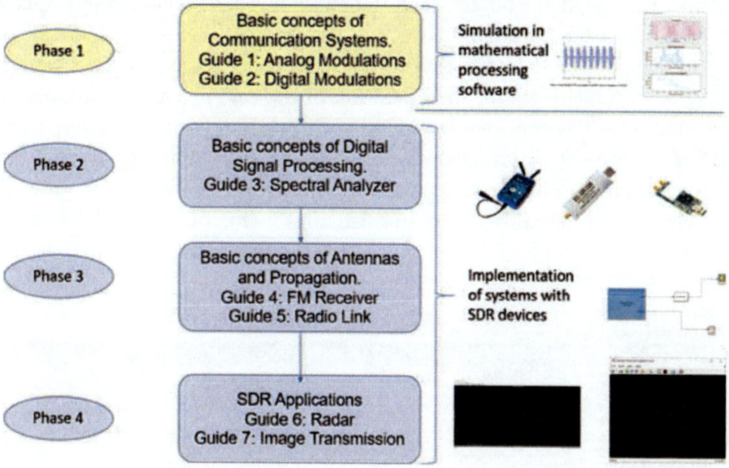

Fig. 1 SDR-supported communications lab architecture

3.1 Tools and Devices Used for the Implementation of the Laboratory

The laboratory is designed so that it can be developed on different platforms such as MATLAB, GNU RADIO or LABVIEW, among others. Said implementation will depend on the availability of equipment or affinity with the tool that one has. This flexibility is important, since it gives the possibility of exploring various tools and their potentialities, in the laboratory the implementations were carried out mainly with Matlab and the Simulink tool, the SDRSharp tool was also used in some guides.

Regarding the specifications of the laboratory, considerations were taken into account that allow the use of devices of low complexity, low cost and that have the necessary characteristics to implement systems. Table 1 shows some characteristics of the devices to be used in the laboratory and with which a solution can be given to the requirements of the guidelines proposed.

Table 1 Specifications devices used in the SDR communications laboratory

Device	Frequency range	Maximum bandwidth	Functionality
RTL2838	0.5–1766 MHz	2.4 MHz	RX
Adalm pluto	325–3.8 GHz	20 MHz	TX/RX
LimeSDR Mini	19–3.75 MHz	30.72 MHz	TX/RX

3.2 Techno-Pedagogical Integration of Processes for Incorporating SDR into Distance Learning Methodology

The communications laboratory based on SDR technology at UNAD, strengthens the development of the practical component in situ of the Communication Systems and Antennas and Propagation courses. This laboratory is available through face-to-face meetings and interaction with the SDR devices in working groups developing the laboratory guides in a collaborative way. The computer rooms of the University are used where the complements for the operation of the equipment are installed and can access said resource. To access the laboratory, it is necessary for the student to enter the MATLAB development environment, this numerical computing system offers an integrated development environment with its own programming language or through the interconnection of functional blocks of Simulink, which will be the interface of the Lab as shown in Fig. 2. In Fig. 3 you can see the devices of the laboratory in operation.

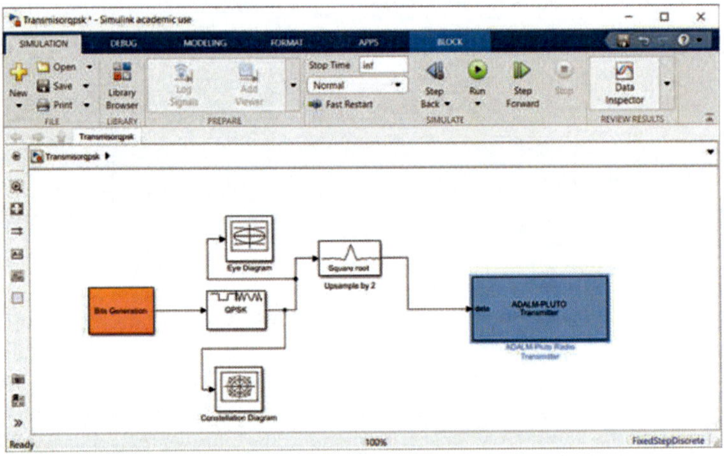

Fig. 2 Matlab Simulink working interface from SDR lab

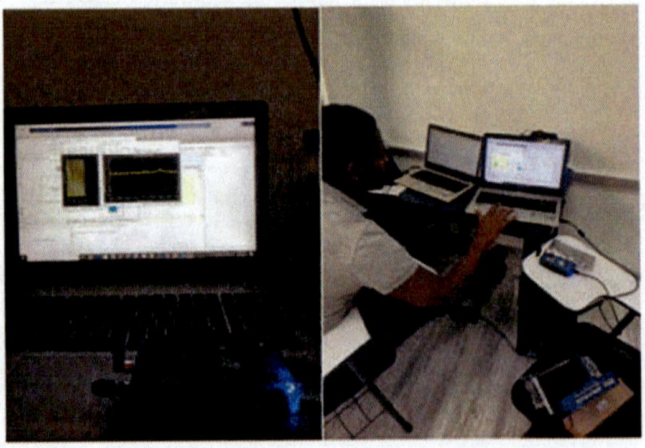

Fig. 3 SDR laboratory in operation

The experimentation with the laboratory had the participation of 54 students. Once a homogeneous universe of experimentation had been determined, two groups of 27 students each were formed, defining group A as the reference group that develops the practical component of the Communication Systems and Antennas and Propagation courses in a conventional way as they have been working at the University, while group B refers to the group of students who interacted with the SDR-based communications laboratory. Group B students interact in the laboratory collaboratively, in groups of three to perform tests with the reconfigurable hardware of the physical devices in current existing facilities of CEAD Palmira. The proposed laboratory evidently allows us to take a step beyond theory, these students accessed the laboratory between two and three sessions, with an approximate of 8 h of work, in which they developed the seven laboratory guides, from the basic phase to the implementation of medium complexity applications. In these sessions, students have the support of the teacher, as well as in the development of a conventional practical component, with the teacher becoming a facilitator within the contextual learning scenario.

3.3 Application of Test of Basic Knowledge of Communication Systems, Antennas and Propagation

Initially, a knowledge test is carried out for antennas and propagation courses and communications systems to two groups, a control group and a test group. Both groups with a sample of 27 students each chosen randomly and with independent samples. In the control group we will call Grupo_A are the students who performed their laboratory practices In situ, and the test group we'll Grupo_B who performed their lab practices using SDR. Each test was evaluated on a scale of 0 to 20 points, for a total of 20 questions. Weighted results for both groups can be seen in Table 2.

Table 2 Results obtained in the control group and test group

Group A	Group A	Group A	Group A	Group B	Group B	Group B	Group B
8	12	18	15	15	18	14	18
20	15	16	16	16	19	19	18
14	16	16	17	15	19	18	16
6	16	14	9	16	18	16	18
19	18	15	19	9	16	7	16
19	5	13	15	17	16	11	14
12	11	17		12	13	16	

It is proposed to establish which of the two teaching methods allow the student to obtain better academic performance based on the assessment of the concepts developed in the knowledge test, one through traditional teaching in a laboratory and the other by incorporating technology through the use of SDR. In order to contrast the null hypothesis of normality of the data. In order to apply this statistical technique, the following conditions must be checked:

1. Normality in the distribution of scores for each of the groups.
2. Homocedasticity or equality of population variances of qualifications according to the group.
3. Independence of observations, which is achieved by extracting independent random samples from the populations. This hypothesis is usually assumed as a result of the design of the experiment, as the different levels of the factor that are randomly assigned (In this specific case, the students are chosen randomly). Therefore, we will only check the other two assumptions. However, before we check to see if the conditions necessary to perform an ANOVA are met, we will explore the data.

They are then shown in Table 3. The descriptive statistics of the data, which collects the information from two samples belonging to Groups A and B, each with a total of 27 samples per group, with an average score obtained in the knowledge test of 14.48 for Group_A and 15.56 for Group_B, and a standard deviation of 3.94 and 3.02 respectively. And with a range of scores that oscillate between 5.00 and 20.00 for the knowledge test of Group A and scores between 7.00 and 19.00 for Group_B. The quartiles indicate, for example, that half of the students in Group A obtained at least a score of 15.00 and half of the students of Group_B obtained 16.00 and that 75% of the students in Group A have a score of more than 17.00 and Group B a score of more than 18.00.

Table 3 Descriptive statistics

Groups	N	Mean	Typical deviation	Minimum	Maximum	Percentiles		
						25	50	75
A	27	14.48	3.94	5.00	20.00	12.50	15.00	17.00
B	27	15.56	3.02	7.00	19.00	14.50	16.00	18.00

According to Fig. 4, it is observed that for the analysis of Group_A, the boxplot parameters in r statistics show: presence of an outlier value in 5, and for Group_B, two outliers in 9 and 7. Also, it is observed that the data is skewed to the left in both Figs. 4 and 5.

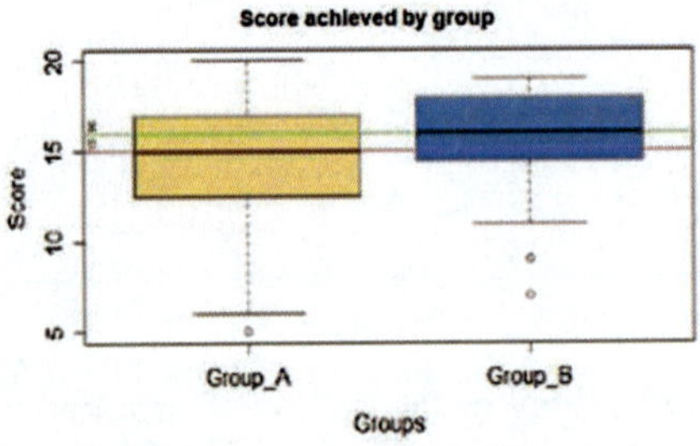

Fig. 4 Box plot score vs groups

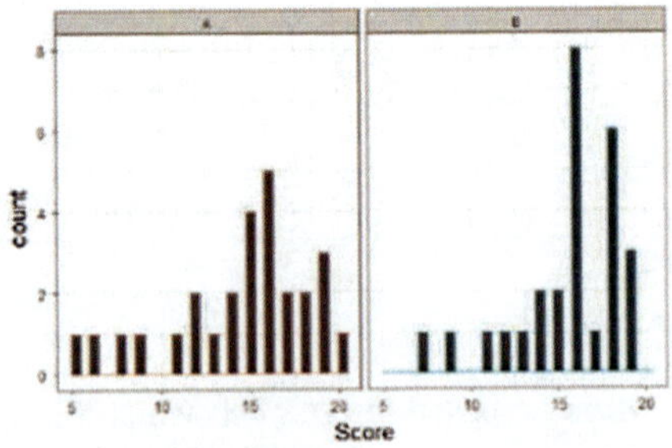

Fig. 5 Histogram of score vs groups

Next, a contrast is performed for normality. In this contrast, the null hypothesis is the normality hypothesis, that is, there are no differences between our distribution and a normal distribution with that mean and that standard deviation. To contrast normality, we use the Shapiro–Wilk test. As the p-value obtained is (0.03391) and this is less than alpha (0.05), the null hypothesis (H0) is rejected, therefore, the score obtained by "Groups_A" does not present a normal or parametric behavior. We also carry out the analysis for "Group_B". As the p-value obtained is (0.00308) is less than $\alpha = 0.05$, the null hypothesis (H0) is rejected, therefore, the score obtained by "Groups_B" does not show normal or parametric behavior either. If we carry out an analysis of the linear regression model of the residual errors, we observe that it is also obtained that the residual errors are not normally distributed for either of the two groups, as evidenced below:

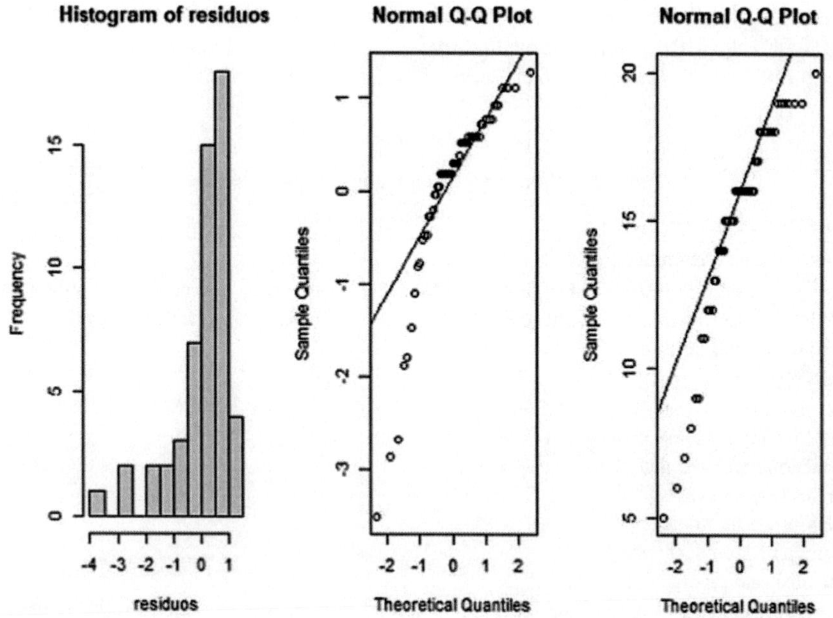

Fig. 6 Representation of residuals, distribution of residuals and distribution of data by group

In Fig. 6 it is observed that the graph on the left is the histogram of the residuals, it is evident that the residuals are skewed to the left, the graph in the center represents the distribution of the residuals and the one on the right the distribution of the data, in the three graphs it is observed that the data do not behave as a normal distribution. Therefore, a test is required to demonstrate whether the null hypothesis that the medians of the scores between groups do not differ from each other. For this, we apply the non-parametric Kruskal—Wallis test, and we obtain the following results: The test indicates that there are no significant differences between both groups, since the p-value is 0.3065 such that it is greater than the significance level of 0.05. Therefore, we do not reject the null hypothesis, and we conclude that the medians between groups A and B do not differ from each other. It is concluded that when using the SDR technological tool in the test group there is no significant difference in their teaching method with respect to the control group that uses the traditional teaching method with the size sample of experiment.

3.4 Laboratory Perception Instrument

A perception instrument was applied, in this case a survey of group B students, a group that had contact with the communications laboratory supported by SDR. 26 students from the target group participated once the pilot test was completed, in addition to 2 teachers from the area who accompanied the piloting. For the survey, participants answered each question on scale of 1 to 5, with 1 (the lowest value) and 5 (the highest value). Table 4 shows the results of the questions asked. The questions were classified into four categories "contribution to training", "Motivation against the tool", "Perception of SDR technology" and "Laboratory assessment".

Table 4 Laboratory perception instrument questions

Questions of survey	1	2	3	4	5
Contribution to learning					
Do you consider that the SDR laboratory encourages and stimulates learning?	Definitely not (0%)	No (0%)	Regular (0%)	Yes (3,6%)	Definitely yes (96,4%)
Do you consider that the SDR laboratory presents a didactic alternative that would help to better appropriate the topics of communication systems, antennas and propagation?	Definitely not (0%)	No (0%)	Regular (0%)	Yes (3,6%)	Definitely yes (96,4%)
In general terms, to what extent does the proposed practical exercise help you to learn effectively?	Definitely not (0%)	No (0%)	Regular (0%)	Yes (71,4%)	Definitely yes (26,6%)
Motivation in front of the tool					
Do you think that the use of SDR technology generates added value in your academic and professional development?	Definitely not (0%)	No (0%)	Regular (0%)	Yes (3,6%)	Definitely yes (96,4%)

(continued)

Table 4 (*continued*)

Questions of survey	1	2	3	4	5
Has the approach with the SDR laboratory motivated you to learn more about telecommunications issues?	Definitely not (0%)	No (0%)	Regular (0%)	Yes (3,6%)	Definitely yes (96,4%)
Perception of SDR technology					
Do you consider that the use of technological tools based on SDR would improve the laboratories in situ at UNAD?	Definitely not (0%)	No (0%)	Regular (0%)	Yes (14,3%)	Definitely yes (85,7%)
The use of SDR devices that you have found in terms of difficulty?	Very difficult (0%)	Difficult (10,7%)	Regular (50%)	Easy (35,7%)	Very Easy (3,6%)
Has the material used as a study guide for the use of the SDR-based laboratory been interesting?	Definitely not (0%)	No (0%)	Regular (0%)	Yes (28,6%)	Definitely yes (71,4%)
Do you consider that the use of SDR-based laboratory-based development tools support the process of updating courses in the Electronics, Telecommunications and networks chain?	Definitely not (0%)	No (0%)	Regular (0%)	Yes (3,6%)	Definitely yes (96,4%)
Laboratory assessment					
What is your overall assessment of this type of SDR laboratories? (1 minor—5 major)	1 (0%)	2 (0%)	3 (0%)	4 (17,9%)	5 (82,1%)
Would you recommend that a colleague take courses where SDR laboratories are organized?	Definitely not (0%)	No (0%)	Regular (0%)	Yes (35,7%)	Definitely yes (64,3%)

4 Conclusions and Future Work

The implementation of the SDR communications laboratory as a contextual pedagogical strategy at UNAD allowed to prove the relevance of the proposal with the pilot test described. UNAD Telecommunications Engineering students acquired practical experience and applied the knowledge acquired theoretically in a contextual setting, managing to apply the basic concepts of Antennas and Propagation. Regarding the instruments applied in the experimentation, it was possible to identify a highly positive perception of the students and teachers who participated in the laboratory, standing out high in the different categories of the instrument. A favorable response was presented regarding the level of difficulty that the students had with the use of the laboratory, since despite the fact that none of the students had experience in the use of SDR technology, only 10.7% expressed that there was difficulty in the management of the equipment, managing to estimate a fast learning curve for the students when using this type of devices. It should be noted that the perception measured by the participants is quite favorable, placing 100% of the participants in the highest qualification levels (4 and 5) in relation to all the questions posed, both in the section on contribution to learning, motivation with the use of the tool and perception of the technology used. Even 82.1% of the participants rate the laboratory with the highest level (5/5) and 17.9% with a (4/5).

On the other hand, the measurement instrument for the appropriation of basic concepts of Communications Systems and Antennas and Propagation, although it is true, gave favorable results in terms of the average of qualifications of group B, a group that had access to the laboratory, the statistical analysis did not allows us to conclude that there were significant differences regarding the comparison of the learning process with and without the laboratory.

As future works, it seeks to replicate the tests carried out at the national level to verify with a much larger sample the existence or not of a significant difference in the use of the laboratory as a pedagogical tool. Given the good perception results, the use of SDR technology is also planned not only in Communication Systems and Antennas and Propagation courses, but also in several additional courses of the program. In addition, the implementation of pilot tests is planned in different programs of the ETR chain and in the different centers nationwide. Due to the flexibility and potential of technology, there is also evidence of an innovative alternative to support various research processes at the University, as well as the strengthening of the laboratory with more equipment that allows supporting postgraduate projects in the modality of applied projects and projects of investigation.

Acknowledgements. We are grateful for the support of the doctorate in modeling and scientific computing of the Faculty of Basic Sciences of the University of Medellín for their contribution in the process of this work.

References

1. Abadía García C, Vela Gonzáles PA, Montero Vargas P (2016) Lineamientos Generales del Currículo de la UNAD, Bogotá
2. UNAD (2011) Proyecto Académico Pedagógico Solidario, Bogotá
3. C. S. U. UNAD (2020) Acuerdo 029 del 12 de Agosto de 2020, Bogotá
4. Abadía García C, Montero Vargas R (2015) Lineamientos Académicos para el Desarrollo del Componente Práctico, Bogotá
5. Wireless Innovation Forum. Wireless Innovation Forum. https://www.wirelessinnovation.org/what_is_sdr

6. Sruthi MB, Abirami M, Manikkoth A, Grandhiraj R, Soman KP (2013) Low cost digital transceiver design for Software
7. Kramarev D, Hong Y, Viterbo E (2014) Software Defined Radio Implementation of a Two-way Relay Network with Digital Network Coding
8. Thaj T, Viterbo E OTFS Modem SDR Implementation and Experimental Study of Receiver Impairment Effects
9. University of Colorado. Communication, data, information and network sciences research area - Research Groups. https://www.colorado.edu/communications-data-science/research-groups. Accessed 2020
10. University of Colorado Department of Aerospace Engineering Sciences (2020) Thinsat Operational Management
11. University of Notre Dame (2020) Wireless Institute University of Notre Dame. https://wireless.nd.edu/about/software-defined-radio-sdr-lab/
12. Ruiz Peñas LE, Crespo Cadenas C (2018) Sistemas Basados en SDR para Equipamiento de Comunicaciones Inalámbricas, Sevilla
13. Pinar Domínguez I (2011) Software Defined Radio: USRP y plataformas de desarrollo
14. Iowa State University Iowa State University - Electrical and Computer Engineering. https://www.ece.iastate.edu/research/centers-institutes-and-laboratories/software-defined-radio-laboratory/
15. Mao S, Huang Y, Li Y (2014) On developing a software defined radio laboratory course for undergraduate wireless engineering curriculum
16. Gustafson II, MR, Nadeau JC, Schaad D (2013) A grand challenge-based framework for contextual learning in engineering
17. CORD (2016) What is Contextual Learning
18. Mathworks. Desplegar y prototipar un sistema de radio definida por software (SDR) con MATLAB y Simulink. https://la.mathworks.com/discovery/sdr.html

Comparison of Two Technologies for Digital Payments: Challenges and Future Directions

Galena Pisoni[1](✉), Bálint Molnár[2], and Ádám Tarcsi[2]

[1] Université Côte d'Azur, Nice, France
galena.pisoni@univ-cotedazur.fr
[2] Eötvös Loránd University, ELTE, IK Pázmány Péter 1/C, Budapest 1117, Hungary

Abstract. Digital payments are dramatically increasing around the world. Global non-cash payments has observed an increased by 60% since 2015. This rapid growth has been a results of technological innovation and increased presence of regulation. Digital payments offer exciting opportunities and many banks and other financial institutions have been innovating in the domain. Established financial institutions, also non-financial institution such as start ups or technology companies, are contributing significantly to financial technology innovation in the payments market. This article provides a comparison of two technological innovations and challenges, provides the basis for a conceptual framework on how to compare innovative digital payments solutions, such as future directions in the evolving payments landscape.

1 Introduction

Technology surrounds us and helps us in most of our daily activities. Smartphones have become essential tools in our everyday lives as they expand our physical and intellectual abilities and they allow us to carry out tasks that were difficult even to imagine not many decades ago. Countless innovative products and services have been developed and are continuously being created in order to let us take advantage from the opportunities provided by innovation and to serve various and ever-increasing purposes.

Payment systems are the cornerstone of every economy, and payments innovation is a critical contributor to worldwide economic development [15]. New technologies and changing needs of final customers have brought to the developments of new payments systems. These innovations have been driven by a variety of players: banks, non-banks and regulators [3]. Other financial institutions (and not banks) have been successful in identifying unmet needs and enhancing existing payment solutions, embracing underserved populations and enhancing payments system efficiency [5].

At the same time, banks also have a long history of payments innovation that is sometimes overlooked. Banks control the core payments infrastructure. Banks also lead the way in serving customers differently, with new products.

M. E. Auer et al. (eds.), *Online Engineering and Society 4.0*, Lecture Notes in Networks and Systems 298, https://doi.org/10.1007/978-3-030-82529-4_46

In addition, new payments types require close collaboration between banks and non-banks to create the underlying infrastructure [4,13].

Given the financial and reputational impact of data breaches, privacy and security are frequently among the top concerns for such developments [6]. GDPR compliance is often viewed as a costly and challenging effort and ultimately something that could obstruct projects. These reservations can lead to reluctance in investment and innovation. Incorporating privacy considerations from the beginning of a project can have huge benefits [8]. Digital industries are centred around consumer trust and in an age where data is a public concern, one of the best ways to build this trust is to be able to show that data handling practices are strong and ethical. For these reasons, businesses that can find inventive solutions to privacy challenges are going to be in the best competitive position in the digital sector.

In this paper we analyze the case of two digital finance companies and their solutions for innovative digital payments. In Sect. 2 we present the case studies we analyze in Sect. 3 we present a comparison between the case studies and future directions.

2 Case Studies Analyzed

2.1 Case Technology 1

With this innovative payment authentication system people can always have their digital documents with them. Customer who choose to adopt this innovation will not need to manual insert any letter or any numbers from their documents, and the dedicated application will do it for them. It is a customer driven technology, and it's final aim is to offer a better service to the final customer.

This case study is an innovative application that allows the user to store his or her identity credentials on the smartphone. Moreover it has the potential of becoming a broadly-used payment authentication system. It may help the clients of the banks to perform transactions more safely. It also may improve the user experience by creating a more secure authentication and providing the functionality for the digital signature. With this application in order to verify the identity it would be enough to scan the passport and then insert a PIN to ensure total security. The bureaucratic paper-loaded procedures of client identification in the banks may be significantly simplified and accelerated, moreover the updates of any data in the personal documents can be done swiftly without the need to visit the branch office of the bank. Apart from that, it addresses the desire of the people to avoid manual compilation of the information needed for the authentication or registration purposes (other than the password or PIN-code).

We have detected four possible application areas in which this technology could be used, that are:

- Authentication of users;
- Digital signature for users;
- Registration of the users;
- Management of the smart ID-account;

Looking more deeply, from the side of the customer, the main advantages are the fact that the application works in the mobile phone or tablet of the user, it increases the security on diverse levels, it accelerates the bureaucracy procedures of identification and simplification of any update of the personal documents, it provides a better customer experience and that fact that customer can avoid manual compilation of the personal information in any registration or authentication phase.

On the other hand, from the point of view of the bank, we have detected three main advantages which are:

- A faster data management (no need of long manual registration),
- Banks will avoid any mistakes in clients' data and the enjoy of a better image for the bank (offering a paper-free service).

2.2 Case Technology 2

The second case study is a new technology that 'authenticates' payments with a help of facial recognition of the customer, the transaction is allowed if the customer smiles at the 'payment toll camera' at checkout. This company aims to revolution the way we think about digital payments. In fact, the main feature of the product consists in a technology allowing for the possibility to pay with biometrics: for now, the focus is on facial recognition; but in the future, it is possible that other connected methods will be considered as well.

Where the possibility to use one's own phone to buy products and services is already considered to be a novelty, this case is already one step further: no cards, no phones—only the user's face (once enrolled). The application is designed to be compliant with the main relevant EU legislation both on biometrics and digital technologies.

Banks will play an important role for the future of the service, for instance they can propose the registration to of such a service when opening a new bank account; in this way, they directly benefit from the added value of the solution, as they are able to offer an additional service which competitors do not have. In addition, merchants using this service increase their sales by offering this innovative payment method.

At the same time, it is fundamental to underline that according to the same Regulation, Member States of the EU have the possibility to introduce stricter laws and/or specifications in order to protect users' personal data as they deem to be the most appropriate modality. Consequently, when launching the product in different Member States, it is necessary to have a full understanding of the legal situation on the matter.

3 Comparison of Payments Systems: Towards a Framework for Payments System Evaluation

Ultimately, payments innovation is about enhancing value for consumers and businesses, through increased speed, convenience, information, etc.

As the case studies above clearly show, different players have different roles to play in payments innovation, and bring different strengths and weaknesses to the innovation space.

Both of the case studies we analyzed focused on unmet needs of the customers, provided better, easier to use, faster way for final customers to pay, lead with value not the technology, and both worked hard to minimize risk and increase protection of personal data. This is consistent with previous research that said raised the importance for privacy and security risks of digital payments [1], while keeping the customer in focus [2,11,14].

We compare the studies on four aspects: convenience, security, simplicity, and availability (Table 1).

Convenience: Payments systems must be more convenient than traditional payments systems. If the system must go thought various procedures to use the payment service, users might lose interest and prefer the traditional payment approaches.

Security (and Privacy): These payments systems keep customers data, security is an important requirement, and the users data should not be exposed to malicious attackers, and such information must not be available to unauthorized third party.

Simplicity: The solution must be lighter and with developments of IoT technology this aspect became even more important for each sphere, payments included. The payment system must be optimized and simple.

Availability: New payments services should be designed in a way that allow for an easy availability in most possible places, that means on various devices or operating systems.

Table 1. Classification of the case studies in relation to their convenience, security, simplicity, and availability affordances

	High	Medium	Low
Convenience	Case study 2	Case study 1	
Security	Case study 1 Case study 2		
Simplicity	Case study 2	Case study 1	
Availability	Case study 1		Case study 2

We exclude other factors from this study that others scholars used [7,9] because we focus on digital payments solutions from innovation point of view and we want to study and understand better their potential to penetrate in established markets and to dominate them. We are aware that other factors may influence the general availability of the payment system in first place, however, we investigate the design logic of the payments system and how they influence to higher or lower possibility for adoption by final customers.

In terms of convenience, our second case study requires that merchants need to have the hardware and software installed on premises, which consist of big disadvantage on availability aspect, while the fact that the first case study is only an application on the cellphone in this respect provides simpler availability for users. On convenience aspect, the second case technology is superior as it provides simpler way to do payments compared to traditional payments system, while the first case seen from this perspective presents a burden, as still the final users need to enter traditional payment details and then use the app to confirm identity, a step that improves security and privacy, still at the moment not commonly present in current payment systems. Both of the case studies have put in place high level of security due to regulation, however also from the value they provide to the customers: the first case improves confirmation of identity and makes payments more secure, while the second provides additional levels of security due to the thorough check thought fingerprint and face recognition. Both of the case studies provide as simple as possible interaction.

For both of the cases it is important to underline the fact that the EU's General Data Protection Regulation is particularly strict on any kind of use and storing of any data (biometric or personal data from ID). At the same time, the use of ID and biometrics in the field of digital payments is not uncommon; as such, the connection has not been widely studied and undoubtedly needs further understand and research, both from technical and legal perspectives.

4 Discussion

We study how new types of digital payments solutions are designed and configured to enter the payment markets. We analyzed two case studies of innovative payments solutions and we glean insights on how the platforms are developed and designed across 4 axes: convince, security, simplicity and availability.

One can argue that banks and merchants prefer payments instruments that are easy to use right-away or highly compatible and accessible with their current payment infrastructure, to reduce adoption costs and this add another possibility or variety to the existing customer and business structure. In this respect, the solution based on mobile payment like in our case study 1, can potentially turn to be a challenge for merchants in terms of incompatibility with existing payment infrastructures.

Controlling and leveraging on the existing user base is a precondition for success of digital payments innovative solutions, so in this respect our case study 2 provided better basis for still serving the customer base of the merchant while diversifying and providing different (if not even) fun approaches to engage them.

Ultimately, it is important to recognise the roles that banks, non-banks and regulators can play in innovation. Both of the case studies came from non-banks institutions, still collaboration between all is important step to guarantee success of the new payment deployment.

5 Conclusions and Future Directions

Our research compared two case studies of innovative digital payments solutions and provided a conceptual for comparing different payments solutions design and configurations to understand successful payment solutions. We performed a comparative-case study in a European setting, identified challenges and opportunities for the case studies, and provided the basis for a conceptual contribution to better describe and understand contemporary digital payments platforms in terms of their market-entry potential. The framework that we propose can be used by merchants or market players to compare and reflect on decisions regarding digital payments adoption.

This study has several limitations: first we don't take hardware complexities, or the platform effect and collaboration between the banks or regulators in the payments ecosystem, both might have impact on market entry for the case studies. Another important aspect to mention is that both of the case studies are in pilot stage, so also their affordances in terms of convenience, security, simplicity and availability may change.

Future research could focus on innovative payments solutions and what they mean for the banks. Payment innovation may for instance lead change in fees and may require transactions regulation. A fee reduction may change the traditional model on which for now for instance banks process payments with cards, and may in turn increase or decrease possibilities for use.

Another future research can explore how the different layers of complexities of the payment platforms, and how for instance the introduction of such payments system influence and shape existing complex business processes [12], or for instance if on strategy side for financial companies to become more digital [10], and how potentially changes in various technology layers impact adoptions by final customers.

Acknowledgement. The project was partially supported by "Application Domain Specific Highly Reliable IT Solutions" project that has been implemented with the support provided from the National Research, Development and Innovation Fund of Hungary, financed under the Thematic Excellence Programme TKP2020-NKA-06 (National Challenges Subprogramme) funding scheme.

References

1. Acharya B (2017) Privacy and security risks of digital payments
2. Ali R, Barrdear J, Clews R, Southgate J (2014) Innovations in payment technologies and the emergence of digital currencies. Bank of England Quarterly Bulletin p Q3
3. Buckley RP, Mas I (2016) The coming of age of digital payments as a field of expertise. U Ill JL Tech & Pol,'y, p 71

4. Hedman J, Henningsson S (2012) Competition and collaboration shaping the digital payment infrastructure. In: Proceedings of the 14th annual international conference on electronic commerce, pp 178–185
5. Henningsson S, Hedman J (2014) Transformation of digital ecosystems: the case of digital payments. In: Information and communication technology-EurAsia conference, pp 46–55. Springer
6. Johnson VL, Kiser A, Washington R, Torres R (2018) Limitations to the rapid adoption of m-payment services: Understanding the impact of privacy risk on m-payment services. Comput Hum Behav 79:111–122
7. Kazan E, Damsgaard J (2013) A framework for analyzing digital payment as a multi-sided platform: A study of three European NFC solutions
8. Oliinyk I, Echikson W (2018) Europe's payments revolution. stimulating payments innovation while protecting consumer privacy. CEPS research report no. 2018/06, September 2018
9. Ondrus J, Gannamaneni A, Lyytinen K (2015) The impact of openness on the market potential of multi-sided platforms: a case study of mobile payment platforms. J Inf Technol 30(3):260–275
10. Pisoni G (2020) Going digital: case study of an Italian insurance company. J Bus Strategy 42(2):106–115
11. Riemer K, Hafermalz E, Roosen A, Boussand N, El Aoufi H, Mo D, Kosheliev A (2017) The fintech advantage: Harnessing digital technology, keeping the customer in focus. University of Sydney, Business School and Capgemini, Technical report
12. Rodrıguez C, Engel R, Kostoska G, Daniel F, Casati F, Aimar M (2012) Eventifier: Extracting process execution logs from operational databases. In: Proceedings of the demonstration track of BPM, vol 940, pp 17–22
13. Tan JD, Purba JT, Widjaya AE (2019) Financial technology as an innovation strategy for digital payment services in the Millenial generation. In: 1st Aceh Global Conference (AGC 2018), Atlantis Press
14. Vives X (2019) Digital disruption in banking. Ann Rev Financ Econ 11:243–272
15. Weichert M (2008) Payments innovation: a comparison of banks and non-banks and how they can learn from each other. J Paym Strategy Syst 2(3):236–249

Machine Learning: A Tool for Formulating Tracking Strategies for Covid-19 Pandemic in Colombia

Hernando Vélez Sánchez(✉) and Luini Leonardo Hurtado Cortes

Universidad Distrital Francisco José de Caldas, Bogotá, Colombia
{hvelezs,llhurtadoc}@udistrital.edu.co

Abstract. The Covid-19 pandemic has greatly affected the health conditions of the population as well as the economies around the world. In some countries, health systems have been overwhelmed, since in most cases they were not prepared to assist the large number of people who were affected. In the absence of vaccines and treatments of proven efficacy, the most obvious measure is the lockdown. An initial lockdown of five months was one of the first measures taken by the Colombian government to impede the spread of the virus. However, this measure did not prevent the country from becoming the fifth in number of infected people worldwide. With a population of about 8 million people and the social dynamics and mobility typical of a capital city, Bogotá has high rates in the statistics of incidence, effective reproduction and mortality. It is assumed that after reaching herd immunity, the incidence of the virus diminishes. But the current priority in the country is to focus on the scarce resources on the sites and the communities most prone to contagion. This document presents the use of machine learning-based models to identify these potential nuclei. A clustering model was constructed using the Simple K-means algorithm to identify clusters of people who are most susceptible to contracting the virus. Four algorithms were used for determining the value of the effective reproduction number. The best performing algorithm was linear regression with a correlation coefficient of 0.753 and artificial neural networks with 0.774.

Keywords: Covid-19 · Machine learning · Tracking

1 Introduction

For the obtaining a tracking as efficient as possible in terms of resources and hits, it is important to identify the trend in the spread of the virus, that is, what kind of communities, people or social groups with high reproduction rate are more likely to contract the virus. To carry out this identification, the number of diagnostic tests must be increased. However, due to the limitation in the availability of the tests and the logistics for their application and processing, especially in highly populated and developing countries, prioritization in the use of these resources is necessary [10, 11].

In this vein, a question arises: On what kind of people and communities should these resources be focused in order to detect existing cases or prevent the appearance of new cases through selective isolation?

M. E. Auer et al. (eds.), *Online Engineering and Society 4.0*, Lecture Notes in Networks and Systems 298, https://doi.org/10.1007/978-3-030-82529-4_47

The purpose of this study is to provide a tool that can help to answer the stated question. For this purpose, a model was developed using machine learning and data mining to establish the effective reproduction number, which is a key parameter to predict the presence of new cases. The model is based on sociodemographic information from the open database of the city of Bogota, which contains data of infections and new cases, as well as their evolution. The probability of new waves in the pandemic requires decision-making tools that allow taking advantage of the lessons learned from the first wave.

1.1 Related Works

Several researchers [4] have developed applications to help in the detection of diseases related to Covid-19, for example, the blue dot software. Facing the Covid-19 outbreak originated in Wuhan, China, the blue dot software uses its AI system to identify an unknown type of pneumonia. In Boston Children's hospital, the use of the HealthMap tool helps in the detection of Covid-19 [2]. The Egida platform of the Universidad de los Andes in Colombia [3] allows understanding the evolution of the pandemic to predict the appearance of new cases.

Through the processing and analysis of news from all countries in the world, together with information related to commercial flight routes and disease outbreaks, BlueDot [5] was able to predict not only the existence of a potentially pandemic disease, but also its epicenter and its most immediate trajectory. A group of Basque scientists reinvented the "Ubikare" app for studying the behavior of the SARS-CoV2 virus. Using this structure, they created an app called "Reg Covid 19". The University of California [6] in San Francisco and Stanford University led epidemiological, clinical and genetic studies that made it possible to customize the measures to be taken based on the results from the Californian population.

In accordance with [7], several U.S. institutions are developing new AI technology or using preexisting technology to monitor and treat the new coronavirus. HealthMap, the AI application run by Boston Children's Hospital, was launched in 2006. It was among the first tracking mechanisms to detect the Covid-19 outbreak in China, and can even be used by the general public to receive real-time information on disease events in their community or where they may travel.

Researcher Jongeun You from the University of Colorado Denver (United States) [9] published a new research article in *The American Review of Public Administration*. In that article, the author reviews the public health policies of South Korea as it is one of the few countries that has not confine the entire population. According to the research, some of the critical factors in South Korea's public health administration and management that led to success include national infectious disease plans, collaboration with the private sector, rigorous contact tracing, a system of adaptive healthcare and government-driven communication.

The University of Singapore developed an algorithm based on artificial intelligence to predict the spread of the pandemic. Finally, John Hopkins University has a very complete database on the development of the pandemic and built models to analyze its dynamics and suggest control measures.

2 Methodology

This section presents the methodology used for the development of the work. First, a data-based model of pandemic behavior is formulated for each of two cases: relationship between variables and virus transmission, and calculation of the effective reproduction number. Then, databases are preprocessed and machine learning algorithms are selected to solve the cases. The models built with the algorithms are trained and tested using standardized metrics. The best performing models are selected. The free software Weka 3.8 from the University of Waikato, New Zealand, is used for preprocessing, training and testing the models.

2.1 Virus Spread Model

A conceptual model of the spread of the pandemic is initially presented based on the existing bibliography. The phases of the pandemic are: importation of cases, containment, community contagion, and sustainment. Virus transmission involves biological and socio-demographic (physiological and social behavior characteristics of individuals), as well as political factors. Human behavior may be very complex to model. According to the majority of the models used in epidemiology, variables such as population density, access to an adequate health system, and public policies in health and epidemiology are factors that have a great influence on the advance of the virus. Due to the transmission characteristics, a physical closeness is required between individuals including the physical spaces that are in contact with them. Other important factors are: virus mutation, acquired immunity and measures taken by society and the state.

Susceptible-Infected-Recovered Model (SIR). The most widely used models are based on the SIR model, formulated in 1927 by Kermack and McKendrick, which for many years has been a reference for the study of the spread of pandemics during the 20th and 21st centuries. The basic assumptions of these models consider a homogeneous and constant population. It is assumed that there are the next states at time t: susceptible (S), latent (L), infected (I), and removed (R). The interaction between individuals is random, since there is no external intervention that changes the contact rate of the population. Not all these assumptions are really fulfilled, for example, the fourth (R). The values of S(t), I(t), and R(t) change over time. The mathematical models establish relationships between the factors described, and use specific parameters of the cases such as: average duration of the disease, interaction rate, probability of contagion, and recovery rate.

An important element is the initial transmission rate R_0. Most models establish that a susceptible person can become infected due to contact with an infected person and then, they go through a latent state first. It can then be removed by isolation, fatality, or immunity. A probability model (Table 1) that meets the above characteristics is:

Table 1 Transition probabilities for the time increment [16]

Transition to	Probability
(s − 1, l + 1, i, r)	(αsi + oh)
(s, l − 1, I + 1, r)	(βh + oh)
(s, l, i −1, r + 1)	γih + oh
No change	Balance of probability

According to [16], a general epidemic model of Markov, continuous in time, establishes the initial conditions (S_0, Lo Io R_0) = (n, a, 0, 0), and a stochastic model for a closed community of size n + a. Durations of latency and infection are independent random variables with exponential distribution. Assuming a long latency period with constant duration and a short infection period, the binomial distribution can be used. These models are stochastic, thus they can approximate deterministic models. The form obtained is;

$$\frac{ds_t}{dt} = -\alpha s_t i_t, \ \frac{dl_t}{dt} = \alpha s_t i_t - \beta l_t \ \frac{di_t}{dt} = \beta l_t - \gamma i_t, \ \frac{dr_t}{dt} = \lambda i_t \tag{1}$$

To determine the number of infected people, it is necessary to know the effective reproduction number $R_e(t)$. For its determination, there are three fundamental aspects: biological factors, like duration of the infection period, probability that an infection passes from an infectious person to a susceptible person, and social factors such as the average number of times per unit of time that an infectious individual meets a susceptible individual during the day. The key is to analyze only the results of the propagation. From this point of view, only factors related to the possibility of contact between people and between them and public use sites will be taken into account. This work addresses both issues, emphasizing the behavioral characteristics that are closely related to the sociodemographic characterization of individuals. The database used in this work provides values of $R_e(t)$ calculated using a modified Bayesian time series model called the state-space model [13]. The expression for its calculation is:

$$R(t) = \sum\nolimits_{\tau=1}^{\infty} \omega(t) \frac{R(t-\tau)}{\cap} \tag{2}$$

with ω(t): weights
∩: number of days the person can infect.

As mentioned above, the tool used for this calculation is data mining and the machine learning is employed to establish the sociodemographic conditions that favor the spread of the disease. Since this is a new situation for most of the countries, the results obtained in the application of the proposed models have a limited scope. Therefore, conclusions in line with the behavior of the pandemic can only be obtained if greater information and variables that have not been collected are included in the data provided by the organizations in charge of it.

The calculation of $R_e(t)$ allows evaluating the effectiveness of the measures adopted to confront the pandemic under time-varying conditions (retrospective), and predicting its spread under constant conditions over a time interval. Values less than one mean that the pandemic is not spreading, while the higher it is above one, the greater is the speed of the spread [12].

A common objection to the kind of mathematical models presented above is that such models involve too many simplifying assumptions, but it is wrong to reject them just on the basis of their simplicity.

Data Model and Databases. The data-driven model has a structure defined by the following data levels: attribute or variable level, attribute class level and attribute value level. For the approach to the data-driven model, the following databases are used;

- Positive cases of Covid-19 in Colombia from the Instituto Nacional de Salud. Open data portal [1].
- Bogotá Mayor's office database [15].

Relationship of Cases with Sociodemographic Variables. To find the relationship between the probability that a susceptible person becomes infected and the sociodemographic variables, the model is built based on the first database. Table 2 shows the structure of the data-based model.

Table 2 Dataset attributes

Attribute	Class	Values
Onset of symptoms date	Nominal	
Diagnostic date	Nominal	Date
City of residence	Nominal	Bogotá, out of Bogotá, unknown,
Town of residence	Nominal	Usaquén, Barrios Unidos, Fontibón, Puente Aranda, Teusaquillo, Candelaria, Ciudad Bolívar, Antonio Nariño, Usme, San Cristóbal, Mártires, Rafael Uribe, Tunjuelito, Suba, Bosa, Kennedy, Chapinero, Engativá,
Age	Numeric	
Gender	Nominal	F, M
Type of case	Nominal	Unknown, under study, imported, related,
Location	Nominal	Home, passed away, passed away does not apply, hospital, hospital UCI,
State	Nominal	Passed away, passed away does not apply, moderate, serious, mild, recovered
Ethnic group	Nominal	Indigenous, black, other
Social stratum	Nominal	1 and 2, 2, 3,4, 3 and 4, 2 and 3, 2-3-5, 3 and 4, 3-4-5
Lock down	Nominal	Total, partial, no

Table 3 Attributes for effective reproduction number

Attribute	Class	Values
Window_Start_Date_	Nominal	date
Winndow_End_Date	Nominal	date
Mean (R)	Numeric	Bogotá, out of Bogotá, no data
Std (R)	Numeric	Data
Cuantil.0.025(R)	Numeric	Data
Cuantil.0.05(R)	Numeric	Data
Cuantil.0.25(R)	Numeric	Data
Median(R)	Numeric	Data
Cuantil.0.75(R)	Numeric	Data
Cuantil.0.95(R)	Numeric	Data
Cuantil.0.975(R)	Numeric	Data
Town	Nominal	
Lockdown	Nominal	No, partial Bogotá, partial, total
Population	Numeric	Data
Population density	Numeric	Data
Infected	Numeric	Data
Susceptible	Numeric	Data
Recovered	Numeric	Data
Contact rate	Numeric	Data

Calculation of the Effective Number of Reproduction. To build this model, the second database is used. $R_e(t)$ is calculated from the previously mentioned space-state time series model [13]. Table 3 shows the attributes, their class and values.

Data Analysis. Figure 1 shows the variation of the mean value of the effective reproduction number over time since the start of the pandemic. The colors represent the quarantine condition. It can be established that the highest values of $R_e(t)$ correspond to the condition when there was no restrictions for people's activities, which is the case of non-quarantine and is represented by the green dots. On the contrary, the points in blue color correspond to the condition of full quarantine and are located at the lowest values of $R_e(t)$, with some exceptions. The partial quarantine conditions are presented in red and show intermediate values of $R_e(t)$. Figure 2 presents the behavior of the effective number of reproduction in Bogotá for different confinement conditions. The trends are similar to those of Fig. 1.

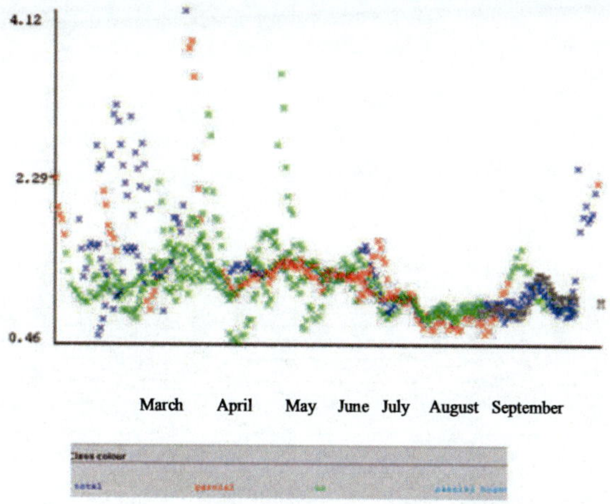

Fig. 1 Variation of the effective reproduction number depending on the date and the mobilization restrictions

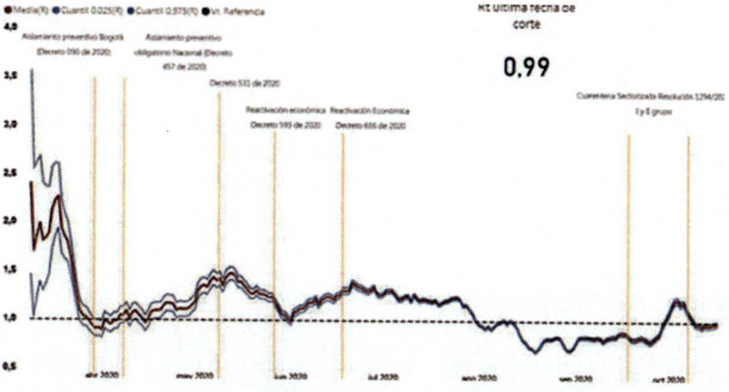

Fig. 2 Evolution of $R_e(t)$ in Bogotá [14]

When $R_e(t)$ is analyzed per localities, a homogeneous behavior is observed in Fig. 3, since the values follow a very similar trend. However, the red points that correspond to the towns of Kennedy, Barrios Unidos and Rafael Uribe are located in the middle and upper part of the figure, which indicates higher values of $R_e(t)$. The figure also shows the relationship between the existence or not of quarantine and the value of $R_e(t)$.

It can be observed that total quarantine values report lower number of points, which denotes the existence of fewer cases, while for partial quarantine the points increase a little and in the case without quarantine, the density of cases is much higher and the number of points of this last case a little above the previous ones.

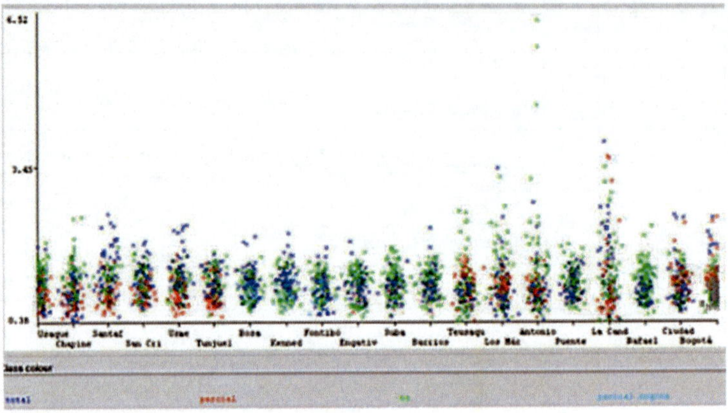

Fig. 3 Behavior of $R_e(t)$ per localities and quarantine

2.2 Grouping and Classification Models

After selecting the databases for the two cases under study and preprocessing them, the grouping algorithms are selected for the identification of common characteristics in the spread of the pandemic and the socio-demographic variables, and then, the effective number of reproduction is calculated. For these purposes, four supervised learning algorithms are selected for this case classification. For training the algorithms used, the split method is used, which consists of dividing each database into two subsets: one of 66% for training and the other of 34% for testing and validation. The goodness of fit of the models is determined using metrics such as MAE, RAE, RMSE and RRSE.

3 Results

The application of the methodology described for the two cases is presented below.

3.1 Case Identification Model for Tracking

A model is built in order to identify relationships between the attributes that may influence the transmission of the virus. The data was taken from the database, whose structure is in Table 2. Unsupervised algorithms of grouping and association are used, on the one hand, to identify common characteristics in the attributes of each group of infected people, and on the other, to establish applicable rules for the formulation of policies for tracking possible new cases [8]. This model may be applied to the strategy that the Colombian government has designed to mitigate the pandemic.

Clustering

To perform the grouping of sociodemographic variables with common characteristics, the unsupervised Simple K-Means algorithm was used as described below.

- Group the data into k different groups based on the position of its centroid. To determine to which cluster a data is added, calculate the Euclidean distance between the data and the centroid of the different clusters and choose the cluster with the minimum distance. The metric used is the mean for numerical data or the mode for nominal data. The steps of the algorithm are the following:
- Choose K examples that act as seeds for k clusters.
- Each example is added to the class that most closely matches it.
- The new seeds are chosen by calculating the centroids of the classes.
- Update the seeds by calculating the centroids of the classes.
- To achieve convergence, the second step is considered.

Cluster Analysis. An analysis of each of the clusters built using the K-Means algorithm is presented. The number of instances is 240 288.

Table 4 evidences that the main differentiating elements are the localities where the people reside. These localities comprise people who belong to strata between 2 and 3 and live in populated neighborhoods. This probably means that, despite the quarantine, they mobilize to produce sustenance for their families. The tracing of more susceptible people to infection can be focused on the localities with the highest number of infections and on people with the characteristics defined in the clusters.

Table 4 Cluster features

Cluster	City	Town	Age	Gender	Type of case	Estratum	Ethnic group	Lock down
0	Bogotá	Bosa	10–20	F	Study	3-4-5	Other	Partial
1	Bogotá	Unknown	21–30	M	Study	3-4-5	Other	Partial
2	Bogotá	Engativá	31–40	M	Study	3	Other	Partial
3	Bogotá	Suba	41–50	F	Study	2-3-5	Other	Partial
4	Bogotá	Fontibón	31–40	M	Study	3-4	Other	Partial
5	Bogotá	Engativá	51–60	M	Study	3	Other	Partial
6	Bogotá	Kennedy	61–70	M	Study	2-3	Other	Partial
7	Bogotá	Tunjuelito	41–50	M	Study	2-3	Other	Partial

Figure 4 shows the eight clusters defined and differentiated by type of cases summarized in the table above, most types of cases have not been identified in terms of their type of transmission and are listed as cases under study. The cluster with the highest number of cases is 0, which is located in Bosa, with a percentage of 36% of cases, followed by clusters 1, 2 and 3. The cluster with the lowest number is 7 with only 4% of the total.

This allows orienting actions for the tracking of cases, which can be initially headed by localities like Bosa, Engativá, Suba and the others identified in the table. The search can be directed to people with the characteristics defined for each cluster.

It should be noted that the Bogotá Mayor's office directed a partial quarantine for the localities with the highest number of infections. The measure included all the

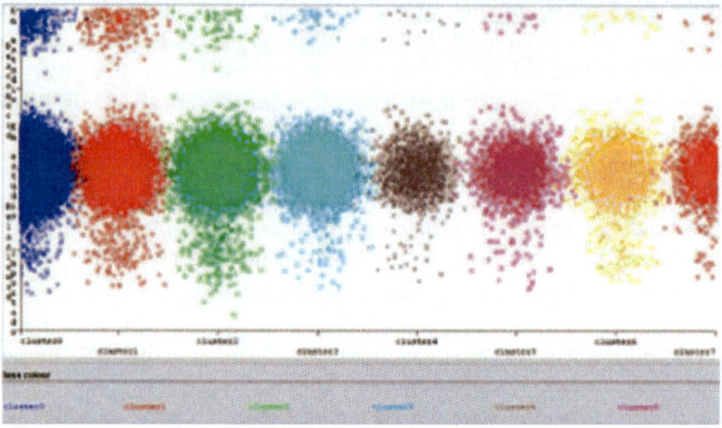

Fig. 4 Clusters by type of case

people who lived in those localities, excluding those who carried out activities of high impact on society. However, the model built allows further specifying the type of population that should be tracked, for example, according to their age, gender, stratum and ethnic group.

Figure 5 presents the clusters compared with the different grouping variables. In the first case, the clusters are associated with strata. It can be observed that cluster 0 is similarly distributed among all strata, while strata 3 and 4 predominate in cluster 4 and strata 1 and 2, 4 and 5 in cluster 4. In this figure, the grouping is compared with the quarantine type. In the partial type, there are few infected in cluster 4 and without quarantine there are almost no counts in clusters 3 and 4. Regarding the type of case, most of them had not been identified and were under study, followed by cases acquired through contact with related persons. When analyzing by gender, it is observed that the distribution is similar in the different clusters, with the female gender prevailing in strata 1 and 5 and the male gender in 0, 1 and 2. Finally, age is distributed in a similar way in the different clusters, except for the case of clusters 2, 3, 4 and 7, which have very few cases in the range of 90 to 104 years.

3.2 Model of Effective Reproduction Number Re

Taking into account that the variable or attribute is used as a continuous classifier, the algorithm selected is regression. All models have been trained with 66% of the data and tested with 34%. For this purpose, four models are built using four different algorithms:

- Gaussian process algorithm
- Linear regression algorithm
- Neural networks algorithm (MLP)
- Support Vector machine algorithm (SMO)

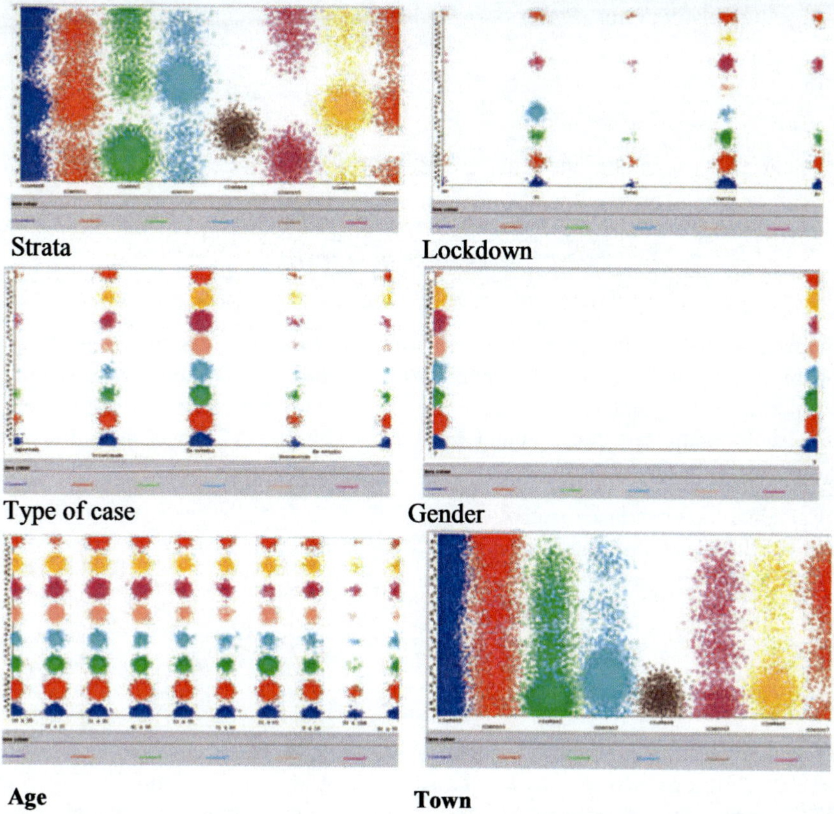

Fig. 5 Clusters by stratum, quarantine, type of case, gender, age, location. Cluster in x axis

Estimation of precision and error in regression algorithms can be done by means of the following metrics:

- Correlation coefficient between the estimated variable and the predictors
- Absolute mean error (MAE)

$$MAE = \frac{1}{n}\sum_{i=1}^{n}|f_i - y_i| = \frac{1}{n}\sum_{i=1}^{n}|e_i| \tag{3}$$

- Square root of the absolute mean error (RMSD)

$$RMSD = \sqrt{\sum_{t=1}^{n}\frac{(\widehat{y}_t - y)^2}{n}} \tag{4}$$

- Absolute relative error (RAE)

$$\frac{\textit{Mean absolute error}}{\textit{Mean abolute error (Zero R)}} \times 100 \tag{5}$$

- Square root of the absolute relative error (RRSE)

$$\frac{\textit{Root mean square error}}{\textit{Root mean square error (Zero R)}} \times 100 \tag{6}$$

Gaussian Process Algorithm. This algorithm allows data classification. It uses probability distributions on finite random variables that have a multivariate Gaussian distribution. The algorithm parameters are: batch size 100, noise 1.0, filter type is normalized training data, polynomial Kernel E 1.0–C-250, sed 1. The steps followed by the algorithm are the following:

- Calculate the covariance matrix.
- Invert the covariance matrix using Cholesky factorization. $LL^T = K$
- Optimization process: determine the values of the hyperparameters.

$$\log p(y|x,\theta) = -\frac{1}{2}y^T K^{-1} - \frac{1}{2}\log|K| - \frac{n}{2}\log 2\pi \tag{7}$$

Performance of the Model. This model presents a correlation coefficient of 0.6656. The absolute mean error values and its square root present low values (0.2084 and 0.2737), while the absolute relative error and its square root present high values (79.58% and 74.88%, respectively). Figure 6 shows the behavior of the error when comparing the estimated and true values: True Mean $R_e(t)$ in x axis and Predicted Mean $R_e(t)$ in y axis.

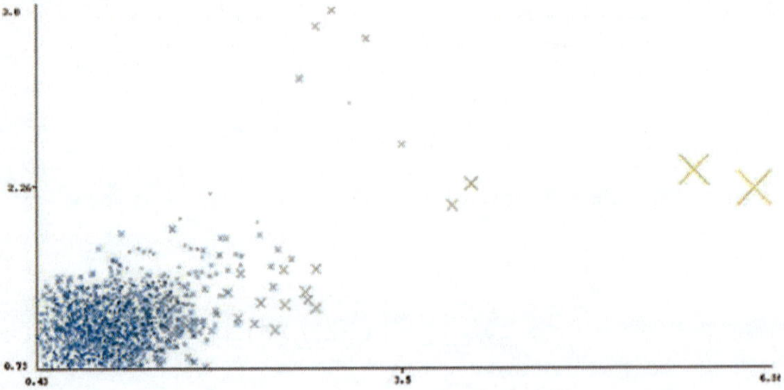

Fig. 6 Gaussian processes classification error. Predicted (y axis) true values (x axis) of $R_e(t)$

Linear Regression Algorithm. This algorithm allows using numeric variables. The model parameters are batch size 100, attribute selection method M5 method and ridge 1.0e−08.

Analysis of Model Results. Figure 7 (a) shows the predicted value of the mean of $R_e(t)$ on the y-axis, as well as the true value on the x-axis. It can be observed that when drawing an imaginary line at 45°, the predicted values are a little above the true ones. Figure 7 (b) shows $R_e(t)$ vs time. The correlation coefficient is 0.7453, which indicates a high relationship between $R_e(t)$ and the attributes selected to build the model.

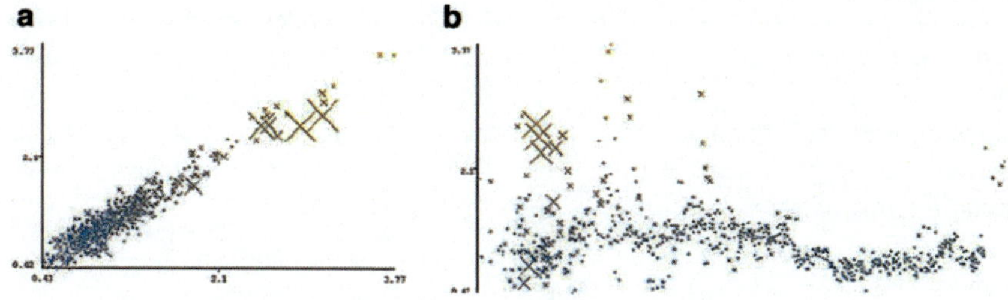

Fig. 7 **a** Classification error $R_e(t)$. **b** $R_e(t)$ vs time

Neural Networks Algorithm (MLP). Considering the data model outlined, the following levels are defined:

S: space where the whole of the database is defined. S = [s1, s2, s3,........ sp]
p: number of instances or records
A: vector representing the set of attributes that describe the contagion. It varies between 1 and *n*.
X: set of values of attributes of vector A. It varies between 1 and *n*.
C: vector containing the values of classes that allow the classification of R(t). It varies between 1 and *m*.
O: vector containing the values of vector class. It varies between 1 and *m*.

The training stage follows the following steps:

- Configure the MLP: set the number of hidden layers, learning rate and momentum.
- Scan the data set: entries X, output C.
- Assign the weights to links between neurons of entry and hidden layers.
- Active neurons of hidden layers.
- Calculate the error between the output and the reference values.

$$e_j(n) = d_j(n) - y_j(n) \tag{8}$$

- Back propagate the error and reassign the weight values and calculate the new output and error. The equations used are:

$$\varepsilon(n) = \frac{1}{2}\sqrt{e_j^2(n)} \tag{9}$$

$$\Delta w_{ji} = -\gamma \partial v_j(n) y_i(n) \tag{10}$$

$$I = \sum w_{ii} + \theta_i \tag{11}$$

The parameters used in the configuration of the neural network are: number of hidden layers 3, learning rate 0.3, momentum 0.2, training time 500 and batch size 100.

Analysis of the Model Results. The application of this model obtains the following performance metrics: correlation coefficient of 0.774, mean absolute error of 0.2001, root of absolute error of 0.2808, absolute relative error of 70.141%, and square root of relative error of 70.95%. Figure 8 (a) shows the relationship between the predicted and true values of $R_e(t)$, and (b) $R_e(t)$ vs time. This is the algorithm that shows the best results in terms of precision.

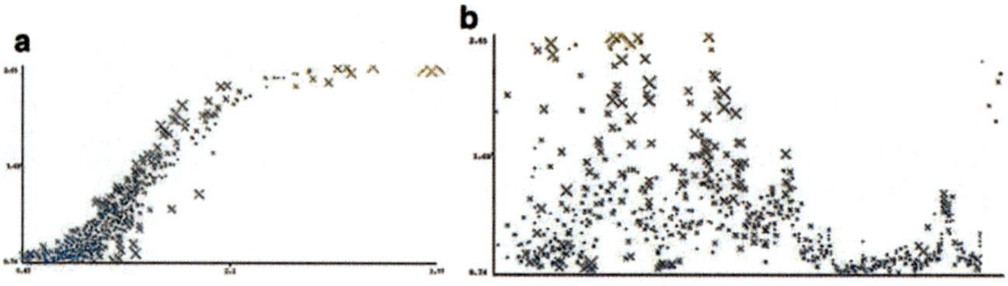

Fig. 8 **a** MLP classification error. **b** Predicted $R_e(t)$ vs time

Figure 9 shows the comparison of $R_e(t)$ values by type of quarantine. It is observed that the highest number of points corresponds to the case in which there is no quarantine. When comparing the mean values of each group, the lowest values correspond

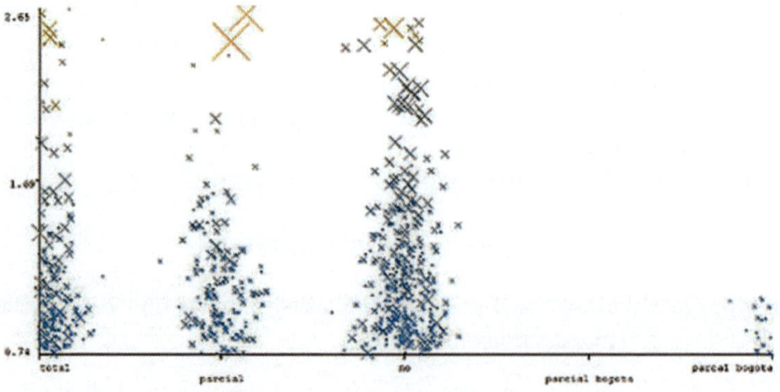

Fig. 9 Value of $R_e(t)$ predicted by class of quarantine

to total quarantine with 0.96, followed by partial quarantine with 1.012 and the highest to no quarantine with 1.18.

Table 5 shows the predicted and true values for some instances shown in Fig. 8. In some cases, the estimated value is greater than the true one, in others it is minor.

Table 5 Instance attribute values: low and high values of R(t) MLP

Instance	Predicted mean (R)	Mean (R)	Location	Lock down	Population	Density
531	0.9579	0.9475	Suba	Total	24,088	11,693
654	1.0832	0.982	Ciudad Bolívar	Partial	24,088	11,693
654	1.17905	1.3331	Barrios Unidos	No	24,088	11,693

Support Vector Machine Algorithm. The minimal sequential optimization algorithm uses a series of regression-based supervised learning methods that include SVMs, which learns from a set of training data and predicts new data. This type of algorithm establishes a model of equations, where the different parameters can be learned using various methods. The parameters are: batch size 100, c 1.0, filter type is normalized training data, polynomial Kernel E 1.0 c. regOptimizer RegSMOimproved T0.001.

Analysis of the Results of the Model. Figure 10 shows the classification error. Most of the values estimated by the classifier are lower than the true ones. Although the scale on the vertical axis (estimated values) starts at a higher value than on the horizontal axis (true values), the shape of the points distribution should be close to a 45° slope so that both values be the most precise as possible. As the true values increase, the estimated values move further away from the hypothetical 45° line, which indicates a greater error in the estimation. However, the amount of values above the viewpoint of 1.2 (indicated by the degradation of the blue color towards gray and yellow) is very small. Most of the values are in the range between 0.8 and 1.35, where the difference between the estimated and the true values is small, which means a smaller error. The correlation

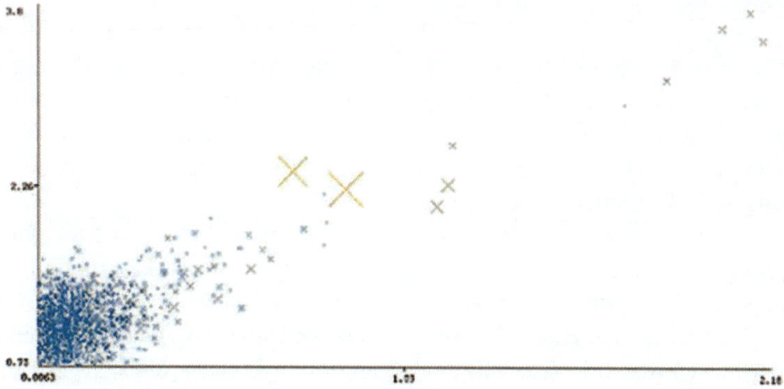

Fig. 10 Error between predicted and true values of $R_e(t)$

coefficient is 0.6398, with a MAE of 0.2059 and RMSE of 0.3063, while the values of RAE and RRSE are 81.44% and 71.41%, respectively.

3.3 Comparison of Results of the Models

The performance comparison of the different models is presented in Table 6, to estimate the number of effective reproduction with the defined metrics. A better performance of the Linear Regression and Neural Networks algorithms (Multilayer Perceptron MLP) can be observed. These algorithms present higher correlation coefficients and lower errors, particularly Linear Regression. To determine the value of $R_e(t)$ under given conditions, the attribute values are entered, the model is loaded, and then, the estimated value is obtained.

Table 6 Comparison of the models

Metric	Gaussian process	Linear regression	Multilayer perceptron	SMO
Correlation coefficient	0.6656	0.7453	0.774	0.6398
Mean absolute error	0.2084	0.1462	0.2001	0.2059
Root mean squared error	0.2757	0.264	0.2808	0.3063
Relative absolute error	79.58	57.8181	79.1415%	81.4459%
Root relative squared error	74.88	66.711	70.9568%	77.411%
Instances	1402	1402	1402	1402

4 Conclusions

The probability of contagion is closely associated with the sociodemographic characteristics of the population that determine the behavior of individuals. The grouping of the sociodemographic variables in clusters allows identifying common characteristics of the population in order to orient the tracking of positive cases.

The models for the calculation of the effective reproduction number $R_e(t)$ show good performance with the highest precision in linear regression and MLP algorithms.

The calculation of the effective reproduction number is of great importance to project the number of infections over time. The traditional models consider several assumptions that limit their application in specific cases. Data based models lack the restrictions of mathematical models.

The inclusion of more detailed information such as the activity of the people, the composition of the family group, the number and type of trips, vaccination progress, would improve the prediction capacity of the models. As more data is included, the accuracy of the model increases. The use of machine learning tools is very useful when predicting the reproduction rate and its distribution in social groups.

The contact rate (β) varies depending on government measures as well as social culture. In the data-based media, this parameter is implicit in the socio-demographic variables state regulations.

References

1. Ins. http://ins.gov.co/Noticias/Paginas/coronavirus-casos.aspx. Accessed 06 Oct 2020
2. http://www.elhospital.com/temas/Datos-mas-confiables-para-la-deteccion-de-casos-infecciosos-por-COVID-19+134568. Accessed 13 Nov 2020
3. http://ingenieria.uniandes.edu.co/es/noticias/egida-plataforma-analitica-inteligencia-artificial-prediccion-evolucion-pandemia-covid-19. Accessed 05 Nov 2020
4. https://www.parcdesalutmar.cat/media/upload/pdf/NP_IA_COVID19_CAST_1587986164.pdf. Accessed 10 Nov 2020
5. https://campussanofi.es/e-professionals/noticias/inteligencia-artificial-covid-19/. Accessed 10 Nov 2020
6. https://www.lavanguardia.com/local/barcelona/20201026/484231555907/biotecnologia-inteligencia-artificial-covid19-coronavirus-barcelona-global.html
7. https://ge.usembassy.gov/tracking-covid-19-with-artificial-intelligence-july-22/#: ~ :text=A%20recent%20study%20by%20researchers,for%20COVID%2D19%20by%20radiologists. Accessed 31 Oct 2020
8. https://www.nature.com/articles/s42256-020-0194-1. Accessed 6 Nov 2020
9. https://www.infosalus.com/asistencia/noticia-sido-tan-exitosa-estrategia-corea-sur-contra-covid-19-20200812171052.html. Accessed 5 Nov 2020
10. http://www.javiercolomo.com/coro.htm.estadísticas.covid19.mundo. Accessed 2 Oct 2020
11. https://www.bbc.com/mundo/noticias52402064#: ~ :text=Sin%20embargo%2C%20la%20estrategia%20del,m%C3%A1s%20de%20140%20pacientes%20recuperados. Vietnam. Accessed 9 Oct 2020
12. Manrique Abril Fred G, Agudelo Calderón Carlos A (2020) SIR model of the Covid 19 pandemic in Colombia. Rev Salud pública 22:1–9
13. Paez Salamanca Gustavo Nicolas, Franco Restrepo Zea Jose Fernando, Estimación del numero efectivo de transmision del covid 19 en Colombia.
14. http://saludata.saludcapital.gov.co/osb/index.php/datos-de-salud/enfermedades-trasmisibles/covid19/. Accessed 15 Oct 2020
15. https://www.datos.gov.co/en/d/gt2j-8ykr/visualization, last accessed 2020/9/21
16. Becker N (1979) The uses of epidemic models. Biometrics 35:295–305

Towards a Universally Accessible Laboratory for the Teaching and Learning of Electronic and Computer Engineering

Ian Grout(✉)

Department of Electronic and Computer Engineering, University of Limerick, Limerick V94 T9PX, Ireland
Ian.Grout@ul.ie

Abstract. In the teaching and learning of electronic and computer engineering concepts, experiential learning underpins much of the theory learnt. Laboratory-based activities are essential to the student learning experience, enabling the development of a more in-depth understanding of specific theoretical concepts. Those involved in the teaching and learning of the engineering subjects will be familiar with the role and requirements of the laboratory, the requirements of the experiment, the "typical" student profile, and the anticipated laboratory experiment outcomes. Within the laboratory setting, it is typical to design the laboratory and the experiment with a single means of "Engagement" (the "why" of learning), a single means of "Representation" (the "what" of learning), and a single means of "Action and Expression" (the "how" of learning). This is however not necessarily supportive for all students and can in many cases limit or prevent access to learning, particularly for those with disabilities. For institutions delivering laboratory experiments for the "typical" student with single means approach, multiple means now need to be supported. This however is not necessarily easy to achieve. The laboratories considered would include face-to-face (F2F), fully online, and blended learning delivery and participation approaches. A better understanding of accessibility requirements, and the adoption of guidelines and "good practice" approaches, can support the individual academic in achieving the goal of universal access, that is to provide access for all. In this paper, the role of the laboratory experiment in electronic and computer engineering experiential learning is considered, and the need to design and develop laboratory experiments that are accessible to all, presented.

Keywords: Electronic engineering · Laboratory · UDL · Accessibility

1 Introduction

In the teaching and learning of electronic and computer engineering concepts, with a focus in this paper on higher education, experiential learning underpins much of the theory learnt. Laboratory-based work is essential to the student experience, enabling the development of a more in-depth understanding of specific theoretical concepts, as well as the development of essential practical skills required by the practicing professional. This is particularly noticeable in the engineering and science subjects. Those

M. E. Auer et al. (eds.), *Online Engineering and Society 4.0*, Lecture Notes in Networks and Systems 298, https://doi.org/10.1007/978-3-030-82529-4_48

involved in the teaching and learning of the engineering and science subjects will be familiar with the role of the laboratory, along with the requirements of the laboratory setting, the requirements of the experiment, the "typical" student profile, and the anticipated laboratory experiment outcomes.

Within the laboratory setting, it is usual to design the laboratory and the experiment for a "typical" student and a single means of *Engagement* (the "*why*" of learning), a single means of *Representation* (the "*what*" of learning), and a single means of *Action and Expression* (the "*how*" of learning). This is however not necessarily supportive for all students and can in many cases limit or prevent access to learning, particularly for students with disabilities. For institutions delivering laboratory experiments for the "typical" student with single means of *Engagement*, *Representation* and *Action & Expression*, where these are defined by CAST [1] and their Universal Design for Learning (UDL) framework [2], multiple means now need to be supported. In this paper, the role of the laboratory experiment in electronic and computer engineering experiential learning is considered, and the need to design and develop laboratory experiments that are accessible to all, presented. These would include face-to-face (F2F) laboratories, fully online laboratories, and blended learning laboratories (a mix of F2F and online laboratories).

This paper will consider the situation experienced by many educators in that there is no longer, in many situations, a "typical" student profile. Students come from different countries (cultural considerations), may be from a socially disadvantaged background, may be mature students returning to education, or may have one or more disabilities, where the disabilities may also be hidden disabilities. Education institutions have a need, underpinned by national legislation, to support all students irrespective of individual circumstances.

To consider the laboratory experiment from an accessibility perspective places the student at the center of the decision making process, and decisions are made on the basis of how the student can access the laboratory based on their particular needs by building on the individuals' strengths and abilities. For the educator to achieve this, UDL principles have been developed and can be realized by considering specific UDL guidelines within the laboratory setting. This requires an understanding of the benefits of UDL and how it can support the educator as well as the student. For example, UDL is supported around the world by range of organizations such as CAST in the United States of America [1, 2], AHEAD in Ireland [3, 4], and the Centre for Universal Design Australia [5, 6].

The paper will be structured as follows. In Sect. 2, the laboratory experiment is considered and the "one size fits all" approach typically adopted in the design, delivery, and assessment of the laboratory activities discussed. Limitations of this approach, particularly where student diversity and laboratory accessibility aspects are taken into account, are identified. In Sect. 3, the approach that supports universal access, UDL, is introduced and how this relates to the design, deployment, and assessment of laboratory-based teaching and learning presented. Section 4 will present a review of work already undertaken in UDL within the laboratory setting in identified education scenarios will be provided. In Sect. 5, consideration will be given into how laboratories can be designed, or redesigned, with universal access in mind. Finally, Sect. 6 will provide conclusions to the work presented.

2 Role of the Laboratory in the Teaching and Learning of Electronic and Computer Engineering Concepts

In higher education, it is typical for teaching and learning to be undertaken in a range of modes of delivery and student participation, depending on the requirements of the study programme and the ability for the provider to support specific modes. In a traditional teaching approach, the institution campus has been the physical focal point for studies, and face-to-face (F2F) teaching undertaken using a mixture of lectures, tutorials, and laboratory experiments. The academic timetable would be developed to interleave the lectures, tutorials, personal study time, and laboratories such that they are undertaken by the student in a manner that supports prior learning.

With the maturing of electronic, computer, and internet technologies, alternative approaches to the F2F teaching approach has led to the introduction of online learning where the student and academic provider are physically distant and connected through a network or internet connection. Fully F2F and fully online programmes each have advantages and disadvantages. Somewhere in between these two approaches, increasing adopted by education providers, is the blended learning approach that attempts to utilize advantages of F2F and online approaches to include F2F teaching and learning only when necessary. Considering the types (classifications) of laboratory set-up possible, see Table 1, this can be either local or remote, and physical or virtual [7].

Table 1 Classifications of laboratories

Experiment type	Local experimenter (same location as the experiment)	Remote experimenter (different location from the experiment)	
Experiment type	Local	Remote	
Real (physical)	At-presence (physical)	Remote laboratory	Hybrid
Virtual (simulated)	At-presence (simulated)	Virtual laboratory	

In many situations, the local and real laboratory is the preferred approach to laboratory provision in that it gives the students the ability to interact with physical laboratory equipment and components/systems to investigate, and direct access for the student to the teacher in order to ask and answer questions F2F. However, in other scenarios then the local and virtual laboratory is preferred when simulation techniques, languages, and the use of tools are required to be learnt. For example, in microelectronics the use of analog, digital, and mixed-signal simulation techniques are required prior to any physical integrated circuit (IC) fabrication and test. Where the student cannot gain access to the laboratory in a local setting, the remote mode of learning would be utilized. For example, with distance learning programmes, to support students who cannot attend a physical laboratory, and where specific scenarios prohibit access to a physical university campus. Whatever the approach taken, the laboratory is a dynamic and complex environment for both the teacher and the learner. It can also require more effort to resource, set-up, run, and assess than a theoretical, lecture-based approach. However, the benefits to the student who successfully completes a well-structured and operated laboratory with a well-defined set of appropriate learning outcomes cannot be underestimated. To fully support the student in their laboratory based experiential learning, the whole process of how the laboratory is accessed and experiment run should be considered. Figure 1 shows key interactions between the people, laboratory environment, and experiment that would typically be encountered in an electronic engineering teaching module. The figure could also be adapted to include more details and to visualize different disciplines.

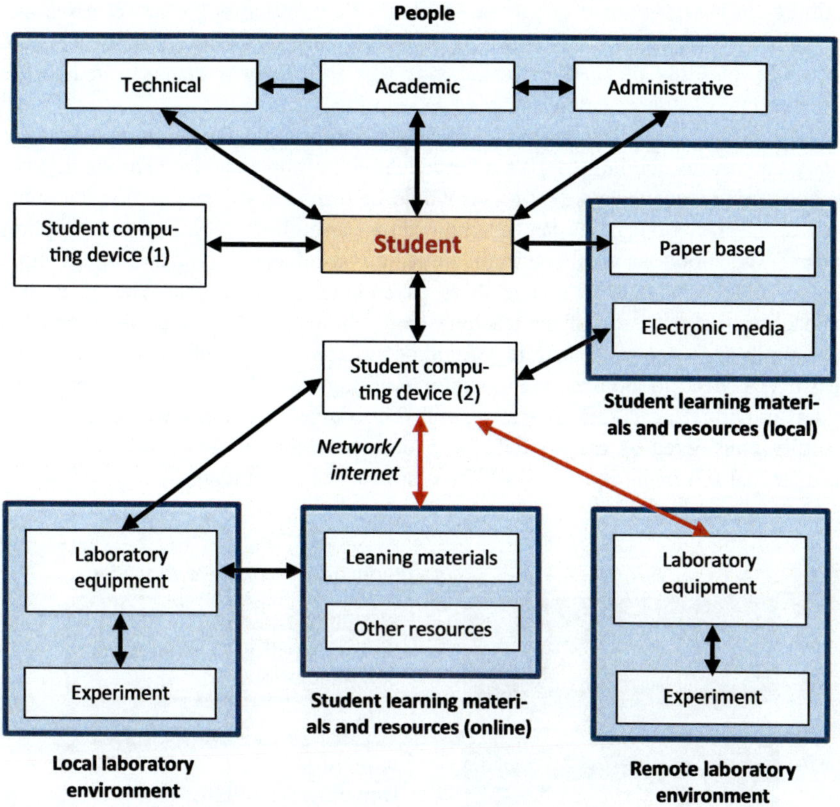

Fig. 1 Student interactions within the laboratory

If analysis of a particular laboratory environment and experiment to experience were to be undertaken, it would be seen that the laboratory is a complex environment with multiple interactions between the different parts that form the laboratory and the experiment to undertake. Attempting to capture such interactions in a formal way so that improvements or alternative approaches could be considered is however a non-trivial task.

3 Universal Design for Learning (UDL)

Universal Design for Learning (UDL) [1, 2] is a set of principles that can realized through UDL guidelines (the guidelines being the principles put in practice) for education providers to structure their teaching and learning in order to give all students equal opportunities. It is based on three core principles, as shown in Fig. 2, of *multiple means of Engagement, Representation*, and *Action & Expression*, with guidelines to provide options in which to realize these principles in practice. This essentially offers flexibility in ways the students can access and engage with learning materials as well as flexibility in how the students can show what they know through the set assessment mechanisms. This flexibility is aimed to enable all students to engage in their learning process by building on their particular strengths in a manner that is suited to their particular circumstances when engaging in learning.

By providing flexibility where possible and appropriate, UDL supports the removal of any barriers to learning to give all students equal opportunities. The flexibility is provided to all students and the options would be provided so that the required quality levels and learning outcomes can be achieved by any option or combination of options. However, what options would be available would be subject specific and would require the individual academic to realize these as and when appropriate. Hence, a single approach to the provision of the teaching would be replaced by multiple approaches, and a single approach to the learning would be replaced by multiple approaches. This is not a trivial scenario to create and operate, requiring effort from the institution as well as the individual academic to realize this. The range of options would need to be carefully considered as there would be a need to resource the options and also to consider that too many options could be confusing to the student.

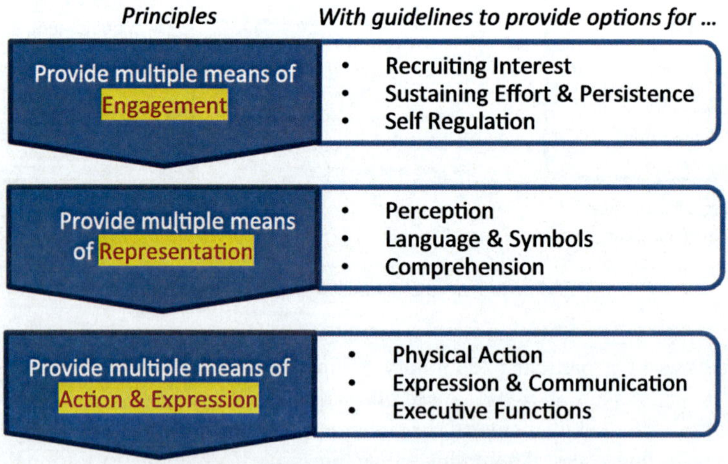

Fig. 2 UDL principles with guidelines to provide options for …

To get this right is however not necessarily easy to achieve. It requires commitment from the institution as well as the individual academic. It does however consider the student as an individual learner, removes the idea that there is a "typical" student, and the student is at the center of decision making. After all, what really is "typical"? Today, the student population consists of individuals from different countries (an increasingly international student population), may be from a socially disadvantaged background, may be mature students returning to education, or may have one or more disabilities. Organizations and education institutions globally are dealing with the need to support the changing student demographic and some institutions already implement UDL as part of the everyday activities. To illustrate this, for example, Table 2 identifies a number of resources that showcase what UDL is, and how UDL can be used.

Table 2 Example online resources on UDL (all sites accessed on 11th November 2020)

Organization	Document	Online resource
CAST	The Universal Design for Learning guidelines	http://udlguidelines.cast.org/binaries/content/assets/udlguidelines/udlg-v2-2/udlg_graphicorganizer_v2-2_numbers-no.pdf
AHEAD	Universal Design for Learning in higher education—license to learn	https://ahead.ie/userfiles/files/shop/free/UDLL%20Online.pdf
University College Dublin [8]	Toolkit for inclusive higher education institutions, from vision to practice	https://www.ucd.ie/all/t4media/0274_UCD_TOOLKIT_1118_ONLINE_LR.pdf
SOLAS [9]	A conceptual framework of Universal Design for Learning (UDL) for the Irish further education and training sector	https://www.solas.ie/f/70398/x/948bcabcc4/udl-for-fet-framework.pdf
Council of Ontario Universities [10]	Creating an accessible science laboratory environment for students with disabilities	http://www.accessiblecampus.ca/wp-content/uploads/2017/01/Creating-an-Accessible-Science-Laboratory-Environment.pdf

Although the focus in this paper is on UDL, this discussion should be considered in context. UDL is one of a number of models in the overall theme of *universal design* (UD) [11]. UD is defined in [12] as:

> *"Universal Design is the design and composition of an environment so that it can be accessed, understood and used to the greatest extent possible by all people regardless of their age, size, ability or disability."*

Universal instructional design (UID) or *universal design for instruction* (UDI) is a set of nine principles [13] that is defined as:

> *"The design of teaching and learning products and environments to be usable by all people, to the greatest extent possible, without the need for adaptation or specialized design."*

UDL should also considered alongside accessibility initiatives such as assistive technologies, and world-wide web accessibility through the W3C WAI ("World Wide Web" "Web Accessibility Initiative") [14]. The W3C WAI is involved in the development of standards and support materials to support the creation and implementation of digital creations such as websites and applications that are ideally accessible to all "*by design*".

4 Applying UDL Principles and Guidelines in the Laboratory Environment

Laboratories are designed to achieve a set of learning outcomes whilst working within the capabilities of the education provider. The individual academic will develop learning material and experiments that can be operated within a set time and in a safe manner. There is no universally adopted procedure to follow to create the laboratory experience, and relies much on the experience of the academic to create experiments that is both useful and interesting to the student. Considering the undertaking of the laboratory requires all the phases from initialization through to completion, the actions of the people involved, and the interactions between people to be considered:

1. **Before**. Preparation of learning materials, preparation of the laboratory environment, laboratory equipment, and experiment. Familiarization of the laboratory requirements and procedures by the student.
2. **During**. Accessing the laboratory environment. Setting-up the laboratory equipment and experiment. Running the experiment and taking results. Experiment completion actions. Leaving the laboratory.
3. **After**. Undertaking the laboratory assessment, and submission of the assessment. Review and feedback actions.

The approaches that are usually adopted tend to follow a single approach or a limited number of alternative approaches to engage with the student. This can work for many students, but not all. Particularly in STEM (Science, Technology, Engineering, and Mathematics), the laboratory can be a problem for access to students with disabilities. This can lead to an underrepresentation of students with disabilities in STEM [12], a situation that is not desirable as education should be an inclusive activity [12, 13]. Applying the three principles of UDL can be considered to the three laboratory phases identified above. This takes into account the understanding that the student population is not homogenous, rather it is heterogenous with much diversity.

The diversity would be a part of the student population make-up with students coming from different countries (cultural considerations), may be from a socially disadvantaged background, or may have one or more disabilities, where the disabilities may also be hidden disabilities. In addition, it is a decision for the student whether to disclose their disability to the institution or not, and this would affect the supports that the institution would be able to provide. However, this requires an institutional level response. For example, support for students with disabilities is usually focused on the *accommodations*. The term *accommodation* is usually used to describe changes made for an individual student, for example alteration of a learning environment, changes to curriculum format, or the provision of equipment that allows an individual with a disability to access learning materials and/or complete assigned tasks. This would be on an individual needs-basis. This leads to institutions providing *reasonable accommodations* that have a basis in legislation. However, UDL considers a different approach. Instead of focusing on accommodations for individual students with disabilities, UDL provides a framework to change the actual learning environment for all [15]. Flexibility and choices allow all students to realize the requirements placed on them but in a manner that best suits their needs. The flexibility and choices are for all students. To give examples of using UDL in further and higher education scenarios, in [16], the authors report on the inclusion of UDL to make laboratory spaces more accessible with a perspective from a Department of Chemical Engineering. In [17], the authors report on a literature review on UDL in postsecondary STEM education for students with disabilities, and in [18], the authors report on the inclusion of UDL to minimize student stress in science laboratories. These show approaches to student support.

To consider now how UDL has been included in a teaching module by the author, a laboratory-based module was analyzed for the inclusion of selected UDL principles and guidelines. This module is a skills development module, with the students following a "typical" engineering type design project flow from initial problem specification to presentation of a working system developed, for third-year students studying an information technology (IT) focused study programme. The module was based on undertaking a project to develop a sensor system considered as a resource within the Internet of Things (IoT), as shown in Fig. 3, a focus that would be beneficial as it would naturally be of interest to many of the students. Where the students have interest, there would be more engagement, and potentially better outcomes.

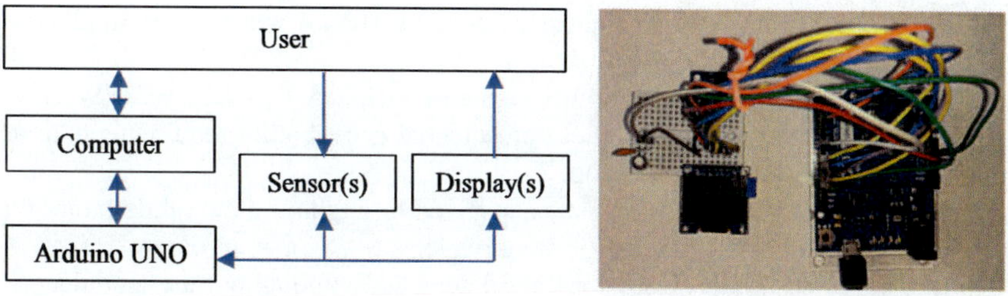

Fig. 3 Sensor system overview and example prototype hardware

In this module, computer-side programming using the Python language was undertaken. For most of the students, this would be their initial introduction to Python programming, so the laboratory work was undertaken to introduce a new programming language, demonstrate to the students new programming concepts, and to allow the students to undertake further personal research activities to obtain a more detailed understanding of these new concepts introduced. On the hardware side, a microcontroller-based sensor system based in the Arduino UNO platform, external sensor peripherals, and visual displays, would introduce the students to the embedded systems design. Again, for most of the students, this would be an introduction to embedded systems. The actual focus of the student work can be modified each year to have a different focus, be in more towards the embedded microcontroller system or the computer-based software development. Hence, by adapting the project on a year-by-year basis, whilst remaining within the requirements of the module learning outcomes, it would be possible to update the student engagement approach. Taking this one stage further, student engagement could be further considered by enabling the students to be given with a number of options for a project focus, within practical limits and to avoid potential confusion, through an initial discussion with the students at the start of the module. With this laboratory approach of teaching and learning, the undertaking of the project work (the "journey") is considered as important as the result (the "destination"). Whilst there would be many possible ways to include UDL principles and guidelines, the key considerations in this iteration of the module were to:

- Provide a set of enhanced aims and objectives, with a new material introduced through an updated plan that supports scaffolding (enhancing learning by building on experiences and knowledge as new skills are learnt).
- Enable students to work individually or in pairs, and where appropriate, ensure that students can undertake their learning activities by building on their strengths, with the potential to make adjustments to specific activities.
- Support multiple means of engagement by developing content in text, image and video formats, with the video formats including narration and captioning. Adopting a consistent style of presentation of content.
- Provide options for submitting assessment material. For example, the final presentation could be undertaken as a live presentation or a prerecorded video. However, it would be important to ensure that whatever option of assessment was selected by the student, all students would be treated equally.
- Ensure content is provided in well in advance of when it will be used to allow for students to prepare at their own pace.
- Introduce the students to accessibility concepts such as UD, UDL, W3C WAI, and assistive technology, to inform them of potential accessibility requirements in any hardware or software they develop.
- Build in multiple and small assessments staged throughout the module rather than one large assessment at the end of the module.
- Allow for guided group discussions and interactions throughout the module.

Much of what can be done to adapt teaching and learning would anyway be "good practice", but by considering aspects of UDL in the context of an individual module, would enable a linkage of actions to benefits. Extending this to the laboratory setting, by undertaking an analysis on how the laboratory can be developed and operated for universal access, moves the discussion towards whether a laboratory can be "universally designed" so that it is inherently accessible by all without the need for additional interventions. However, there would be practicality and resource considerations to also consider. This would also be a continuous process of teaching adaptation.

5 Towards a Universally Designed Laboratory?

In order to consider the idea of a universally designed laboratory, it is important to identify what is actually the definition of this term. In the context of this work:

> *"A universally designed laboratory is a laboratory accessible to all irrespective of individual circumstances through the application of Universal Design for Learning (UDL) principles and guidelines."*

The design will consider all aspects of a laboratory:

1. The design of, facilities within, and access to & from, a physical laboratory.
2. The design of, facilities within, and access to & from, a remote laboratory.
3. The design and deployment of the teaching and learning materials.
4. The design and deployment of the experiment(s).
5. The design and deployment of the assessment approaches.
6. The provision of suitable dialog and feedback mechanisms.
7. The integration of the laboratory into the teaching module.
8. The integration of the teaching module into the institutional level CMS (Content Management System) or LMS (Learning Management System).

The approach does not come without a cost, so cost versus potential benefits need to be carefully considered. Initially, it might be that there would be a high cost associated with this, but long-term costs could potentially be lower as the *accommodations* can be reduced as the need for specific accommodations for individual students could be reduced *by design*. How to formally capture ideas and relevant information in a non-intrusive manner could be considered by modelling the interactions between the people and the facilities, both local and remote. Figure 4 shows an approach that identifies different levels of the laboratory, and the interactions between these levels.

The four levels in Fig. 4 can be mapped to the three UDL principles that will inform UDL guidelines which can be used within the laboratory environment. Considering the people, these would be the students, the academics leading the teaching, the technical support within the laboratory, and the administrative support staff. Based on initially considering the academic and the student, it would be possible to develop a planner to formally capture the use, or planned use, of UDL principles within the laboratory. This would come from initially developing a plan on how laboratories fit into the context of the overall teaching module, and would be a standardized way to capture the inclusion of UDL principles in the laboratory. How much level of detail is actually required is an important question to ask. Too much detail requested in an information capture form means that it could become a significant administrative overhead with potentially no noticeable academic benefit, and may simply become a time consuming "box ticking exercise". This would miss the important point of attempting to capture information to improve the student learning experience. It can also consider a single laboratory, the suite of laboratory experiments as a whole, or the laboratory within the overall context of the module.

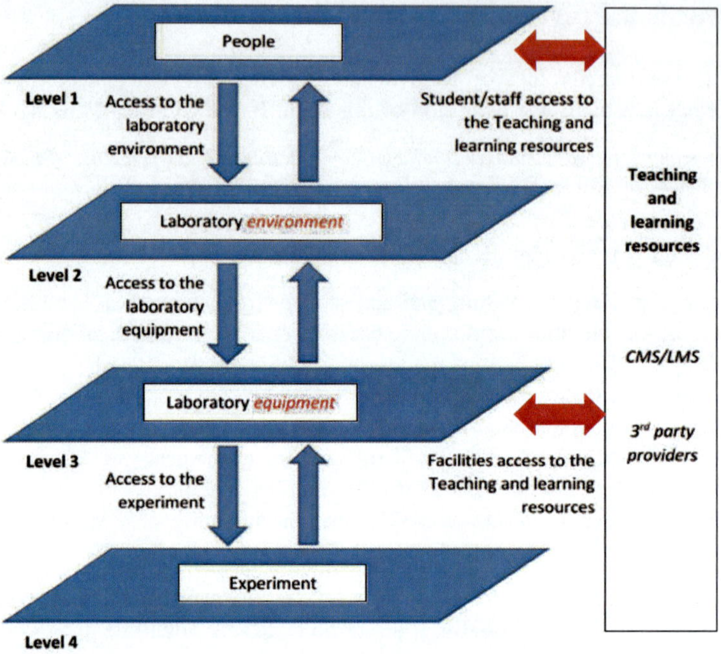

Fig. 4 Laboratory interactions at different levels in the laboratory operation

6 Conclusions

In this paper, the role of the laboratory experiment in electronic and computer engineering experimentation has been considered, and the need to design and develop laboratory experiments that are accessible to all, elaborated. The laboratory is a complex, dynamic and varied environment, and there would be no "one size fits all" approach in how the laboratory is set-up, deployed, and evaluated. These laboratories could include face-to-face (F2F) laboratories, fully online laboratories, and blended learning laboratories. Laboratories and experiments now need to consider UDL principles to provide students with multiple means of *Engagement* (the "why" of learning), multiple means of *Representation* (the "what" of learning), and multiple means of *Action and Expression* (the "how" of learning). This can be achieved through the use of UDL guidelines. A better understanding of accessibility requirements, and the adoption of guidelines and "good practice" approaches, can support the individual academic in achieving the goal of universal access.

References

1. CAST. About Universal Design for Learning. http://www.cast.org/impact/universal-design-for-learning-udl. Accessed 6 Nov 2020

2. CAST. The UDL Guidelines. http://udlguidelines.cast.org/. Accessed 6 Nov 2020
3. AHEAD. Home. https://www.ahead.ie/. Accessed 6 Nov 2020
4. AHEAD. Universal Design for Learning. https://www.ahead.ie/udl. Accessed 6 Nov 2020
5. Centre for Universal Design Australia. http://universaldesignaustralia.net.au/. Accessed 6 Nov 2020
6. Centre for Universal Design Australia, DI and UDL: Is there a difference?. http://universaldesignaustralia.net.au/category/policy-and-policy-development/policy-udl/. Accessed 6 Nov 2020
7. Palloff R, Pratt K (2003) The virtual student, a profile and guide to working with online learners, United States of America, Jossey-Bass. ISBN 0-7879-6474-3
8. University College Dublin, Toolkit for Inclusive Higher Education Institutions, From Vision to Practice. https://www.ucd.ie/all/t4media/0274_UCD_TOOLKIT_1118_ONLINE_LR.pdf . Accessed 6 Nov 2020
9. SOLAS. A Conceptual Framework of Universal Design for Learning (UDL) for the Irish Further Education and Training Sector. https://www.solas.ie/f/70398/x/948bcabcc4/udl-for-fet-framework.pdf. Accessed 6 Nov 2020
10. Council of Ontario Universities, Creating an Accessible Science Laboratory Environment for Students with Disabilities. http://www.accessiblecampus.ca/wp-content/uploads/2017/01/Creating-an-Accessible-Science-Laboratory-Environment.pdf. Accessed 6 Nov 2020
11. AHEAD. Universal Design Principles. https://www.ahead.ie/ra_appendix_udprinciples. Accessed 6 Nov 2020
12. Centre for Excellence in Universal Design, What is Universal Design. http://universaldesign.ie/What-is-Universal-Design/. Accessed 6 Nov 2020
13. Disabilities, Opportunities, Internetworking, and Technology (DO-IT), University of Washington, Universal Design of Instruction (UDI): Definition, Principles, Guidelines, and Examples. https://www.washington.edu/doit/universal-design-instruction-udi-definition-principles-guidelines-and-examples. Accessed 6 Nov 2020
14. W3C Web Accessibility Initiative. https://www.w3.org/WAI/. Accessed 6 Nov 2020
15. Council of Ontario Universities Creating an Accessible Science Laboratory Environment for Students with Disabilities. http://www.accessiblecampus.ca/wp-content/uploads/2017/01/Creating-an-Accessible-Science-Laboratory-Environment.pdf. Accessed 6 Nov 2020
16. Division of Teaching Excellence and Innovation, University of California, Making Lab Spaces More Accessible to the Variety of Students We Serve. https://dtei.uci.edu/news/making-lab-spaces-more-accessible-to-the-variety-of-students-we-serve/. Accessed 6 Nov 2020
17. Schreffler J, Vasquez E III, Chini J et al (2019) Universal design for learning in postsecondary STEM education for students with disabilities: a systematic literature review. IJ STEM Ed 6:8. https://doi.org/10.1186/s40594-019-0161-8 Accessed 6 Nov 2020
18. Miller D, Lang PL (2016) Using the universal design for learning approach in science laboratories to minimize student stress. J Chem Educ 93:1823–1828 https://doi.org/10.1021/acs.jchemed.6b00108

Promoting Microelectronic Through Remote FPGA Based Laboratory

Pedro Plaza[1(✉)], Manuel Castro[1], Elio Sancristobal[1], Pablo Orduña[2], Luis Rodríguez[2], Javier Garcia[3], Félix García-Loro[1], Rosario Gil[1], Antonio Menacho[1], Manuel Blázquez[1], Blanca Quintana[1], Sergio Martin[1], Francisco Mur[1], Alejandro Macho[1], Pablo Baizán[1], Ramón Carrasco[1], German Carro[1], África López-Rey[1], Clara Pérez[1], and Slavka Tzanova[4]

[1] UNED, Madrid, Spain
{pplaza, mcastro, elio, fgarcialoro, rgil, amenacho, mblazquez, bquintana, smartin, fmur, amacho, pbaizan, fmur, amacho, pbaizan, clarapm}@ieec.uned.es
[2] LabsLand, Madrid, Spain
{pablo, luis}@labsland.com
[3] Deusto, Spain
zubia@deusto.es
[4] Technical University of Sofia, Sofia, Bulgaria
slavka.tzanova@tu-sofia.bg

Abstract. Educational continuity has been made possible by the fact that teacher education and teacher collaboration systems have evolved rapidly to provide teachers with the knowledge and skills they need both to adopt new pedagogies and to support students and their families during the pandemic. Therefore, it is recommended that the ongoing professional development of teachers be more integrated into the organization of their educational work. Now, what tools can we offer students to continue doing practices from home in the same way as if they were in the university's laboratories? It seems that more than ever remote laboratories are the solution or perhaps the best alternative. A remote laboratory could be defined as a set of hardware and software technologies that allow the user, through the Internet, to carry out an experiment in the same way as if they were in the face-to-face laboratory. There are a lot of benefits of using remote laboratories in the learning process. From our side, and what is the object of this contribution, we have developed a series of seminars through which we develop educational content around microelectronics. During these remote access seminars, attendees have been able to use LabsLand's remote FPGA (Field Programmable Gate Array) laboratories to experiment while developing the theoretical contents and challenges proposed.

Keywords: Collaboration · FPGA · Learning technology · Microelectronic · Remote laboratories

M. E. Auer et al. (eds.), *Online Engineering and Society 4.0*, Lecture Notes in Networks and Systems 298, https://doi.org/10.1007/978-3-030-82529-4_49

1 Context

We are facing an exceptional situation, never experienced before, which has highlighted the enormous difficulties of all our systems, and especially the educational one. Since the closure of the schools, as a preventive measure, in the face of the COVID-19 pandemic we are facing a scenario of uncertainty in which the entire educational community is trying to provide the most appropriate responses to its students [1–3].

Schools, local educational administrations, and States will have to develop dynamic strategies for educational continuity that can be quickly adjusted to the different situations that may arise. This includes the design of strategies to offer feedback and accompaniment to students, their families, teachers, and the societies that surround them.

During lockdown, many countries have experienced deficiencies and inequalities in access to technology and skills for online teaching and learning. Therefore, it is a priority to address these deficiencies in order to face the next course, both from governments and from the educational community, because it is essential to help students develop the skills they need to face tomorrow.

2 Purpose

Educational continuity has been made possible by the fact that teacher education and teacher collaboration systems have developed rapidly to provide teachers with the knowledge and skills they need both to adopt new pedagogies and to support students and their families during the pandemic. Therefore, it is recommended that the ongoing professional development of teachers be more integrated into the organization of their educational work.

Now, what tools can we offer students to continue doing practices from home in the same way as if they were in the university's laboratories? It seems that more than ever remote laboratories are the solution or perhaps the best alternative. A remote laboratory could be defined as a set of hardware and software technologies that allow the user, through the Internet, to carry out an experiment in the same way as if they were in the face-to-face laboratory. Or, explained differently, what is not a remote laboratory? It is not a simulator or a virtual laboratory where the answer of the experiment is designed, developed, and coded by a programmer. The authors in the following papers provide empirical results of the benefits of using remote laboratories in the learning process [4–10].

From our side, and what is the object of this contribution, we have developed a series of seminars through which we develop educational content around microelectronics. During these remote access seminars, attendees have been able to use LabsLand's remote FPGA (Field Programmable Gate Array) laboratories (https://labsland.com/) to experiment while developing the theoretical contents and challenges proposed.

3 Approach

This paper presents a pilot test framed in microelectronic where attendees were using a remote FPGA based laboratory. FPGAs are devices that allow us to describe a digital circuit using a specific language, being the most common are VHDL (Very High-Speed Integrated Circuit Hardware Description Language) and Verilog. After the process of synthesis, placement and routing we can record our description in the memory connected to the FPGA or directly to the FPGA (if it has the memory integrated). When working with this type of devices it is necessary to install the software and have a development kit available for use. In addition, it is usually necessary a PC with quite high requirements. Thanks to the remote FPGA based laboratory, students do not need to install any software, nor have a development kit to perform the necessary experiments.

The following sub-sections show the different sessions that were held as part of the activities of the IEEE Student Branch at UNED. Other examples of this type of activities can be found at [11–19].

3.1 Introduction to FPGAs, First Steps in Microelectronic Design

In a first session the different types of programmable logic devices and their historical appearance were shown: PAL (Programmable Array Logic) in 1978, GAL (Generic Array Logic) in 1985, FPGA (Field Programmable Gate Array) in 1985, CPLD (Complex Programmable Logic Device) in 1988 and SoC (System on Chip) in 2012. Later, FPGAs were described as those devices that allow the description of digital circuits. The two most common languages are: VHDL and Verilog. Next, the development flow normally used was shown: Digital Design, HDL Description, Synthesis, Placement and Routing, and loading into the FPGA.

The most common elements were also shown: Logic Gates, Memory Blocks, Clocks, DSP and SerDes. In a next step, a series of recommendations were indicated: (1) Parallel execution, and (2) Synchronous and asynchronous processes. The work with this type of devices should be oriented to: Modular design and Reuse. The main manufacturers were indicated: Xilinx, Altera (currently Intel), Lattice and Atmel (currently Microchip). It was also indicated a series of development platforms that can be thought for makers: (1) Arty A7 (Xilinx), (2) Icezum Alhambra (Lattice,), (3) Terasic DE0-Nano (Intel), and (4) Arduino MKR Vidor 4000 (Intel). In addition, different application development tools with FPGA and circuit simulation were indicated. To complete the content of the session, a series of sectors where the use of FPGA is quite extended were shown: Aerospace, Audio, Automotive, Broadcast, Electronics, Data Centers, High Performance Computing, and Industrial and Medical.

Finally, a series of applications made with FPGA were shown: (1) servo control (https://youtu.be/UwN5NicwRTo), (2) humanoid robot control based on servos (https://youtu.be/aCz6ryLf3ss and https://youtu.be/QxWW5MXiIDI) and (3) Dual Educational Robotics Laboratory [20–22]. Finally, a challenge was proposed to the attendees: (1) Imagine an application to develop with FPGA, (2) Acquire the components, and (3) Assemble and program your solution.

3.2 First Steps with FPGA, Describing in VHDL

In another session it was shown how digital circuits are described using the VHDL language. At the beginning of this session some differences between FPGAs and microprocessors were discussed. Microprocessors compared to FPGAs: they are sequential, the code is always executed, and they usually have fewer input/output ports. Some precautions to be taken when developing with FPGAs were also indicated, such as being careful with the loops, paying attention to the running conditions, and it is highly recommended to take advantage of the modular design.

In addition, several features of VHDL were indicated:

- Originally developed by the United States Department of Defense in the early 1980s
- Language standard defined by the IEEE
- To program PLD, FPGA, ASIC and similar

Later, the structure of.vhd files and the differences between sequential logic and combinational logic were described.

In a next step, the LabsLand remote FPGA lab used during the session was detailed and the attendees were able to access it to work with the examples shown. Figure 1 shows LabsLand's remote FPGA lab.

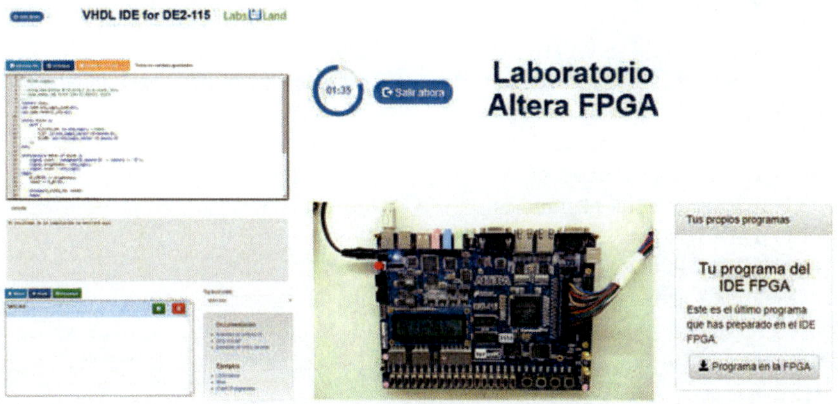

Fig. 1 LabsLand's remote FPGA lab

The first description of a circuit was a BCD-7segment decoder:

- Inputs: The first four switches from the bottom, V_SW(3 downto 0) are the BCD input
- Outputs: the 7 segments on the right, G_HEX0(6 downto 0)
- Logic with case sentence

The challenge for the attendees was to make an E appear when there is an invalid entry. They were also given another challenge to make the 7 segments represent hexadecimal codes.

As a description of more complex circuits, we worked with the description of a counter, a frequency divider, a state machine and finally, basic filters.

All the examples shown were given to the attendees during the session. At the end of the session, the following challenges were proposed to the attendees so that they could work on the content shown:

- Using the example "ADDER.vhd":
  - Make nothing visible in the 7-segment if at least one of the two entries is not BCD
  - Use the G_LEDs to indicate whether the first summation input is GREATER than the second summation input, whether the first summation input is LESS than the second summation input, or whether they are equal
- Using the example "COUNTER_MOD.vhd":
  - Add a V_SW switch that makes the counter count or not, with switch = 1 counts, and with 0 it does not count
  - Add a V_SW switch that makes the counter go from 0 to 9 if it is 0 and go from 9 to 0 if it is 1. The chosen switch is 0, the rightmost one
  - Same as the previous one but when changing the V_SW switch from 0 to 1, or vice versa the counter goes to the beginning of the other sequence. That is, if it was going up (switch = 0) and it goes down (switch = 1), then the counter goes to 9 and then it continues
- Using the example "STATE_MACHINE_MOD":
  - Use everything you have learned to make a clock with the 7 segments:
  - G_HEX0 as units of seconds
  - G_HEX1 as tens of seconds
  - G_HEX1 as tens of seconds
  - G_HEX2 as units of minutes
  - G_HEX3 as tens of minutes
  - G_HEX4 as units of hours
  - G_HEX5 as tens of hours
- Use everything you have learned to perform a pong with the 7 G_HEX segments and the V_BT buttons for two players.

3.3 Digital Signal Processing with FPGA

In a third session, the topic was Digital Signal Processing (DSP). Digital signal processing is one of the most powerful technologies that will shape science and engineering in the twenty-first century. Some fields of application of DSP are: Communications, Medical imaging, Radar and sonar, High fidelity music, and Oil exploration, among others. In the framework of FPGA there are two types of processing: discrete signals and digital values.

Throughout this session, it was shown how digital circuits of a series of digital filters were described using the VHDL language. The result of the description was also developed on an FPGA using LabsLand's remote FPGA laboratory.

The first activity of the session was focused on basic filtering. To do this, work was done on the effect of the bounces of the signals coming from the buttons. The challenge was proposed to the attendees to answer the following questions:

- How many bounces are detected when using a button?
- How could the bounces be filtered?

Examples of how to count the pushbutton bounces and how to filter those bounces were shown.

In a second step, we worked on an example of filtering the signal received from a sensor. A sensor emulator with noise and its corresponding filter was used for this purpose. The challenge was proposed to the attendees to answer the following questions:

- What is the effect of noise on the measurements received?
- How could the sensor values be filtered?

Figure 2 shows the sensor emulator with noise (top) and the filtering effect of the example shown (bottom).

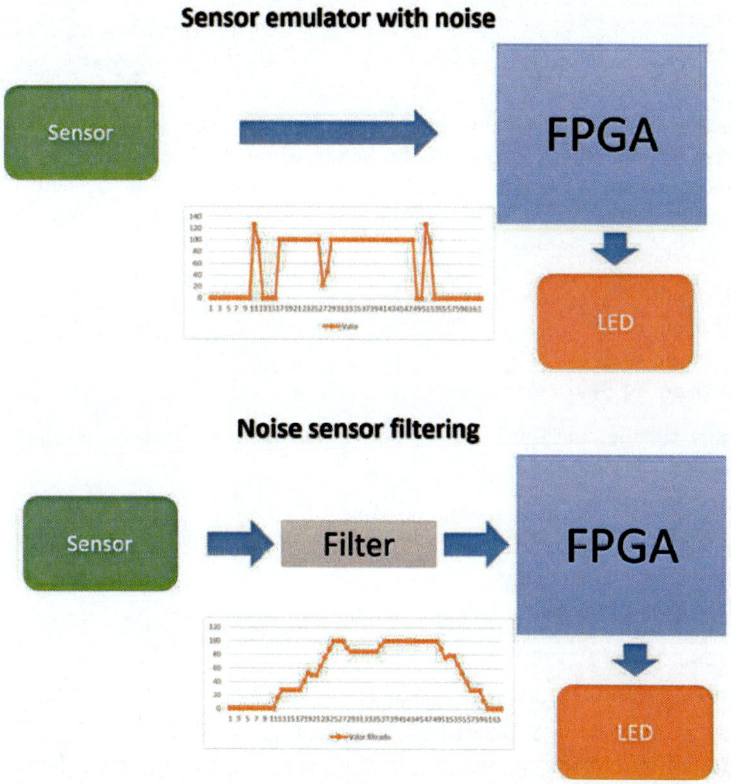

Fig. 2 Sensor emulator with noise (top) and the filtering effect of the example shown (bottom)

All the examples shown were given to the attendees during the session. At the end of the session, the following challenges were proposed to the attendees so that they could work on the content shown:

- Using the example "BUTTON_BOUNCING": Make another type of filtering to add to your circuit to the bounces of the buttons
- Using the example "SENSOR_NOISE": Describe a filtering that improves immunity to noise generated by the sensor emulator compared to the one used in the example "NOISE_FILTER".

3.4 Safety Critical Applications with FPGA

In a last session, the basis of the development of Safety Critical applications with FPGA was shown. Safety Critical systems are those systems protected to perform functions that:

- Avoid risks to people
- Avoid risks to the environment

These systems consider:

- Hardware
- Software
- Human aspects

Companies dedicated to this type of application have a RAMSS department (Reliability, Availability, Maintainability, Safety, and cyberSecurity).

A Safety Critical function must detect a fault and direct the system to a safe situation.

The architectures can be:

- 1oo1 (1 out of 1)
- 1oo2 (1 out of 2)
- 2oo2 (2 out of 2)
- 2oo3 (2 out of 3)

To this session, the 2oo2 architecture is detailed. A 2oo2 architecture can be designed:

- With diversity in the people who design it
- With hardware diversity
- With a variety of software
- With diverse firmware

Firmware diversity can be:

- Positive logic < > Logic in complement
- Upstream counters < > Downstream counters
- Coding of state machines

The applications of this type of development can be.

- Infrastructure
- Medicine
- Nuclear Engineering
- Entertainment
- Transport: Railway, Automotive, Aviation and Aerospace

As a practical example, during this session a counter was shown representing your count in a 7-segment and the indication of discrepancy in each of the FPGAs. Figure 3 shows a block diagram of the system to be implemented.

Fig. 3 Block diagram of the system to be implemented

The example was developed using LabsLand's remote FPGA lab. The examples shown were given to the attendees during the session. At the end of the session, the following challenges were proposed to the attendees so that they could work on the content shown:

- Do you dare to create your own Safety Critical application?
- To be Safety Critical you should have:
  - Positive logic < > Logic in complement
  - Upstream counters < > Downstream counters
  - Coding of status machines

4 Outcomes

Throughout the paper we present the learning methodology used and different educational resources to be used in the framework of microelectronic education based on FPGA. During our educational experiences, a total of 65 attendees were adhered to the sessions:

- 29 attendees were connected to Introduction to FPGAs, first steps in microelectronic design
- 18 attendees were connected to First steps with FPGA, describing in VHDL
- 10 attendees were connected to Digital Signal Processing with FPGA

- 8 attendees were connected to Safety Critical Applications with FPGA

The countries of connection were France, Germany, India, Mexico, Peru, Spain, United States and Venezuela.

All attendees were attending the sessions from their corresponding country and they were able to use the FPGA LabsLand's remote FPGA laboratories during the session and during one week after the corresponding session.

5 Conclusions

As a main conclusion, it has been possible to demonstrate that the methodologies used, and the LabsLand's remote FPGA laboratories are a good option for promoting collaborative work between students. Furthermore, it has been demonstrated how microelectronics can be developed remotely in an easy way. Therefore, the results are useful for the teachers who wants to initiate their group of students around microelectronics remotely.

The use of LabsLand's remote FPGA laboratories in microelectronics teaching environment promotes the development of skills and competencies (basic use of FPGA, DSP with FPGA, Safety Critical applications with FPGA) in an affordable and easy-to-use way. In addition, LabsLand's remote FPGA laboratories are valid for the development of scalable educational content and is ideal for collaborative learning. We have also found that LabsLand's remote FPGA laboratories are suitable for developing skills in participants.

Given that the number of participants is not extremely high, it would be advisable to work with the same contents and a greater number of samples to be able to consolidate the conclusions reflected here.

Acknowledgements. This work has been co-funded by the Madrid Regional Government through the e-Madrid-CM Project under Grant S2018/TCS-4307, co-funded by the European Structural Funds (FSE and FEDER). The authors acknowledge the support provided by the UNED Industrial School of Engineering, the UNED Doctoral School, and the project "Techno-Museum: Discovering the ICTs for Humanity" (IEEE Foundation Grant #2011-118LMF).

Authors also acknowledge the support of the e-LIVES. e-Learning InnoVative Engineering Solutions-Erasmus+ Capacity Building in Higher Education 2017-585938-EPP-12017-1-FR-EPPKA2-CBHE-J, I4EU -Key Competences for an European Model of Industry 4.0, Erasmus+ Strategic Partnership nº 2019-1-FR01-KA202-06296 and ECoVEM, European Centre of Vocational Excellence in Microelectronics, Erasmus + KA3 - Dedicated VET Tools, Centres of Vocational Excellence nº 620101-EPP-1-2020-1-BG-EPPKA3-VET-COVE.

References

1. Gaudiot J-L, d Kasahara H (2020) Computer education in the age of COVID-19. Computer 53(10):114–118

2. Ocaña JM, Morales-Urrutia EK, Pérez-Marín D et al (2020) Can a learning companion be used to continue teaching programming to children even during the covid-19 pandemic? IEEE Access 8:157840–157861
3. Nassr RM, Aborujilah A, Aldossary DA et al (2020) Understanding education difficulty during COVID-19 lockdown: reports on malaysian university students' experience . IEEE Access 8:186939–186950
4. Rodríguez Artacho M, Castro Gil M, Díaz G et al (2016) eMadrid project: authoring, reuse and remote labs. In: 2016 international symposium on computers in education (SIIE), Salamanca, pp 1–3
5. Dávila L, Sancristobal E, Castro M (2010) Hardware implementation of remote laboratory for digital electronics. In: IEEE EDUCON 2010 conference, Madrid, 2010, pp 357–362.
6. German CF, Borrego RC, Sancristobal Ruiz E et al (2015) Usingcommon elements to explain electromagnetism to children: remote laboratory of electromagnetic crane. In: 2015 IEEE frontiers in education conference (FIE), El Paso, TX, pp 1–5
7. Macho A, Baizán P, Blazquez M, et al (2019) Work in progress: Proof of concept: Remote Laboratory Raspberry Pi + FPAA, 2019 IEEE World Conference on Engineering Education (EDUNINE), Lima, Peru, 2019, pp. 1–4
8. Garcia-Loro F, Sancristobal E, Alves G, et al (2016) New experiences and strategies in remote laboratories and apps for electronics: proposal for a special session. In: 2016 13th international conference on remote engineering and virtual instrumentation (REV), Madrid, pp 3–3
9. Martin S, Sancristobal E, Gil R et al (2011) Engineering education: Importance and relevance of ubiquitous technologies. In: 2011 IEEE GCC conference and exhibition (GCC), Dubai, pp 245–246
10. García-Loro F, Baizán P, Blázquez-Merino M et al (2019) Spreading remote laboratory scope through a federation of nodes: VISIR case. IEEE Revista Iberoamericana de Tecnologias del Aprendizaje, 14(4):107–116
11. Caeiro-Rodríguez M, Llamas M, Sancristobal E et al (2020) Activities of the Spanish chapter of the IEEE education society. In: 2020 IEEE global engineering education conference (EDUCON), Porto, Portugal, pp 1932–1936
12. Plaza P, Sancristobal E, Carro G et al (2019) Scratch as driver to foster interests for STEM and educational robotics. IEEE Revista Iberoamericana de Tecnologias del Aprendizaje 14 (4):117–126
13. Plaza P, Sancristobal E, Carro G et al: Scratch day to introduce robotics. In: 2018 IEEE global engineering education conference (EDUCON), Tenerife, pp 208-216
14. Plaza P, Sancristobal E, Carro G et al (2019) STEM and educational robotics using scratch. In: 2019 IEEE global engineering education conference (EDUCON), Dubai, United Arab Emirates, pp 330-336
15. Plaza P, Peixoto A, Sancristobal E et al (2020) Visual block programming languages and their use in educational robotics. In: 2020 IEEE global engineering education conference (EDUCON), Porto, Portugal, pp 457-464
16. Plaza P, Sancristobal E, Carro, G et al (2017) Home-made robotic education, a new way to explore. In: 2017 IEEE global engineering education conference (EDUCON), Athens, pp 132-136
17. Plaza P, Sancristobal E, Carro G et al (2018) Traffic lights through multiple robotic educational tools. In: 2018 IEEE global engineering education conference (EDUCON), Tenerife, pp 2015-2020
18. Plaza P, Sancristobal E, Carro G et al (2018) Multiplatform educational robotics course to introduce children in robotics. In: 2018 IEEE Frontiers in education conference (FIE), San Jose, CA, USA, pp 1-9.

19. Plaza P, Sancristobal E, Carro, G.: Lighting through educational robotics, et al (2018) XIII technologies applied to electronics teaching conference (TAEE). La Laguna 2018:1–7
20. Plaza P, Sancristobal E, Carro G et al (2018) European robotics week to introduce robotics and promote engineering. Comput Appl Eng Educ 26:1068–1080
21. Plaza P, Sancristobal E, Carro G et al (2016): A wireless robotic educational platform approach. In: 13th International conference on remote engineering and virtual instrumentation (REV), Madrid, pp 145–152
22. Plaza P, Sancristobal E, Carro, G.: Collaborative robotic educational tool based on programmable logic and Arduino, et al (2016) Technologies applied to electronics teaching (taee). seville 2016:1–8
23. Plaza P, Sancristobal E, Carro G, et al (2016) Robotic Educational Tool to engage students on Engineering, IEEE frontiers in education conference (FIE). Erie, PA, USA 2016:1–4

Data Collection: Use and Transformation in Predictive Maintenance Models

Hernando Vélez Sánchez(✉) and Luini Leonardo Hurtado Cortés

Universidad Distrital Francisco José de Caldas, Bogotá, Colombia
{hvelezs,llhurtadoc}@udistrital.edu.co

Abstract. Operational Reliability of the equipment for the operation of industrial plants is an indicator of great importance in industrial processes, which must be measured as continuously as possible with a minimum of interruptions. To avoid the occurrence of failures, the appropriate time to intervene the equipment must be established. The prediction of this indicator is based on data collected from the equipment condition and its proper analysis. However, failure data are not always available. The key issue is how to obtain and process good quality fault data to obtain high precision results. This document studies the use of Industry 4.0 and big data principles in the identification of potential faults that leadinglead to a functional failure in a multistage injection pump, which is a critical equipment for the disposal of wastewater in the hydrocarbon industry. The values of torque, amperage and rpm of the equipment, as well as temperature, and pressure and flow of the water were generated using the specialized simulation software Simscape™ from the MATLAB tool kit. Various fault types were defined for the pump and some models were built using machine learning with the simulation data to detect the presence of faults and identify their type. The model with the highest precision was selected and corresponds to a tree classification algorithm with an accuracy of 95%.

Keywords: Data generation · Fault detection · Machine learning

1 Introduction

The evaluation of the mechanical, electrical and structural integrity of the equipment and components of industrial processes and civil works in general is an issue that has evolved recently in an important way. The development of techniques for monitoring the assets condition, including non-destructive tests (NDE), allows determining the real condition of such systems, their failure probabilities and residual life with different levels of precision. A large amount of real-time information about the operation of the mentioned systems can be obtained, but frequently data on fault conditions are unknown and difficult to collect from the actual installation. The objective of the use of this information is to diagnose the existence of faults in assets, using simulated data. A model that represents the degradation process is the known P-F curve, as shown in Fig. 1. The potential fault is detected at point P and monitoring is carried out in the P-F interval.

M. E. Auer et al. (eds.), *Online Engineering and Society 4.0*, Lecture Notes in Networks and Systems 298, https://doi.org/10.1007/978-3-030-82529-4_50

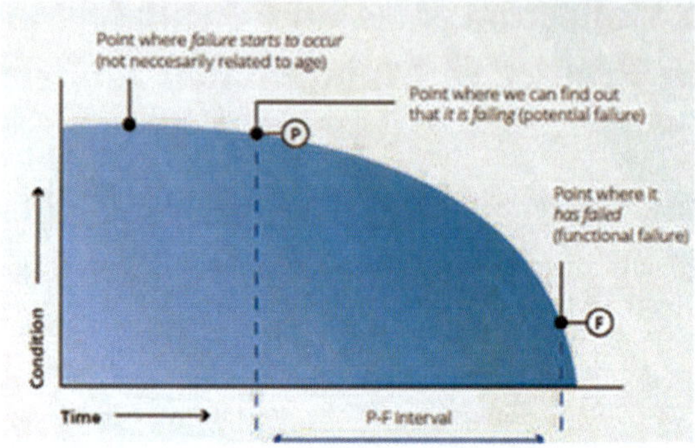

Fig. 1 P-F curve. Source: www.maintworld.com

Two recurring questions from these issues are, first, if the equipment has an abnormal condition and, second, how long can it work before presenting a fault. Answering these questions allows making decisions to preserve the integrity of the equipment. However, establishing the existence of a fault or the residual life of a system without making an observation generates uncertainty due to the probabilistic nature of the phenomena of its degradation. This is the problem studied in this document. There are numerous approaches that attempt to decrease the uncertainty of these estimates. Additionally, the development of concepts within the Internet of Things, digital twin, big data and, more generally, Industry 4.0, introduces new tools to achieve the objectives defined above. In this sense, predictive maintenance can be considered as a part of the Internet of Things.

The characterization of the materials used in industrial processes and in the manufacture of equipment and machines is generally carried out in the laboratory, under standardized conditions. However, the conditions existing in the work environment or operational context are generally different. For this reason, the evaluation of the actual performance of a material in its particular operational context must be done on-site during service. In any case, the degradation process of the material is stochastic due to the large number of variables that can affect its behavior. From the point of view of the service stage, the traditional works for PHM (prediction health management) are based on the interpretation of the data taken from the physical model. The use of monitoring techniques for the detection of failures consists in the processing and analysis of data with operating or production variables of the assets. These data can be obtained from the sensors installed in the equipment or simulated with specialized software when sensors fault data are not available.

1.1 Related Works

Wiian Ren et al. [1] used wavelet packet decomposition and Elman neural network for the extraction of features that allow identifying the faults in progressing cavity pumps in some oilfields located in China. Iserman [2] presented model-based fault detection methods using data from many measured variables to extract information of possible changes caused by faults. The relationships raised are analytical in terms of equations. Twin mesh [3] used digital twin based on CAE to model the hydraulic behavior of rotary pumps. Multistage pumps are not included. Mathworks [4] employed a diagnostic feature designer tool for identifying faults in a triplex reciprocating pump. It generates simulated pump fault data, for example, multiclass fault detection.

Mathworks [5] also made a list with the applications of different descriptive statistics for the characterization of signals in predictive maintenance.

Mathworks [6] developed a methodology to extract the features of the data signals from the equipment and identify the condition indicators that serve to detect and identify their failures.

Matworks [7] found signal-based statistical metrics that can be applied to any type of signal, including vibration signals. Changes in these metrics may indicate changes in the condition of the asset.

In [8], a descriptive essay is presented about the application of digital twin in the development of products, particularly the design of centrifugal pumps and the simulation of the phenomenon of cavitation.

Mathworks [9] developed an example of fault detection and identification in a triplex pump with data generated in Simscape™, which is a Simulink® module. The digital model defines and generates the data of three different faults, calculates the fault condition indicators and builds a machine learning model to carry out the identification of the faults. Finally,

Fowler [10] obtained a patent for the development of a controller method to create a digital model of a centrifugal pump. The model is connected to the sensors of the equipment and updates the operation and production data of the pumping system.

2 Methodology

To analyze the behavior of physical systems, models are frequently used to represent them. The methodology used here is summarized in the diagram in Fig. 2. The first stage is gathering the data available from the real installation or the simulated using specialized software. In the second stage, the data are conditioned and a first exploration (preprocessing) is performed by analyzing the signals from sensors or simulating the characterization of the installation behavior with and without faults in the time and frequency domains. Next, a search for features that can be used to identify the presence and type of faults is carried out. These features are called condition indicators. The failure modes to be studied were previously defined. Next, a data-based model is built (trained and tested) to detect and identify the present faults. Finally, the algorithm is deployed to the asset control system.

The basis for the development of any prediction algorithm is the available data which, once filtered, constitutes its input. The data can be obtained mainly from sensors installed in the equipment and then transmitted and recorded. Simulation is another alternative, especially when fault data are not enough. This can be achieved by physical and mathematical models or by generating data that represents the behavior of the system. Alternatively, both ways can be used.

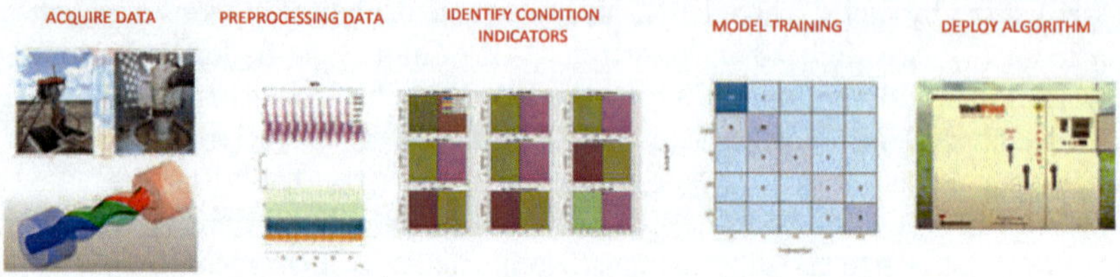

Fig. 2 Workflow diagram of data-based predictive maintenance

2.1 Multistage Pumps Failure Modes

The pump considered for this work is a Schlumberger-manufactured multistage pump, used in different applications, including the hydrocarbon industry. The hydraulic behavior of the pump corresponds to the principle of Euler equation. Due to the rotary movement of the pumping element, characteristic components of the fault may appear at mild-high frequencies. The multistage pump can fail in several ways. Based on the experience and data taken from these assets, the following faults are identified:

- Casing wear
- Pipeline leak
- Pump leak
- Impeller corrosion
- Impeller wear (code 11)
- Bearing wear or break (code 10)
- Seals leakage (code 100)

Additionally, there may be faults in the motor, coupling, speed meter and control system. The fault modes listed with codes were included in the data processing program. A brief description of the fault modes taken into account in this study (impeller wear, seals wear, and bearing fault) are briefly described in the following paragraphs. In addition to these fault conditions, the normal operation of the equipment is also considered and assigned to the code 0.

Normal Operation: Although several parameters are directly related to the operation of the pump, pressure and flow are perhaps the most representative from the hydraulic point of view. Due to its construction, this type of pump has uniform discharge pressure and flow characteristics with a minimum pulsation when operating under normal conditions.

Impeller Wear: The graph in the time domain of this type of fault is characterized by a decrease in pressure with values close to 0.

Seals Leakage: It is characterized by pressure swings with irregular peaks well above the normal operating pressure.

Bearing Wear or Break: It is characterized by the appearance of vibrations that increase in amplitude and frequency and are transmitted to the casing.

2.2 Data Model and Databases

There are data taken by the sensors of the equipment, but no failure data are known. The database with the measurements generated from the Simscape™ consists of 4000 records of the operation of four pumps functioning at different fields and pumping wastewater at different conditions of temperature and density. Fluid flow, temperature and pressure values are simulated. These data constitute the input to the predictive maintenance algorithm. Some concepts of the variables and parameters that describe the condition of the equipment are defined as follows.

Fault Parameters: Impeller wear, seal wear, bearing wear break.

Fault Variable Values: Values in which a fault condition exists (fault limits) for each type of fault. These values are defined by the user.

Condition Variables: Normal, impeller wear, seal leaks, bearing wear break.

Data Variables of the Process: Temperature, Total Dynamic Head (TDH), flow (Q).

Data Variables of Feature and Fault: fPeak, pLow, pMid, pHigh, pKurtosis, qMean, qVar, qSquewness, qKurtosis, qPeak, 2Peak, qCrest, qRMS, qMAD and qCSRange.

Table 1 presents the properties of the fluid and the operation and production of the pump.

Table 1 System characteristics

Conditions	Parameter	Values
Fluid properties	Density	
	Viscosity	
	Temperature	
Production conditions	TDH (variable) Psi	Variable 300–600
	Flow (variable) (BFPD)	7000
Operating conditions	RPM	Variable
	VAC (volts)	220

2.3 Pump Model Creation

The Mathworks® Simscape™ tool is employed to generate the data used to extract the features that serve as input in the classification model. This tool kit includes models for different kinds of systems, such as fluid, electrical, mechanical and hydraulics. In this case, the fluid and hydraulic modules are used to build a model of the system. Regarding the pump, the program has the possibility of importing models built in tools such as AutoCAD or any drawing program. Figure 3 shows a multistage pump drawing.

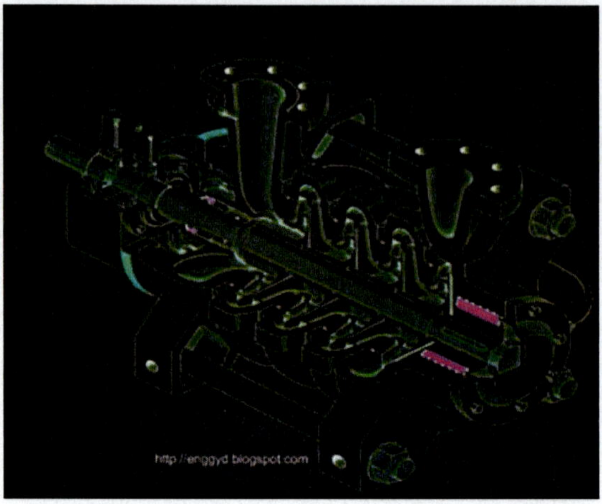

Fig. 3 Multistage pump [6]

First, the model of the multistage pump is created. For this purpose, the CAD file is imported, including its geometric parameters. Then, a library is created in Simscape™ to configure the hydraulic model from the program. The model is configured for three types of faults. These faults are parameterized as variables in the MATLAB workspace through the dialog menu in the pump parameter block. Table 2 presents the centrifugal pump settings in Simscape™.

Table 2 Parameter block settings

Parameter		
Model parametrization	Polynomial aproximation	
Finit aproximating coefficient	326.8	Pa
Second aproximating coefficient	3.104e4	Pa-s/kg
Third aproimatting coefficient	1.097e7	Pa-s^2/(kg-m^3)
Fourth approximating coefficient	2.136e-5	Pa-s^2/(kg-m^3)
Correction factor	0.5	
Pump design delivery	*130*	lpm
Reference angular velocity	*1770*	rpm
Reference density	*920*	Kg/m^3
Leak resistance	*1e6*	Pa
Drive shaft torque	*0.1*	N-m
Torque-pressure coefficient	*0.1e−5*	N-m/Pa
Angular speed treshold for flow reversal:	*1e−9*	Rad/s

Figure 4 shows the model of the centrifugal pump built with Simscape™. This model can be connected with the equipment instrumentation to record operating data from the sensors.

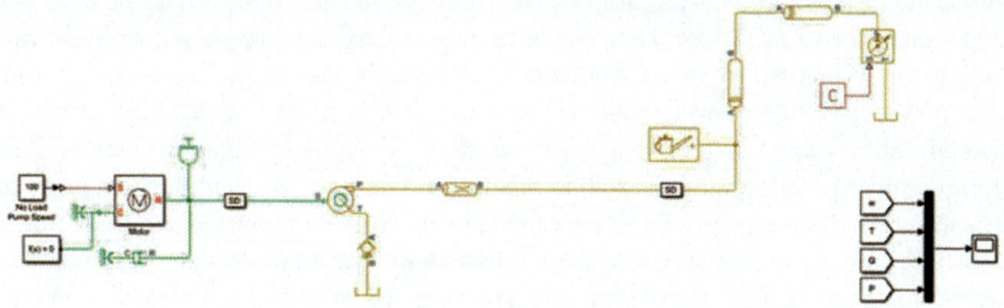

Fig. 4 Simscape™ model of the multistage pump

The failure models were included in the pumping circuit. The faults in each component will be reflected in the variables that can be calculated and evaluated: angular velocity, head P, discharge Q and torque on the motor-pump shaft T. In this case, only the variation in the head P and discharge Q are considered [11].

Generating Data. Data are then simulated for normal and fault conditions. For each of the three types of fault, a range of values is defined that represents the severity of the fault. In this case, the range consists of ten values between the interval limits. Values must allow the classification model to distinguish between a faultless condition and a significant value in each type of fault.

Simulation Settings with Different Scenarios. Combinations of failure parameters are built to create objects for simulation. With the simulation objects defined, the Simulink command is used for running the simulations. For each simulation input, the random seed is different to obtain different results. The results are stored in a subfolder and then assembled into an array of vectors. The same failure conditions and their severity generate different simulation results, which are grouped into vectors, according to the different combinations of fault parameter values defined above. For each of these groups, different sets of fault values (severity) are created. The dataset from the sensors is used to tune the simulated data.

2.4 Development of Predictive Model

Once the data are generated, the input data are processed and selected for model training.

Preprocess Training Data. The pre-processing of the input signals is performed, in this case, the pressure developed by the pump (TDH). The signals are shown graphically in the time and frequency domains. The pressure spectrum allows identifying peaks at the fundamental frequency that corresponds to the rotation speed of the pump.

Extracting Features from Data. When analyzing data in the time domain, all fault signals can be mixed. On the contrary, in the frequency domain, they are separated according to the characteristic frequency of the component and the associated fault class. An important part in the development of the algorithm is the identification of the signal features that can be useful for detecting fault conditions in the equipment, as well as the type of fault it is. These features can be obtained by analyzing the data in the time or frequency domains, or in a combination of both. In the frequency domain, bandwidth power, average power, peak values, peak frequencies and harmonics can be obtained. In a mixed analysis, spectral entropy, spectral kurtosis and conditional spectral moment are found. For each member of the data set, the features related to the variable pump discharge pressure are calculated from the pre-processed data. Not all the features of the signal are useful to differentiate the state of the equipment and identify existing faults. These features are selected using the Diagnosis Feature Designer tool. Table 2 presents the main features used for analyzing the signals. After identifying the condition indicators, they are grouped into a feature vector that will be used to classify the input data.

Condition Indicators. These are features (metrics, statistics) whose behavior changes in a predictable way as the equipment produces a fault and its condition degrades as a consequence of it. They are extracted in the data processing. When analyzing raw data, it is difficult to distinguish between normal and faulty operation because the signals may be almost superimposed.

Detection and Classification Model. The objective is to assign labels (no fault, fault) to the condition of the equipment, given some known condition indicators associated with the data variables (pressure, flow). These condition indicators are the input to the model. Therefore, this is a classification problem.

This condition indicators are used to build the classification model that allows detecting and identifying equipment faults. The dataset for training the classification model is configured with the data variables, that is, the condition indicators and the condition variables for the fault variables values. A program function is used to calculate the features and another command to add the value of the feature and fault variable to the assembly. The aim is to find a machine learning algorithm that detects and identifies faults with good accuracy. To test the model, a confusion matrix or other proved metrics can be used.

2.5 Deploy Algorithm

Once the classification algorithm is validated, it is deployed into the equipment through its control system. Then, it continues to function and modify its parameters according to the new information obtained.

3 Results

The following subsections present the results obtained by the different models built with the Predictive maintenance tool from Mathworks®.

3.1 Processing and Extracting Features From the Database

The data generated with Simscape™ is processed to obtain information about the behavior of the TDH variable in the time domain (s), as shown in Fig. 5.

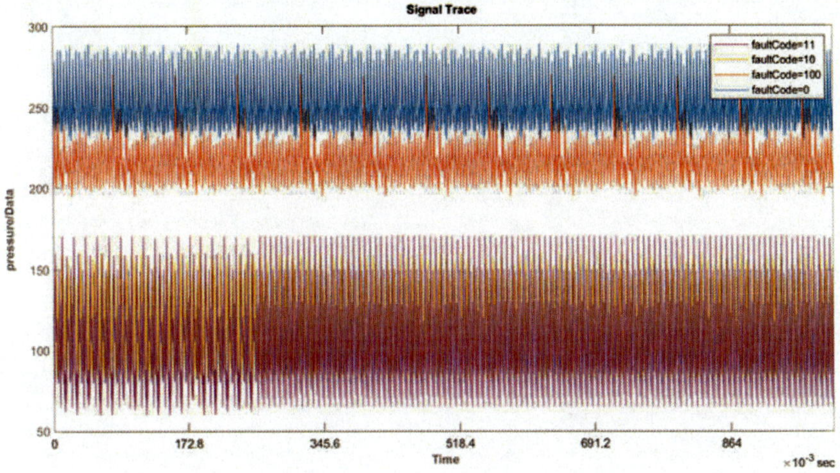

Fig. 5 Signal of the pressure from Simscape™ in the time domain

However, the displayed signal shows no clear differences for the fault codes indicated by colors. Therefore, the features in the time domain are extracted, such as the mean, the standard deviation, the root mean square (RMS) and the form factor or the kurtosis, which is used to obtain the following histograms. It is not possible to clearly identify the pressure peaks for failure modes 100 (seal leakage), 11 (impeller wear) and 10 (bearing wear break). Due to the difficulty in identifying patterns in the signals, the analysis of their features can be carried out.

Figure 6 shows the different features of the data in the form of histograms for the condition variables represented in colors. Their interpretation is a bit difficult, since it is necessary to analyze which colors are perfectly distinguished from the others in each feature. For example, in the first column bottom (RMS), the light blue color corresponding to the condition code 0 (faultless) is clearly different from the colors of the other failure modes. However, the feature shown barely discriminates the other colors, since they are mixed.

For a better definition of faults, a conversion to the frequency domain is performed. An algorithm for an order-20 autoregressive model is used. This method fits the model to the data and then calculates the spectrum of that estimated model. The faultless behavior of the pressure developed by the pump shows peaks in frequencies related to the rotation speed of the pump shaft. In the frequency domain, the peaks for the different faults are observed. The plot of the spectra calculated on a logarithmic scale shows some resonant peaks (Fig. 7).

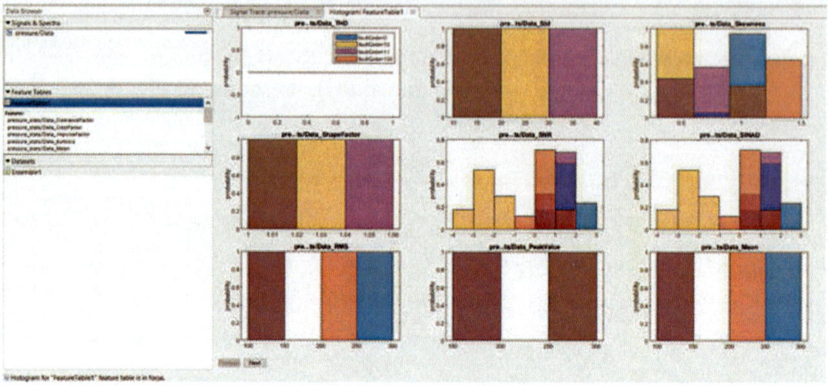

Fig. 6 Feature histograms for different fault modes and condition indicators

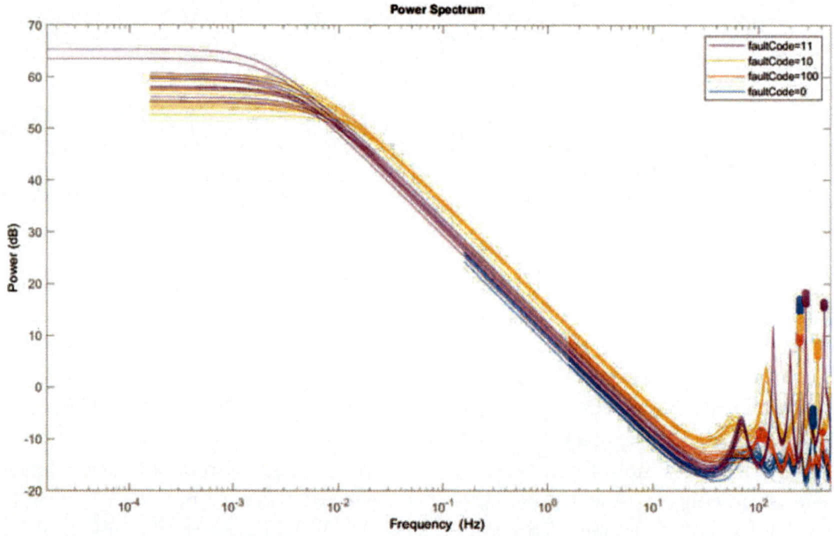

Fig. 7 Data generated in the frequency domain

Spectral features, such as peaks, and modal coefficients, such as damping factor, natural frequency and band power, can be calculated in the area with the greatest peaks. In this case, these features are extracted in a frequency band between 100–450 Hz. There are five spectral peaks for each signal, as shown in Fig. 8.

The clearance factor provides information on the occurrence of faults by evaluating the magnitude of the peaks. This magnitude is obtained by dividing the peak value into the root mean square value of the absolute amplitudes. In this case, the greater this factor is, the farther the fault appears. The analysis of Table 3 evidences that the impeller wear (11) is the farthest fault from the start, while the seal leak (100) is the closest.

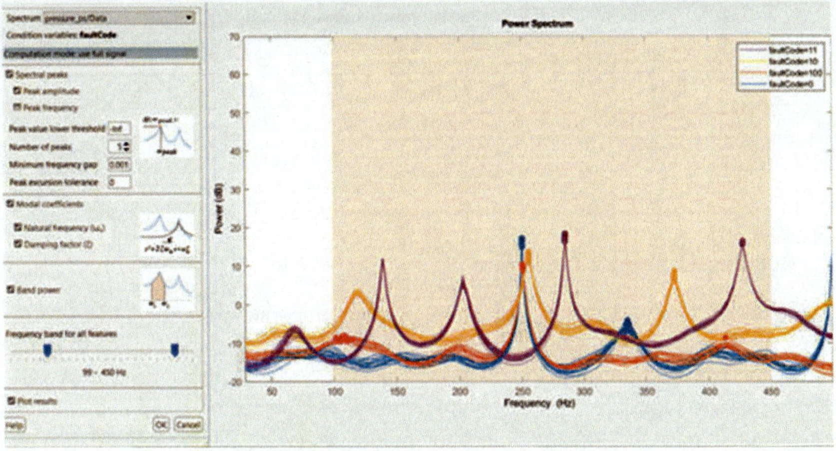

Fig. 8 Spectral peaks of each signal

Table 3 Signal features

Time domain	Frequency domain	Time frequency domain
Basic statistics	Power bandwith	Spectral entropy
Mean	Mean frequency	Spectral Kurtosis
Standard deviation	Peak values	Conditional spectral moment
Shape factor	Peak frequencies	Conditional temporal moment
Root mean square	Harmonics	
High order statistics		
Skewness		
Kurtosis		
Signal to noise ratio		
Impulsive metrics		
Peak factor		
Crest factor		
Impulse factor		

About the crest factor, it indicates the appearance of a fault and is obtained by dividing the peak into the RMS. In this case, the seal leakage fault in the pipe presents this condition. The impulse factor compares the height of the maximum peak with the average value of the signal. Higher values indicate impeller wear and bearing wear in the pump.

Next, metrics related to high-order statistics, such as third moment (kurtosis) and impulsive, are analyzed. The highest kurtosis value and highest mean pressure occur for impeller wear (Table 4).

Table 4. Values of impulsive signal features and kurtosis for failure modes

Fault code	Clearance factor	Crest factor	Impulse factor	Kurtosis
10	1.3435	1.3164	1.3342	1.5265
100	1.2391	1.2337	1.2373	4.9613
11	1.7326	1.5775	1.6734	2.2599

3.2 Defining Condition Indicators

The diagnostic feature designer tool of the predictive maintenance module performs a ranking of the features extracted from the data to select the most appropriate to serve as input in the classification model. In this case, all features with a one-way ANOVA metric > 5e4 are exported. Figure 9 shows the ranking of features and evidences that the most suitable for the construction of the model are mean, RMS, standard deviation, peak value and %shape factor.

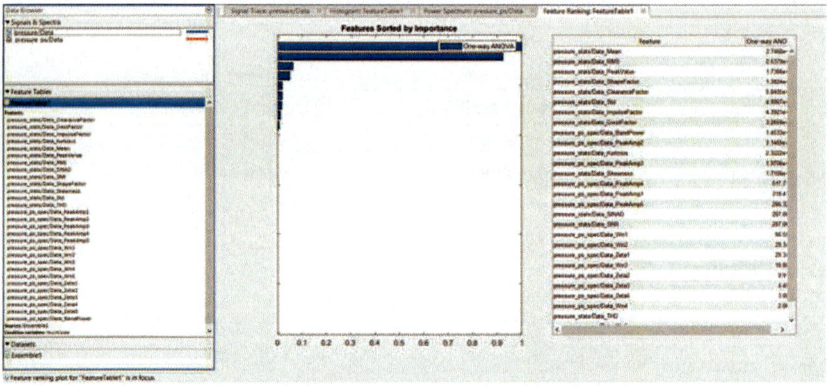

Fig. 9 Features ranking showing condition indicators

3.3 Training and Testing the Models

After selecting the condition indicators, the next step is to train a classification model based on these features. A classifier is created with the condition indicators as inputs and the status of the equipment (normal or faulty) as output. The vector support machine is trained using a second-order polynomial kernel. The dataset is divided into a set for training and another for testing the model. The program selects several classification algorithms and builds a model with each of them. The models are trained and tested, thus obtaining the metrics that demonstrate the goodness of fit of each one of them.

Figure 10 presents the behavior of the crest and clearance factors. There is a clear linear relationship between the two condition indicators.

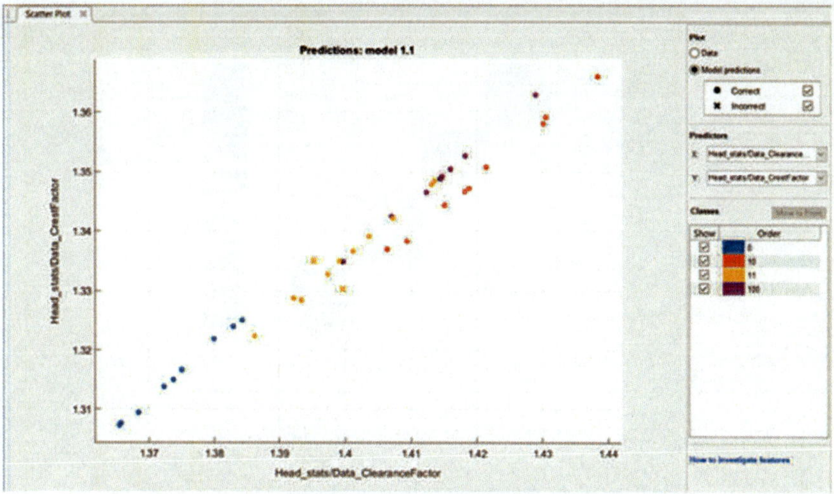

Fig. 10 Metrics of the built models. Crest factor vs clear factor

The confusion matrix in Fig. 11 shows the number of true positives and false positives corresponding to the percentages in Fig. 12. The total number of observations is 10 for each state of the equipment (no fault, fault). Each table presents the number of observations in each classification. For example, the failure modes 10 and 100 had 10 observations classified correctly (true positives), that is, 100%.

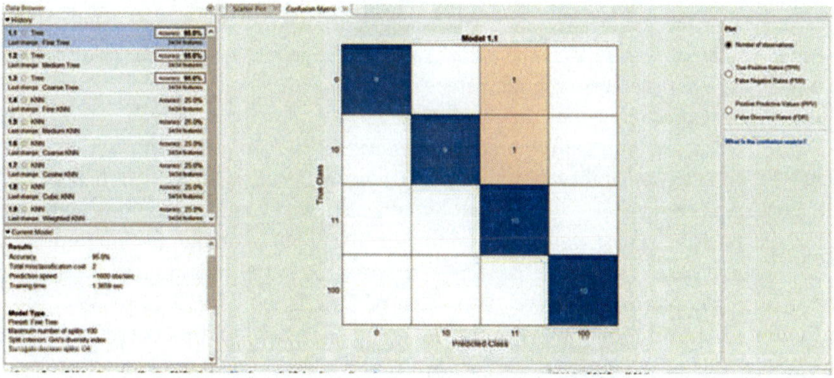

Fig. 11 Confusion matrix: True positive rates (TPR) and false negative rates (FNR)

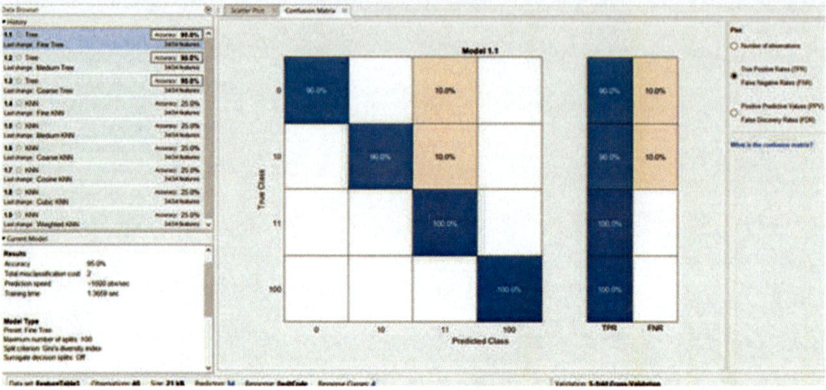

Fig. 12 Number of true positives and false negatives

Figure 12 shows the rate of the equipment conditions that were correctly classified (true positives rate) on the diagonal of the matrix, whereas the percentage of incorrectly classified conditions (false negatives rate) is outside the diagonal. It can be observed that the highest number of correct answers corresponds to failure modes 11 (impeller wear) and 100 (seal leakage) with 100%. The accuracy obtained for this model is 95%. The algorithm can be improved by performing a more detailed analysis of fault behavior in more specific frequency ranges and applying band-pass filter in cases where failures were not predicted but did exist (row 1 of the diagonal).

The algorithms shown in the graph are classification trees and KNN. The tree algorithms achieved an accuracy of 95% that corresponds to the number of true positives or hits of the model. The confusion matrix shows that the failure mode 10 was classified incorrectly as 11 in some cases, which is impeller wear. When failure modes are not clearly identified because the values of the ranges are very close, ranges of faults values can be varied for better differentiation.

Figure 13 shows the positive predictive values (PPV) and the fault discovery rates (FDR). PPV is the probability that a failure or a normal state will be identified as such. FDR comprises the type 1 errors in null hypothesis tests of multiple comparisons; it is like a false positive. These values are complementary. For example, for the fault mode 11, all positive cases (10) were identified in this way, but also 1 in class 10 and 1 in class 0 were identified as positive. The value of PPR is then 10/12 or 83.3%.

Figure 14 shows a parallel coordinates graph. For the faultless condition in blue color, the values of the indicators are shown in the vertical lines: clearance factor -1.2, mean 1.2, peak value 0.75, RMS 1.2, shape factor -1.8. This allows differentiating the values of the condition indicators in the different conditions of the pump.

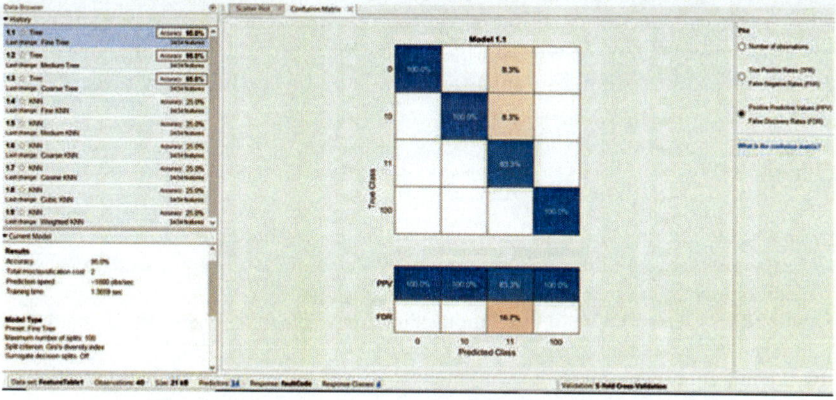

Fig. 13 True prediction values (TPV) and false discovered rates (FDR)

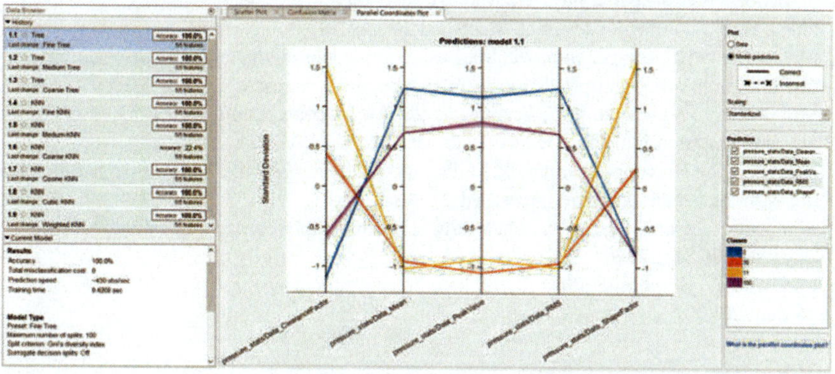

Fig. 14 Parallel coordinates plot

4 Summary

A computational tool was employed to build an algorithm for failure detection and classification in multistage pumps for wastewater reinjection in oil fields. Three characteristic faults of this type of equipment were defined.

When data on equipment faults is not available, it can be generated using specialized computational tools. In this case, the module Simscape™, a simulation tool of the software Matlab, was used.

With the data generated, statistics were calculated to select condition indicators. The features of the indicators were extracted to select the most appropriate ones. The criteria for ranking them was Anova one way.

Condition indicators were used for training a classification algorithm based on machine learning. The confusion matrix and the accuracy of the algorithm are employed to determine the fit of the classification models. The model with the best fit obtained an accuracy of 95%.

The values of the indicators based on the pressure state variable present unclear differences between some of the fault modes. In this case, the ranges of fault variables may be modified to distinguish faults from the faultless condition more easily.

References

1. Ren, W, Lu Y, Dong H (2010) Fault diagnosis of progressing cavity pumps well based on wavelet package and Elman neural network. In: Proceedings of the 8th world congress on intelligent control and automation July 6–9, Jinan China
2. Fault IR, Systems D (2006) An Introduction from default detection to fault tolerance. Springer Verlag, Berlin
3. CADFEM Homepage. https://www.cadfem.in/products/additional-software/twinmesh.html
4. Mathworks® Homepage. http://enngydblogspot.com. Accessed 20 Oct 2020
5. Mathworks® Homepage. https://la.Mathworks.com/help/stats/principal-component-analysis-pca.html. Accessed 11 Oct 2020
6. Mathworks® Homepage. https://la.Mathworks.com/help/predmaint/ug/analyze-and-select-features-for-pump-diagnostics.html. Accessed 30 Oct 2020
7. Mathworks® Homepage. https://la.Mathworks.com/help/predmaint/ug/data-ensembles-for-condition-monitoring-and-predictive-maintenance.html. Accessed 18 Oct 2020
8. Mathworks® Homepage. https://opencadd.com.br/wp-content/uploads/feature-extraction-identifying-condition-indicators-with-matlab.pdf. Accessed 06 Nov 2020
9. Mathworks® Homepage. https://la.Mathworks.com/help/predmaint/ug/file-ensemble-datastore-with-measured-data.html. Accessed 21 Aug 2020
10. Mathworks® Homepage. https://la.Mathworks.com/help/predmaint/ug/signal-features.html. Accessed 17 Sept 2020
11. Mathworks® Homepage https://la.Mathworks.com/products/simscape-fluids.html. Accessed 25 Sept 2020

Motivation Management of Domestic and International Engineering Students in the Digital Era

Irina Makarova[1(✉)], Gulnar Nadirova[2], Azhar Serikkaliyeva[3], Polina Buyvol[1], Vadim Mavrin[1], Eduard Mukhametdinov[1], and Larisa Fatikhova[1]

[1] Kazan Federal University, Syuyumbike prosp., 10a, Naberezhnye Chelny 423812, Russia
{IrVMakarova, PABujvol, VGMavrin, EMMuhametdinov}@kpfu.ru, LEFatihova@ksu.ru

[2] Khoja Akhmet Yassawi International Kazakh-Turkish University, Eurasian Research Institute, Mametova Street, 48, 050004 Almaty, Kazakhstan
g.nadirova@eurasian-research.org

[3] Al-Farabi Kazakh National University, Al-Farabi Avenue, 71, 050040 Almaty, Kazakhstan
Azhar.Serykkalieva@kaznu.kz

Abstract. Growth sphere of engineering and high technology creates conditions for the rapid economy development, but to realize these opportunities, highly qualified specialists are needed, who are subject to increased requirements both in terms of professional skills and competencies, and in personal qualities terms. This creates new challenges to the educational system, in particular to the engineering education system, which must overcome inertia, reduced student motivation, insufficient funding for the re-equipment of laboratories and increased requirements for lecturers. At the same time, both the system of student motivation, which allows them to reveal their talents, and a change in the paradigm of the educational process organization. Engineering education should combine not only the modern methods use and teaching aids, but also an orientation towards the needs of real production and business for create in student's ability to self-study and subsequently his rapid adaptation to production conditions. The experience gained shows that the use of new teaching methods in engineering education contributes to the development of engineering competencies required for high-tech industries. This requires creating a learning environment that motivates students to pursue engineering professions.

Keywords: Economy digitalization · Engineering education · Motivation system

1 Introduction

Natural resources depletion and ecosystems critical conditions exacerbated the world community problems in the twenty-first century. Accelerating urbanization, along with positive effects on the economy, has a negative impact on the environment, which leads

M. E. Auer et al. (eds.), *Online Engineering and Society 4.0*, Lecture Notes in Networks and Systems 298, https://doi.org/10.1007/978-3-030-82529-4_51

to climate change. The fourth industrial revolution and the economy digitalization (experts also position as "technonomic" [1, 2]), as a result of technological breakthroughs, led to the intellectualization of all activity spheres, thereby changing employment patterns [3–5]. This reality, according to analysts' forecasts, will lead to a rethinking of the attitude to work in general and to the work process in particular [6]. All this leads to an increase in the requirements for the engineers' competence and the engineering education system as a whole, that should ensure the training quality of engineers necessary for the business and society development.

Human capital, its creativity and ability to innovate will be the driving force for companies to thrive in the digital era [7], what greatly enhance the requirements for the engineering education quality and sets a twofold goal for it. On the one hand, it is necessary to train personnel for the digital economy that will meet the digital business needs in all activity areas, especially in the high-tech industry. On the other hand, education system digitalization itself is necessary, including new teaching technologies, as well as innovating content. A student's competitiveness is determined not only by his knowledge and skills in the chosen field of activity, but also by the ability to make independent decisions in the design process. At the same time, many countries note that the interest of young people in obtaining an engineering education is declining, which requires significant efforts in finding ways to increase the motivation for obtaining engineering professions. This applies not only to domestic students, but also to those who come to study from abroad on quotas and grants, but do not yet have a clear understanding of the needs of the labor market.

Despite the progress in the engineering education system achieved with the help of e-learning systems, a decrease in motivation for mastering digital technologies is characteristic of both students and professors, since it requires significant efforts. The main study goal is to identify effective ways to increase motivation in the engineering education system. For this, firstly, it is necessary to determine the reasons for the decline in young people's interest in engineering education. Secondly, to determine the possibilities of influencing the motivational signs. Thirdly, to create a tool for monitoring the educational process based on the "student–teacher" feedback, which will allow to correct training courses that are poorly perceived by students. The changes should contribute to the creation of systemic strategy aimed at increasing the engineering education system sustainability, the possibility of its continuous improvement and development, which will meet the economy real sector needs. Particular attention should be paid to changing the paradigm of the educational process organization, which should combine not only the use of modern methods and means of teaching, but also an orientation towards the real production and business needs in order to form the student's ability to self-study and subsequently his rapid adaptation to production conditions.

2 Background: Motivation, Its Characteristics and Incentives

Motivation is defined as the internal energy that ensures a person's activity at work and in life. The motivational sphere contains a number of motives: ideals of value orientations, needs, motives, goals, interests, etc. Taking this into account, the students' motivation characterizes the processes, methods and means of their stimulation to cognitive activity, and to active study of educational content. The authors of numerous scientific and practical researches identify the reasons for low motivation, ways to increase it and the mechanisms of influencing the student in order to increase their motivation.

2.1 Reasons for Low Motivation of Engineering Students

The authors of the article [8] assume that the majority of students do not attend their classes, mainly out of fear or lack of real interest in the academic disciplines. It is proposed to use a variety of technological tools and rewards to achieve greater motivation and self-confidence development, which will allow to work out new teaching methods and tools, as well as monitoring student progress.

Student dropout is an important topic, especially for engineering students, a significant proportion of whom drop out for various reasons. The article [9] suggests that learning motivation and current emotional state are very important for academic progress in studies as well as student retention. The authors proposed a set of teaching methods to improve the students' emotional condition, reduce negative stress and enable them to organize individual learning behaviour.

In the study [10] it was found that only 37% of students studying in the UCH electronic engineering program completed it, many refused the program due to poor academic performance. To increase the intrinsic motivation of students, additional and cross-cutting scenarios are considered, such as: project fair, project competition and research congresses at the university itself, and also take into account the condition of students, age range, motivational characteristics, availability of time.

The authors of paper [11] analyses the relationship between two variables: the usefulness of the online community of practice (OCP), that students perceived and their satisfaction with the tasks given to them. It should be noted that the perceived usefulness of OCP by students is in direct proportion to cooperation between students and the active role they play in the development of their own learning and leads to greater student satisfaction at the university level.

The study [12] results show that motivation to learn and academic self-efficacy can effectively predict the students autonomous learning behaviour of, which is characteristic of blended learning. In addition, academic self-efficacy has a positive effect on motivation to learn. Students with a high self-efficacy sense are more confident in their abilities and dare to solve difficult learning problems, have a stronger motivation to learn, and are more likely to receive learning satisfaction and better academic performance after achieving a goal.

The article [13] goal was to test the effectiveness of learning results self-reflection to increase student confidence and to acquire skills. The results of the correlation analysis between self-reflection and the overall performance of students show that students' coursework and final grades improved significantly, which allowed both students and the professor to adjust the educational process by using the self-reflection journal.

2.2 Motivation Enhancement: New Methods and Feedback

The study [14] presents research that was conducted to examine the degree of improvement of motivation received by first-year engineering students after completing an introductory engineering course using the How People Learn (HPL) framework structure as the basis for design. The authors note the success of this experience in increasing students' motivation.

Encouraging internal motivation in the classroom is a central focus of many researchers, since students' individual differences can affect their level of internal motivation. The authors of the study [15] examines the influence of personality on the level of students' internal motivation in a gaming and self-determined online environment. The obtained results showed that only extraversion, consciousness and openness of the personality significantly affect the level of students' internal motivation for learning. These results can be explored by developers of the game learning environment to provide more elements of game design to enhance student's internal motivation based on their personality.

Article [16] is an empirical study of the motivation of computer science students. Their motivational powers are mainly related to intrinsic aspects such as achievement motivation, willingness to improve, a sense of achievement, self-confidence, and so on. However, motivation gaps were noted related to willingness to attend classes, assessment processes, perceived usefulness of their efforts, attention received by their teachers, and resources provided by college. differences are related to achievement motivation, willingness to improve, intrinsic motivation and perception of usefulness.

The document [17] presented a framework that uses personalized motivation in the form of feedback, tips, and reminders to increase student engagement in online learning. The main focus is on the capabilities of modelling and monitoring the involvement level as a likelihood indicator that a student will successfully complete the course. This is especially true for the development of online learning courses such as MOOCs, which place more emphasis on self-regulation and require each student to manage their learning more than traditional classrooms. When analyzing detailed student feedback, a number of potential improvements were identified.

The gamification principles using can strengthen and expand student participation in learning activities and, therefore, contribute to more effective acquiring of knowledge. The authors of the articles [18, 19] assessed the effectiveness and suitability of gamification and reward for education, its relevance to learning goals, and the promotion of both engagement and internal motivation. The educational process has been improved through progress monitoring and a rating system.

As shown in [20], the using of teaching media also contributes to the increase of students' motivation, which is classified as very high in terms of such aspects as student engagement, awards, because it provides realistic expectations, creates incentives and directs their behaviour. Virtual reality technology has a positive effect on students' motivation too. The article [21] shows that the use of virtual reality leads to an increase in involvement level and a decrease in cognitive efforts.

Recognizing the advantage of the Internet of Things in transforming learning into a competency-based model that requires students to play a more active role, the authors of [22] examine its impact on student motivation. Analysis of results shows that, although students are generally motivated towards science, however, the IoT using through the implementation of dynamic, contextualized and real actions provides the development of various skills that are useful for analysing actions, interactions, preference trends and changes in the student skills' level.

The authors of the research [23] showed that although feedback from peers or mentors does not affect any of the motivational factors, feedback contributes to greater interaction between students. Autonomy is significantly affected only by feedback from the teacher, and in blended learning it can contribute to rigor and basic needs.

The article [24] proposes a salary forecasting system to increase students' motivation for learning. To create a forecasting model, a decision tree method with seven functions is used. The result shows that the system can effectively increase student motivation to learn, as well as show them a positive outlook on their future.

Many factors that are classified as internal and/or external affect student motivation, engagement and, as a result, their academic advancement. In the article [25], to highlight positive and negative factors, multifactorial motivation analysis was used, which will allow controlling the motivation of students and increasing their involvement by controlling these factors. In addition, motivation is influenced by different driving forces, monitoring and evaluation methodology.

Paper [26] presents a student motivation study to learn programming. The authors analysed several aspects of motivation and then compared this information with the grades received by students in introductory programming courses. The results show the need for pedagogical strategies that can help most students overcome the natural difficulties of learning to code, improve student motivation, confidence, and perception, making them more prepared for the effort of learning to code.

The students' performance is impacted by the academic context, their ability and motivation, especially in engineering disciplines. The articles [27] presents an empirical study conducted with the participation of 149 students studying computer science at the university, discusses the benefits and results of innovative educational methods, as well as their impact on student motivation.

2.3 Teachers' Motivation and Its Impact on Student Motivation and Success

The research [28] is aimed at studying the influence of teachers' competences on the students' motivation to achieve. Educator's competencies are external factors that keep students motivated to perform well. The results showed that the influence of teachers' competence on the students' motivation progress was 34.4%, while a great influence was exerted on the motivation for performance, consisting of social, pedagogical and personal competencies. Students assessed that personal and pedagogical competencies affect the students' motivation to learn in their academic and social environment and play a more significative role in comparison with other competencies.

Article [29] explores the links between classroom climate, psychological need satisfaction and motivation. In a learning setting, these three basic psychological needs are met when students experience efficiency and mastery; supporting communication with others; as well as choice and control. Research shows that educators play an important role in creating environments that support these three needs through their pedagogical choices, interactive style, culture, and classroom learning climate. The authors found that student needs satisfaction and a positive assessment of the learning climate are positively correlated with autonomous motives, however, students' positive motivation is unstable over time, but there are important links between motivations and course variables that teachers can influence when choosing educational activities and learning methods.

The document [30] shows that professors play a key role in motivating students. The overall goal of most professors is to help students acquire knowledge and competencies relevant to their future life. These research results are used to create new lecture formats to activate students, while it is very important to take into account students' motivation when developing curricula and choosing lecture topics, since the way of lecturing is the second most important source of motivation and demotivation.

They directly affect motivation through their role models and good lecture didactics, and the indirect impact is even greater in curriculum and lecture design. Understanding students' motivation is an important aspect of effective course design [31]. Since motivation is linked to learning outcomes, from critical thinking to creativity and lifelong learning, helping students develop positive learning motivation is critical to engagement and success.

As shown in the article [32], motivation plays an important role in the innovative experience of engineering students, especially in solving the unique problems that innovation presents. It was found that students in general expressed more examples of internal and integrated regulation in comparison with introjected and external motivation, while the teacher must determine the range of potential internal motivators and consider each of them in the project description. By inspiring in students a sense of ownership and responsibility for their projects results (for example, through autonomy and connection), professors could stimulate greater motivation for innovation for students of various engineering.

Developments in the information technology sphere such as multimedia technology, multimedia teaching software contribute to the teaching quality and to unleash the students' creative talents in accordance with their abilities through a stimulating strategy of intelligent problem solving, because, as shown in the document [33], students have a strong desire to learn.

3 Case Study: Exploring Ways to Increase Students' Motivation in the Digitalization Era

3.1 The System of Motivation in the Educational System

The educational system can be analysed both in terms of processes and participants. The impact on the process is carried out through organizational and management structures, which have different departments, each responsible for their own process and, accordingly, managing it. For these purposes, there are regulatory documents and instructions, as well as feedback. As a rule, management is carried out in accordance with the activities types. Traditionally, the learning process was carried out by transferring educational content in the classroom: in the lectures form, laboratory studies and practical exercises. With the advent of computers and the information technology's development, new types of educational content have appeared, as well as new teaching methods. To expand the possibilities in implementing the concept of "life-long learning", technologies for the distance education implementation have appeared.

As can be seen from the above review, these technologies currently provide the opportunity to receive a quality education, although they have problems. Despite the fact that education goal is to learning personnel for work in various, including new, areas, but pedagogical techniques and methods use technologies and developments from other economy areas, therefore the educational system is rather inert compared to large systems of other types (for example, industrial, organizational and technical, etc.), This contradiction is most acutely reflected in engineering education, since it is the engineer who generates new ideas, which are subsequently implemented in the form of new products and technologies. That is why, in the digitalization era, society's demand for the engineering education system has intensified, where the student's abilities for technical creativity, creativity of thinking and skills in implementing technologies in various industries should develop. Thus, the goal of engineering education is to prepare an engineer "for the future" [34, 35].

This goal can be realized only if the student has developed the ability to be creative, which is regulated through the motivation system. It is necessary not only to classify motives and impulses, but also a priorities system, which will allow to influence certain qualities of the student's personality, as well as reasonably form pedagogical goals. A flexible universal mechanism is also needed to: track changes in the external environment; adequately respond to them, adapting individual elements of the educational system to new conditions; introduce new technological advances into the educational process.

To realize the motive, to master the self-education methods, the students need to set and fulfil many auxiliary goals: to learn to see the long-term results of their educational activity, to subordinate to them academic work stages, to set goals for the learning activities implementation and their self-examination and etc.

3.2 Educational Motives and Their Features

Motives are one of the mobile systems that can be influenced. However, even the most positive and diverse motives only create a potential for the student's development, since the motives realization depends on the goal-setting processes, i.e. the ability to set goals and achieve them in learning. Figure 1 shows the stages of the formation of educational motivation. The difference between the concepts of "internal motivation" and "external motivation" requires the establishment of some unique criteria. To increase the motivation of students for engineering education, it is necessary to formalize the motivation criteria, to stimulate the positive and reduce the impact of the negative. By purposefully forming a sustainable system of motives for activity, you can help a future specialist in professional adaptation and career growth. The educational process effectiveness is directly related to how high the motivation and incentives to master the future profession.

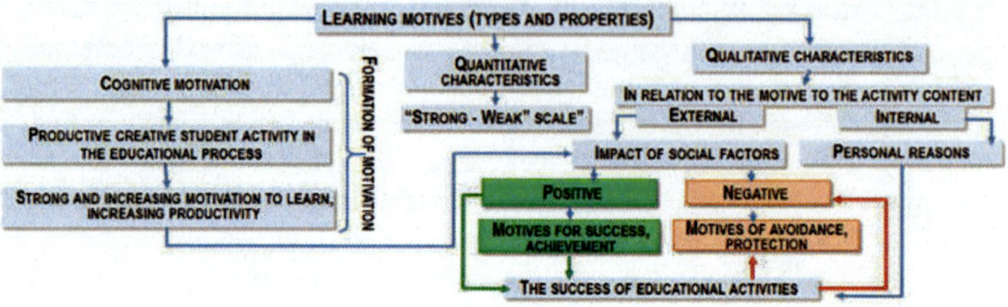

Fig. 1 Impact of motivation on learning activities

In the system of educational motives, external and internal motives are intertwined. Internal motives include their own desire to learn and their own development in the learning process. External motives come from parents, teachers, a group in which a student is studying, environment or society. The decisive importance should be given not to the external influence, but to the internal motive forces (Fig. 2).

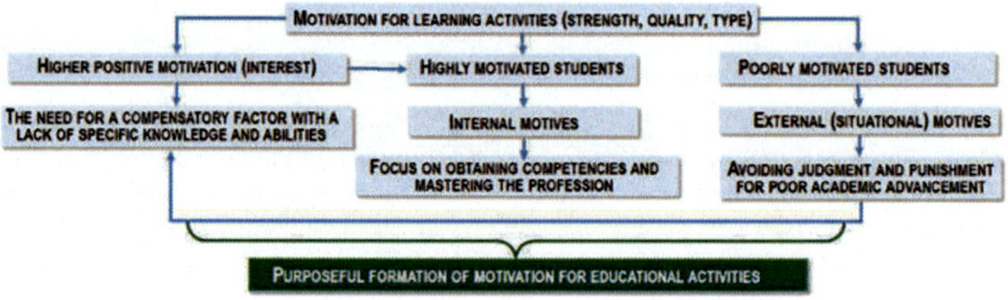

Fig. 2 Formation of motivation for learning activities

There are different ways to increase the motivation of students in higher educational institutions, but the main methods of influence can be divided into three areas: understanding the relevance of the acquired knowledge in professional activities; formation, along with interest in discipline, understanding of the possibility of knowledge practical use; providing feedback with the teacher for advice and assistance during the educational process.

3.3 Studying the Teachers' Motivation When it is Necessary to Adjust the Educational Process

Since the transition to distance learning was carried out in accordance with the order of the Ministry of Education and Science of Russia, that at the end of the academic year, the Ministry of Education and Science of Russia, together with the Institute of Social Analysis and Forecasting of the Russian Academy of National Economy and Public Administration, conducted a mass survey of the universities teaching staff on development of the online environment in the coronavirus pandemic context. It summarizes the attitude of teachers to distance education in the context of a sharp change in their professional activities and the organization of their personal lives.

This survey has become a good opportunity to assess the readiness of the educational system to function in new conditions, since it allows us to analyze the readiness of teachers to use new tools, which, judging by previous studies (Chap. 2), affect the increase in student motivation. In addition, in view of the importance of student-professor feedback in stimulating personal motivation and interest in learning, it is important to create a comfortable environment for student–teacher interaction.

33,987 teachers, or about 15% of the total of the teaching staff, were taken part in the survey. The results of the study showed that teachers are organizationally ready for the transition to distance learning formats, but psychologically they do not accept such a sharp break with traditional full-time education. A skeptical attitude to what is happening is due to both the peculiarities of the disciplines taught (for example, technical and experimental), and conservative views on the nature of teaching. Although teachers formally accept the transition to distance education, they have an internal, latent rejection of it in all areas of training, regardless of gender, age, social and professional status. The survey recorded two positive social characteristics of the professional community of university teachers: the adoption of a state policy on countering coronavirus infection and the availability of skills and abilities to work in an online environment.

46.1% of the respondents disagree with the change in education towards individualization and adjustment for each student. The main threats associated with the impossibility of liberalizing education, the transition to a distance format, are called: the decline in students' motivation to learn; lack of students' skills and abilities to maintain discipline and diligence in distance learning; emotional breakdowns of both students and teachers; increased workload on teachers; lack of an individual approach in the education system, impersonality; inability to control the level of knowledge; restriction in some spheres (for example, in engineering) on the remote knowledge transfer; educational processes formalization, a tendency to stereotyped, unified solutions. In general, the results of the survey in Russia coincide with the opinion of the teachers in our university (Fig. 3).

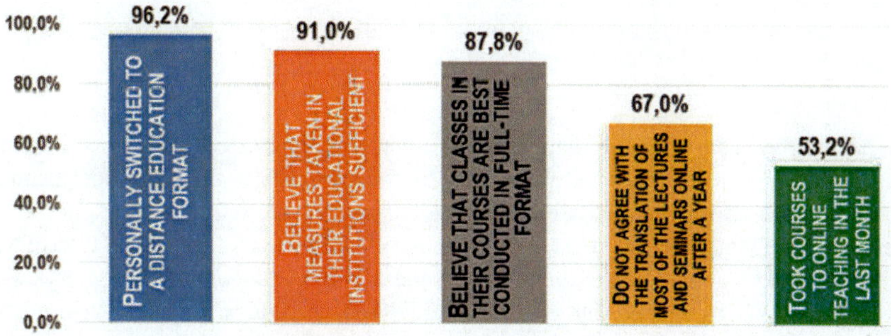

Fig. 3 Results of the teacher survey

According to the greater part of both the professional educational community and students, a mixed form of education will be the most rational, since not all types of distance learning are effective.

3.4 Influence of External Factors on Students' Motivation

To identify educational trends and assess overall performance after distance learning from March 30 to June 30, 2020, the summer session results of 1–4 courses engineering students of KFU (1,191,107 -Materials Science and Materials Technology; 1,191,109 - Mechanical engineering; 1,191,111 - Design and technological support of machine-building industries; 1,191,113 - Vehicle designer; 1,191,115 - Vehicles' maintenance and repair; 1,191,116 - Vehicles and automotive economy; 1,191,121 - Transport process technology; 1,191,133 - Car service) were analysed. In particular, according to the examination testing results in natural sciences, the average score in the theoretical part turned out to be higher than according to the results of completing practical tasks in the semester. The resulting dependencies by disciplines: mathematics (M), structural materials technology (SMT), information technology (IT), descriptive geometry and computer graphics (DG & CG), are presented in the scattering diagrams (Fig. 4 - 5).

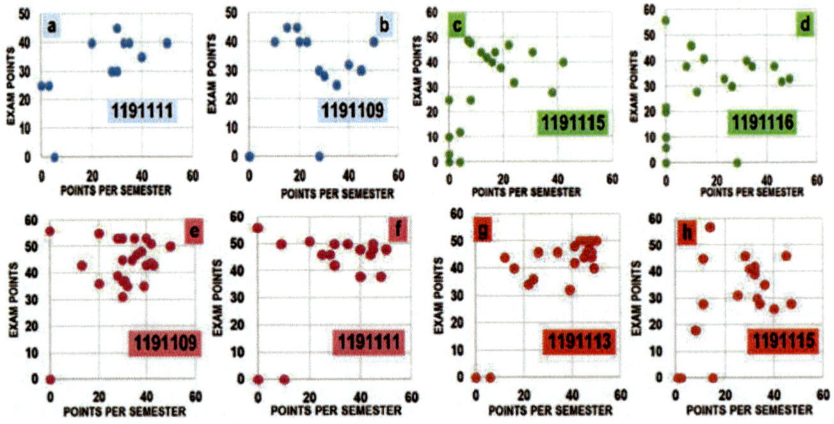

Fig. 4 Exam results in maths (a, b - teacher 1, c, d - teacher 2) and in SMT (e, f - teacher 3, g, h - teacher 4)

As can be seen from the diagrams, the revealed pattern does not depend on the teacher personality, but is determined only by the curriculum, therefore, the distance technologies' expediency and effectiveness are characteristic only a lecture course. At the same time, obtaining practical skills requires closer interaction between the student and the teacher. A small points' number near the origin, corresponding to students with low academic performance, indicates the individual perceptions' specific features of this cycle disciplines by these students, and possible problems with basic learning. The questionnaire survey results of these students made it possible to identify such

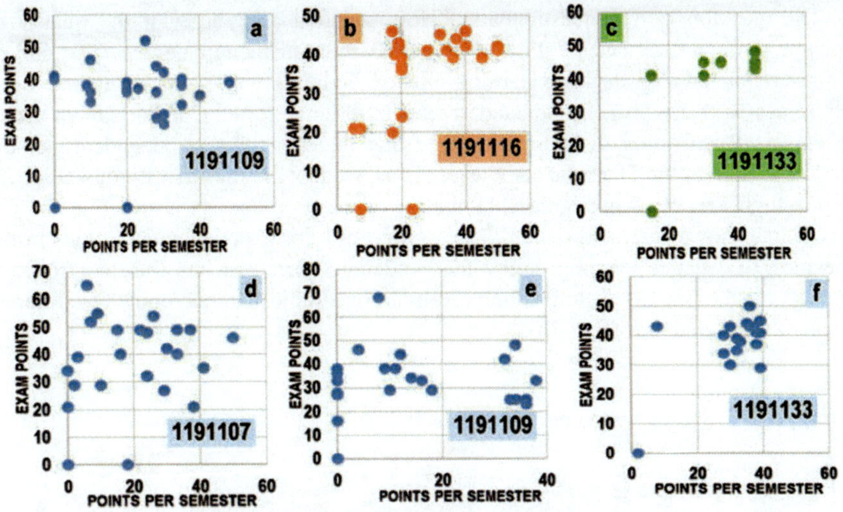

Fig. 5 Exam results in IT (a - teacher 1; b - teacher 2; c - teacher 3) and in DG & CG (d, e, f - teacher 4)

problems in learning as difficulties in self-control during independent work and a lack of free time to complete tasks.

At the same time, the scatter diagram of the physics session results (Fig. 5 a, b) clearly shows two clusters: students who scored low scores both in the semester and on the exam, and students with a score above the average. The questionnaire survey showed that the first group students had problems with the use of educational-methodological materials developed by the teacher, which indicates the need to address the effectiveness issues of distance technologies using in teaching disciplines that include the practical skills acquisition.

Similar results typical for foreign languages (Fig. 6 c–e), are explained by the complexity of organizing paired language practice when using some distance platforms (MS Teams), or the lack of such an opportunity (Virtual classroom).

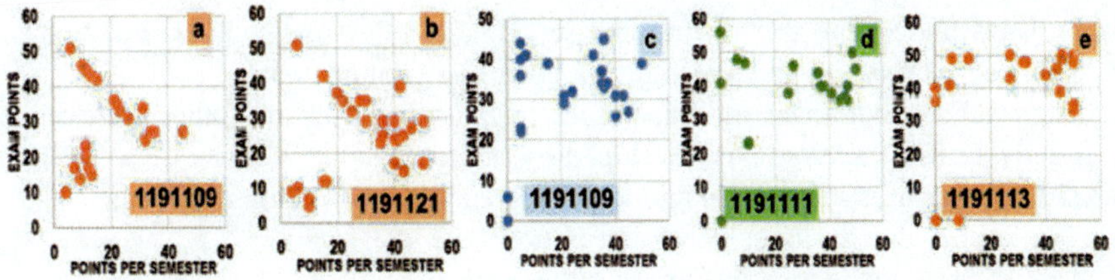

Fig. 6 Exam results in physics (a, b - teacher 1) and foreign languages (c - teacher 2; d - teacher 3; e - teacher 4)

The question of the questionnaire "How satisfied are you with the quality of teaching foreign languages" allowed to reveal the satisfaction degree. The survey results conducted after distance learning showed slightly lower grades compared to the fall semester 2019–2020, conducted in the traditional form. Comparison of exam results in IT and other disciplines showed that there is a direct dependence of the session results on the IT proficiency level. Thus, students in the fourth group (Fig. 7a) have a high average score both in IT and in other disciplines (see also Fig. 4). The survey has shown that students from this group did not experience difficulties in the transition to distance learning during the pandemic. Moreover, the third control loop framework, the passing examinations results in one subject in one course to different teachers were analysed (Fig. 7b).

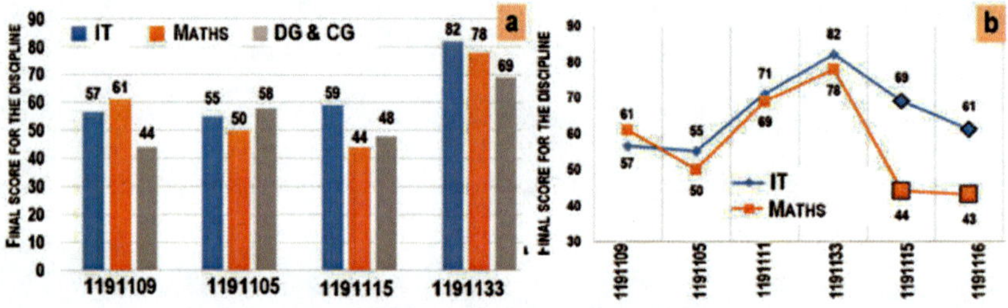

Fig. 7 Average scores based on exams' results (a); comparison of average scores in IT and Maths (b)

It was found that students receive approximately the same scores in M and IT, with the exception of the last two groups. In these groups, M classes were taught by another teacher who scored slightly lower points as a survey result "A teacher through the student eyes". Thus, it was found that academic performance depends not only on the individual student's qualities, but also on the teaching methodology.

In the course of the questionnaire "Teacher through the eyes of a student" for each student, according to the disciplines of the curriculum, a list of teachers was generated and 10 criteria were formulated. Each criterion could be evaluated on a 10-point scale (minimum - 1 point, maximum - 10 points):

1. I always understand the lesson purpose;
2. The teacher presents the material in a clear and creative way;
3. The teacher emphasizes the importance of the subject for the future profession;
4. I actively use teaching aids developed by the teacher;
5. The teacher uses visual aids and technical teaching aids during classes;
6. The teacher demonstrates general erudition;
7. The teacher is objective in assessing the work of students;
8. The teacher is friendly and tactful, treats students with respect;
9. I am impressed by the appearance and demeanor, organization and discipline of the teacher, his culture of speech;
10. The teacher is available for consultation.

The questions were conventionally divided into three groups. When answering the first group questions, students had to rate their attitude on a five-point scale (1 - the lowest quality, 5 - the highest). In the second group of questions, the student's attitude was assessed in accordance with the formulated criteria. The third group of questions provided an opportunity to formulate your own answer. The results of processing the questionnaires are shown in the diagrams (Fig. 8). On the whole, students were satisfied with the organization of the educational process.

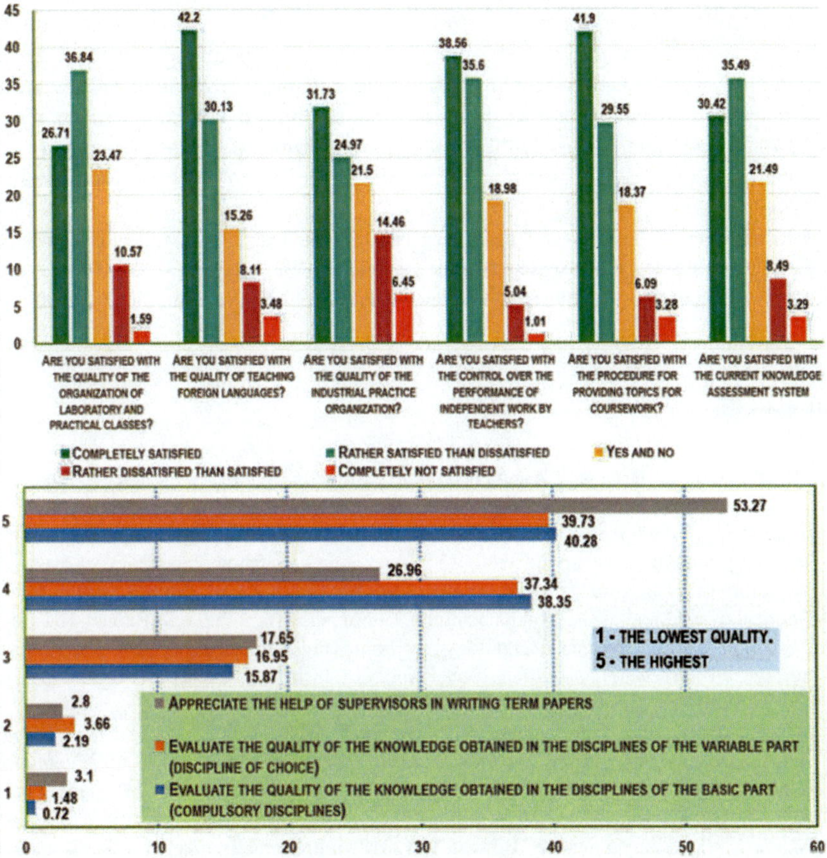

Fig. 8 Results of the student survey

4 Conclusion

To create the possibility of training engineers "for the future", conditions are needed for changing the people consciousness, since the need for engineers with creative thinking becomes more and more acute. We need a transition to a creator's society, which in the context of the intellectualization of all activity spheres requires the creation of prerequisites for a conscious choice of the engineering profession. The changes should contribute not only to the creation of a systemic strategy, but influence the new tools and teaching methods application. We investigated the engineering students motivation for automotive industry. During the experiment, we found that the creation of a positive motivational sphere should include two mechanisms for the formation of motives through activity: "bottom-up" and "top-down". It should be borne in mind that the motivation problems are also the problems of internal cognitive processes, motives, choice. We compared the learning outcomes in the traditional form as well as in the forced transition to a distance format during the COVID 19 pandemic. To make informed decisions, we studied the motivation system, compiled questionnaires for interviewing students and teachers, and then a questionnaire was conducted, the results of which were generalized and used to adjust the teaching methods. We compared the results of the examination session for students of different specialties, courses, groups, which allowed us to draw conclusions about the effectiveness of the proposed method.

Acknowledgements. This research has been funded by the Science Committee of the Ministry of Education and Science of the Republic of Kazakhstan (Grant AP08957435 - The impact of ethnic Kazakh students (kandastar) on the educational space of Kazakhstan: academic, social and cultural aspects).

References

1. `Technonomy' to bring tech leaders to Wayne State (2012). https://www.michiganradio.org/post/technonomy-bring-tech-leaders-wayne-state.
2. Techonomy: Watch Andrew Anagnost Discuss the Future of Making Things (2017). https://labs.blogs.com/its_alive_in_the_lab/2017/11/techonomy-watch-andrew-anagnost-discuss-the-future-of-making-things.html.
3. Shubenkova K, Valiev A, Mukhametdinov E, Shepelev V, Tsiulin S, Reinau KH (2018) Possibility of digital twins technology for improving efficiency of the branded service system. In: Proceedings - 2018 global smart industry conference (GloSIC), pp 1–7. Chelyabinsk
4. Shepelev V, Glushkov A, Almetova Z, Mavrin V (2020) A study of the travel time of intersections by vehicles using computer vision. In: Proceedings of the 6th international conference on vehicle technology and intelligent transport systems (iMLTrans), vol 1. SCITEPRESS, Portugal, pp 653–658
5. Makarova I, Shubenkova K, Buyvol P (2018) Intellectualization of the management processes at the enterprise of automotive industry. In: Proceedings of the international conference on geoinformatics and data analysis (ICGDA 2018). Association for Computing Machinery, New York, NY, USA, pp 177–181
6. 'Talentism' defines success in new capitalism, says Davos chief (2020). https://asia.nikkei.com/Editor-s-Picks/Interview/Talentism-defines-success-in-new-capitalism-says-Davos-chief
7. The social enterprise at work: Paradox as a path forward 2020 Deloitte Global Human Capital Trends (2020). https://www2.deloitte.com/content/dam/Deloitte/cn/Documents/human-capital/deloitte-cn-hc-trend-2020-en-200519.pdf

8. Soto Muñoz JG et al (2019) Experience in the use of rewards to help increase motivation and participation of university students. In: XIV Latin American conference on learning technologies (LACLO). San Jose Del Cabo, Mexico, pp 142–148
9. Fritzsche ES, Schlingensiepen J, Kordts-Freudinger R (2018) Study motivation and academic emotions in engineering students: A case study in german higher education. In: IEEE global engineering education conference (EDUCON), pp 563–570. Tenerife
10. Tirado-Mendoza G et al (2019) Work-in-Progress: Application of academic strategies to strengthen intrinsic motivation and reduce the number of students who abandon engineering programs at universities in Lima – North. In: IEEE world conference on engineering education (EDUNINE). Lima, Peru, pp 1–4
11. Garay-Ruiz U et al (2016) Students' motivation and satisfaction in an online practice community. In: International symposium on computers in education (SIIE). Salamanca, pp 1–3
12. Bai X et al (2020) College students' autonomous learning behavior in blended learning: learning motivation, self-efficacy, and learning anxiety. In: 2020 international symposium on educational technology (ISET). Bangkok, Thailand, pp 155–158
13. Chikhalsouk M et al (2019) The effect of self reflection as an assessment tool for improving universities' students performance: case study. In: 2019 advances in science and engineering technology international conferences (ASET). Dubai, United Arab Emirates, pp 1–6
14. Samsuri NS, Mohd-Yusof K, Aziz AA (2017) Enhancing the first year engineering student motivation through an introductory engineering course. In: 7th world engineering education forum (WEEF). Kuala Lumpur, pp 349–354
15. Tlili A et al (2019) Personality effects on students' intrinsic motivation in a gamified learning environment. In: IEEE 19th International Conference on Advanced Learning Technologies (ICALT). Brazil, pp 100–102
16. López-Fernández D et al (2019) Motivation of computer science engineering students: analysis and recommendations. In: 2019 IEEE frontiers in education conference (FIE). Covington, KY, USA, pp 1–8
17. Staikopoulos A et al (2015) Enhancing student engagement through personalized motivations. In: IEEE 15th international conference on advanced learning technologies. Hualien, pp 340–344
18. Pombo N, Garcia N, Alves P (2019) How to get a badge? unlock your mind: motivation through student empowerment. In: 2019 IEEE global engineering education conference (EDUCON). Dubai, United Arab Emirates, pp 115–119
19. Hasan ML et al (2017) Improving students' motivation to learn through gamification. In: 2017 7th world engineering education forum (WEEF). Kuala Lumpur, pp 642–647
20. Budiman E et al (2018) The effect of mobile learning media on student learning motivation in data structure courses. In: 3rd international conference on information technology, information system and electrical engineering. Yogyakarta, Indonesia, pp 60–64
21. Gargrish S, Mantri A, Singh G, Harun (2020) Measuring students' motivation towards virtual reality game-like learning environments. In: Indo–Taiwan 2nd international conference on computing, analytics and networks. Rajpura, Punjab, India, pp 164–169
22. Magalhães A, Andrade A, Alves JM (2020) Impact of the Internet of Things on 3rd cycle students motivation in an interdisciplinary approach to science. In: 15th iberian conference on information systems and technologies (CISTI). Sevilla, Spain, pp 1–7
23. Bombaerts G, Nickel PJ (2017) Feedback for relatedness and competence: can feedback in blended learning contribute to optimal rigor, basic needs, and motivation? In: IEEE global engineering education conference (EDUCON). Athens, pp 1089–1092

24. Khongchai P, Songmuang P (2016) Improving students' motivation to study using salary prediction system. In: 13th international joint conference on computer science and software engineering. Khon Kaen, pp. 1–6
25. Kbar G, Alazab A, Agbinya J (2019) Multi-factor based enhancing students' motivations. In: 2019 IEEE international conference on industrial technology (ICIT). Melbourne, Australia, pp 1054–1059
26. Gomes A et al (2018) Student motivation towards learning to program. In: 2018 IEEE frontiers in education conference (FIE). San Jose, CA, USA, pp 1–8
27. López-Fernández D et al (2019) Motivation of computer science students at universities organized around small groups. In: IEEE global engineering education conference (EDUCON). Dubai, United Arab Emirates, pp 1128–1135
28. Kusumajati DA, Ruman YS, Oktriono K (2017) The influence of lecturers' competencies towards students' performance motivation: a case study at higher education. In: International symposium on educational technology. Hong Kong, pp 173–176
29. Stolk JD et al (2018) Learners' needs satisfaction, classroom climate, and situational motivations: evaluating self-determination theory in an engineering context. In: 2018 IEEE frontiers in education conference (FIE). San Jose, CA, USA, pp 1–5
30. Fuhrmann T (2018) Motivation centered learning. In: IEEE frontiers in education conference. San Jose, CA, USA, pp 1–5
31. Stolk JD et al (2014) To what extent can instructors influence student motivation in the classroom? In: IEEE Frontiers in Education Conference Proceedings. Madrid, pp 1–3
32. Fila ND, Purzer Ş (2017) Using self-determination theory to understand engineering student motivation during innovation projects. In: IEEE frontiers in education conference (FIE). Indianapolis, IN, pp 1–8
33. Fan D (2016) An empirical study on the motivation strategies of students' learning motivation based on intelligent decision analysis. In: 2016 international conference on intelligent transportation, big data & smart city (ICITBS). Changsha, pp 186–189
34. Makarova I et al (2018) Improving the quality of engineering education by developing the system of increasing students' motivation. In: Advances in intelligent systems and computing, vol 716. Springer, Cham, pp 150–161
35. Makarova I et al (2018) Interaction between education and business in digital era. In: IEEE industrial cyber-physical systems (ICPS). St. Petersburg, pp 503–508

Virtual Learning of Engineering Students from the Low-Resource Environment During the Pandemic

Sasi Sekhar Mallampalli(✉) and Shriya Goyal

Uttaranchal University, Dehradun, India

Abstract. Learning in the virtual environment was a new experience for many of the students from the rural areas of Andhra Pradesh, India. The abrupt closure of educational institutions across the county during the lockdown made many engineering institutions in India explore virtual teaching and learning, a plausible option. It threw unprecedented challenges both for the teachers and students to tune to virtual learning even without any Learning Management System (LMS). Many engineering students who did not have even a computer had to attend the classes on their mobile devices and complete the given tasks using the same. The study aimed at understanding the learning experiences and investigating the challenges and issues faced by engineering undergraduate students (n = 25) from small towns and villages where the students had to rely on mobile data for attending the classes. The participants were divided into five teams with five members each ensuring that at least one student has a desktop or laptop in each team. Two collaborative writing tasks were assigned to the students to be completed as a part of the 'Technical Writing for Engineers' course. The mixed-methods approach was adopted and the data were collected through virtual interaction with individual students and teams. Students were able to complete the given tasks though they were not up to the expectations of the required standards. The results have implications for teachers teaching writing skills and students who write using mobile applications.

Keywords: Virtual learning · Mobile Assisted Language Learning · Collaborative writing · Learning Management System · Low-resource environment

1 Introduction

Writing effectively in a global language is not only one of the most challenging aspects of second language learning but also one of the most sought-after skills for employability and career success [1]. Communicating effectively using different technological tools is one of the essential skills in the digital age of communication. In order to prepare the students for the global workforce, it is imperative to train students in using different mobile applications individually and collaboratively.

M. E. Auer et al. (eds.), *Online Engineering and Society 4.0*, Lecture Notes in Networks and Systems 298, https://doi.org/10.1007/978-3-030-82529-4_52

1.1 Importance of Collaborative Writing for Engineering Students

Communication, Collaboration, Critical thinking, and Creativity are the most sought-after 'Learning Skills' of the twenty-first century for all learners to succeed in college, at work, and in life [2]. Good writing skills are paramount for effective communication as well as in keeping oneself ahead of others. The ability to articulate ideas and information clearly and effectively through digital tools is one of the most desired skills for engineering students for seizing the best job opportunities.

1.2 Mobile Assisted Language Learning (MALL)

Mobile Assisted Language Learning (MALL) is one of the most widely recognized fields of mobile learning. It has attracted the attention of researchers and practitioners as one of the most promising fields of research for enabling millions of learners to learn and acquire skills in second language around the world. The user-friendly features and functions of mobile devices make them the most useful gadgets. The affordances of smartphones led educators and researchers to explore and experiment with the uses of these handheld devices in language pedagogy.

The availability of a good number of applications for learning a foreign or second language makes MALL the most convenient means of learning for millions of learners who can afford a smartphone. Learners can access rich learning materials anytime and from anywhere using the application that suits their learning needs. Affordability and accessibility are the two important factors that play a crucial role in choosing a mobile application.

1.3 Virtual Learning during Pandemic in a Low-Resource Environment

The abrupt closure of educational institutions across the country at the outset of the pandemic had a disruptive impact on the studies of engineering students as was the case within all aspects of life. As the national lockdown led to the complete closure of all human activity, educational institutions were no exception. All the institutions of Higher Education were closed and they had to find different ways and means of reaching out to students to save their precious academic year and studies.

Despite huge developments in mobile connectivity and the use of mobile devices, the infrastructural facilities in many institutions were not equipped with enough facilities to deal with the problem effectively. However, online learning classes and virtual learning were initiated across many engineering colleges in the country. More than 80% of engineering students in rural areas had smartphones but only less than 35% had desktop or laptop computers. Moreover, many institutions did not have any Learning Management Systems (LMS).

1.4 Technical Writing for Engineering Students Using Mobile Applications

Informal learning plays a crucial role in developing second language learning. Learners when they start interacting with others learn many things incidentally. Educators around the world are on the lookout for more innovative ways and seek to provide an interactive environment that engages students and motivates them to collaborate and share their ideas and experiences [3]. Online chatting applications offer immense scope for developing second language skills. Text messaging applications provide opportunities for developing writing skills in an informal learning setting.

Oxford [4] and Savignon [5] opine that communicative competence in a foreign/second language could be developed through realistic interactions using meaningful and contextualized language. The limited-time that is allotted for the classroom is hardly enough for practicing language skills. Furthermore, classroom teaching is often driven by course books and provides little room for contextualized language use and interactions. Hence, it is highly imperative to create a space for students to practice and interact in the target language as much as possible to acquire a good level of communicative competence in the target language.

The present study attempted to investigate how the engineering students with low resources worked collaboratively using the mobile devices for completing the two tasks assigned to them as a part of their curriculum during the pandemic. The study was based on the proposition that students' writing skills in particular could be improved in an effective and engaging way through virtual learning environments using the mobile applications.

2 Related Literature

This brief literature review presents a relevant definition of online or virtual learning for the present context, stakeholders, current theories of online learning, and the most vital factors for success in online learning.

The definition of e-learning undergoes a modification with the evolution of technologies (from Web 0 to Web 4.0). E-learning can be defined as the development of knowledge and skills through the use of information and communication technologies (ICTs), particularly to support interactions for learning, interactions with content, learning activities and tools, and other people [6]. For the present paper, online learning, virtual learning, and e-learning have been used synonymously. Virtual learning or online learning was thrust upon the system as an alternative method by the pandemic.

Literature advocates various parties that comprise the world of e-learning as stakeholders. Anyone who is an integral part of an organization is a stakeholder [7]. Broadly, four stakeholders can be identified in the world of e-learning. They are: (a) learners, (b) instructors, (c) designers and d) implementers [8]. Education institutes and employers come under implementers.

The majority of the learning theories centered on the learners' needs and some on the learner and instructor together. Virtual programs can be far more effective if the program is tailored and executed keeping the interest of the learner and his learning style in mind. Instructional strategies and the choice of technology play a vital role in virtual learning [9]. Many learning theories emphasize a three-dimensional interaction of the learner with the course content, co-learners, and the instructor [10]. From the literature on e-learning theories, it is noticed that the concept of learning is associated not only with the aspects of cognition but also various psychological, environmental, and technical aspects which have a great influence on the world of digital learning [11].

Though there has been a lot of research on virtual learning during the pandemic, only a few studies studied the problem concerning the low-resource environment and collaborative learning tasks during COVID-19 and how the learners without any adequate infrastructure support responded to virtual learning and adapted to the new ways of virtual learning and teaching in the absence of previous experience.

3 Research Objectives

The present study investigates how the engineering students with low resources worked collaboratively using the mobile devices for completing the two tasks assigned to them as a part of their curriculum during the pandemic. The two collaborative tasks were:

(1) Preparing a PowerPoint Presentation using *Google Slides* on a selected topic
(2) Writing a project report collaboratively using the *Google Documents*

Hence, the study aims at investigating the following research questions (RQ):

RQ1. How did the participants prepare the presentation collaboratively using the limited resources they had in the virtual learning environment?
RQ2. How did the participants accomplish the task of writing a project report collaboratively in the virtual learning environment?
RQ3. What challenges and issues did they face during virtual learning during the pandemic?
RQ4. What was the overall impact of virtual learning on engineering students' performance in the Technical Writing Course?

4 Methodology

The study adopted the qualitative method of collecting and analyzing the data. Participants were divided into five groups with five students each. Each group was assigned two tasks: (a) creating a PowerPoint presentation using *Google Slides* and (b) writing a project report using *Google Documents*. The quality of the presentations and the report writing were adjudicated with the support of rubrics. The focus was more on the process of accomplishing the task collaboratively and on their response to the virtual learning experience.

4.1 Participants

The participants in the study were first-year engineering students who were pursuing Electronics and Communication Engineering course in an accredited college which is affiliated to JNTU, Kakinada. The participants were from rural areas and small towns with no proper broadband connectivity. They were from low-income group families hardly with any facilities for learning at home and limited internet access.

4.2 Procedure

The participants were taught technical writing skills for eight weeks to complete the syllabus which was abruptly stopped when the national lockdown was announced. All the participants were given two separate tasks to be done collaboratively at the end of eight weeks of instruction.

All the team members have to participate in the collaborative tasks by sharing the work among themselves and communicating their work regularly using *WhatsApp* where the teacher was also a member to support the students. Two weeks were allocated for the completion of the tasks. Students were required to complete the work using *Google Slides* for presentation and Google Document for report writing.

4.3 Materials/Tools Used

- *WhatsApp* group for each group separately for communication among the group members
- *Google Docs and Google Slides* apps

4.4 Data Collection

Data were collected from the questionnaire and the virtual personal interviews and interaction with each group separately.

5 Results

The results of the study are presented in the order of the research questions studied.

Research Question 1: How did the participants prepare the presentation collaboratively using the limited resources they had in the virtual learning environment?

The learners worked in small teams and shared their work based on their skill-set and the availability of the resources. As each group had one member either a laptop or desktop computer, he was assigned the final designing part of the presentation. The other members contributed by gathering relevant information and suitable images for each slide. They communicated among themselves using *WhatsApp* and completed the presentation in the given time.

Research Question 2: How did the participants accomplish the task of writing a project report collaboratively in the virtual learning environment?

Similar to the practice of making the presentation, each group shared the work among the team members while writing the project report. Once again, the member with the laptop/ desktop had to play a key role in the final documentation of the report. The other team members made thorough research on the given topic and contributed to the project of report writing.

The study found out that the students felt preparing a presentation using *Google Slides* was comparatively easier than writing the report using *Google Documents* collaboratively as shown in Fig. 1. It was also found that the quality of the presentations was better than the project reports.

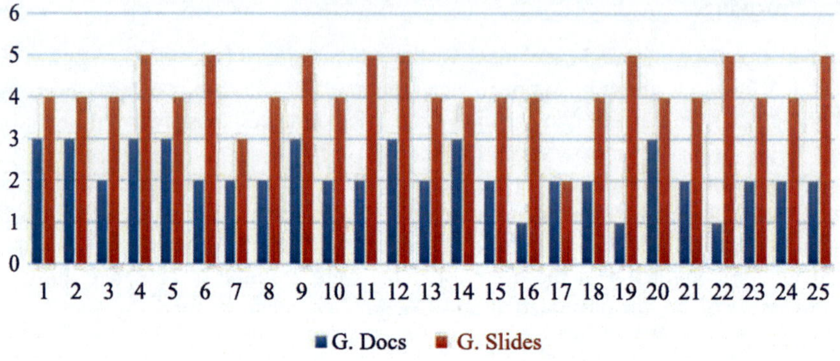

Fig. 1. Participants' performance in using G.Slides and G.Docs.

Research Question 3: What challenges and issues did they face during virtual learning during the pandemic?

There were several issues and challenges faced both by the faculty and the students alike. A majority of faculty members were technologically-challenged and experienced a tough time in delivering lessons online. Very often the teachers had to continue lecturing without the video due to internet bandwidth issues which was an unusual experience for many students.

Many a time online classroom turned very noisy with everyone keeping the mike on as the teachers had little hands-on experience in using the online platforms like Google class or Zoom which offered free access during the pandemic.

Lack of exposure and training in using Learning Management Systems posed a big challenge for the teachers as well as to the learners. Students were not given any orientation class or training in using the technology. They learnt using the tools on a trial-and-error basis.

Teachers and students had to rely on free resources like Google classrooms and *WhatsApp* for conducting online sessions in the absence of any LMS.

Very few institutions had their own LMS. Though Moodle was in place in some institutions – the faculty do not have the required technical knowledge and skills. the lack of technical skills is the biggest hurdle for faculty.

Further even students did not have the expected and required technical know-how. For instance, when *Google Slides* was introduced to the class for a PowerPoint Presentation on the project, 80% of the students admitted that they never made a PowerPoint Presentation using Google-slides. This was exactly the state of affairs of a reputed engineering college that is affiliated to JNTU Kakinada, Andhra Pradesh. The students were first-year engineering students and the majority of them from small towns and rural areas. 80% of the students attending the classes using their mobile phones and the rest using laptops and desktops.

The main challenge of using a mobile was the frequent disturbance in the form of calls and messages from social media. In some cases, the entire family depended on a single mobile for all their communication needs and, as a result, students had to leave the class in the middle and rejoin later. The students had to listen to the class by keeping the video camera switched off and the teacher had to teach without seeing the students due to the network bandwidth issues. For teachers, it is like teaching in a vacuum as the teacher is not able to see the students and had to depend on the text chat option for interacting with the students.

A few teachers also who were not equipped with a desktop or laptop computer had to depend on mobile devices for conducting the classes. A survey among the engineering faculty in low-resource engineering colleges revealed that they did not have the technical know-how in using the mobile devices effectively.

Research Question 4. What was the overall impact of virtual learning on engineering students' performance in the Technical Writing Course?

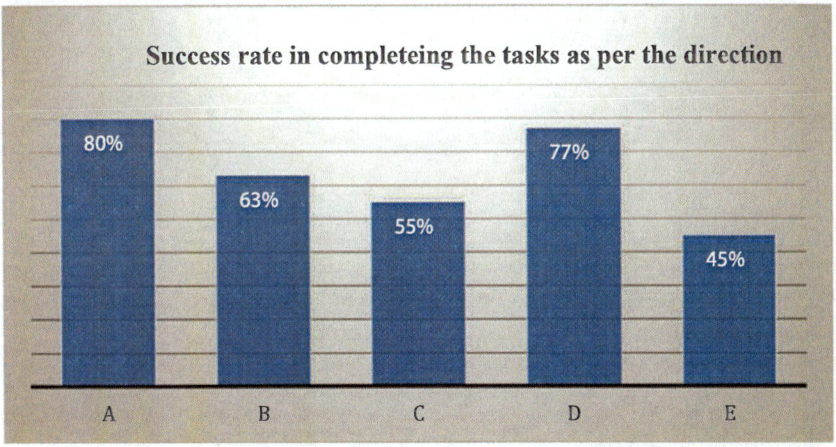

Fig. 2. Participants' success rate in completing the tasks as directed.

From Fig. 2. it is evident that Group A and Group D completed both the tasks with 80% and 77% of success rates respectively and Group B and Group C completed the tasks with 63% and 55% respectively while Group E recorded a 45% success rate in completing the given tasks.

The reasons for the good performance are better internet connectivity, participants' better exposure to the use of technology, and better coordination with the participants of the group members. While the reasons for poor performance can be attributed to poor and unreliable internet connectivity, inefficient team coordination, and a serious lack of technical skills of the participants.

6 Discussion and Conclusion

The experimental study examined the performance of the participants' collaborative writing tasks using Google Docs and *Google Slides* for enhancing the writing skills of the participants.

Two collaborative tasks that were assigned to the students were completed successfully by all the five groups though they varied in the quality of presentation and in following the instruction in detail. Finally, the 'Technical Writing for Engineers' course was completed and feedback at the end of the course revealed the mixed satisfaction levels of the students. The intermittent network issues, the small screen, the absence of LMS affected the learning experience adversely. In short, the issues the learners faced could be summarized into five points: (1) poor connectivity; (2) lack of institutional infrastructure to support faculty and students; (3) lack of LMS; (4) Lack of technical know-how, and (5) poor team working skills.

COVID-19 has forced higher education institutions to rethink how they could reach out to the students in a better way. One of the positive influences on the pandemic on the education field was the universal acceptance of mobile devices as a tool for learning and teaching.

References

1. Hyland K (2003) Writing and teaching writing. Second Lang Writ 1–30
2. Stauffer B (2020) What Are 21st Century Skills? https://www.aeseducation.com/blog/what-are-21st-century-skills
3. Batianeh AM (2014) The effect of text chat assisted with word processors on Saudi English major students' writing accuracy and productivity of authentic texts. Int J Emerg Technol Learn (IJET) 9(9):32
4. Oxford RL (1991) Language learning strategies: what every teacher should know. Heinle, Boston
5. Savignon SJ (1997) Communicative competence: theory and classroom practice, 2nd edn. McGraw-Hill, New York
6. Tîrziu AM, Vrabie C (2015) Education 2.0: E-learning methods. Procedia Soc Behav Sci 186:376–380
7. Wagner N, Hassanein K, Head M (2008) Who is responsible for e-learning success in higher education? Astakeholder's analysis. Educ Technol Soc 11(3):26–36
8. Choudhury S, Pattnaik S (2020) Emerging themes in e-learning: a review from the stakeholders' perspective. Comput Educ 144:103657
9. Clark RE (1994) Media will never influence learning. Educ Technol Res Dev 42(2):21–29
10. Mayer RE, Moreno R (2003) Nine ways to reduce cognitive load in multimedia learning. Educ Psychol 38(1):43–52
11. Venkatesh V, Morris MG, Davis GB, Davis FD (2003) User acceptance of information technology: toward a unified view. MIS Q. 27:425–478

Concept for a Teaching Medium for Immersive Forms of Teaching in Automation Technology

Jenny Schaffrath(✉)

Phoenix Contact GmbH & Co. KG, TechEducation, Blomberg, Germany
jenny.schaffrath@phoenixcontact.com

Abstract. Current industrial transformation processes in the field of automation technology entail new learning media if the teaching is intended to take place in a practice-oriented and industry-oriented manner. Immersive learning environments offer the opportunity not only to adopt new methods for knowledge transfer but also to use training scenarios that would be difficult or impossible to depict in laboratory environments:

Actions that involve high costs or even danger in the event of misconduct (such as incorrect control of a machine), understanding of the sequence of individual production steps in large industrial plants.

In addition, an important aspect of virtual learning environments is that process knowledge can be made accessible to students. Virtual teaching media and hybrid training devices can thus contribute to the democratization of education by making it possible to acquire practical knowledge without the hurdle of large investments in industrial technology.

Finally, virtual teaching media enable technical qualification in state-of-the-art technologie. The virtual learning environments can be adapted to technological leaps with significantly less effort, and due to advancing digitization high-quality data sets for generating 3D models are available anyway.

In order to be able to generate advantages of virtual worlds in the area of teaching media for automation technology in a success-oriented manner, concepts and technological arrangements are required. These must precisely capture and describe the current industrial status, anticipate technological trend topics and accompany the teaching person into new learning.

Keywords: Teaching automation technology · Immersive education forms · Augmented reality

1 Introduction and Background

Digitalization is leading to a transformation of the electrical and automation industry prompting, new qualification requirements. Due to the highly automated processes, the work in cyber-physical systems requires mental models of complex process-related interrelationships in addition to other core needs that are also changing [4]. The learner must develop a steadily growing and more comprehensive understanding of energy and data flows already during training and studies and need to successively expand this understanding. In addition, the automation industry must increasingly shift its thinking (and learning) away from technical components towards complex solutions. In the

M. E. Auer et al. (eds.), *Online Engineering and Society 4.0*, Lecture Notes in Networks and Systems 298, https://doi.org/10.1007/978-3-030-82529-4_53

context of strategies for decarbonization and sector coupling, holistic approaches must also be established in technical qualification.

At the same time, practice-oriented training of future technical specialists and experts is essential. An integral part of the acquisition of technical competencies is work in the laboratory, or in other words:

"Learning is the process whereby knowledge is created through the transformation of experience." [2].[1]

However, providing an authentic learning environment and solution, close to the real industrial reality for the automation industry, requires immense investments.

Currently, investments in the hundreds of thousands of euro range are necessary even for miniaturized plants so that students can constitute an understanding of machine interrelationships and acquire process knowledge from learning on real plants or demonstration plants.[2] Real demo industrial plants are far more expensive. In addition to the high initial investment costs, space requirements and lack of mobility are further obstacles. Furthermore, faulty control of plants, which is not only conceivable in the learning process but even desirable in the sense of experience-based learning, can lead to expensive damage to the demonstrators, sometimes even posing a danger to the learner.

Virtual worlds offer new opportunities for both teachers and learners in an industrial context.

Augmented reality (AR) and virtual reality (VR) already enable faster trainings of employees in industry and digital support of maintenance work through AR assistance systems. Complex components can be explained in a comprehensible and vivid way. An understanding of technological interrelationships is for example supported through the visualization of data streams.[3]

In learning environments, immersive learning worlds are motivational catalysts and technologies and, for example, historical facts become vivid through virtual worlds.[4] In technical education, the advantages of immersive learning worlds are obvious:

"they are not dangerous,[5] they are usually mobile,[6] there is no material wear of expensive training equipment, no damage to expensive equipment due to improper handling, practical and realistic training situations can be mapped, and it is possible to modify environmental variables (weather, lighting conditions, error messages from devices)" [1].

[1] Kolb defines a learning cycle that starts from concrete experience through critical observation and analysis. Here, things that the learner cannot (yet) explain to himself have a special meaning. The existing (theoretical) concept is adapted by insights of the reflection of the concrete experience and is finally applied.

[2] Cf. iLernfabrik christiani.

[3] In the company Phoenix Contact, AR applications are primarily used in the areas of sales and marketing, sporadically in the area of production and in the area of building automation.

[4] Compare google result "shark" which can be projected as a 3D AR model in the living room.

[5] Except for motion sickness in VR environments.

[6] Here is the restriction related to the necessary computing power depending on the design of the 3D models.

Despite numerous applications of virtual reality and augmented reality in industry, there are hardly any teaching aids for the field of education (academic or vocational) in the automation industry.[7] Due to its scalability and direct operational capability without any further technical devices,[8] AR in particular is currently utilized far behind its potential. The connection of virtual systems with AR could also make the advantages of this technology accessible for the teaching of automation technology.

Nevertheless, it is also important to plan didactically supportive measures for learning in simulated worlds. Teaching materials must be accompanied by the production of suitable (virtual) documents and didactic concepts.

Jolie et al. define the framework.
for such conceptions:
"Clear learning objectives, work orders and instructions.
Permanently available background information.
References and exercises that stimulate reflection, for example setting a specific state of the simulation" [1].

Based on these considerations, the TechEducation team of the Phoenix Contact company developed a concept for a hybrid teaching tool that links a virtual industrial world to a real control system and embedded it into a didactic concept.

If programmed correctly, the real controller controls the virtual system. The training device allows various learning settings (individual work/group work), can be used interdisciplinary and is cost-effective.

2 Materials and Methods

2.1 Hardware

The physical basis of the teaching medium "EDU PLCnext augmented" is an Eduline PLCnext starter kit. This technology bundle, designed in a specialized form factor and standardized for teaching, serves to demonstrate the PLCnext Ecosystem. It is suitable for teaching control engineering, industrial network engineering including cloud connectivity. Students plan and create control programs, execute them on the teaching aid and map complex automation systems via networks. In addition to the IEC 61, 131 programming languages, the PLC can also be programmed using high-level languages such as C#, C++ or Matlab Simlink. The open source platform also enables maximum compatibility with other communication protocols via add-ons. This enables interdisciplinary teaching at the interface of OT and IT. The aspects IOT and industrial communication can also be addressed with the teaching material. For example, actuators and sensors can be connected via MQTT or Node-RED and an OPC UA server can be configured via the PLCnext Engineer. The Proficloud of Phoenix Contact can also connect both local and globally distributed network participants.

[7] There are numerous teaching aids in the field of automotive, welding.

[8] Unlike VR, which requires special glasses.

Fig. 1 Hardware rack of the learning carrier with the physical controller

2.2 First Expansion Stage

The virtual system of the learning medium has two stages of expansion based on the two levels described by Potkonjak et al. [8].[9]

In the first expansion stage, a virtual industrial plant is controlled by the real Eduline PLC next starter kit.

The Eduline PLCnext Starter Kit is a special arrangement of real industrial components. The special design of the training board allows AR tracking via object recognition. This also means that any starter kit that exists at a university can use the AR learning scenarios if they download[10] the corresponding data packages from the company's TechEducation Knowledge Hub.[11] Additional physical makers are not necessary. The application is downloaded to the cell phone and used via app.[12]

The AR connection has several layers. In a first layer, the physical Eduline starter kit can be scanned. Components are explained on the training board (text, links and video tutorials[13]). This also enables a self-directed learning process. The directly experienced reality is thus expanded by additional information.

[9] The first level includes the task of being able to work with a technical system. The second level deals in greater depth with the question of how a technical system works exactly in order to be able to solve more complex tasks.

[10] Phoenix Contact operates a university network and provides eduline starter kits free of charge to universities as part of its membership. Currently, 140 universities in 30 countries are members of the Edunet network.

[11] Company-owned knowledge base for universities and vocational schools.

[12] Modular structure, via UI web browser, app can be downloaded from Technology Hub via QR code (on package insert), learning scenarios are individually selectable.

[13] On the PLCnext Community of the company Phoenix Contact.

In a second layer users can choose between different industrial learning scenarios.

To illustrate the range of possible immersive learning environments, two different learning scenarios will be presented:

Learning scenario I "Industrial plant"

This scenario represents a classic industrial plant. Individual plant components (conveyor belt, pick & place, sensors, actuators) are controlled by the controller.

Learning scenario II "Wind farm"

The second learning scenario represents a wind turbine. A single turbine has to be programmed correctly to enable the rotation of the rotor blades.

Both learning scenarios can hardly be simulated in training centers. This is only possible in miniaturized form.

Scaling the virtual world is already possible in the first expansion stage. This enables the representation of higher-level contexts. In the example of the learning scenario "Wind turbine", additional decentralized feed-in units might be seen by changing the zoom factor and the image angle. This makes it possible to address further topics, e.g. charging current control, grid-compliant feed-in and enhanced grid. An energy system coupled between the electricity, heat, mobility and industry sectors will be the future – not only in the automation industry. Thus, the Eduline Starter Kit offers to visualize opportunities for sector coupling and the All Electric Society even in the first expansion stage. However, at this stage the influence is limited to interactions with one virtual asset per scenario. In the future, the integration of higher-level platforms (such as a cloud-based parking management system) is conceivable.

The simulated reality is represented by optimized 50 k polygons of industrial plants (Mesh network when distributed over several polygons, if there is no need for virtual intervention on the special system parts). The aim of the low resolution is that the own mobile device can be used with a viewer app.[14] At the same time, a realistic image should be achieved despite a small polygon number to keep the degree of immersion as high as possible. Therefore, the very common solution Vuforia[15] as an engine is not used. Instead, a gaming engine solution that is more versatile and highly performant is used, to realize object-based tracking, as well as to integrate animations.

What is the task for learners?

The learners first interact with a real physical control system. They analyze the controller and learn how to program it. This can be done with the help of teachers, but also in the self-learning process with the help of the first layer of the AR simulation. The task is presented in a next step: Program the controller so that the virtual system runs.

Didactic added value: The classical training of control technology and PLC programming is significantly enriched, and the learner can see the success of his work in a realistic 3D scenario. Additional virtual information enables the self-learning process. The virtual system contextualizes the functions of the controller.

[14] Compliant with data protection (BYOD, IOS and Android capable).

[15] https://developer.vuforia.com/.

2.3 Second Expansion Stage

In a second stage of the learning medium, the intentional programming of errors/disruptions of the real world is provided. Practice-oriented technology didactics should always aspire to prepare for professional/occupational fields of activity with the requirements for high problem-solving skills. An "ideal" virtual world is not effective here. The aim is to enable the learner to react to unforeseen events. The didactic approach here is problem-based learning, whereby a special attention of the learner is generated by a conflict. The target state of the system is disturbed by external influences. These need to be analyzed and the learner must make decisions and implement them on the physical learning medium. The behavior of the virtual attachment allows him to see if he has done it successfully.

Possible errors in learning scenario I

The pick and place unit has too little suction pressure during the product suction process due to great wear of the suction head. Here, the suction device must be readjusted. The industrial reference is the topics of maintenance. The learner must identify the fault on the virtual plant, correct it and redefine the suction pressure.

Possible errors in learning scenario II

The rotor blades are iced up. The machine must not be started without de-icing (otherwise, chunks of ice could pose a danger to life or knock out the bearing shat due to a weight rebalance). Here the focus is on the evaluation of additional sensor data from the detectors. The machine may only be operated after it has been de-iced.

A gust of wind is coming up. The energy requirements of the grid must be analyzed, the strength of the wind must be analyzed, and the learner must decide whether and how much he puts the rotor blades in the wind.

All scenarios provide explanatory information about the exact potential of danger. Nevertheless, the worst-case scenarios are also implemented in the AR environment to create an impressive learning experience.

Task of the learner:

Analyze real error scenarios and make decisions on the control of the system. Programming the controller according to the respective error. There must be no danger to the plant and no danger to humans. Nevertheless, it should work as efficiently as possible.

2.4 Third Expansion Stage: AI-Integration

The Eduline PLCnext Starter Kit is suitable for interdisciplinary use in the field of information technology due to the possibility of addressing the controller in high-level languages. Thus, it is obvious to map industrial applications of current IT topics on the teaching medium and to make them teachable and understandable. AI[16] has gained a lot of interest because it has the potential allow this.[17] The topic of artificial intelligence is of great importance in the field of education. Learning Analytics is already a well-

[16] AI = artificial intelligence.

[17] Currently, there are numerous textbooks in Germany for machine learning, deep learning, AI, etc. but no special textbooks for the training area.

received topic in the context of “AI in Study and Teaching”. AI is already successfully used when it comes to optimizing individual learning processes [5]. There is also promising research regarding the use of intelligent tutoring systems [6]. The field of AI as a teaching tool, however, has been less systematically addressed. Research on teaching AI from a competence-oriented point of view has not been published yet. There is currently no empirically validated competence model, as well as competence tests for the field of AI as a prerequisite for the systematic development in competences of students. Therefore, necessary basic research is needed which examines how humans learn AI [3]. Unesco has commissioned studies to identify necessary competences in the field of AI [7, 9]. De la Higuera [7] suggests five pillars for AI teaching and learning:

“Data Awareness”, “Randomness“, “Computational Thinking and Coding”, “Post AI Humanism” und “Critical Thinking for AI”.

In developing the AI learning setting for the Eduline Starter Kit, the Phoenix Contact TechEducation development team was guided by these basic pillars. The team has developed an arrangement that makes concepts accessible behind neural networks, classification, matrix multiplication, neuronal output nodes, etc. These are particularly addressed in accompanying didactic materials. The growing importance of AI for the industry is clearly discussed. A discussion of the topic is essential. Beyond theoretical AI teaching, another aim of the teaching aid bundle is also to enable action-oriented teaching of the topic and competence training through the solution of real problems.

Core mission for learners:

A set of real, anonymized plant data is available to the learners, in accordance with the basic learning scenario chosen in the first and second expansion stages. These data are based on real disturbance events. The learners’ task is to find anomalies by means of a self-programmed AI, to determine these anomalies (with the help of additional information from the teaching materials and other information sources such as databases, weather services, etc.) and to program the controller according to the real requirements.

Learning contents are:

Program a neural network.
Initialize the network.
Create link weights, initialize them differentially.
Query the network/create output from the neuronal network.
Analyze the results.
Formulate recommendations for action based on predictions.
If necessary, program the control system according to the recommended action.

In the didactic setting, real data is deliberately work with. These are available in an industrial company like Phoenix Contact and can be used for training purposes after anonymization. This also gives learners the opportunity to deal with the behavior of real data, which is clearly different from constructed data. In addition, the relevance of the industrial relationship and the relevance of the learning content are modeled by real problems and scenarios.

3 Results and Conclusion

AR applications provide the opportunity to create industrial proximity in classrooms. The augmented reality can transport learning content that would be difficult or impossible to transfer only through hardware. This implementation concept illustrates possibilities for a hybrid teaching tool in the field of automation. It is both a training board and a demonstrator and enables different forms of teaching.

From frontal inputs to self-directed learning, from group work (several students access a control/virtual system) to flipped classroom learning, all common forms of learning are possible with this teaching tool.

Learning fields of control technology (planning and creating control programs with elementary and complex data formats, developing library-capable, norm-compliant function modules for automation devices also in higher programming languages, evaluation of intelligent sensors, analysis of process-related data, control of networked automation systems) and information technology (data processing, data security, anomaly detection) and even plant safety, work safety and trouble-free operation are addressed. The hybridity of the teaching equipment allows all this with the smallest investment effort, no set-up time and a maximum of mobility.

Acknowledgements. Thanks to all the colleagues who shared their experiences in different technology areas with me and who supported me during the preparation of this paper.

References

1. Jolie S, Katzky, U, Bredl, K, Kappe F, Krause D (2011) Simulationen und simulierteWelten. In: Ebner M, Schön S (Hrsg.) Lehrbuch für Lernen und Lehren mit Technologien. epuli GmbH, Berlin
2. Kolb DA (1984) Experiential learning: experience as the source of learning and development. Prentice Hall, Upper Sadle River
3. Shapiro B, Norvig P, Fiebrink R (2018) How machine learning impacts the undergraduate computing curriculum. Commun ACM 61:27–29
4. Zinke G, Schenk H, Wasiljew E (2014) Berufsfeldanalyse zu industriellen Elektroberufen als Voruntersuchung zur Bildung einer möglichen Berufsgruppe (Abschlussbericht). BiBB, Bonn
5. Olivé DM, Huynh DQ, Reynolds M, Dougiamas M (2018) Wiese D A supervised learning framework for learning management systems. Commun. ACM 8:631–652
6. du Boulay B (2019) Escape from the Skinner Box: The case for contemporary intelligent learning environments. Br J Edu Technol 50(6):2902–2919. https://doi.org/10.1111/bjet.12860
7. de la Higuera, C (2019) A report about education, training teachers and learning artificial intelligence: overview of key issues. https://www.k4all.org/wp-content/uploads/2019/11/Teaching_AI-report_09072019.pdf
8. Potkonjak V, Vukobratovic M, Jovanovic K, Medenica M (2010) Virtual mechatronic/robotic laboratories-a step further in distance learning. Comput Educ 55(2):465–475
9. Pedro F, Subosa, M (2019) Artificial intelligence in education: Challenges and opportunities for sustainable development (Working Papers on Education Policy Nr. 07). UNESCO, Paris